Lecture Notes in Computer Science 14612

The series Lecture Notes in Computer Science (LNCS), including its subseries Lecture Notes in Artificial Intelligence (LNAI) and Lecture Notes in Bioinformatics (LNBI), has established itself as a medium for the publication of new developments in computer science and information technology research, teaching, and education.

LNCS enjoys close cooperation with the computer science R & D community, the series counts many renowned academics among its volume editors and paper authors, and collaborates with prestigious societies. Its mission is to serve this international community by providing an invaluable service, mainly focused on the publication of conference and workshop proceedings and postproceedings. LNCS commenced publication in 1973.

Nazli Goharian · Nicola Tonellotto · Yulan He ·
Aldo Lipani · Graham McDonald ·
Craig Macdonald · Iadh Ounis
Editors

Advances in Information Retrieval

46th European Conference on Information Retrieval, ECIR 2024
Glasgow, UK, March 24–28, 2024
Proceedings, Part V

Editors
Nazli Goharian
Georgetown University
Washington, WA, USA

Nicola Tonellotto
University of Pisa
Pisa, Italy

Yulan He
King's College London
London, UK

Aldo Lipani
University College London
London, UK

Graham McDonald
University of Glasgow
Glasgow, UK

Craig Macdonald
University of Glasgow
Glasgow, UK

Iadh Ounis
University of Glasgow
Glasgow, UK

ISSN 0302-9743 ISSN 1611-3349 (electronic)
Lecture Notes in Computer Science
ISBN 978-3-031-56068-2 ISBN 978-3-031-56069-9 (eBook)
https://doi.org/10.1007/978-3-031-56069-9

This Springer imprint is published by the registered company Springer Nature Switzerland AG
The registered company address is: Gewerbestrasse 11, 6330 Cham, Switzerland

Paper in this product is recyclable.

Preface

The 46th European Conference on Information Retrieval (ECIR 2024) was held in Glasgow, Scotland, UK, during March 24–28, 2024, and brought together hundreds of researchers from the UK, Europe and abroad. The conference was organised by the University of Glasgow, in cooperation with the British Computer Society's Information Retrieval Specialist Group (BCS IRSG) and with assistance from the Glasgow Convention Bureau.

These proceedings contain the papers related to the presentations, workshops, tutorials, doctoral consortium and other satellite tracks that took place during the conference. This year's ECIR program boasted a variety of novel work from contributors from all around the world. In addition, we introduced a number of novelties in this year's ECIR. First, ECIR 2024 included for the first time a new "Findings" track, which was offered to some full papers that were deemed to be solid, but which could not make the main conference track. Second, ECIR 2024 ran a new special IR4Good track that presented high-quality, high-impact, original IR-related research on societal issues (such as algorithmic bias and fairness, privacy, and transparency) at the interdisciplinary level (e.g., philosophy, law, sociology, civil society), which go beyond the purely technical perspective. Third, ECIR 2024 featured a new innovation called the "Collab-a-thon", intended to provide an opportunity for participants to foster new collaborations that could lead to exciting new research, and forge lasting relationships with like-minded researchers. Finally, ECIR 2024 introduced a new award to encourage and recognise researchers who have made significant contributions in using theory to develop the information retrieval field. The award was named after Professor Cornelis "Keith" van Rijsbergen (University of Glasgow), a pioneer in modern information retrieval, and a strong advocate of the development of models and theories in information retrieval.

The ECIR 2024 program featured a total of 578 papers from authors in 61 countries in its various tracks. The final program included 57 full papers (23% acceptance rate), an additional 18 finding papers, 36 short papers (24% acceptance rate), 26 IR4Good papers (41%), 18 demonstration papers (56% acceptance rate), 9 reproducibility papers (39% acceptance rate), 8 doctoral consortium papers (57% acceptance rate), and 15 invited CLEF papers. All submissions were peer-reviewed by at least three international Program Committee members to ensure that only submissions of the highest relevance and quality were included in the final ECIR 2024 program. The acceptance decisions were further informed by discussions among the reviewers for each submitted paper, led by a Senior Program Committee member. Each track had a final PC meeting where final recommendations were discussed and made, trying to reach a fair and equal outcome for all submissions.

The accepted papers cover the state-of-the-art in information retrieval and recommender systems: user aspects, system and foundational aspects, artificial intelligence & machine learning, applications, evaluation, new social and technical challenges, and other topics of direct or indirect relevance to search and recommendation. As in previous years, the ECIR 2024 program contained a high proportion of papers with students as first authors, as well as papers from a variety of universities, research institutes, and commercial organisations.

In addition to the papers, the program also included 4 keynotes, 7 tutorials, 10 workshops, a doctoral consortium, an IR4Good event, a Collab-a-thon and an industry day. Keynote talks were given by Charles L. A. Clarke (University of Waterloo), Josiane Mothe (Université de Toulouse), Carlos Castillo (Universitat Pompeu Fabra), and this year's Keith van Rijsbergen Award winner, Maarten de Rijke (University of Amsterdam). The tutorials covered a range of topics including explainable recommender systems, sequential recommendation, social good applications, quantum for IR, generative IR, query performance prediction and PhD advice. The workshops brought together participants to discuss narrative extraction (Text2Story), knowledge-enhanced retrieval (KEIR), online misinformation (ROMCIR), understudied users (IR4U2), graph-based IR (IRonGraphs), open web search (WOWS), technology-assisted review (ALTARS), geographic information extraction (GeoExT), bibliometrics (BIR) and search futures (SearchFutures).

The success of ECIR 2024 would not have been possible without all the help from the strong team of volunteers and reviewers. We wish to thank all the reviewers and meta-reviewers who helped to ensure the high quality of the program. We also wish to thank: the reproducibility track chairs Claudia Hauff and Hamed Zamani, the IR4Good track chairs Ludovico Boratto and Mirko Marras, the demo track chairs Giorgio Maria Di Nunzio and Chiara Renso, the industry day chairs Olivier Jeunen and Isabelle Moulinier, the doctoral consortium chairs Yashar Moshfeghi and Gabriella Pasi, the CLEF Labs chair Jake Lever, the workshop chairs Elisabeth Lex, Maria Maistro and Martin Potthast, the tutorial chairs Mohammad Aliannejadi and Johanne R. Trippas, the Collab-a-thon chair Sean MacAvaney, the best paper awards committee chair Raffaele Perego, the sponsorship chairs Dyaa Albakour and Eugene Kharitonov, the proceeding chairs Debasis Ganguly and Richard McCreadie, and the local organisation chairs Zaiqiao Meng and Hitarth Narvala. We would also like to thank all the student volunteers who worked hard to ensure an excellent and memorable experience for participants and attendees. ECIR 2024 was sponsored by a range of learned societies, research institutes and companies. We thank them all for their support. Finally, we wish to thank all of the authors and contributors to the conference.

March 2024

Nazli Goharian
Nicola Tonellotto
Yulan He
Aldo Lipani
Graham McDonald
Craig Macdonald
Iadh Ounis

Organization

General Chairs

Craig Macdonald	University of Glasgow, UK
Graham McDonald	University of Glasgow, UK
Iadh Ounis	University of Glasgow, UK

Program Chairs – Full Papers

Nazli Goharian	Georgetown University, USA
Nicola Tonellotto	University of Pisa, Italy

Program Chairs – Short Papers

Yulan He	King's College London, UK
Aldo Lipani	University College London, UK

Reproducibility Track Chairs

Claudia Hauff	Spotify & TU Delft, Netherlands
Hamed Zamani	University of Massachusetts Amherst, USA

IR4Good Chairs

Ludovico Boratto	University of Cagliari, Italy
Mirko Marras	University of Cagliari, Italy

Demo Chairs

Giorgio Maria Di Nunzio	Università degli Studi di Padova, Italy
Chiara Renso	ISTI - CNR, Italy

Industry Day Chairs

Olivier Jeunen ShareChat, UK
Isabelle Moulinier Thomson Reuters, USA

Doctoral Consortium Chairs

Yashar Moshfeghi University of Strathclyde, UK
Gabriella Pasi Università degli Studi di Milano Bicocca, Italy

CLEF Labs Chair

Jake Lever University of Glasgow, UK

Workshop Chairs

Elisabeth Lex Graz University of Technology, Austria
Maria Maistro University of Copenhagen, Denmark
Martin Potthast Leipzig University, Germany

Tutorial Chairs

Mohammad Aliannejadi University of Amsterdam, Netherlands
Johanne R. Trippas RMIT University, Australia

Collab-a-thon Chair

Sean MacAvaney University of Glasgow, UK

Best Paper Awards Committee Chair

Raffaele Perego ISTI-CNR, Italy

Sponsorship Chairs

Dyaa Albakour	Signal AI, UK
Eugene Kharitonov	Google, France

Proceeding Chairs

Debasis Ganguly	University of Glasgow, UK
Richard McCreadie	University of Glasgow, UK

Local Organisation Chairs

Zaiqiao Meng	University of Glasgow, UK
Hitarth Narvala	University of Glasgow, UK

Senior Program Committee

Mohammad Aliannejadi	University of Amsterdam, Netherlands
Omar Alonso	Amazon, USA
Giambattista Amati	Fondazione Ugo Bordoni, Italy
Ioannis Arapakis	Telefonica Research, Spain
Jaime Arguello	The University of North Carolina at Chapel Hill, USA
Javed Aslam	Northeastern University, USA
Krisztian Balog	University of Stavanger & Google Research, Norway
Patrice Bellot	Aix-Marseille Université CNRS (LSIS), France
Michael Bendersky	Google, USA
Mohand Boughanem	IRIT University Paul Sabatier Toulouse, France
Jamie Callan	Carnegie Mellon University, USA
Charles Clarke	University of Waterloo, Canada
Fabio Crestani	Università della Svizzera italiana (USI), Switzerland
Bruce Croft	University of Massachusetts Amherst, USA
Maarten de Rijke	University of Amsterdam, Netherlands
Arjen de Vries	Radboud University, Netherlands
Tommaso Di Noia	Politecnico di Bari, Italy
Carsten Eickhoff	University of Tübingen, Germany
Tamer Elsayed	Qatar University, Qatar

Liana Ermakova	HCTI/Université de Bretagne Occidentale, France
Hui Fang	University of Delaware, USA
Nicola Ferro	University of Padova, Italy
Norbert Fuhr	University of Duisburg-Essen, Germany
Debasis Ganguly	University of Glasgow, UK
Lorraine Goeuriot	Université Grenoble Alpes (CNRS), France
Marcos Goncalves	Federal University of Minas Gerais, Brazil
Julio Gonzalo	UNED, Spain
Jiafeng Guo	Institute of Computing Technology, China
Matthias Hagen	Friedrich-Schiller-Universität, Germany
Allan Hanbury	TU Wien, Austria
Donna Harman	NIST, USA
Claudia Hauff	Spotify, Netherlands
Jiyin He	Signal AI, UK
Ben He	University of Chinese Academy of Sciences, China
Dietmar Jannach	University of Klagenfurt, Germany
Adam Jatowt	University of Innsbruck, Austria
Gareth Jones	Dublin City University, Ireland
Joemon Jose	University of Glasgow, UK
Jaap Kamps	University of Amsterdam, Netherlands
Jussi Karlgren	SiloGen, Finland
Udo Kruschwitz	University of Regensburg, Germany
Jochen Leidner	Coburg University of Applied Sciences, Germany
Yiqun Liu	Tsinghua University, China
Sean MacAvaney	University of Glasgow, UK
Craig Macdonald	University of Glasgow, UK
Joao Magalhaes	Universidade NOVA de Lisboa, Portugal
Giorgio Maria Di Nunzio	University of Padua, Italy
Philipp Mayr	GESIS, Germany
Donald Metzler	Google, USA
Alistair Moffat	The University of Melbourne, Australia
Yashar Moshfeghi	University of Strathclyde, UK
Henning Müller	HES-SO, Switzerland
Julián Urbano	Delft University of Technology, Netherlands
Marc Najork	Google, USA
Jian-Yun Nie	Université de Montreal, Canada
Harrie Oosterhuis	Radboud University, Netherlands
Iadh Ounis	University of Glasgow, UK
Javier Parapar	University of A Coruña, Spain
Gabriella Pasi	University of Milano Bicocca, Italy
Raffaele Perego	ISTI-CNR, Italy

Benjamin Piwowarski	CNRS/ISIR/Sorbonne Université, France
Paolo Rosso	Universitat Politècnica de València, Spain
Mark Sanderson	RMIT University, Australia
Philipp Schaer	TH Köln (University of Applied Sciences), Germany
Ralf Schenkel	Trier University, Germany
Christin Seifert	University of Marburg, Germany
Gianmaria Silvello	University of Padua, Italy
Fabrizio Silvestri	University of Rome, Italy
Mark Smucker	University of Waterloo, Canada
Laure Soulier	Sorbonne Université-ISIR, France
Torsten Suel	New York University, USA
Hussein Suleman	University of Cape Town, South Africa
Paul Thomas	Microsoft, USA
Theodora Tsikrika	Information Technologies Institute/CERTH, Greece
Suzan Verberne	LIACS/Leiden University, Netherlands
Marcel Worring	University of Amsterdam, Netherlands
Andrew Yates	University of Amsterdam, Netherlands
Shuo Zhang	Bloomberg, UK
Min Zhang	Tsinghua University, China
Guido Zuccon	The University of Queensland, Australia

Program Committee

Amin Abolghasemi	Leiden University, Netherlands
Sharon Adar	Amazon, USA
Shilpi Agrawal	Linkedin, USA
Mohammad Aliannejadi	University of Amsterdam, Netherlands
Satya Almasian	Heidelberg University, Germany
Giuseppe Amato	ISTI-CNR, Italy
Linda Andersson	Artificial Researcher IT GmbH TU Wien, Austria
Negar Arabzadeh	University of Waterloo, Canada
Marcelo Armentano	ISISTAN (CONICET - UNCPBA), Argentina
Arian Askari	Leiden University, Netherlands
Maurizio Atzori	University of Cagliari, Italy
Sandeep Avula	Amazon, USA
Hosein Azarbonyad	Elsevier, Netherlands
Leif Azzopardi	University of Strathclyde, UK
Andrea Bacciu	Sapienza University of Rome, Italy
Mossaab Bagdouri	Walmart Global Tech, USA

Ebrahim Bagheri	Ryerson University, Canada
Seyedeh Baharan Khatami	University of California San Diego, USA
Giacomo Balloccu	Università degli Studi di Cagliari, Italy
Alberto Barrón-Cedeño	Università di Bologna, Italy
Alvaro Barreiro	University of A Coruña, Spain
Roberto Basili	University of Roma Tor Vergata, Italy
Elias Bassani	Independent Researcher, Italy
Christine Bauer	Paris Lodron University Salzburg, Austria
Alejandro Bellogin	Universidad Autonoma de Madrid, Spain
Alessandro Benedetti	Sease, UK
Klaus Berberich	Saarbruecken University of Applied Sciences, Germany
Arjun Bhalla	Bloomberg LP, USA
Sumit Bhatia	Adobe Inc., India
Veronika Bogina	Tel Aviv University, Israel
Alexander Bondarenko	Friedrich-Schiller-Universität Jena, Germany
Ludovico Boratto	University of Cagliari, Italy
Gloria Bordogna	National Research Council of Italy - CNR, France
Emanuela Boros	EPFL, Switzerland
Leonid Boytsov	Amazon, USA
Martin Braschler	Zurich University of Applied Sciences, Switzerland
Pavel Braslavski	Ural Federal University, Russia
Timo Breuer	TH Köln (University of Applied Sciences), Germany
Sebastian Bruch	Pinecone, USA
Arthur Câmara	Zeta Alpha Vector, Netherlands
Fidel Cacheda	Universidade da Coruña, Spain
Sylvie Calabretto	LIRIS, France
Cesare Campagnano	Sapienza University of Rome, Italy
Ricardo Campos	University of Beira Interior, Portugal
Iván Cantador	Universidad Autónoma de Madrid, Spain
Alberto Carlo Maria Mancino	Politecnico di Bari, Italy
Matteo Catena	Siren, Italy
Abhijnan Chakraborty	IIT Delhi, India
Khushhall Chandra Mahajan	Meta Inc., USA
Shubham Chatterjee	The University of Edinburgh, UK
Despoina Chatzakou	Information Technologies Institute, Greece
Catherine Chavula	University of Texas at Austin, USA
Catherine Chen	Brown University, USA
Jean-Pierre Chevallet	Grenoble Alpes University, France
Adrian-Gabriel Chifu	Aix Marseille University, France

Evgenia Christoforou	CYENS Centre of Excellence, Cyprus
Abu Nowshed Chy	University of Chittagong, Bangladesh
Charles Clarke	University of Waterloo, Canada
Stephane Clinchant	Naver Labs Europe, France
Fabio Crestani	Università della Svizzera Italiana (USI), Switzerland
Shane Culpepper	The University of Queensland, Australia
Hervé Déjean	Naver Labs Europe, France
Célia da Costa Pereira	Université Côte d'Azur, France
Maarten de Rijke	University of Amsterdam, Netherlands
Arjen De Vries	Radboud University, Netherlands
Amra Deli	University of Sarajevo, Bosnia and Herzegovina
Gianluca Demartini	The University of Queensland, Australia
Danilo Dess	Leibniz Institute for the Social Sciences, Germany
Emanuele Di Buccio	University of Padua, Italy
Gaël Dias	Normandie University, France
Vlastislav Dohnal	Masaryk University, Czechia
Gregor Donabauer	University of Regensburg, Germany
Zhicheng Dou	Renmin University of China, China
Carsten Eickhoff	University of Tübingen, Germany
Michael Ekstrand	Drexel University, USA
Dima El Zein	Université Côte d'Azur, France
David Elsweiler	University of Regensburg, Germany
Ralph Ewerth	Leibniz Universität Hannover, Germany
Michael Färber	Karlsruhe Institute of Technology, Germany
Guglielmo Faggioli	University of Padova, Italy
Fabrizio Falchi	ISTI-CNR, Italy
Zhen Fan	Carnegie Mellon University, USA
Anjie Fang	Amazon.com, USA
Hossein Fani	University of Windsor, UK
Henry Field	Endicott College, USA
Yue Feng	UCL, UK
Marcos Fernández Pichel	Universidade de Santiago de Compostela, Spain
Antonio Ferrara	Polytechnic University of Bari, Italy
Komal Florio	Università di Torino - Dipartimento di Informatica, Italy
Thibault Formal	Naver Labs Europe, France
Eduard Fosch Villaronga	Leiden University, Netherlands
Maik Fröbe	Friedrich-Schiller-Universität Jena, Germany
Giacomo Frisoni	University of Bologna, Italy
Xiao Fu	University College London, UK
Norbert Fuhr	University of Duisburg-Essen, Germany

Petra Galuščáková	University of Stavanger, Norway
Debasis Ganguly	University of Glasgow, UK
Eric Gaussier	LIG-UGA, France
Xuri Ge	University of Glasgow, UK
Thomas Gerald	Université Paris Saclay CNRS SATT LISN, France
Kripabandhu Ghosh	ISSER, India
Satanu Ghosh	University of New Hampshire, USA
Daniela Godoy	ISISTAN (CONICET - UNCPBA), Argentina
Carlos-Emiliano González-Gallardo	L3i, France
Michael Granitzer	University of Passau, Germany
Nina Grgic-Hlaca	Max Planck Institute for Software Systems, Germany
Adrien Guille	Université de Lyon, France
Chun Guo	Pandora Media LLC, USA
Shashank Gupta	University of Amsterdam, Netherlands
Matthias Hagen	Friedrich-Schiller-Universität Jena, Germany
Fatima Haouari	Qatar University, Qatar
Maram Hasanain	Qatar University, Qatar
Claudia Hauff	Spotify, Netherlands
Naieme Hazrati	Free University of Bozen-Bolzano, Italy
Daniel Hienert	Leibniz Institute for the Social Sciences, Germany
Frank Hopfgartner	Universität Koblenz, Germany
Gilles Hubert	IRIT, France
Oana Inel	University of Zurich, Switzerland
Bogdan Ionescu	Politehnica University of Bucharest, Romania
Thomas Jaenich	University of Glasgow, UK
Shoaib Jameel	University of Southampton, UK
Faizan Javed	Kaiser Permanente, USA
Olivier Jeunen	ShareChat, UK
Alipio Jorge	University of Porto, Portugal
Toshihiro Kamishima	AIST, Japan
Noriko Kando	National Institute of Informatics, Japan
Sarvnaz Karimi	CSIRO, Australia
Pranav Kasela	University of Milano-Bicocca, Italy
Sumanta Kashyapi	University of New Hampshire, USA
Christin Katharina Kreutz	Cologne University of Applied Sciences, Germany
Abhishek Kaushik	Dublin City University, Ireland
Mesut Kaya	Aalborg University Copenhagen, Denmark
Diane Kelly	University of Tennessee, USA

Jae Keol Choi	Seoul National University, South Korea
Roman Kern	Graz University of Technology, Austria
Pooya Khandel	University of Amsterdam, Netherlands
Johannes Kiesel	Bauhaus-Universität, Germany
Styliani Kleanthous	CYENS CoE & Open University of Cyprus, Cyprus
Anastasiia Klimashevskaia	University of Bergen, Italy
Ivica Kostric	University of Stavanger, Norway
Dominik Kowald	Know-Center & Graz University of Technology, Austria
Hermann Kroll	Technische Universität Braunschweig, Germany
Udo Kruschwitz	University of Regensburg, Germany
Hrishikesh Kulkarni	Georgetown University, USA
Wojciech Kusa	TU Wien, Austria
Mucahid Kutlu	TOBB University of Economics and Technology, Turkey
Saar Kuzi	Amazon, USA
Jochen L. Leidner	Coburg University of Applied Sciences, Germany
Kushal Lakhotia	Outreach, USA
Carlos Lassance	Naver Labs Europe, France
Aonghus Lawlor	University College Dublin, Ireland
Dawn Lawrie	Johns Hopkins University, USA
Chia-Jung Lee	Amazon, USA
Jurek Leonhardt	TU Delft, Germany
Monica Lestari Paramita	University of Sheffield, UK
Hang Li	The University of Queensland, Australia
Ming Li	University of Amsterdam, Netherlands
Qiuchi Li	University of Padua, Italy
Wei Li	University of Roehampton, UK
Minghan Li	University of Waterloo, Canada
Shangsong Liang	MBZUAI, UAE
Nut Limsopatham	Amazon, USA
Marina Litvak	Shamoon College of Engineering, Israel
Siwei Liu	MBZUAI, UAE
Haiming Liu	University of Southampton, UK
Yiqun Liu	Tsinghua University, China
Bulou Liu	Tsinghua University, China
Andreas Lommatzsch	TU Berlin, Germany
David Losada	University of Santiago de Compostela, Spain
Jesus Lovon-Melgarejo	Université Paul Sabatier IRIT, France
Alipio M. Jorge	University of Porto, Portugal
Weizhi Ma	Tsinghua University, China

Joel Mackenzie	The University of Queensland, Australia
Daniele Malitesta	Polytechnic University of Bari, Italy
Antonio Mallia	New York University, USA
Behrooz Mansouri	University of Southern Maine, USA
Masoud Mansoury	University of Amsterdam, Netherlands
Jiaxin Mao	Renmin University of China, China
Stefano Marchesin	University of Padova, Italy
Giorgio Maria Di Nunzio	University of Padua, Italy
Franco Maria Nardini	ISTI-CNR, Italy
Mirko Marras	University of Cagliari, Italy
Monica Marrero	Europeana Foundation, Netherlands
Bruno Martins	University of Lisbon, Portugal
Flavio Martins	Instituto Superior Técnico, Lisbon
David Massimo	Free University of Bolzano, Italy
Noemi Mauro	University of Turin, Italy
Richard McCreadie	University of Glasgow, UK
Graham McDonald	University of Glasgow, UK
Giacomo Medda	University of Cagliari, Italy
Parth Mehta	IRSI, India
Ida Mele	IASI-CNR, Italy
Chuan Meng	University of Amsterdam, Netherlands
Zaiqiao Meng	University of Glasgow, UK
Tristan Miller	University of Manitoba, Canada
Alistair Moffat	The University of Melbourne, Australia
Jose Moreno	IRIT/UPS, France
Gianluca Moro	University of Bologna, Italy
Josiane Mothe	Univ. Toulouse, France
Philippe Mulhem	LIG-CNRS, France
Cataldo Musto	University of Bari, Italy
Suraj Nair	University of Maryland, USA
Hitarth Narvala	University of Glasgow, UK
Julia Neidhardt	TU Wien, Austria
Wolfgang Nejdl	L3S and University of Hannover, Germany
Thong Nguyen	University of Amsterdam, Netherlands
Diana Nurbakova	INSA Lyon, France
Hiroaki Ohshima	Graduate School of Information Science, Japan
Harrie Oosterhuis	Radboud University, Netherlands
Salvatore Orlando	Università Ca' Foscari Venezia, Italy
Panagiotis Papadakos	Information Systems Laboratory - FORTH-ICS, Greece
Andrew Parry	University of Glasgow, UK
Pavel Pecina	Charles University, Czechia

Harrisen Scells	Leipzig University, Germany
Philipp Schaer	TH Köln (University of Applied Sciences), Germany
Jörg Schlötterer	University of Marburg, Germany
Ferdinand Schlatt	Friedrich-Schiller-Universität Jena, Germany
Christin Seifert	University of Marburg, Germany
Giovanni Semeraro	University of Bari, Italy
Procheta Sen	University of Liverpool, UK
Ismail Sengor Altingovde	Bilkent University, Türkiye
Vinay Setty	University of Stavanger, Norway
Mahsa Shahshahani	Accenture, Netherlands
Zhengxiang Shi	University College London, UK
Federico Siciliano	Sapienza University of Rome, Italy
Gianmaria Silvello	University of Padua, Italy
Jaspreet Singh	Amazon, USA
Sneha Singhania	Max Planck Institute for Informatics, Germany
Manel Slokom	Delft University of Technology, Netherlands
Mark Smucker	University of Waterloo, Canada
Maria Sofia Bucarelli	Sapienza University of Rome, Italy
Maria Soledad Pera	TU Delft, Germany
Nasim Sonboli	Brown University, USA
Zhihui Song	University College London, UK
Arpit Sood	Meta Inc, USA
Sajad Sotudeh	Georgetown University, USA
Laure Soulier	Sorbonne Université-ISIR, France
Marc Spaniol	Université de Caen Normandie, France
Francesca Spezzano	Boise State University, USA
Damiano Spina	RMIT University, Australia
Benno Stein	Bauhaus-Universität, Germany
Nikolaos Stylianou	Information Technologies Institute, Greece
Aixin Sun	Nanyang Technological University, Singapore
Dhanasekar Sundararaman	Duke University, UK
Reem Suwaileh	Qatar University, Qatar
Lynda Tamine	IRIT, France
Nandan Thakur	University of Waterloo, Canada
Anna Tigunova	Max Planck Institute, Germany
Nava Tintarev	University of Maastricht, Germany
Marko Tkalcic	University of Primorska, Slovenia
Gabriele Tolomei	Sapienza University of Rome, Italy
Antonela Tommasel	Aarhus University, Denmark
Helma Torkamaan	Delft University of Technology, Netherlands
Salvatore Trani	ISTI-CNR, Italy

Giovanni Trappolini	Sapienza University, Italy
Jan Trienes	University of Duisburg-Essen, Germany
Andrew Trotman	University of Otago, New Zealand
Chun-Hua Tsai	University of Omaha, USA
Radu Tudor Ionescu	University of Bucharest, Romania
Yannis Tzitzikas	University of Crete and FORTH-ICS, Greece
Venktesh V	TU Delft, Germany
Alberto Veneri	Ca' Foscari University of Venice, Italy
Manisha Verma	Amazon, USA
Federica Vezzani	University of Padua, Italy
João Vinagre	Joint Research Centre - European Commission, Italy
Vishwa Vinay	Adobe Research, India
Marco Viviani	Università degli Studi di Milano-Bicocca, Italy
Sanne Vrijenhoek	Universiteit van Amsterdam, Netherlands
Vito Walter Anelli	Politecnico di Bari, Italy
Jiexin Wang	South China University of Technology, China
Zhihong Wang	Tsinghua University, China
Xi Wang	University College London, UK
Xiao Wang	University of Glasgow, UK
Yaxiong Wu	University of Glasgow, UK
Eugene Yang	Johns Hopkins University, USA
Hao-Ren Yao	National Institutes of Health, USA
Andrew Yates	University of Amsterdam, Netherlands
Fanghua Ye	University College London, UK
Zixuan Yi	University of Glasgow, UK
Elad Yom-Tov	Microsoft, USA
Eva Zangerle	University of Innsbruck, Austria
Markus Zanker	University of Klagenfurt, Germany
Fattane Zarrinkalam	University of Guelph, Canada
Rongting Zhang	Amazon, USA
Xinyu Zhang	University of Waterloo, USA
Yang Zhang	Kyoto University, Japan
Min Zhang	Tsinghua University, China
Tianyu Zhu	Beihang University, China
Jiongli Zhu	University of California San Diego, USA
Shengyao Zhuang	The University of Queensland, Australia
Md Zia Ullah	Edinburgh Napier University, UK
Steven Zimmerman	University of Essex, UK
Lixin Zou	Wuhan University, China
Guido Zuccon	The University of Queensland, Australia

Additional Reviewers

Pablo Castells
Ophir Frieder
Claudia Hauff
Yulan He
Craig Macdonald
Graham McDonald
Iadh Ounis
Maria Soledad Pera
Fabrizio Silvestri
Nicola Tonellotto
Min Zhang

Contents – Part V

Industry Papers

Doctoral Consortium Papers

Tutorials

Workshops

Conference and Labs of the Evaluation Forum (CLEF)

IR for Good Papers

Measuring Bias in a Ranked List Using Term-Based Representations

Amin Abolghasemi[1(✉)], Leif Azzopardi[2], Arian Askari[1], Maarten de Rijke[3], and Suzan Verberne[1]

[1] Leiden University, Leiden, The Netherlands
{m.a.abolghasemi,a.askari,s.verberne}@liacs.leidenuniv.nl

[2] University of Strathclyde, Glasgow, UK
leif.azzopardi@strath.ac.uk

[3] University of Amsterdam, Amsterdam, The Netherlands
m.derijke@uva.nl

Abstract. In most recent studies, gender bias in document ranking is evaluated with the NFaiRR metric, which measures bias in a ranked list based on an aggregation over the unbiasedness scores of each ranked document. This perspective in measuring the bias of a ranked list has a key limitation: individual documents of a ranked list might be biased while the ranked list as a whole balances the groups' representations. To address this issue, we propose a novel metric called TExFAIR (term exposure-based fairness), which is based on two new extensions to a generic fairness evaluation framework, attention-weighted ranking fairness (AWRF). TExFAIR assesses fairness based on the term-based representation of groups in a ranked list: (i) an explicit definition of associating documents to groups based on probabilistic term-level associations, and (ii) a rank-biased discounting factor (RBDF) for counting non-representative documents towards the measurement of the fairness of a ranked list. We assess TExFAIR on the task of measuring gender bias in passage ranking, and study the relationship between TExFAIR and NFaiRR. Our experiments show that there is no strong correlation between TExFAIR and NFaiRR, which indicates that TExFAIR measures a different dimension of fairness than NFaiRR. With TExFAIR, we extend the AWRF framework to allow for the evaluation of fairness in settings with term-based representations of groups in documents in a ranked list.

Keywords: Bias · Evaluation · Document Ranking

1 Introduction

Ranked result lists generated by ranking models may incorporate biased representations across different societal groups [7,11,39]. Societal bias (unfairness) may reinforce negative stereotypes and perpetuate inequities in the representation of groups [21,50]. A specific type of societal bias is the biased representation of genders in ranked lists of documents. Prior work on binary gender bias

N. Goharian et al. (Eds.): ECIR 2024, LNCS 14612, pp. 3–19, 2024.
https://doi.org/10.1007/978-3-031-56069-9_1

Query: Who is the best football player

1	... currently he plays for Ligue 1 club Paris Saint-Germain ...	1	... currently he plays for Ligue 1 club Paris Saint-Germain ...
2	... she previously played for Espanyol and Levante ...	2	... He is Real Madrid's all-time top goalscorer, scoring 451 ...
3	... She became the first player in the history of the league ...	3	... he was named the Ligue 1 Player of the Year, selected to ...
4	... he returned to Manchester United in 2021 after 12 years ...	4	... he returned to Manchester United in 2021 after 12 years ...

Fig. 1. Two ranked lists of retrieved results for "who is the best football player". Documents in blue contain only female-representative terms and documents in red contain only male-representative terms. In terms of NFaiRR, fairness of both ranked result lists is zero (minimum fairness).

in document ranking associates each group (*female*, *male*) with a predefined set of gender-representative terms [6,39,40], and measures the inequality of representation between the genders in the result list using these groups of terms. While there have been efforts in optimizing rankers for mitigating gender bias [39,44,57], there is limited research addressing the metrics that are used for the evaluation of this bias. The commonly used metrics for gender bias evaluation are *average rank bias* (which we refer to as ARB) [40] and *normalized fairness in the ranked results* (NFaiRR) [39]. These metrics have been found to result in inconsistent fairness evaluation results [22].

There are certain characteristics of ARB and NFaiRR that limit their utility for bias evaluation of ranked result lists: ARB provides a signed and unbounded value for each query [40], and therefore the bias (unfairness) values are not properly comparable across queries. NFaiRR evaluates a ranked list by aggregating over the unbiasedness score of each ranked document. This approach may result in problematic evaluation results. Consider Fig. 1, which shows two rankings for a single query where the unbiasedness score of all documents is zero (as each document is completely biased to one group). The fairness of these two rankings in terms of NFaiRR is zero (i.e., both have minimum fairness), while it is intuitively clear that the ranking on the left is fairer as it provides a more balanced representation of the two groups. There are metrics, however, that are not prone to the kind of problematic cases shown in Fig. 1, but are not directly applicable to fairness evaluation based on term-based group representation off-the-shelf. In particular, *attention-weighted rank fairness* (AWRF) [12,37,43] works based on soft attribution of items (here, documents) to multiple groups. AWRF is a generic metric; for a specific instantiation it requires definitions of: (i) the association of items of a ranked list with respect to each group, (ii) a weighting schema, which determines the weights for different rank positions, (iii) the target distribution of groups, and (iv) a distance function to measure the difference between the target distribution of groups with their distribution in the ranked list. We propose a new metric *TExFAIR* (term exposure-based fairness) based on the AWRF framework for measuring fairness of the representation of different groups in a ranked list. TExFAIR extends AWRF with two adaptations: (i) an explicit definition of the association of documents to groups based on probabilistic term-level associations, and (ii) a ranked-biased discounting factor

(RBDF) for counting non-representative documents towards the measurement of the fairness of a ranked list.

Specifically, we define the concept of *term exposure* as the amount of attention each *term* receives in a ranked list, given a query and a retrieval system. Using term exposure of group-representative terms, we estimate the extent to which each group is represented in a ranked result list. We then leverage the discrepancy in the representation of different groups to measure the degree of fairness in the ranked result list. Moreover, we show that the estimation of fairness may be highly impacted by whether the non-representative documents (documents that do not belong to any of the groups) are taken into account or not. To count these documents towards the estimation of fairness, we propose a rank-biased discounting factor (RBDF) in our evaluation metric. Finally, we employ counterfactual data substitution (CDS) [30] to measure the gender sensitivity of a ranking model in terms of the discrepancy between its original rankings and the ones it provides if it performs retrieval in a counterfactual setting, where the gender of each gendered term in the documents of the collection is reversed, e.g., "he" → "she," "son" → "daughter."

In summary, our main contributions are as follows:

- We define an extension of the AWRF evaluation framework with the metric *TExFAIR*, which explicitly defines the association of each document to the groups based on a probabilistic term-level association.
- We show that non-representative documents, i.e., documents without any representative terms, may have a high impact in the evaluation of fairness with group-representative terms and to address this issue we define a rank-biased discounting factor (RBDF) in our proposed metric.
- We evaluate a set of ranking models in terms of gender bias and show that the correlation between TExFAIR and NFaiRR is not strong, indicating that TExFAIR measures a different dimension of fairness than NFaiRR.

2 Background

Fairness in Rankings. Fairness is a subjective and context-specific constraint and there is no unique definition when it comes to defining fairness for rankings [2,17,31,45,56]. The focus of this paper is on measuring fairness in the representation of groups in rankings [13,32,36,39,55], and, specifically, the setting in which each group can be represented by a predefined set of group-representative terms. We particularly investigate gender bias in document ranking and follow prior work [5,16,39,40,57] on gender bias in the binary setting of two groups: female and male. In this setup, each gender is defined by a set of gender-representative terms (words), which we adopt from prior work [39].

Previous studies on evaluating gender bias [7,39,40,57] mostly use the ARB [40] and NFaiRR [39] metrics. Since the ARB metric has undesirable properties (e.g., being unbounded), for the purposes of this paper we will focus on comparing our newly proposed metric to NFaiRR as the most used and most recent of the two metrics [39,57]. Additionally, there is a body of prior work addressing the evaluation of fairness based on different aspects [4,10,15,45,53,54]. The metrics used in these works vary in different dimensions including (i) the goal of fairness, i.e., what does it mean to be fair, (ii) whether the metric considers relevance score as part of the fairness evaluation, (iii) binary or non-binary group association of each document, (iv) the weighting decay factor for different positions, and (v) evaluation of fairness in an individual ranked list or multiple rankings [36,37]. In light of the sensitivity of gender fairness, which poses a constraint where each ranked list is supposed to represent different gender groups in a ranked list equally [7,39,57], we adopt attention-weighted rank fairness (AWRF) [43] as a framework for the evaluation of group fairness in an *individual ranked list* with soft attribution of documents to multiple groups.

Normalized Fairness of Retrieval Results (NFaiRR). In the following, q is a query, $\mathrm{tf}(t, d)$ stands for the frequency of term t in document d, G is the set of N groups where G_i is the i-th group with $i \in \{1, \dots, N\}$, V_{G_i} is the set of group-representative terms for group G_i, d_q^r is the retrieved document at rank r for query q, and k is the ranking cut-off. $M^{G_i}(d)$ represents the magnitude of group G_i, which is equal to the frequency of G_i's representative terms in document d, i.e., $M^{G_i}(d) = \sum_{t \in V_{G_i}} \mathrm{tf}(t, d)$. τ sets a threshold for considering a document as neutral based on $M^{G_i}(d)$ of all groups in G. Finally, J_{G_i} is the expected proportion of group G_i in a balanced representation of groups in a document, e.g., $J_{G_i} = \frac{1}{2}$ in equal representation for $G_i \in \{\textit{female}, \textit{male}\}$ [7,39,57].

Depending on $M^{G_i}(d)$ for all $G_i \in G$, document d is assigned with a neutrality (unbiasedness) score $\omega(d)$:

$$\omega(d) = \begin{cases} 1, & \text{if } \sum\limits_{G_i \in G} M^{G_i}(d) \leq \tau \\ 1 - \sum\limits_{G_i \in G} \left| \dfrac{M^{G_i}(d)}{\sum_{G_x \in G} M^{G_x}(d)} - J_{G_i} \right|, & \text{otherwise.} \end{cases} \tag{1}$$

To estimate the fairness of the top-k documents retrieved for query q, first, the neutrality score of each ranked document d_q^r is discounted with its corresponding position bias, i.e., $(\log(r+1))^{-1}$, and then, an aggregation over top-k documents is applied (Eq. 2). The resulting score is referred to as the *fairness of retrieval results* (FaiRR) for query q:

$$\mathrm{FaiRR}(q, k) = \sum_{r=1}^{k} \frac{\omega(d_q^r)}{\log(r+1)}. \tag{2}$$

As FaiRR scores of different queries may end up in different value ranges (and consequently are not comparable across queries), a background set of documents S is employed to normalize the fairness scores with the *ideal FaiRR* (IFaiRR) of S for query q [39]. $\mathrm{IFaiRR}(q, S)$ is the best possible fairness result that can be achieved from reordering the documents in the background set S [39]. The NFaiRR score for a query is formulated as follows:

$$\mathrm{NFaiRR}(q, k, S) = \frac{\mathrm{FaiRR}(q, k)}{\mathrm{IFaiRR}(q, S)}. \tag{3}$$

Attention-Weighted Rank Fairness (AWRF). Initially proposed by Sapiezynski et al. [43], AWRF measures the unfairness of a ranked list based on the difference between the exposure of groups and their target exposure. To this end, it first computes a vector E_{L_q} of the accumulated exposure that a list of k documents L retrieved for query q gives to each group:

$$E_{L_q} = \sum_{r=1}^{k} v_r a_{d_q^r}. \tag{4}$$

Here, v_r represents the attention weight, i.e., position bias corresponding to the rank r, e.g., $(\log(r+1))^{-1}$ [12,43], and $a_{d_q^r} \in [0,1]^{|G|}$ stands for the alignment vector of document d_q^r with respect to different groups in the set of all groups G. Each entity in the alignment vector $a_{d_q^r}$ determines the association of d_q^r to one group, i.e., $a_{d_q^r}^{G_i}$. To convert E_{L_q} to a distribution, a normalization is applied:

$$nE_{L_q} = \frac{E_{L_q}}{\|E_{L_q}\|_1}. \tag{5}$$

Finally, a distance metric is employed to measure the difference between the desired target distribution $\hat{E}$ and the nE_{L_q}, the distribution of groups in the ranked list retrieved for query q:

$$\text{AWRF}(L_q) = \Delta(nE_{L_q}, \hat{E}). \tag{6}$$

3 Methodology

As explained in Sects. 1 and 2, NFaiRR measures fairness based on document-level unbiasedness scores. However, in measuring the fairness of a ranked list, individual documents might be biased while the ranked list as a whole balances the groups' representations. Hence, fairness in the representation of groups in a ranked list should not be defined as an aggregation of document-level scores.

We, therefore, propose to measure group representation for a top-k ranking using term exposure in the ranked list as a whole. We adopt the weighting approach of AWRF, and explicitly define the association of documents on a term-level. Additionally, as we show in Sect. 5, the effect of documents without any group-representative terms, i.e., non-representative documents, could result in under-estimating the fairness of ranked lists. To address this issue, we introduce a rank-biased discounting factor in our metric. Other measures for group fairness exist, and some of these measures also make use of exposure [10,45].[1] However, these measures are not at the term-level, but at the document-level. In contrast, we perform a finer measurement and quantify the amount of attention a *term* (instead of document) receives.

[1] Referring to the amount of attention an item (document) receives from users in the ranking.

Term Exposure. In order to quantify the amount of attention a specific term t receives given a ranked list of k documents retrieved for a query q, we formally define *term exposure* of term t in the list of k documents L_q as follows:

$$\mathrm{TE}\,@k(t, q, L_q) = \sum_{r=1}^{k} p_o(t \mid d_q^r) \cdot p_o(d_q^r). \tag{7}$$

Here, d_q^r is a document ranked at rank r in the ranked result retrieved for query q. $p_o(t \mid d_q^r)$ is the probability of observing term t in document d_q^r, and $p_o(d_q^r)$ is the probability of document d at rank r being observed by user. We can perceive $p_o(t \mid d_q^r)$ as the probability of term t occurring in document d_q^r. Therefore, using maximum likelihood estimation, we estimate $p_o(t \mid d_q^r)$ with the frequency of term t in document d_q^r divided by the total number of terms in d_q^r, i.e., $\mathrm{tf}(t, d_q^r) \cdot |d_q^r|^{-1}$. Additionally, following [31,45], we assume that the observation probability $p_o(d_q^r)$ only depends on the rank position of the document, and therefore can be estimated using the position bias at rank r. Following [39,45], we define the position bias as $(\log(r+1))^{-1}$. Accordingly, Eq. 7 can be reformulated as follows:

$$\mathrm{TE}\,@k(t, q) = \sum_{r=1}^{k} \frac{\frac{\mathrm{tf}(t, d_q^r)}{|d_q^r|}}{\log(r+1)}. \tag{8}$$

Group Representation. We leverage the term exposure (Eq. 8) to estimate the representation of each group using the exposure of its representative terms as follows:

$$p(G_i \mid q, k) = \frac{\sum_{t \in V_{G_i}} \mathrm{TE}\,@k(t, q)}{\sum_{G_x \in G} \sum_{t \in V_{G_x}} \mathrm{TE}\,@k(t, q)}. \tag{9}$$

Here, G_i represents the group i in the set of N groups indicated with G (e.g., $G = \{female, male\}$), and V_{G_i} stands for the set of terms representing group G_i. The component $\sum_{G_x \in G} \sum_{t \in V_{G_x}} \mathrm{TE}\,@k(t, q)$ can be interpreted as the total amount of attention that users spend on the representative terms in the ranking for query q. This formulation of the group representation corresponds to the normalization step in AWRF (Eq. 5).

Term Exposure-Based Divergence. To evaluate the fairness based on the representation of different groups, we define a fairness criterion built upon our term-level perspective in the representation of groups: in a fairer ranking – one that is less biased – each group of terms receives an amount of attention proportional to their corresponding desired target representation. Put differently, a divergence from the target representations of groups can be used as a means to measure the bias in the ranking. This divergence corresponds to the distance function in Eq. 6. Let $\hat{p}_{Gi}$ be the target group representation for each group G_i (e.g., $\hat{p}_{G_i} = \frac{1}{2}$ for $G_i \in \{female, male\}$ for equal representation of male and female), then we can compute the bias in the ranked results retrieved for the query q as the absolute divergence between the groups' representation and their corresponding target representation. We refer to this bias as the *term exposure-based divergence* (TED) for query q:

$$\text{TED}(q,k) = \sum_{G_i \in G} |p(G_i \mid q,k) - \hat{p}_{G_i}|. \tag{10}$$

Rank-Biased Discounting Factor (RBDF). With the current formulation of group representation in Eq. 9, non-representative documents, i.e., the documents that do not include any group-representative terms, will not contribute to the estimation of bias in TED (Eq. 10). To address this issue, we discount the bias in Eq. 10 with the proportionality of those documents that count towards the bias estimation, i.e., documents which include at least one group-representative term. To take into account each of these documents with respect to their position in the ranked list, we leverage their corresponding position bias, i.e., $(\log(1+r))^{-1}$ for a document at rank r, to compute the proportionality. The resulting proportionality factor which we refer to as *rank-biased discounting factor* (RBDF) is estimated as follows:

$$\text{RBDF}(q,k) = \frac{\sum_{r=1}^{k} \frac{\mathbb{1}[d_q^r \in S_R]}{\log(1+r)}}{\sum_{r=1}^{k} \frac{1}{\log(1+r)}}. \tag{11}$$

Here, S_R stands for the set of representative documents in top-k ranked list of query q, i.e., documents that include at least one group-representative term. Besides, $\mathbb{1}[d_q^r \in S_R]$ is equal to 1 if $d_q^r \in S_R$, otherwise, 0. Accordingly, we incorporate RDBF(q,k) into Eq. 10 and reformulate it as:

$$\text{TED}(q,k) = \sum_{G_i \in G} |p(G_i \mid q,k) - \hat{p}_{G_i}| \cdot \frac{\sum_{r=1}^{k} \frac{\mathbb{1}[d_q^r \in S_R]}{\log(1+r)}}{\sum_{r=1}^{k} \frac{1}{\log(1+r)}}. \tag{12}$$

Alternatively, as TED(q,k) is bounded, we can leverage the maximum value of TED to quantify the fairness of the rank list of query q. We refer to this quantity as *term exposure-based fairness* (TExFAIR) of query q:

$$\text{TExFAIR}(q,k) = \max(\text{TED}) - \text{TED}(q,k). \tag{13}$$

In the following, we use TExFAIR to refer to TExFAIR with proportionality (RBDF), unless otherwise stated. With $\hat{p}_{G_i} = \frac{1}{2}$ for $G_i \in \{\textit{female}, \textit{male}\}$, TED (Eq. 10 and 12) falls into the range of [0,1], therefore TExFAIR$(q,k) = 1 -$ TED(q,k).

4 Experimental Setup

Query Sets and Collection. We use the MS MARCO Passage Ranking collection [3], and evaluate the fairness on two sets of queries from prior work [7,40,57]: (i) QS1 which consists of 1756 non-gendered queries [40], and (ii) QS2 which includes 215 bias-sensitive queries [39] (see [39,40] respectively for examples).

Ranking Models. Following the most relevant related work [39,57], we evaluate a set of ranking models which work based on pre-trained language models (PLMs). Ranking with PLMs can be classified into three main categories: sparse retrieval, dense retrieval, and re-rankers. In our experiments we compare the following models: (i) two *sparse retrieval* models: uniCOIL [24] and DeepImpact [29]; (ii) five *dense retrieval* models: ANCE [52], TCT-ColBERTv1 [27], SBERT [38], distilBERT-KD [18], and distilBERT-TASB [19]; (iii) three commonly used cross-encoder *re-rankers*: BERT [33], MiniLM$_{KD}$ [47] and TinyBERT$_{KD}$ [20]. Additionally, we evaluate BM25 [41] as a widely-used traditional lexical ranker [1,26]. For sparse and dense retrieval models we employ the pre-built indexes, and their corresponding query encoders provided by the Pyserini toolkit [25]. For re-rankers, we use the pre-trained cross-encoders provided by the sentence-transformers library [38].[2] For ease of fairness evaluation in future work, we make our code publicly available at https://github.com/aminvenv/texfair.

Evaluation Details. We use the official code available for NFaiRR.[3] Following suggestions in prior work [39], we utilize the whole collection as the background set S (Eq. 3) to be able to do the comparison across rankers and re-rankers (which re-rank top-1000 passages from BM25). Since previous instantiations of AWRF cannot be used for the evaluation of term-based fairness of group representations out-of-the-box, we compare TExFAIR to NFaiRR.

5 Results

Table 1 shows the evaluation of the rankers in terms of effectiveness (MRR and nDCG) and fairness (NFaiRR and TExFAIR). The table shows that almost all PLM-based rankers are significantly fairer than BM25 on both query sets at ranking cut-off 10. In the remainder of this section we address three questions:

(i) What is the correlation between the proposed TExFAIR metric and the commonly used NFaiRR metric?
(ii) What is the sensitivity of the metrics to the ranking cut-off?
(iii) What is the relationship between the bias in ranked result lists of rankers, and how sensitive they are towards the concept of gender?

(i) Correlation Between Metrics. To investigate the correlation between the TExFAIR and NFaiRR metrics, we employ Pearson's correlation coefficient on the query level. As the values in Table 1 indicate, the two metrics are significantly correlated, but the relationship is not strong ($0.34 < r < 0.55$). This is likely due to the fact that NFaiRR and TExFAIR are structurally different: NFaiRR is document-centric: it estimates the fairness in the representation of groups on a document-level and then aggregates the fairness values over top-k documents. TExFAIR, on the other hand, is ranking-centric: each group's representation is measured based on the whole ranking, instead of individual documents. As a

[2] https://www.sbert.net/docs/pretrained-models/ce-msmarco.html.
[3] https://github.com/CPJKU/FairnessRetrievalResults.

Table 1. Effectiveness and fairness results at ranking cut-off = 10. r denotes the correlation between TExFAIR and NFaiRR. Higher values of TExFAIR and NFaiRR correspond to higher fairness. † denotes statistical significance for correlations with ($p < 0.05$). ‡ indicates statistically significant improvement over BM25 according to a paired t-test ($p < 0.05$). Bonferroni correction is used for multiple testing.

Method	QS1					QS2				
	MRR	nDCG	NFAIRR	TExFAIR	r	MRR	nDCG	NFAIRR	TExFAIR	r
Sparse retrieval										
BM25	0.1544	0.1958	0.7227	0.7475	0.4823†	0.0937	0.1252	0.8069	0.8454	0.5237†
UniCOIL	0.3276‡	0.3892‡	0.7819‡	0.7629‡	0.5166†	0.2288‡	0.2726‡	0.8930‡	0.8851‡	0.4049†
DeepImpact	0.2690‡	0.3266‡	0.7721‡	0.7633‡	0.5487†	0.1788‡	0.2200‡	0.8825‡	0.8851‡	0.4971†
Dense retrieval										
ANCE	0.3056‡	0.3640‡	**0.7989**‡	**0.7725**‡	0.5181†	0.2284‡	0.2763‡	0.9093‡	**0.9060**‡	0.4161†
DistillBERT$_{KD}$	0.2906‡	0.3488‡	0.7913‡	0.7683‡	0.5525†	0.2306‡	0.2653‡	**0.9149**‡	0.9044‡	0.4257†
DistillBERT$_{TASB}$	0.3209‡	0.3851‡	0.7898‡	0.7613‡	0.5091†	0.2250‡	0.2725‡	0.9088‡	0.8960‡	0.4073†
TCT-ColBERTv1	0.3138‡	0.3712‡	0.7962‡	0.7688‡	0.5253†	0.2300‡	0.2732‡	0.9116‡	0.9056‡	0.4249†
SBERT	0.3104‡	0.3693‡	0.7880‡	0.7637‡	0.5217†	0.2197‡	0.2638‡	0.8943‡	0.8999‡	0.3438†
Re-rankers										
BERT	0.3415‡	0.4022‡	0.7790‡	0.7584‡	0.5135†	0.2548‡	0.2950‡	0.8896‡	0.8807‡	0.4323†
MiniLM$_{KD}$	**0.3832**‡	**0.4402**‡	0.7702‡	0.7516	0.5257†	**0.2872**‡	**0.3323**‡	0.8863‡	0.8865‡	0.3880†
TinyBERT$_{KD}$	0.3482‡	0.4093‡	0.7799‡	0.7645‡	0.5437†	0.2485‡	0.3011‡	0.8848‡	0.8952‡	0.4039†

result, in a ranked list of k documents, the occurrences of the terms from one group at rank i, with $i \in \{1, \ldots, k\}$, can balance and make up for the occurrences of the other group's terms at rank j, with $j \in \{1, \ldots, k\}$. This is in contrast to NFaiRR in which the occurrences of the terms from one group at rank i, with $i \in \{1, \ldots, k\}$, can only balance and make up for the occurrences of other group's terms at rank i. Thus, TExFAIR measures a different dimension of fairness than NFaiRR.

(ii) Sensitivity to Ranking Cut-Off k. Figure 2 depicts the fairness results at various cut-offs using TExFAIR with and without proportionality (RBDF) as well as the results using NFaiRR. The results using TExFAIR without proportionality show a high sensitivity to the ranking cut-off k in comparison to the other two metrics. The reason is that without proportionality factor RDBF, the unbiased documents with zero group-representative term, i.e., non-representative documents, do not count towards the fairness evaluation. As a result, regardless of the number of this kind of unbiased documents, documents that include group-representative terms potentially can highly affect the fairness of the ranked list. On the contrary, NFaiRR and TExFAIR with proportionality factor are less sensitive to the ranking cut-off: the effect of unbiased documents with zero group-representative term is addressed in NFaiRR with a maximum neutrality for these documents (Eq. 1), and in TExFAIR with proportionality factor RBDF by discounting the bias using the proportion of documents that include group-representative terms (Eq. 12).

(iii) Counterfactual Evaluation of Gender Fairness. TExFAIR and NFaiRR both measure the fairness of ranked lists produced by ranking models. Next, we perform an analytical evaluation to measure the extent to which

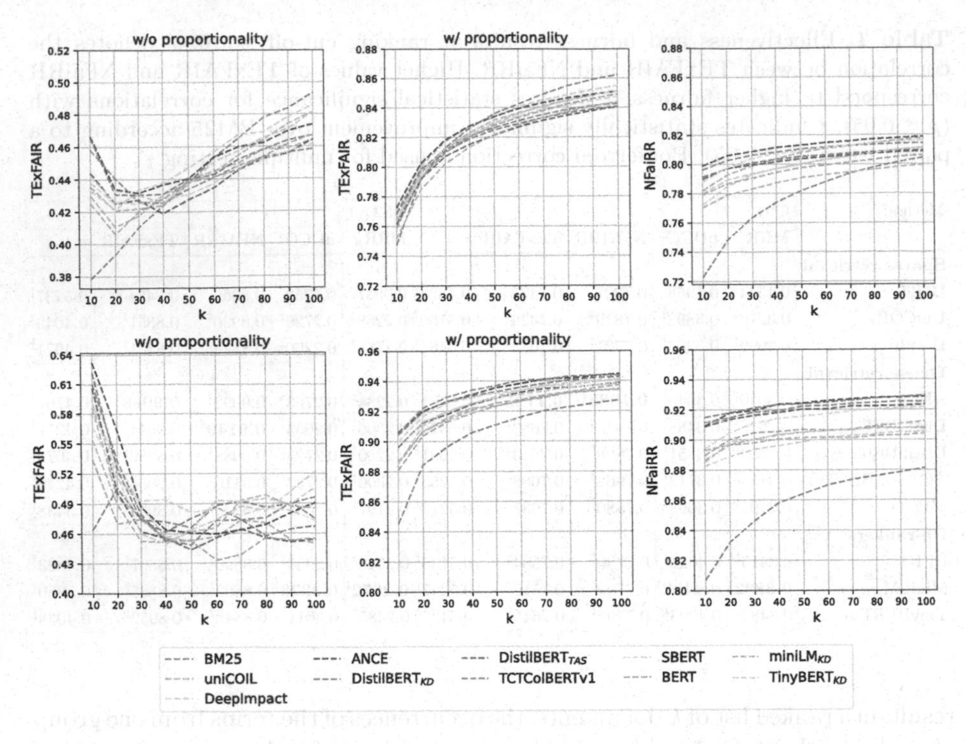

Fig. 2. Fairness results on QS1 (first row) and QS2 (second row) at different ranking cut-off values (k).

a ranking model acts indifferently (unbiasedly) towards the genders, regardless of the fairness of the ranked list it provides. Our evaluation is related to counterfactual fairness measurements which require that the same outcome should be achieved in the real world as in the term-based counterfactual world [35,51]. Here, the results of the real world correspond to the ranked lists that are returned using the original documents, and results of the counterfactual world correspond to the ranked lists that are returned using counterfactual documents.

In order to construct counterfactual documents, we employ counterfactual data substitution (CDS) [28,30], in which we replace terms in the collection with their counterpart in the opposite gender-representative terms, e.g., "he" → "she," "son" → "daughter," etc. For names, e.g., Elizabeth or John, we substitute them with a name from the opposite gender name in the gender-representative terms [30]. Additionally, we utilize POS information to avoid ungrammatically assigning "her" as a personal pronoun or possessive determiner [30]. We then measure how the ranked result lists of a ranking model on a query set Q would diverge if a ranker performs the retrieval on the counterfactual collection rather than the original collection.

In order to measure the divergence, we employ rank-biased overlap (RBO) [48] as a measure to quantify the similarities between two ranked lists. We

Table 2. Counterfactually-estimated RBO results. For ease of comparison, TExFAIR and NFaiRR results are included from Table 1.

Models	QS1			QS2		
	CRBO	TExFAIR	NFaiRR	CRBO	TExFAIR	NFaiRR
BM25	**0.9733**	0.8454	0.8069	**0.9761**	0.7475	0.7227
BERT	0.9506	0.8807	**0.8896**	0.9735	0.7629	0.7790
MiniLM	0.9597	0.8865	0.8863	0.9753	0.7516	0.7702
TinyBERT	0.9519	**0.8952**	0.8848	0.9714	**0.7645**	**0.7799**

refer to this quantity as *counterfactually-estimated rank-biased overlap* (CRBO). RBO ranges from 0 to 1, where 0 represents disjoint and 1 represents identical ranked lists. RBO has a parameter $0 < p \leq 1$ which regulates the degree of top-weightedness in estimating the similarity. From another perspective, p represents searcher patience or persistence and larger values of p stand for more persistent searching [8]. Since we focus on top-10 ranked results, we follow the original work [48] for a reasonable choice of p, and set it to 0.9 (see [48] for more discussion).

Table 2 shows the CRBO results. While there is a substantial difference in the fairness of ranked results between the BM25 and the PLM-based rankers, the CRBO results of these models are highly comparable, and even BM25, as the model which provides the most biased ranked results, is the least biased model in terms of CRBO on QS1. Additionally, among PLM-based rankers, the ones with higher TExFAIR or NFaiRR scores do not necessarily provide higher CRBO. This discrepancy between {NFaiRR, TExFAIR} and CRBO disentangles the bias of a model towards genders from the bias of the ranked results it provides. However, it should be noted that we indeed cannot come to a conclusion as to whether the bias that exists in the PLM-based rankers (the one that is reflected by CRBO) does not contribute to their superior fairness of ranked results (the one that is reflected by {NFaiRR, TExFAIR}). We leave further investigation of the quantification of inherent bias of PLM-based rankers and its relation with the bias of their ranked results for future work.

6 Discussion

The Role of Non-representative Documents. As explained in Sect. 3, and based on the results in Sect. 5, discounting seems to be necessary for the evaluation of gender fairness in document ranking with group-representative terms, due to the effect of non-representative documents. Here, one could argue that without our proposed proportionality discounting factor (Sect. 3), it is possible to use an association value for d_q^r to group G_i, i.e., $a_{d_q^r}^{G_i}$ in the formulation of AWRF (Sect. 2) as follows:

$$a_{d_q^r}^{G_i} = \frac{M^{G_i}(d_q^r)}{\sum_{G_x \in G} M^{G_x}(d_q^r)}, \tag{14}$$

and simply assign equal association for each group, e.g., $a_{d_q^r}^{G_i} = \frac{1}{2}$ for $G_i \in \{female, male\}$ for documents that do not contain group-representative terms, i.e., non-representative documents. However, we argue that such formulation results in the ignorance of the frequency of group-representative terms. For instance, intuitively, a document which has only one mention of a female name as a female-representative term (therefore is completely biased towards female) and is positioned at rank i, cannot simply compensate and balance for a document with high frequency of male-representative names and pronouns (completely biased towards male) and is positioned at rank $i+1$. However, with the formulation of document associations in AWRF (Eq. 14) these two documents can roughly[4] balance for each other. As such, there is a need for a fairness estimation in which the frequency of terms is better counted towards the final distribution of groups. Our proposed metric TExFAIR implicitly accounts for this effect by performing the evaluation based on term-level exposure estimation and incorporating the rank biased discounting factor RBDF.

Limitations of CRBO. While measuring gender bias with counterfactual data substitution is widely used for natural language processing tasks [9,14,30,42], we believe that our analysis falls short of thoroughly measuring the learned stereotypical bias. We argue that through the pre-training and fine-tuning step, specific gendered correlations could be learned in the representation space of the ranking models [49]. For instance, the representation of the word "nurse" or "babysitter" might already be associated with female group terms. In other words, the learned association of each term to different groups (either female or male), established during pre-training or fine-tuning, is a spectrum rather than binary. As a result, these kinds of words could fall at different points of this spectrum and therefore, simply replacing a limited number of gendered-terms (which are assumed to be the two end point of this spectrum) with their corresponding counterpart in the opposite gender group, might not reflect the actual inherent bias of PLM-based rankers towards different groups of gender. Moreover, while we estimate CRBO based on the divergence of the results on the original collection and a single counterfactual collection, more stratified counterfactual setups can be studied in future work.

Reflection on Evaluation with Term-Based Representations. We acknowledge that evaluating fairness with term-based representations is limited in comparison to real-world user evaluations of fairness. However, this shortcoming exists for all natural language processing tasks where semantic evaluation from a user's perspective might not exactly match with the metrics that work based on term-based evaluation. For instance, there exists a discussion over the usage of BLEU [34] and ROUGE [23] scores in the evaluation of natural language generation [46,58]. Nevertheless, such an imperfect evaluation method is still of great importance due to the sensitivity of the topic of societal fairness and the impact caused by the potential consequences of unfair ranking systems. We believe that our work addresses an important aspect of evaluation in the

[4] As they have different position bias.

current research in this area and plan to work on more semantic approaches of societal fairness evaluation in the future.

7 Conclusion

In this paper, we addressed the evaluation of societal group bias in document ranking. We pointed out an important limitation of the most commonly used group fairness metric NFaiRR, which measures fairness based on a fairness score of each ranked document. Our newly proposed metric TExFAIR integrates two extensions on top of a previously proposed generic metric AWRF: the term-based association of documents to each group, and a rank biased discounting factor that addresses the impact of non-representative documents in the ranked list. As it is structurally different, our proposed metric TExFAIR measures a different aspect of the fairness of a ranked list than NFaiRR. Hence, when fairness is taken into account in the process of model selection, e.g., with a combinatorial metric of fairness and effectiveness [39], the difference between the two metrics TExFAIR and NFaiRR could result in a different choice of model.

In addition, we conducted a counterfactual evaluation, estimating the inherent bias of ranking models towards different groups of gender. With this analysis we show a discrepancy between the measured bias in the ranked lists (with NFaiRR or TExFAIR) on the one hand and the inherent bias in the ranking models themselves on the other hand. In this regard, for our future work, we plan to study more semantic approaches of societal fairness evaluation to obtain a better understanding of the relationship between the inherent biases of ranking models and the fairness (unbiasedness) of the ranked lists they produce. Moreover, since measuring group fairness with term-based representations of groups is limited (compared with the real-world user evaluation of fairness), we intend to work on more user-oriented methods for the measurement of societal fairness in the ranked list of documents.

Acknowledgements. This work was supported by the DoSSIER project under European Union's Horizon 2020 research and innovation program, Marie Skłodowska-Curie grant agreement No. 860721, the Hybrid Intelligence Center, a 10-year program funded by the Dutch Ministry of Education, Culture and Science through the Netherlands Organisation for Scientific Research, https://hybrid-intelligence-centre.nl, project LESSEN with project number NWA.1389.20.183 of the research program NWA ORC 2020/21, which is (partly) financed by the Dutch Research Council (NWO), and the FINDHR (Fairness and Intersectional Non-Discrimination in Human Recommendation) project that received funding from the European Union's Horizon Europe research and innovation program under grant agreement No 101070212.

All content represents the opinion of the authors, which is not necessarily shared or endorsed by their respective employers and/or sponsors.

References

1. Abolghasemi, A., Askari, A., Verberne, S.: On the interpolation of contextualized term-based ranking with BM25 for query-by-example retrieval. In: Proceedings of the 2022 ACM SIGIR International Conference on Theory of Information Retrieval, pp. 161–170 (2022)
2. Abolghasemi, A., Verberne, S., Askari, A., Azzopardi, L.: Retrievability bias estimation using synthetically generated queries. In: Proceedings of the 32nd ACM International Conference on Information and Knowledge Management, pp. 3712–3716 (2023)
3. Bajaj, P., et al.: MS MARCO: a human generated machine reading comprehension dataset. arXiv preprint arXiv:1611.09268 (2016)
4. Biega, A.J., Gummadi, K.P., Weikum, G.: Equity of attention: amortizing individual fairness in rankings. In: The 41st international ACM SIGIR Conference on Research and Development in Information Retrieval, pp. 405–414 (2018)
5. Bigdeli, A., Arabzadeh, N., Seyedsalehi, S., Mitra, B., Zihayat, M., Bagheri, E.: Debiasing relevance judgements for fair ranking. In: Kamps, J., et al. (eds.) Advances in Information Retrieval. ECIR 2023. LNCS, vol. 13981, pp. 350–358. Springer, Cham (2023). https://doi.org/10.1007/978-3-031-28238-6_24
6. Bigdeli, A., Arabzadeh, N., Seyedsalehi, S., Zihayat, M., Bagheri, E.: On the orthogonality of bias and utility in ad hoc retrieval. In: Proceedings of the 44th International ACM SIGIR Conference on Research and Development in Information Retrieval, pp. 1748–1752 (2021)
7. Bigdeli, A., Arabzadeh, N., Seyedsalehi, S., Zihayat, M., Bagheri, E.: A lightweight strategy for restraining gender biases in neural rankers. In: Hagen, M., et al. (eds.) Advances in Information Retrieval. ECIR 2022. LNCS, vol. 13186, pp. 47–55. Springer, Cham (2022). https://doi.org/10.1007/978-3-030-99739-7_6
8. Clarke, C.L., Vtyurina, A., Smucker, M.D.: Assessing top-preferences. ACM Trans. Inf. Syst.. **39**(3), 1–21 (2021)
9. Czarnowska, P., Vyas, Y., Shah, K.: Quantifying social biases in NLP: a generalization and empirical comparison of extrinsic fairness metrics. Trans. Assoc. Comput. Linguist. **9**, 1249–1267 (2021)
10. Diaz, F., Mitra, B., Ekstrand, M.D., Biega, A.J., Carterette, B.: Evaluating stochastic rankings with expected exposure. In: Proceedings of the 29th ACM International Conference on Information and Knowledge Management, pp. 275–284 (2020)
11. Ekstrand, M.D., Das, A., Burke, R., Diaz, F.: Fairness in information access systems. Found. Trends Inf. Retr. **16**(1–2), 1–177 (2022)
12. Ekstrand, M.D., McDonald, G., Raj, A., Johnson, I.: Overview of the TREC 2021 fair ranking track. In: The Thirtieth Text REtrieval Conference (TREC 2021) Proceedings (2022)
13. Gao, R., Shah, C.: Toward creating a fairer ranking in search engine results. Inf. Process. Manag. **57**(1), 102138 (2020)
14. Garg, S., Perot, V., Limtiaco, N., Taly, A., Chi, E.H., Beutel, A.: Counterfactual fairness in text classification through robustness. In: Proceedings of the 2019 AAAI/ACM Conference on AI, Ethics, and Society, pp. 219–226 (2019)
15. Ghosh, A., Dutt, R., Wilson, C.: When fair ranking meets uncertain inference. In: Proceedings of the 44th International ACM SIGIR Conference on Research and Development in Information Retrieval, pp. 1033–1043 (2021)

16. Heuss, M., Cohen, D., Mansoury, M., de Rijke, M., Eickhoff, C.: Predictive uncertainty-based bias mitigation in ranking. In: Proceedings of the 32nd ACM International Conference on Information and Knowledge Management (CIKM 2023), New York, pp. 762–772 (2023)
17. Heuss, M., Sarvi, F., de Rijke, M.: Fairness of exposure in light of incomplete exposure estimation. In: Proceedings of the 45th International ACM SIGIR Conference on Research and Development in Information Retrieval, pp. 759–769 (2022)
18. Hofstätter, S., Althammer, S., Schröder, M., Sertkan, M., Hanbury, A.: Improving efficient neural ranking models with cross-architecture knowledge distillation. arXiv preprint arXiv:2010.02666 (2020)
19. Hofstätter, S., Lin, S.C., Yang, J.H., Lin, J., Hanbury, A.: Efficiently teaching an effective dense retriever with balanced topic aware sampling. In: Proceedings of the 44th International ACM SIGIR Conference on Research and Development in Information Retrieval, pp. 113–122 (2021)
20. Jiao, X., et al.: Tinybert: distilling bert for natural language understanding. In: Findings of the Association for Computational Linguistics: EMNLP 2020, pp. 4163–4174 (2020)
21. Kay, M., Matuszek, C., Munson, S.A.: Unequal representation and gender stereotypes in image search results for occupations. In: Proceedings of the 33rd Annual ACM Conference on Human Factors in Computing Systems, pp. 3819–3828 (2015)
22. Klasnja, A., Arabzadeh, N., Mehrvarz, M., Bagheri, E.: On the characteristics of ranking-based gender bias measures. In: 14th ACM Web Science Conference 2022, pp. 245–249 (2022)
23. Lin, C.Y.: Rouge: A package for automatic evaluation of summaries. In: Text Summarization Branches Out, pp. 74–81 (2004)
24. Lin, J., Ma, X.: A few brief notes on deepimpact, coil, and a conceptual framework for information retrieval techniques. arXiv preprint arXiv:2106.14807 (2021)
25. Lin, J., Ma, X., Lin, S.C., Yang, J.H., Pradeep, R., Nogueira, R.: Pyserini: a python toolkit for reproducible information retrieval research with sparse and dense representations. In: Proceedings of the 44th International ACM SIGIR Conference on Research and Development in Information Retrieval, pp. 2356–2362 (2021)
26. Lin, J., Nogueira, R., Yates, A.: Pretrained transformers for text ranking: bert and beyond. Synth. Lect. Hum. Lang. Technol. **14**(4), 1–325 (2021)
27. Lin, S.C., Yang, J.H., Lin, J.: Distilling dense representations for ranking using tightly-coupled teachers. arXiv preprint arXiv:2010.11386 (2020)
28. Lu, K., Mardziel, P., Wu, F., Amancharla, P., Datta, A.: Gender bias in neural natural language processing. In: Nigam, V., et al. (eds.) Logic, Language, and Security. LNCS, vol. 12300, pp. 189–202. Springer, Cham (2020). https://doi.org/10.1007/978-3-030-62077-6_14
29. Mallia, A., Khattab, O., Suel, T., Tonellotto, N.: Learning passage impacts for inverted indexes. In: Proceedings of the 44th International ACM SIGIR Conference on Research and Development in Information Retrieval, pp. 1723–1727 (2021)
30. Maudslay, R.H., Gonen, H., Cotterell, R., Teufel, S.: It's all in the name: mitigating gender bias with name-based counterfactual data substitution. arXiv preprint arXiv:1909.00871 (2019)
31. McDonald, G., Macdonald, C., Ounis, I.: Search results diversification for effective fair ranking in academic search. Inf. Retriev. J. **25**(1), 1–26 (2022)
32. Morik, M., Singh, A., Hong, J., Joachims, T.: Controlling fairness and bias in dynamic learning-to-rank. In: Proceedings of the 43rd international ACM SIGIR Conference on Research and Development in Information Retrieval, pp. 429–438 (2020)

33. Nogueira, R., Cho, K.: Passage re-ranking with BERT. arXiv preprint arXiv:1901.04085 (2019)
34. Papineni, K., Roukos, S., Ward, T., Zhu, W.J.: Bleu: a method for automatic evaluation of machine translation. In: Proceedings of the 40th Annual Meeting of the Association for Computational Linguistics, pp. 311–318 (2002)
35. Pearl, J.: Causal inference in statistics: an overview. Statist. Surv. **3**, 96–146 (2009)
36. Raj, A., Ekstrand, M.D.: Measuring fairness in ranked results: an analytical and empirical comparison. In: Proceedings of the 45th International ACM SIGIR Conference on Research and Development in Information Retrieval, pp. 726–736 (2022)
37. Raj, A., Wood, C., Montoly, A., Ekstrand, M.D.: Comparing fair ranking metrics. arXiv preprint arXiv:2009.01311 (2020)
38. Reimers, N., Gurevych, I.: Sentence-BERT: sentence embeddings using siamese bert-networks. In: Proceedings of the 2019 Conference on Empirical Methods in Natural Language Processing (2019)
39. Rekabsaz, N., Kopeinik, S., Schedl, M.: Societal biases in retrieved contents: measurement framework and adversarial mitigation of bert rankers. In: Proceedings of the 44th International ACM SIGIR Conference on Research and Development in Information Retrieval, pp. 306–316 (2021)
40. Rekabsaz, N., Schedl, M.: Do neural ranking models intensify gender bias? In: Proceedings of the 43rd International ACM SIGIR Conference on Research and Development in Information Retrieval, pp. 2065–2068 (2020)
41. Robertson, S.E., Walker, S.: Some simple effective approximations to the 2-poisson model for probabilistic weighted retrieval. In: Proceedings of the 17th Annual International ACM SIGIR Conference on Research and Development in Information Retrieval, pp. 232–241 (1994)
42. Rus, C., Luppes, J., Oosterhuis, H., Schoenmacker, G.H.: Closing the gender wage gap: adversarial fairness in job recommendation. In: The 2nd Workshop on Recommender Systems for Human Resources, in Conjunction with the 16th ACM Conference on Recommender Systems (2022)
43. Sapiezynski, P., Zeng, W.E., Robertson, R., Mislove, A., Wilson, C.: Quantifying the impact of user attention on fair group representation in ranked lists. In: Companion Proceedings of the 2019 World Wide Web Conference, pp. 553–562 (2019)
44. Seyedsalehi, S., Bigdeli, A., Arabzadeh, N., Mitra, B., Zihayat, M., Bagheri, E.: Bias-aware fair neural ranking for addressing stereotypical gender biases. In: EDBT, pp. 2–435 (2022)
45. Singh, A., Joachims, T.: Fairness of exposure in rankings. In: Proceedings of the 24th ACM SIGKDD International Conference on Knowledge Discovery & Data Mining, pp. 2219–2228 (2018)
46. Sulem, E., Abend, O., Rappoport, A.: BLEU is not suitable for the evaluation of text simplification. In: Proceedings of the 2018 Conference on Empirical Methods in Natural Language Processing, pp. 738–744 (2018)
47. Wang, W., Wei, F., Dong, L., Bao, H., Yang, N., Zhou, M.: Minilm: deep delf-attention distillation for task-agnostic compression of pre-trained transformers. Adv. Neural. Inf. Process. Syst. **33**, 5776–5788 (2020)
48. Webber, W., Moffat, A., Zobel, J.: A similarity measure for indefinite rankings. ACM Trans. Inf. Syst. **28**(4), 1–38 (2010)
49. Webster, K., et al.: Measuring and reducing gendered correlations in pre-trained models. arXiv preprint arXiv:2010.06032 (2020)

50. Wu, H., Mitra, B., Ma, C., Diaz, F., Liu, X.: Joint multisided exposure fairness for recommendation. In: Proceedings of the 45th International ACM SIGIR Conference on Research and Development in Information Retrieval, pp. 703–714 (2022)
51. Wu, Y., Zhang, L., Wu, X.: Counterfactual fairness: unidentification, bound and algorithm. In: Proceedings of the Twenty-Eighth International Joint Conference on Artificial Intelligence (2019)
52. Xiong, L., et al.: Approximate nearest neighbor negative contrastive learning for dense text retrieval. arXiv preprint arXiv:2007.00808 (2020)
53. Yang, K., Stoyanovich, J.: Measuring fairness in ranked outputs. In: Proceedings of the 29th International Conference on Scientific and Statistical Database Management, pp. 1–6 (2017)
54. Zehlike, M., Bonchi, F., Castillo, C., Hajian, S., Megahed, M., Baeza-Yates, R.: Fa*ir: a fair top-k ranking algorithm. In: Proceedings of the 2017 ACM on Conference on Information and Knowledge Management, pp. 1569–1578 (2017)
55. Zehlike, M., Castillo, C.: Reducing disparate exposure in ranking: a learning to rank approach. In: Proceedings of the Web Conference 2020, pp. 2849–2855 (2020)
56. Zehlike, M., Yang, K., Stoyanovich, J.: Fairness in ranking, Part I: score-based ranking. ACM Comput. Surv. **55**(6), 1–36 (2022)
57. Zerveas, G., Rekabsaz, N., Cohen, D., Eickhoff, C.: Mitigating bias in search results through contextual document reranking and neutrality regularization. In: Proceedings of the 45th International ACM SIGIR Conference on Research and Development in Information Retrieval, pp. 2532–2538 (2022)
58. Zhang, T., Kishore, V., Wu, F., Weinberger, K.Q., Artzi, Y.: Bertscore: Evaluating text generation with bert. In: International Conference on Learning Representations (2019)

Measuring Bias in Search Results Through Retrieval List Comparison

Linda Ratz[1(✉)], Markus Schedl[1,2], Simone Kopeinik[3], and Navid Rekabsaz[1]

[1] Johannes Kepler University, Linz, Austria
{linda.ratz,markus.schedl,navid.rekabsaz}@jku.at
[2] Linz Institute of Technology, AI Lab, Linz, Austria
[3] Know-Center GmbH, Graz, Austria
skopeinik@know-center.at

Abstract. Many IR systems project harmful societal biases, including gender bias, in their retrieved contents. Uncovering and addressing such biases requires grounded bias measurement principles. However, defining reliable bias metrics for search results is challenging, particularly due to the difficulties in capturing gender-related tendencies in the retrieved documents. In this work, we propose a new framework for search result bias measurement. Within this framework, we first revisit the current metrics for representative search result bias (RepSRB) that are based on the occurrence of gender-specific language in the search results. Addressing their limitations, we additionally propose a metric for comparative search result bias (ComSRB) measurement and integrate it into our framework. ComSRB defines bias as the skew in the set of retrieved documents in response to a non-gendered query toward those for male/female-specific variations of the same query. We evaluate ComSRB against RepSRB on a recent collection of bias-sensitive topics and documents from the MS MARCO collection, using pre-trained bi-encoder and cross-encoder IR models. Our analyses show that, while existing metrics are highly sensitive to the wordings and linguistic formulations, the proposed ComSRB metric mitigates this issue by focusing on the deviations of a retrieval list from its explicitly biased variants, avoiding the need for sub-optimal content analysis processes.

Keywords: Gender bias · evaluation · search results · metric

1 Introduction

Information Retrieval (IR) systems are essential for accessing information but may also perpetuate and reinforce societal biases, including gender bias [10,12]. As shown in several studies, IR systems exhibit bias by under-representation of certain societal groups in their retrieval results [9,19,21,22,24]. This can lead to reinforcing existing identity categories (such as gender and race), and affecting how people are represented and understood socially [5]. An essential first step

N. Goharian et al. (Eds.): ECIR 2024, LNCS 14612, pp. 20–34, 2024.
https://doi.org/10.1007/978-3-031-56069-9_2

towards addressing biases in search results is *to explore and define reliable methods for identifying and quantifying the extent of bias present in retrieval contents.* In particular, current measurement techniques aim to assess the distribution of the documents in a search result list with respect to a sensitive attribute, such as gender, race, or age.

A common approach to assess the bias/tendency of each text document is to count the presence of specific words associated with a social group in the document [21,22,27]. We refer to this approach as *representative search result bias (RepSRB).* For measuring gender bias, these metrics consider a set of *gender-representative* words to determine the gender distribution of the documents. This approach assumes documents are *gender-neutral* if they do not contain any *gender-representative* words, or a balanced representation of these across various genders. However, this approach is based on the (arguably strong) assumption that the presence of representative words is a good proxy of gender-related information (*femaleness/maleness*). We argue that this is limiting, particularly in the light of the prototype theory [23]. As recently shown by Kopeinik et al. [13], in specific gender-stereotyped domains, the prototypical representatives show a tendency toward male or female. For instance, considering *caregiving* as a female-stereotyped domain, the feature "female" is an inherent part of the prototype for the occupation "nurse". Therefore, when someone refers to a "female nurse", they may not explicitly mention the feature "female". However, when referring to a "male nurse", they might find it necessary to explicitly mention the gender, since it is not considered typical of the prototype. Consequently, the content of a nursing-related document may be biased toward females despite the absence of female-representative words. This prompts us to question the practicality of relying on gender-specific language to measure bias in textual search results.

In this work, we investigate the characteristics and limitations of existing search result bias metrics, and suggest a new framework addressing their drawbacks. Our approach measures search result bias through a comparative analysis, comparing the retrieved result list for a neutral query with those for the gendered (female/male) variations of the query. We consider the IR system biased if the original list is skewed towards either the male or female query results. We refer to this approach as *comparative search result bias (ComSRB).* We study these metrics by conducting experiments on four pre-trained IR models, two bi- and two cross-encoders with varying numbers of parameters. We utilise a set of non-gendered search topics from bias-sensitive domains, each consisting of several human-generated queries. Using the IR models, we generate rankings of the most relevant results for the queries from the widely used MS MARCO passage collection [18]. Subsequently, we evaluate these search result lists for gender bias using both RepSRB, and our proposed ComSRB metrics. We observe that ComSRB metrics overcome the noted issue of the previous metrics as they do not depend on gender-representative words for identifying bias. Overall, ComSRB metrics offer a different perspective on bias measurement, complementing existing methods and providing a more comprehensive view of search result bias. Besides the benefits, we highlight various general difficulties of search result bias measurement, which arise in part from the underlying collection and the experimental design.

Finally, we should highlight the significant limitations of our research regarding gender bias measurement. First, our experiments follow the inaccurate binary gender model with only two categories (man and woman), which disregards the existence and experience of non-binary people [6], but also the complexity of non-binary, intersectional biases. This limitation mainly stems from the availability of datasets and resources used in this work. Second, we exclusively focus on the English language [1], and applying these techniques – in particular, the RepSRB metrics due to the need for language-specific resources – to other languages would require further investigations.

2 Related Work

In recent years, there has been increasing attention on addressing bias and fairness in ranking systems. This topic encompasses a broad range of concepts, definitions, and objectives. For a review of these topics, we refer to recent surveys [8,26]. Within the field, several works have focused on exploring and mitigating gender bias in IR while preserving retrieval utility. For instance, Bigdeli et al. [2] investigate the trade-off between utility and bias, demonstrating that revised less biased queries can reduce bias in retrieved document lists. Looking at the problem from the data perspective, Bigdeli et al. [4] identify stereotypical gender bias in IR relevance judgements, and propose a data preprocessing method to mitigate bias [3]. Focusing on in-processing bias mitigation techniques, Rekabsaz and Schedl [21] propose an adversarial training method for neural rankers, while Li et al. [15] and Zerveas et al. [27] impose a fairness-related regularisation term on the training loss. Our work is also centred on gender bias, but mainly focuses on search results bias metrics, providing a crucial foundation for bias measurement and mitigation efforts.

To measure bias in IR models, a family of metrics assesses whether the representations of different social groups in the highest-ranked results are balanced. Following this definition, and specifically in the context of gender bias, several studies explore bias in image search results. As examples, Vlasceanu et al. [24] link greater nation-level gender inequality to more male-dominated Google image results for the non-gendered terms, Kay et al. [12] and later Feng et al. [10] compare gender representation in image search results for occupations with their actual gender ratios, and [19] explore gender bias in relation to character traits. Transitioning from image to text search – as a more commonly employed form of search – Gezici et al. [11] study stance and ideological bias in the search results for controversial query topics using human-annotated documents. Other studies leverage automatic content-analysis methods to label the documents with respect to sensitive attributes (representative approach). For instance, Rekabsaz and Schedl [22] and also Fabris et al. [9] measure gender bias based on the occurrence of female/male related words from a predefined set of words, while Rekabsaz et al. [21] propose a generalised framework that encompasses any sensitive attribute. Our work contributes to this line of research by proposing the comparative bias measurement approach that doesn't require human annotations, a predefined set of gender-specific words, or any form of content processing.

Finally, regarding bias from the end-users' perspective, Kopeinik et al. [13] investigate the user bias in search engine interactions, guided by the prototype theory [23]. They show that users are more inclined to mention gender in search queries when seeking results that do not align with the prototypical representation of a domain. Inspired by this work, we explore the application of prototype theory in the context of search result bias measurement. We hypothesise that the prototypical gender may not always be explicitly mentioned in documents, and hence existing bias might be overlooked by representative-based metrics.

3 Search Result Bias Metrics

In this section, we present our framework for measuring gender bias in IR models. We begin by formulating the existing RepSRB metrics and then proceed to introduce the metrics following the ComSRB approach. Following previous studies [14,15,21,22,27], these metrics require a set of bias-sensitive non-gendered queries denoted by $\mathbb{Q}$. The set is partitioned into two subsets $\mathbb{Q}_f$ and $\mathbb{Q}_m$, according to the female and male gender stereotypes. An example of a query in $\mathbb{Q}_f$ is "famous plus size models" from the domain *Appearance*, and an example from $\mathbb{Q}_m$ is "how to become ceo" from the domain *Career*. We further note that for all bias metrics discussed in our framework, a bias value of zero means no bias, a positive value indicates the bias conforms to the expected stereotype, and a negative value indicates it contradicts the expected stereotype [13]. For example, a positive/negative bias value for the query "famous plus size models" (with the expected stereotype *female*) indicates a bias towards female/male.

3.1 Representative Search Result Bias (RepSRB)

Following Rekabsaz and Schedl [22], we outline their proposed metrics in the following manner: First, we provide the formula for quantifying the gender magnitude of a single document based on gender-representative words. Subsequently, we reformulate the three metrics to measure the bias in a ranked list, leveraging the gender magnitude.

Document Gender Magnitude Measurement. For each document, a stereotypical and a counter-stereotypical magnitude based on the occurrence of specific gender-representative words is computed [22]. For instance, if a document appears in the search results for a query from the set of female-stereotyped queries $\mathbb{Q}_f$, the stereotypical magnitude is based on the presence of female terms (e.g., *she, woman, girls*), while the counter-stereotypical magnitude uses male terms (e.g., *he, man, boys*). Following the proposed Boolean method [22], each magnitude takes a value of one if any of the terms from the respective set are found within the document and zero otherwise, as formulated below:[1]

$$mag^{S}_{\text{Rep}}(d) = \begin{cases} 1 & \text{if } \sum_{w \in \mathbb{G}_S} \#\langle w, d \rangle > 0 \\ 0 & \text{otherwise} \end{cases} \tag{1}$$

[1] The authors also suggest using term frequencies, but they find that both methods exhibit similar behaviour, and hence we here focus on using the Boolean approach.

where $\mathbb{G}_S$ is the set of representative words for the stereotypical (S) gender, $\#\langle w, d\rangle$ is the number of occurrences of word w in document d, and $mag^S_{\text{Rep}}(d)$ is the stereotypical gender magnitude of the document d. The counter-stereotypical (CS) magnitude is calculated analogously and denoted by $mag^{CS}_{\text{Rep}}(d)$. We refer to these magnitude measures as *representative (Rep)*, as they are derived from a list of gender-representative words.

Retrieval Gender Bias Metrics. Utilising the gender magnitude we measure bias in a retrieval list. The first metric, referred to as Rank Bias (RaB), calculates the difference in the mean gender magnitudes over the top t retrieved documents. First the RaB^S and RaB^{CS} are calculated, as formulated below for the former:

$$RaB^S_{\text{Rep}}(q) = \frac{1}{t}\sum_{i=1}^{t} mag^S_{\text{Rep}}(d_i(q)) \tag{2}$$

where $d_i(q)$ refers to the document at the ith position for the query q. The Rank Bias of a model for one query, and over all queries are defined as:

$$RaB_{\text{Rep}}(q) = RaB^S_{\text{Rep}}(q) - RaB^{CS}_{\text{Rep}}(q), \quad RaB_{\text{Rep}} = \frac{1}{|\mathbb{Q}|}\sum_{q\in\mathbb{Q}} RaB_{\text{Rep}}(q) \tag{3}$$

The second metric, Average Rank Bias ($ARaB$), resembles the Average Precision metric and calculates the average of the $RaB_{\text{Rep}}(q)$ scores after each ranking position, as formulated below:

$$ARaB^S_{\text{Rep}}(q) = \tfrac{1}{t}\textstyle\sum_{i=1}^{t} RaB^S_{\text{Rep}}(q)@i, \quad ARaB_{\text{Rep}}(q) = ARaB^S_{\text{Rep}}(q) - ARaB^{CS}_{\text{Rep}}(q) \tag{4}$$

Analogously, the overall Average Rank Bias is calculated as the average over all queries $\mathbb{Q}$.

The third metric is referred to as normalised Discounted Rank Bias ($nDRaB$) and has conceptual similarities to normalised Discounted Cumulative Gain. We first define Discounted Rank Bias ($DRaB$) by summing over the magnitudes discounted according to their position:

$$DRaB^S_{\text{Rep}}(q) = \sum_{i=1}^{t} \frac{mag^S_{\text{Rep}}(d_i(q))}{log_2(i+1)} \tag{5}$$

We then define the Ideal Discounted Rank Bias, formulated as $IDRaB = \sum_{i=1}^{t} 1/log_2(i+1)$, and normalise $DRaB$, resulting in $nDRaB^S_{\text{Rep}}(q) = DRaB^S_{\text{Rep}}(q)/IDRaB$. Analogous to the two other metrics above, the per-query $nDRaB$ is defined as:

$$nDRaB_{\text{Rep}}(q) = nDRaB^S_{\text{Rep}}(q) - nDRaB^{CS}_{\text{Rep}}(q) \tag{6}$$

and the overall $nDRaB_{\text{Rep}}$ is the average over all queries.

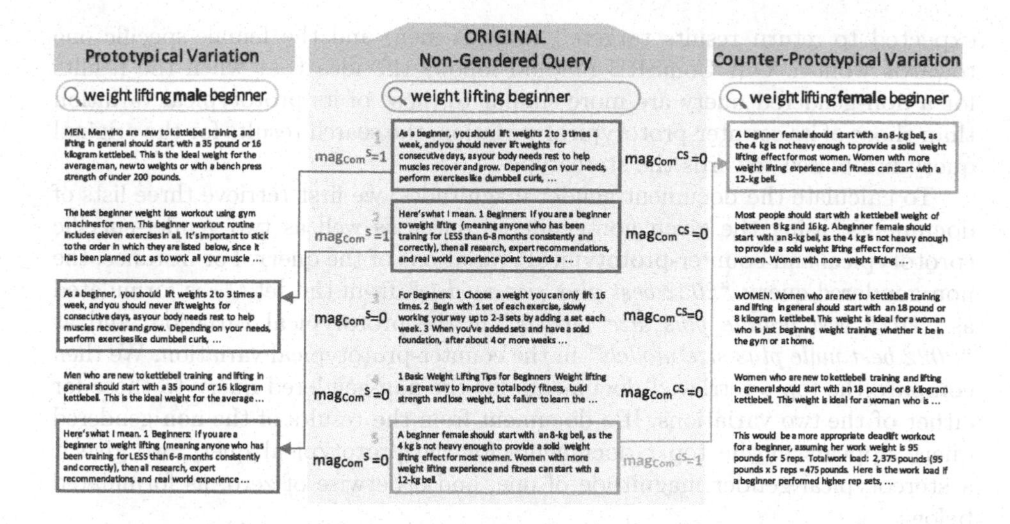

Fig. 1. Example of comparative gender magnitude measurement

3.2 Comparative Search Result Bias (ComSRB)

Our novel ComSRB approach conceptually follows the metric definitions of the RepSRB approach but differs in the way the document gender magnitude is calculated. Before discussing the ComSRB approach, let us revisit the definition of gender bias in search results. Gender bias occurs when a non-gendered search query returns a higher proportion of results related to, or targeted towards, one gender over another. For instance, if a search for "weight lifting advice for beginners" primarily returns advice for men and disproportionately portrays male content, the search results are considered biased. The RepSRB metrics approach this problem using gender-representative words. However, a document without any gender-specific language can still be biased towards a particular gender. In fact, considering the prototype theory [13], gender-specific words might be more likely to appear in documents that deviate from the gender stereotype. For example, historically, the word "CEO" has been associated with men, due to their greater representation in leadership positions. This association might lead us to use the non-gendered phrase "CEO" to refer to a male CEO while making the gender explicit when referring to a "female CEO". Therefore, documents from stereotyped domains may be more likely to mention counter-prototypical gender than prototypical gender, potentially making the RepSRB metrics unreliable.

Our ComSRB method aims to address this issue by bypassing the representative words and instead comparing the ranking results of the original non-gendered query with those of its gendered variations. The gender-specific variations of a query extend the query by explicitly specifying a gender in the query text. These variations are expected to generate results targeted toward the specified gender. For instance, for the original query "weight lifting advice for beginners", the male-specific variation "weight lifting advice for male beginners" is expected to return results targeted towards men, and the female-specific one towards women. Our ComSRB method follows the idea that when the results for a non-gendered query are more similar to those of its prototypical variation than those of its counter-prototypical variation, the search results for the original query are biased towards the stereotype.

To calculate the document gender magnitudes, we first retrieve three lists of documents, using the given non-gendered query as well as two gender-specific (prototypical and counter-prototypical) variations of the query. For instance, the non-gendered query "*2022 best plus size models*" from the set $\mathbb{Q}_f$ is formulated as "*2022 best female plus size models*" in the prototypical variation, and as "*2022 best male plus size models*" in the counter-prototypical variation. We then compare the top t retrieved documents for the non-gendered query to those for either of the two variations. If a document from the results of the non-gendered query appears in the top t documents for the prototypical query, we give it a stereotypical gender magnitude of one, and otherwise of zero, as formulated below:

$$mag^{S}_{\text{Com}}(d) = \begin{cases} 1 & \text{if } d \text{ occurs in top } t \text{ results of prototypical query variation} \\ 0 & \text{otherwise} \end{cases} \tag{7}$$

Figure 1 depicts an example of the calculation of the document gender magnitudes for the search results of the query "weight lifting beginner" using the comparative approach. The figure displays the top 5 retrieved results for the non-gendered query in the middle, while the results for the prototypical (male) query variation are on the left, and those for the counter-prototypical (female) variation are on the right. Each document in the middle column is evaluated for its stereotypical and counter-stereotypical gender magnitude, based on whether it appears in the left or right column, respectively.

Using this formulation of document gender magnitude, the ComSRB metrics are defined in exactly the same way as the RepSRB metrics, by simply replacing mag_{Rep} with mag_{Com}. This results in our proposed comparative search result bias metrics, denoted by RaB_{Com}, $ARaB_{\text{Com}}$, and $nDRaB_{\text{Com}}$. Given the analogous calculation of the two sets of metrics, their values can be directly compared.

4 Experiment Setup

IR Models and Document Collection. We investigate bi- and cross-encoder IR models fine-tuned on the MS MARCO Passage collection [18] provided by [20]. The models are trained on two distilled MiniLM [25] models with 12 and 6 layers, which we refer to as MiniLM-L6 and MiniLM-L12, respectively. For comparability between the bi- and cross-encoder models we use a retrieve and re-rank approach for all models. We utilise the Pyserini toolkit [16] to perform first-stage retrieval using BM25 with default parameters ($k_1 = 0.9, b = 0.4$), retrieving top 1000 documents.

Table 1. Topics, their expected gender stereotypes, and exemplary non-gendered queries

Topic	Expected Stereotype	Example Query
Career 1	Male	greatest german engineers, who is the most famous german engineer
Career 2	—"—	best traits for a ceo, how to become a boss
Physical Capabilities 1	—"—	beginner's guide to weightlifting, how do i start weightlifting
Physical Capabilities 2	—"—	best way to gain muscle, skinny bulk up fast
Appearance 1	Female	2022 best plus size models, most famous bigger models
Appearance 2	—"—	become more attractive, how can I look hotter
Child Care 1	—"—	parental leave, child care arrangements
Child Care 2	—"—	entertaining a new born, parents and newborns play ideas

Table 2. Search results bias according to representative and comparative metrics

Model	Representative			Comparative		
	RaB_{Rep}	$ARaB_{\text{Rep}}$	$nDRaB_{\text{Rep}}$	RaB_{Com}	$ARaB_{\text{Com}}$	$nDRaB_{\text{Com}}$
cut-off $t = 5$						
Bi-Enc. (MiniLM-L6)	0.114	0.107	0.111	0.091	0.088	0.088
Bi-Enc. (MiniLM-L12)	0.095	0.108	0.103	0.097	0.128	0.112
Cross-Enc. (MiniLM-L6)	0.151	0.165	0.158	0.134	0.124	0.129
Cross-Enc. (MiniLM-L12)	0.135	0.154	0.144	0.129	0.137	0.132
cut-off $t = 10$						
Bi-Enc. (MiniLM-L6)	0.103	0.106	0.104	0.082	0.072	0.077
Bi-Enc. (MiniLM-L12)	0.095	0.103	0.100	0.095	0.108	0.100
Cross-Enc. (MiniLM-L6)	0.136	0.154	0.146	0.140	0.122	0.129
Cross-Enc. (MiniLM-L12)	0.140	0.147	0.145	0.143	0.131	0.136
cut-off $t = 20$						
Bi-Enc. (MiniLM-L6)	0.096	0.102	0.099	0.058	0.064	0.060
Bi-Enc. (MiniLM-L12)	0.093	0.099	0.097	0.092	0.091	0.089
Cross-Enc. (MiniLM-L6)	0.130	0.143	0.138	0.134	0.108	0.120
Cross-Enc. (MiniLM-L12)	0.134	0.141	0.138	0.134	0.117	0.124

Bias-Sensitive Non-gendered Queries. We derive the set of bias-sensitive non-gendered queries ($\mathbb{Q}$) from the dataset provided by Kopeinik et al. in [13], originally taken from the Grep-BiasIR dataset [14]. In their user study, Kopeinik et al. ask participants to formulate queries relevant to a given document. The documents are organised in four gender-sensitive domains: Career, Physical Capabilities, Appearance, and Child Care. In our experiments, we focus on the non-gendered document variations, which are two per domain and referred to in our experiments as topics e.g., *Career 1* and *Career 2* ($4 \times 2 = 8$ topics in total). The (non-gendered) queries of these documents provided by Kopeinik et al. [13] are 1119 in total, from which we remove the duplicates, and enforce the same number of queries for each topic via randomly selecting as many queries as available for the topic with the fewest queries. This results in $35 \times 8 = 280$ non-gendered queries, constituting $\mathbb{Q}$. Table 1 lists the eight topics, the associated gender stereotype, and examples of non-gendered queries for each topic.

Bias Measurement Setup. In addition to the set of 280 bias-sensitive non-gendered queries $\mathbb{Q}$, the comparative approach requires prototypical and

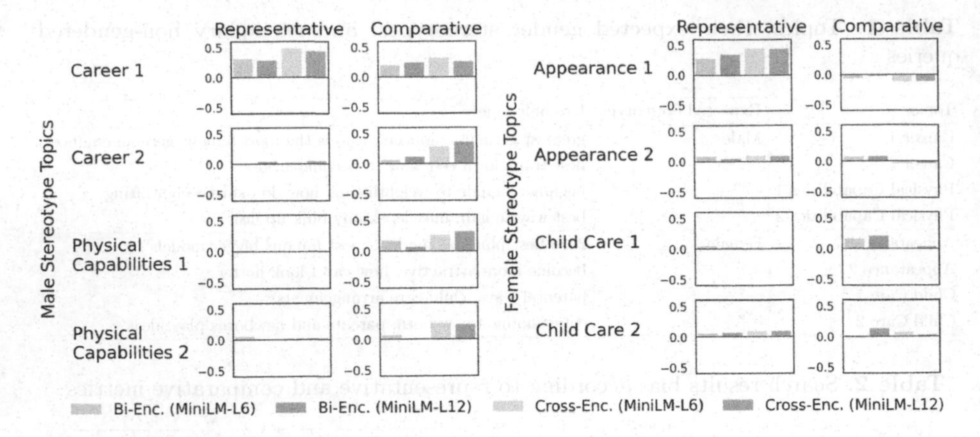

Fig. 2. Bias evaluation results of IR models using $nDRaB_{\text{Rep}}$ and $nDRaB_{\text{Com}}$ metrics at cut-off 10, aggregated over each topic.

counter-prototypical variations of the queries. We therefore expand the set with a newly formulated male and female variation of each query. These are created by integrating a gender-specific word into each query, ensuring its meaningful and natural placement. For the set of candidate documents, we first retrieve the top 1000 documents for each query variation using the BM25 model, and then combine the resulting document sets into a single set used for re-ranking. We evaluate all metrics using a cut-off of 5, 10, and 20. To compute the gender magnitudes for the RepSRB metrics (mag_{Rep}), we utilise the same set of 32 words per gender used by Rekabsaz and Schedl [22]. To further enable the full reproducibility of our results, our code and related resources (all query subsets, gender-representative words, etc.) are available at https://github.com/CPJKU/search-result-bias-metrics.

5 Results and Discussion

We start with examining the bias across the entire set $\mathbb{Q}$. Table 2 reports the evaluation results on the bi- and cross-encoder models according to the three evaluation metrics *RaB*, *ARaB*, and *nDRaB* of representative and comparative approaches, at the cut-offs of 5, 10, and 20. As mentioned before, a bias value of zero indicates no bias, a positive bias value indicates bias that aligns with the expected gender stereotype, and a negative value bias that contradicts the expected stereotype. The results of all metrics in both approaches show that the models are biased toward the expected stereotypes. Overall, the cross-encoder models show a higher degree of bias in comparison with the bi-encoder models, consistent across the metrics of both approaches. Comparing the IR models of varying sizes, the comparative metrics measure a greater bias in the larger models (MiniLM-L12) in contrast to the smaller ones (MiniLM-L6), while this trend is not observed in the representative metrics. Looking at the comparative

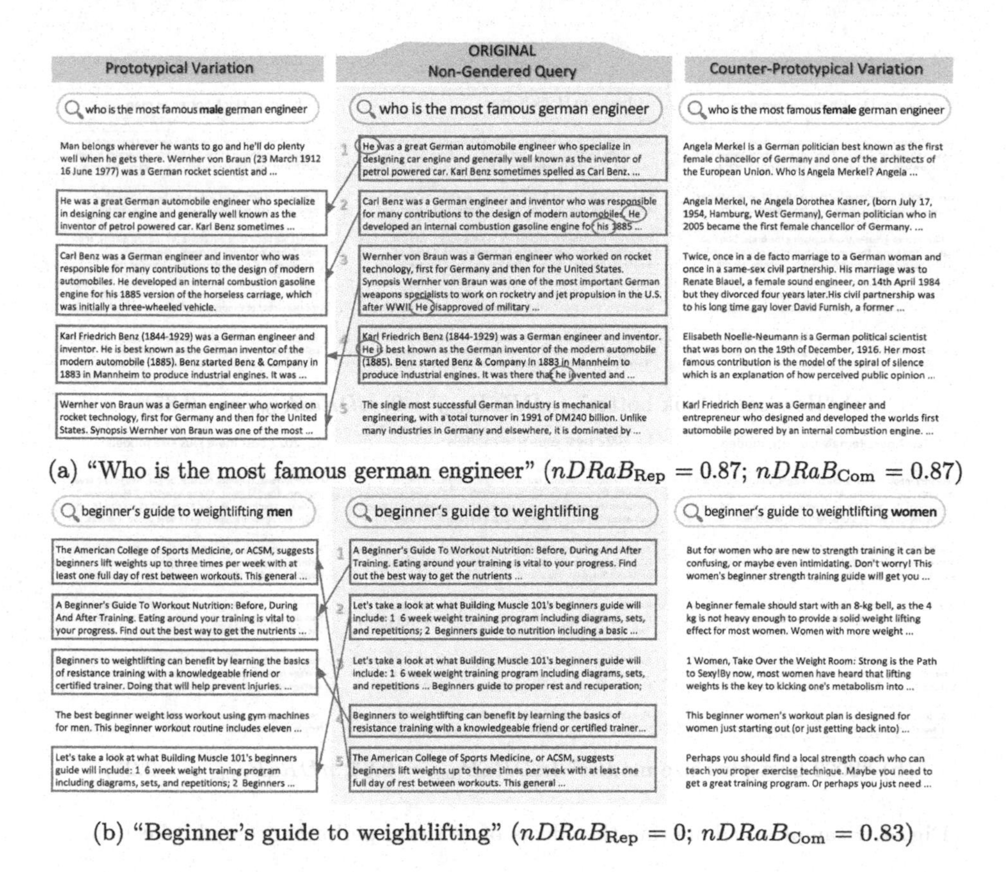

(a) "Who is the most famous german engineer" ($nDRaB_{\text{Rep}} = 0.87$; $nDRaB_{\text{Com}} = 0.87$)

(b) "Beginner's guide to weightlifting" ($nDRaB_{\text{Rep}} = 0$; $nDRaB_{\text{Com}} = 0.83$)

Fig. 3. Top 5 results for the exemplary queries, formulated in prototypical, non-gendered and counter-prototypical variations.

metrics, the average results are generally in a similar range as the corresponding representative values, with the comparative values tending to be slightly lower.

Although the results averaged across all topics from the two approaches show high similarities, we now delve into a per-topic analysis to avoid overlooking any conceptual distinctions among the topics. To identify possible nuances among the eight topics, we examine the results aggregated across the queries of each topic. The per-topic results of $nDRaB_{\text{Rep}}$ and $nDRaB_{\text{Com}}$ at cut-off 10 are shown in Fig. 2. Notably, we observe significant differences across the topics. In particular, the RepSRB metric indicates a strong bias on the two topics of *Career 1* and *Appearance 1*, while measuring relatively small or near-zero bias for all models on the other six topics. In contrast, the ComSRB approach detects more pronounced bias in most topics and models, particularly those with male stereotypes (the four on the left side). To better understand these differences, we analyse how the metrics assess the search result list of specific characteristic examples. We focus on the results of the top five documents ($t = 5$), retrieved by the cross-encoder MiniLM-L12, as it is the model with the best ranking performance [20].

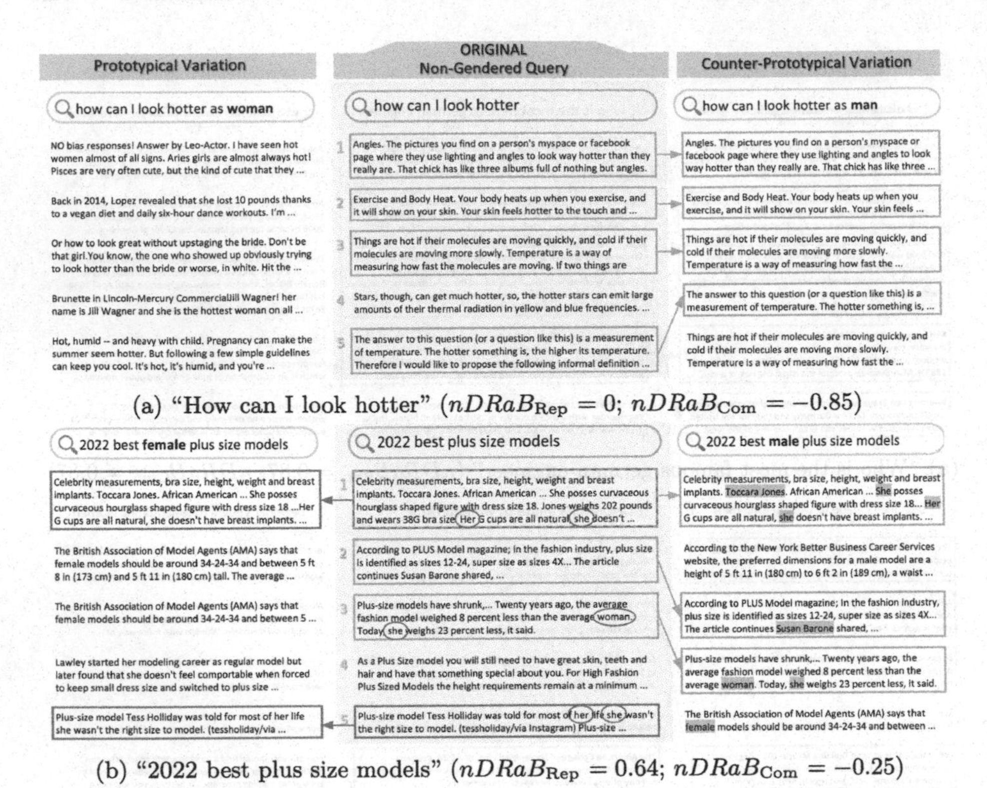

(a) "How can I look hotter" ($nDRaB_{\text{Rep}} = 0$; $nDRaB_{\text{Com}} = -0.85$)

(b) "2022 best plus size models" ($nDRaB_{\text{Rep}} = 0.64$; $nDRaB_{\text{Com}} = -0.25$)

Fig. 4. Examples highlighting the effects of retrieval performance and collection.

We first look at the case where both metrics report similar results. Figure 3a shows a representative example of this scenario with a query from the topic *Career 1*. Looking at the top five retrieved documents for the non-gendered query (middle part), we observe that four of the documents contain male-representative words (like *he*, *his*), resulting in a male (prototypical) bias according to the RepSRB metric. Four of the documents appear in the results for the male query formulation (left part), and hence, we observe the same measurement according to the ComSRB metric (both with $nDRaB = 0.87$).

The next example studies the case with considerably different results, specifically with high positive ComSRB but zero (or even slightly negative) RepSRB. Figure 3b shows the retrieved documents for a representative query of the topic *Physical Capabilities 1*. In this example, the ComSRB metric measures a high bias (0.83) due to a substantial overlap between the search results of the non-gendered, and the ones of the male query (middle and left parts). In contrast, the RepSRB is zero as retrieved documents lack gender-representative words. This example reflects the impact of prototype theory, as several documents biased towards the prototypical gender do not contain gender-specific language. Unlike RepSRB, the ComSRB metric correctly identifies bias, as it does not rely on gender-representative words.

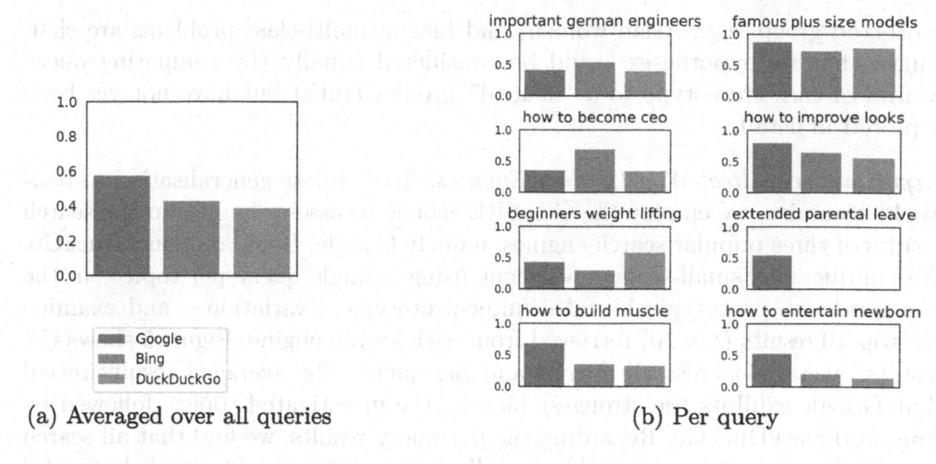

(a) Averaged over all queries (b) Per query

Fig. 5. Search results bias of three popular search engines calculated using our comparative approach. The vertical axes in all plots refer to $nDRaB_{\text{Com}}$.

Finally, we highlight a fundamental limitation, shared by both metrics and rooted in the performance of retrieval, as retrieving non-relevant documents potentially distorts the evaluation and leads to unreliable bias measurement outcomes. Figure 4a shows a representative query from the topic *Appearance 2*, where the retrieved results contain numerous non-relevant documents, particularly for the non-gendered and counter-prototypical (male) queries, leading to non-reliable results by both metrics. This highlights the importance of pre-screening the retrieval results to ensure proper retrieval performance before conducting bias measurement. The following example from the topic *Appearance 1* is shown in Fig. 4b and represents the case where the collection is overpopulated with stereotypical documents (here female), lacking the documents with counter-stereotypical contents. This imbalance also affects both metrics, as the model retrieves stereotypical documents even for the counter-stereotypical query.

In summary, the results and analyses lead us to the following lessons learned: (1) Our comparative approach captures underlying biases more effectively by addressing the issue of the gender-prototypical documents; (2) For broad concepts such as gender bias, analysing the results by topic is more informative compared to solely relying on average results; (3) When interpreting the results of both approaches, it is important to consider the effect of weak retrieval performance and a lack of counter-prototypical contents in the underlying collection. We believe these provide a deeper and more nuanced understanding of bias measurement in search results, paving the way for future studies in this area.

While not included in these experiments, we argue that given a representative set of search queries, the metric could easily be adapted towards other "sensitive attributes" prone to cause discrimination. However, similar to common bias and fairness metrics, as for instance, discussed in classification problems (e.g. [7,17]), the here proposed gender bias metrics are limited to binary class labels (female/male). Assessing intersectional bias (i.e., bias against a subgroup of a protected group, e.g., Asian women) and bias in multi-class problems are challenges that we hypothesise could be considered equally (by comparing query results of each stereotype to a "neutral" ground truth) but have not yet been explored in detail.

Experiments on Real-World Search Engines. To estimate generalisation to real-world scenarios, we employ the ComSRB metric to assess the bias in the search results of three popular search engines, namely Google, Bing, and DuckDuckGo. We conduct this small-scale experiment using a single query per topic – in the non-gendered, prototypical, and counter-prototypical variation – and examine the top 10 results ($t = 10$) retrieved from each search engine. Figure 5 shows the results, averaged across all queries and per-query. The averaged results reveal that Google exhibits the strongest bias on the investigated topics, followed by Bing and DuckDuckGo. Regarding the per-query results, we find that all search engines demonstrate a positive bias for all queries except one ("extended parental leave"). This experiment highlights the practical benefits of the comparative approach, particularly for the systems with expectedly high retrieval performance (such as the studied search engines). The ComSRB metrics are easy to compute and apply to such scenarios, and in contrast to the RepSRB metrics, do not require (possibly error-prone) content analyses nor particular bias-representative resources.

6 Conclusion

We introduce a novel framework for measuring gender bias in search results, addressing the fundamental challenges of current representative-based approaches (RepSRB) in capturing gender-related tendencies. Our proposed comparative search result bias (ComSRB) approach quantifies bias based on the deviations of the set of retrieved documents in response to a non-gendered query compared to the corresponding sets of the query's gender-specific variations. We conduct experiments on the MS MARCO dataset using several neural IR models, utilising a range of gender-sensitive topics, each represented by several queries. Our results demonstrate that the ComSRB metrics offer a practical and more robust approach to bias measurement in comparison with RepSRB metrics, alleviating the need for a predefined set of gender-specific words and content processing.

Acknowledgments. This research was funded in whole or in part by the Austrian Science Fund (FWF): DFH-23 and P36413; and by the State of Upper Austria and the Federal Ministry of Education, Science, and Research, through grants LIT-2020-9-SEE-113 and LIT-2021-10-YOU-215. Other funds include the COMET - Programme, funded by the Austrian Federal Ministry for Transport, Innovation and Technology (bmvit), the Austrian Federal Ministry for Digital and Economic Affairs (bmdw), the Austrian Research Promotion Agency (FFG), the province of Styria (SFG) and partners from industry and academia. The COMET Programme is managed by FFG. For open access purposes, the author has applied a CC BY public copyright license to any author accepted manuscript version arising from this submission.

References

1. Bender, E.M.: On achieving and evaluating language-independence in NLP. Linguist. Issues Lang. Technol. **6** (2011)
2. Bigdeli, A., Arabzadeh, N., Seyedsalehi, S., Zihayat, M., Bagheri, E.: On the orthogonality of bias and utility in ad hoc retrieval. In: Proceedings of the 44th International ACM SIGIR Conference on Research and Development in Information Retrieval, pp. 1748–1752 (2021)
3. Bigdeli, A., Arabzadeh, N., Seyedsalehi, S., Zihayat, M., Bagheri, E.: A light-weight strategy for restraining gender biases in neural rankers. In: Advances in Information Retrieval, pp. 47–55 (2022)
4. Bigdeli, A., Arabzadeh, N., Zihayat, M., Bagheri, E.: Exploring gender biases in information retrieval relevance judgement datasets. In: Hiemstra, D., Moens, M.-F., Mothe, J., Perego, R., Potthast, M., Sebastiani, F. (eds.) ECIR 2021. LNCS, vol. 12657, pp. 216–224. Springer, Cham (2021). https://doi.org/10.1007/978-3-030-72240-1_18
5. Crawford, K.: The trouble with bias. In: Keynote at Annual Conference on Neural Information Processing Systems (NIPS) (2017)
6. Devinney, H., Björklund, J., Björklund, H.: Theories of "gender" in NLP bias research. In: 2022 ACM Conference on Fairness, Accountability, and Transparency, pp. 2083–2102 (2022)
7. Dixon, L., Li, J., Sorensen, J., Thain, N., Vasserman, L.: Measuring and mitigating unintended bias in text classification. In: Proceedings of the 2018 AAAI/ACM Conference on AI, Ethics, and Society, pp. 67–73 (2018)
8. Ekstrand, M.D., Das, A., Burke, R., Diaz, F.: Fairness in information access systems. Found. Trends® Inf. Retriev. **16**, 1–177 (2022)
9. Fabris, A., Purpura, A., Silvello, G., Susto, G.A.: Gender stereotype reinforcement: measuring the gender bias conveyed by ranking algorithms. Inf. Process. Manag. **57**(6), 102377 (2020)
10. Feng, Y., Shah, C.: Has CEO gender bias really been fixed? Adversarial attacking and improving gender fairness in image search. In: Proceedings of the AAAI Conference on Artificial Intelligence, vol. 36, no. 11, pp. 11882–11890 (2022)
11. Gezici, G., Lipani, A., Saygin, Y., Yilmaz, E.: Evaluation metrics for measuring bias in search engine results. Inf. Retrieval J. **24**(2), 85–113 (2021)
12. Kay, M., Matuszek, C., Munson, S.A.: Unequal representation and gender stereotypes in image search results for occupations. In: Proceedings of the Annual ACM Conference on Human Factors in Computing Systems, pp. 3819–3828 (2015)
13. Kopeinik, S., Mara, M., Ratz, L., Krieg, K., Schedl, M., Rekabsaz, N.: Show me a "Male Nurse"! how gender bias is reflected in the query formulation of search engine users. In: Proceedings of the CHI Conference on Human Factors in Computing Systems (CHI 2023), pp. 1–15 (2023)
14. Krieg, K., Parada-Cabaleiro, E., Medicus, G., Lesota, O., Schedl, M., Rekabsaz, N.: Grep-BiasIR: a dataset for investigating gender representation bias in information retrieval results. In: Proceedings of the 2023 Conference on Human Information Interaction and Retrieval (CHIIR 2023), pp. 444–448 (2023)
15. Li, Y., et al.: Debiasing neural retrieval via in-batch balancing regularization. In: Proceedings of the 4th Workshop on Gender Bias in Natural Language Processing (GeBNLP), pp. 58–66 (2022)

16. Lin, J., Ma, X., Lin, S.C., Yang, J.H., Pradeep, R., Nogueira, R.: Pyserini: a Python toolkit for reproducible information retrieval research with sparse and dense representations. In: Proceedings of the 44th Annual International ACM SIGIR Conference on Research and Development in Information Retrieval (SIGIR 2021), pp. 2356–2362 (2021)
17. Maughan, K., Near, J.P.: Towards a measure of individual fairness for deep learning. arXiv e-prints pp. arXiv-2009 (2020)
18. Nguyen, T., et al.: MS MARCO: a human generated machine reading comprehension dataset. In: Proceedings of the Workshop on Cognitive Computation: Integrating Neural and Symbolic Approaches 2016 co-located with the 30th Annual Conference on Neural Information Processing Systems (NIPS 2016), Barcelona, Spain, 9 December 2016. CEUR Workshop Proceedings, vol. 1773 (2016)
19. Otterbacher, J., Bates, J., Clough, P.: Competent men and warm women: gender stereotypes and backlash in image search results. In: Proceedings of the 2017 CHI Conference on Human Factors in Computing Systems, pp. 6620–6631 (2017)
20. Reimers, N., Gurevych, I.: Sentence-BERT: sentence embeddings using Siamese BERT-networks. In: Proceedings of the 2019 Conference on Empirical Methods in Natural Language Processing, pp. 3982–3992 (2019)
21. Rekabsaz, N., Kopeinik, S., Schedl, M.: Societal biases in retrieved contents: measurement framework and adversarial mitigation of BERT rankers. In: Proceedings of the 44th International ACM SIGIR Conference on Research and Development in Information Retrieval, pp. 306–316 (2021)
22. Rekabsaz, N., Schedl, M.: Do neural ranking models intensify gender bias? In: Proceedings of the 43rd International ACM SIGIR Conference on Research and Development in Information Retrieval, pp. 2065–2068 (2020)
23. Rosch, E.H.: Natural categories. Cogn. Psychol. **4**(3), 328–350 (1973)
24. Vlasceanu, M., Amodio, D.M.: Propagation of societal gender inequality by internet search algorithms. Proc. Natl. Acad. Sci. **119**(29), e2204529119 (2022)
25. Wang, W., Wei, F., Dong, L., Bao, H., Yang, N., Zhou, M.: MiniLM: deep self-attention distillation for task-agnostic compression of pre-trained transformers. In: Advances in Neural Information Processing Systems, vol. 33, pp. 5776–5788 (2020)
26. Zehlike, M., Yang, K., Stoyanovich, J.: Fairness in ranking: a survey. arXiv preprint arXiv:2103.14000 (2021)
27. Zerveas, G., Rekabsaz, N., Cohen, D., Eickhoff, C.: Mitigating bias in search results through contextual document reranking and neutrality regularization. In: Proceedings of the International ACM SIGIR Conference on Research and Development in Information Retrieval, pp. 2532–2538 (2022)

SALSA: Salience-Based Switching Attack for Adversarial Perturbations in Fake News Detection Models

Chahat Raj[✉], Anjishnu Mukherjee, Hemant Purohit, Antonios Anastasopoulos, and Ziwei Zhu

George Mason University, Fairfax, VA, USA
{craj,amukher6,hpurohit,antonis,zzhu20}@gmu.edu

Abstract. Despite advances in fake news detection algorithms, recent research reveals that machine learning-based fake news detection models are still vulnerable to carefully crafted adversarial attacks. In this landscape, traditional methods, often relying on text perturbations or heuristic-based approaches, have proven insufficient, revealing a critical need for more nuanced and context-aware strategies to enhance the robustness of fake news detection. Our research identifies and addresses three critical areas: creating subtle perturbations, preserving core information while modifying sentence structure, and incorporating inherent interpretability. We propose SALSA, an adversarial **Sal**ience-based **S**witching **A**ttack strategy that harnesses salient words, using similarity-based switching to address the shortcomings of traditional adversarial attack methods. Using SALSA, we perform a two-way attack: misclassifying real news as fake and fake news as real. Due to the absence of standardized metrics to evaluate adversarial attacks in fake news detection, we further propose three new evaluation metrics to gauge the attack's success. Finally, we validate the transferability of our proposed attack strategy across attacker and victim models, demonstrating our approach's broad applicability and potency. Code and data are available here at https://github.com/iamshnoo/salsa.

Keywords: Adversarial attacks · Robustness · Interpretability · Fake news · Transformers

1 Introduction

The proliferation of misinformation and its potential societal impact has fueled significant research in data mining and information retrieval, particularly in developing methods to detect and mitigate false information. Fake news detection presents a complex challenge due to the nuanced and evolving nature of misinformation, requiring sophisticated algorithms and an understanding of both linguistic subtleties and contextual cues. Recent breakthroughs in language models, particularly transformers [24], capable of understanding subtle contexts, have

C. Raj and A. Mukherjee—Contributed equally to this work.

N. Goharian et al. (Eds.): ECIR 2024, LNCS 14612, pp. 35–49, 2024.
https://doi.org/10.1007/978-3-031-56069-9_3

proven remarkably adept in fake news detection. But, are these systems robust? Adversarial attacks designed to cause input perturbations can depreciate their performances significantly, highlighting underlying vulnerabilities [13,26].

Existing studies on adversarial attacks present certain limitations in their methodologies. Prior work largely depended on heuristics that involve methods such as negation [5], adverb intensity modification [5], and fact distortion [29]. These techniques tend to compromise the fundamental information of the text, creating alterations that might be significant and potentially detectable. Moreover, many of these previous strategies do not provide inherent interpretability in their attack approach, leaving a need for post-hoc explanations to understand the model's altered behavior [25]. We identify three pivotal elements central to our approach in the context of fake news detection. First, the adversarial perturbations must be crafted so that they are subtle, but still lead to significant changes in the model outputs [7]. Second, we want to preserve the text's core information, instead of making significant changes to the content of the sentence. Third, we advocate for an approach where the attack itself is inherently interpretable, negating the need for post-hoc explanations. Towards this, we propose leveraging model interpretability to precisely discern the influence of individual tokens in a news article on the model's predictions. This insight enables us to manipulate the model into producing incorrect outputs by strategically altering the identified critical tokens. In this manner, we create a subtle, content-preserving, and interpretable attack strategy. Our study makes the following contributions:

1. We conceptualize the issue of adversarial attack in fake news detection as a multifaceted problem, addressing two distinct sub-problems: the misclassification of real news articles as fake, and the misclassification of fake news articles as real. This dual-sided approach contrasts with traditional methods predominantly focusing on the misclassification of false information only.
2. We propose to leverage model interpretation as a powerful weapon to reveal the vulnerabilities of a fake news detection model and generate strong attacks. Specifically, we introduce SALSA as an implementation of our idea, which generates input perturbations by implementing similarity-based switching of words according to their salience. This approach subtly modifies the text without necessarily altering the underlying information, effectively overcoming limitations in previous methods.
3. We propose three metrics comprehensively assessing different and complementary aspects of the "success" of an attack strategy.
4. We conduct extensive experiments to evaluate our proposed method on 2 real-world datasets and observe significant improvement in attack success compared to previous methods. Besides, we also perform a study concerning the transferability of our attack strategy, further highlighting our approach's adaptability and potential applications within the landscape of fake news detection.

2 Related Work

The field of fake news detection has witnessed an exponential growth of methods aimed at identifying and mitigating false information [15,20–22,27,28]. While these techniques offer promising solutions, they also expose key limitations that underline the need for further refinement and innovation [8].

There have been numerous advancements toward adversarial attacks in text classification [3,4,16,23]. Methods such as DeepWordBug [6] and TextBugger [11] introduce noise into the data through input perturbations, disrupting the original context of the text and leading to unrealistic manipulations [17]. These perturbations, also observed by Ali et al. [2], present challenges in enhancing robustness, particularly as adversarial examples often rely on fewer words.

Another critical issue lies in the modification of factual information. Flores and Hao [5] highlighted vulnerabilities by attacking fake news detectors through changes in compositional semantics and lexical relations. However, their attacks, along with those by Zhou et al. [29] may distort the factual content of news articles, thus risking the integrity of the information.

Some existing methods, such as Probability Weighted Word Saliency (PWWS) [18], reduce classification accuracy through random heuristic-based word substitutions. While effective, they may alter the context or introduce artificial noise.

Moreover, most of the research until now has focused on fake-attack only, i.e., misclassifying fake news as true [10]. This one-sided view overlooks the nuanced complexity of adversarial behavior. In addition, the reliance on evaluation metrics like flip rate [5], does not necessarily capture all the elements of the attack, pointing to a need for more comprehensive attack evaluation metrics.

In response to these limitations, there is a clear need for an approach that preserves the authenticity and subtlety of the original content while effectively challenging fake news detectors. The existing methods often suffer from changing the context of the news article, introducing noise, or distorting factual information. Recognizing these challenges, we propose SALSA, a similarity-switching approach based on salience scores. Unlike previous methods, SALSA targets a subtle yet efficient attack by simply switching target tokens with candidate words using cosine similarities, replacing the words with their most similar counterparts in the embedding space. This strategy aims to maintain the content and integrity of the news article while introducing a nuanced challenge to the existing detection mechanisms. Moreover, we present a two-faceted view to our approach by modifying information both ways to perform an attack-misclassifying real news as fake and misclassifying fake news as real. Recognizing the limitations of existing metrics, we also propose three new attack evaluation metrics, designed to provide a more in-depth and accurate assessment of the attack landscape.

3 Methodology

To overcome the limitations of the previous approaches, we first define the attack in terms of our desired goals, before proposing our approach for the different components of our algorithm as illustrated in Fig. 1.

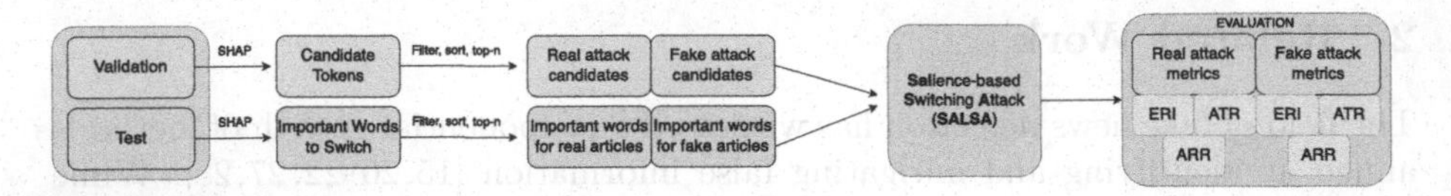

Fig. 1. The pre-requisites for SALSA involves extracting particular words of interest from the validation split to get a pool of attack candidates, and words with high salience scores for switching in the test split. These inputs are then fed into SALSA, along with the article to be perturbed to perform an attack.

3.1 Formalization and Evaluation

Given a news article a with a ground truth label y and predicted label before attack $\hat{y}$, where y and $\hat{y}$ can be either real (1) or fake (0), we have two attack goals. The objective of the attack on a real news article is to manipulate the content in such a way that the manipulated article a' is classified as fake, resulting in a new predicted label $\hat{y}' = 0$. Conversely, the objective of the attack on a fake news article is to alter the content so that the manipulated article a' is classified as real, leading to a new predicted label $\hat{y}' = 1$.

We aim to flip the labels (i.e., $\hat{y}$ should be different from $\hat{y}'$ after an attack procedure T is performed), but specifically for those labels that contribute towards misclassification. We can define this formally as two cases using the example of a "real attack" scenario:

Case 1: The original classification is correct (i.e., $y = 1, \hat{y} = 1$), we want to flip the labels such that: $T(a, y = 1, \hat{y} = 1) \rightarrow a', \hat{y}' = 0$

Case 2: The original classification is incorrect (i.e., $y = 1, \hat{y} = 0$), we do not seek to flip the labels, and thus: $T(a, y = 1, \hat{y} = 0) \rightarrow a', \hat{y}' = 0$

Here, a' is the modified article after procedure T, and $\hat{y}'$ is the final predicted label. T is designed to selectively flip the labels of the first type of articles that were originally getting correctly classified, leading to a misclassification.

We propose three metrics designed to measure the effectiveness of adversarial attacks in the context of fake news classification: Error Rate Increase (ERI), Attack Turning Rate (ATR), and Attack Reserving Rate (ARR). These are percentage values with the maximum being 100, indicating an ideal attack. For all three metrics, the higher the value of the metric, the more the effectiveness of the attack method.

Error Rate Increase: ERI quantifies the percentage growth in the number of errors committed by the model as a result of the input perturbations introduced through the attack.

$$ERI_{\text{real}} = Error_{\text{real}_{(\text{after})}} - Error_{\text{real}_{(\text{before})}}, \qquad Error_{\text{real}} = FN/(FN + TP) \quad (1)$$

$$ERI_{\text{fake}} = Error_{\text{fake}_{(\text{after})}} - Error_{\text{fake}_{(\text{before})}}, \qquad Error_{\text{fake}} = FP/(FP + TN) \quad (2)$$

where FN, FP, FN, TP denote the counts of false negatives, false positives, true positives, and true negatives, respectively.

Attack Transition Rate: ATR quantifies the count of news items that undergo a "transition" from being correctly classified prior to the attack to being misclassified subsequently.

$$ATR_{real} = \frac{|y = 1, \hat{y} = 1, \hat{y}' = 0|}{|y = 1, \hat{y} = 1|}, \quad ATR_{fake} = \frac{|y = 0, \hat{y} = 0, \hat{y}' = 1|}{|y = 0, \hat{y} = 0|} \tag{3}$$

Attack Reserving Rate: ARR quantifies the number of news items that remain misclassified after the attack, where the input perturbations do not reverse the predicted labels, but instead "reserve" them in their incorrect state.

$$ARR_{real} = \frac{|y = 1, \hat{y} = 0, \hat{y}' = 0|}{|y = 1, \hat{y} = 0|}, \quad ARR_{fake} = \frac{|y = 0, \hat{y} = 1, \hat{y}' = 1|}{|y = 0, \hat{y} = 1|} \tag{4}$$

Motivation. To perturb the article a to obtain a' in alignment with the required goals of attack, we emphasize the strategic use of model interpretability. Our approach identifies the model's vulnerabilities, which can then be exploited to manipulate the output. In line with this approach, we present SALSA as an implementation to demonstrate our interpretability-based attack strategy. This method functions by substituting important words w, which contribute to the model's confidence towards the predicted label $\hat{y}$, within the article a, with words w' from a collection of candidate attack tokens. The operation is defined by:

$$\textbf{SALSA}(a, w', w) \rightarrow a' \tag{5}$$

Section 3.2 describes the process by which we determine the candidate attack tokens w', Sect. 3.3 details our process of choosing important words w and finally Sect. 3.4 gives an outline of the algorithm for replacing w with w' based on semantic similarity.

3.2 Candidate Token Generation

The process of adversarial attack on the fake news detection model entails generating suitable candidate tokens to deceive the model into misclassifying real news as fake (real attack) and fake news as real (fake attack).

Table 1. Given a predicted label, we selectively choose attack candidates based on their salience scores for each type of attack.

	Real Attack Candidates	Fake Attack Candidates
Predicted Label: ($\hat{y} = 0$)	+ Salience Scored Tokens	− Salience Scored Tokens
Predicted Label: ($\hat{y} = 1$)	− Salience Scored Tokens	+ Salience Scored Tokens

Salience Score Assignment Using SHAP. To find words of importance in the news articles in the validation split, we use interpretability as a tool to drive our token generation procedure. Primarily, we employ SHAP [12] to find words with high salience scores for every news article. To do so, SHAP uses the model m trained on the training split, and also the labels L predicted by that model for the items in the validation split. Formally,

$$SHAP(a, m, \hat{y}) \rightarrow [+, -] \quad \forall \hat{y} \in L \tag{6}$$

Significance of Salience Scores. For a given model prediction $\hat{y}$, a positive salience score for a word indicates that the presence of the word in the article increases the confidence of the model towards that specific prediction. If these words were to be removed from the article, the model's confidence for the prediction being $\hat{y}$ would reduce significantly, and at some point, the model would find more confidence in predicting the flipped label for the same article.

Global Candidate Pool. We aim to find tokens contributing most to the model's decision for both fake and real news articles so that we can later use them to guide the decision making process towards incorrect predictions on articles from the test split. We will create a candidate set for real attack and another set for fake attack. To do so, we use Eq. 6 to generate a pool of all words with positive salience scores and all words with negative salience scores in the validation split. There are however many articles in this split, resulting in a very large initial pool size. So, we develop a strategy to locate the most important words overall, across all articles. Considering the case of generating real attack candidate tokens, we utilize words with positive salience scores from news items that were predicted as fake ($\hat{y} = 0$) and words with negative salience scores from news items that were predicted as real ($\hat{y} = 1$), both indicate words instrumental for the sentence being classified as fake. Table 1 summarizes the type of salience-scored words used for real and fake attack candidate generation. Only a few tokens per sentence contribute towards the model's decision making process, while the bulk of the remaining tokens make up a very small combined effect. To reflect this on a global scale, across all the articles, we sort the candidate pool based on the absolute values of the salience scores and choose the top N, where N is treated as a hyperparameter that we study extensively in Table 3.

3.3 Target Token Selection

The next step of SALSA is the selection of important target tokens for switching. These are words that contribute significantly to the confidence of the model in making its predictions.

First, we use SHAP with our trained model m on articles in the test split to get corresponding positive and negative salience scores for each word in an article similar to candidate token generation.

However the strategy of choosing important words from the initial pool is different. Here, we select a specific number, M, of words for each article. We study different possible values of the hyperparameter M, in conjunction with the hyperparameter N for candidate token generation in Table 3. The following outlines the approach for each scenario based on our goal for the attack:

1. **y = 0 | ŷ = 0 | Fake Attack:** The prediction needs to change from fake to real. Hence, +ve salience words supporting fake prediction are switched.
2. **y = 0 | ŷ = 1 | Fake Attack:** The prediction should remain as real. Negative salience-scored words opposing the real prediction are removed.
3. **y = 1 | ŷ = 0 | Real Attack:** The prediction should remain as fake. Negative salience words and words with the smallest absolute +ve scores are removed.
4. **y = 1 | ŷ = 1 | Real Attack:** The prediction needs to change from real to fake. Hence, +ve salience words that support the real prediction are switched.

3.4 SALSA

Given the set of candidate tokens, and the target tokens for each article in the test set, we can now formulate our algorithm for input perturbations.

The SALSA algorithm is designed to semantically alter a given article by selectively substituting significant tokens with closely related alternatives from a set of candidates, utilizing word embeddings and cosine similarity for guidance. It starts by transforming words in the target set w and candidate tokens in w' into embeddings (E_w and $E_{w'}$) using a transformer model, capturing their semantic essence in a high-dimensional space. A similarity matrix κ is then derived from the cosine similarity between $E_{w'}$ and E_w. Subsequently, for each vital word w_i in w, tokens in w' are ranked based on their similarity to w_i, excluding w_i itself to avoid redundancy, and stored in sorted order. The algorithm proceeds to replace w_i with the highest-ranked, distinct candidate token $\bar{w}$ in the article, iterating this process for all words in w to generate a semantically similar, yet modified version of the original article.

Algorithm 1. Switch Words Using Salience-based Switching Attack (SALSA)

1: **procedure** SALSA(a, w', w) ▷ a: Article, w': Candidate Tokens, w: Important Words
2: $E_w \leftarrow$ Embeddings of w
3: $E_{w'} \leftarrow$ Embeddings of w'
4: $\kappa \leftarrow$ Cosine Similarity between $E_{w'}$ and E_w
5: **for** each w_i in w **do**
6: $z \leftarrow$ Sorted indices of words similar to w_i in κ
7: $\bar{w} \leftarrow$ The highest ranked $w'_k \in w'$ in z where $w'_k \neq w_i$
8: Replace w_i with $\bar{w}$ in a
9: **end for**
10: **return** a'
11: **end procedure**

4 Experiments

We benchmark our approach on two datasets: NELA-GT-2018 [14] and TFG [1]. A lightweight DistilBERT [19] model fine-tuned for four epochs, achieves 91.65%

and 99.04% classification accuracies on the test splits for both the datasets. For each dataset, we then implement state-of-the-art fake news detection attack baselines and compare them with our proposed algorithm SALSA for both real and fake attacks. We report the results using our three novel evaluation metrics ERI, ATR, and ARR (higher values indicating more effective attack). Further, a hyperparameter study is presented by varying the number of target tokens for switching and the number of candidate tokens for attack. Additionally, we conduct an attack-transferability study by interchangeably using DistilBERT and RoBERTa for different stages of our algorithm. Our goal is to investigate the following research questions:

- **RQ1:** How does SALSA perform compared to state-of-the-art baselines?
- **RQ2:** How do hyper-parameters influence the attack performance?
- **RQ3:** How effective is SALSA in transfer of attacks across different classification models?

Table 2. SALSA outperforms baselines for both real and fake attack on the NELA-GT-2018 dataset. For TFG, it is better for real attack, but for fake attack, the performance is comparable to the negation attack and adverb intensity baselines.

Attack Strategy	NELA-GT-2018 Dataset						TFG Dataset					
	Real Attack			Fake Attack			Real Attack			Fake Attack		
	ERI	ATR	ARR	ERI	ATR	ARR	ERI	ATR	ARR	ERI	ATR	ARR
Negation	24.97	31.14	67.98	16.18	20.75	56.96	11.69	11.85	91.43	2.24	3.13	25.58
Adverb Intensity	25.13	31.30	68.08	16.47	21.04	57.28	11.39	11.55	91.43	2.34	3.21	27.91
Random Injection	33.55	39.82	75.71	17.25	21.97	56.10	16.13	16.38	85.71	8.50	9.35	34.88
Head Injection	35.98	42.68	74.11	24.82	29.66	**61.99**	**72.39**	**73.02**	**94.29**	**24.35**	**25.26**	**46.51**
Tail Injection	34.36	40.68	76.08	19.72	24.72	54.93	13.43	13.58	94.29	6.13	6.93	37.21
Similarity Switching	34.48	41.03	74.01	20.30	25.44	53.64	14.44	14.62	91.43	11.83	12.64	41.86
SALSA	**51.01**	**57.92**	**87.19**	**26.17**	**31.18**	61.24	67.55	68.21	85.71	7.51	8.36	34.88

4.1 Experimental Setup

Datasets. The NELA-GT-2018 dataset, [14], short for "News Landscape Ground Truth 2018", is a rich and expansive resource for news content analysis. Comprised of around 713,000 articles collected from 194 news sources, it covers an entire year of news reporting, from February 2018 to February 2019. The dataset is suitable for a range of tasks in media bias analysis, misinformation detection, and content understanding.

The TFG dataset [1] provides a valuable resource for research on automatic fake news detection. The dataset contains approximately 40k text news articles, evenly divided into reliable and unreliable/fake classes. The data covers a wide range of news topics, sourced from known publishers of mainstream news as well as propaganda and fake news sites.

We use both of these to frame fake news detection as a classification problem, training a classifier to predict whether the input text is "real" or "fake", using a 70:15:15 train-validation-test split.

Baselines. We explore two existing baselines for this challenge, as introduced in the paper by Flores and Hao [5]. The first approach of **negation attack**, identifies third-person singular subjects and negates the associated first-person verbs. The second method, **adverb intensity attack**, eliminates the polarizing adverbs from the articles. Given that both NELA and TFG often feature lengthy articles, these straightforward alterations to the semantic content of the article generally do not significantly influence the model's predictions.

Further, we try two different heuristic approaches that can be considered simplified versions of our final algorithm SALSA. The first approach of **injection attacks** only creates a candidate set similar to SALSA, but instead of switching any words in the article, it simply injects a fixed number of words (10) into the article. The injection location can be chosen randomly, or all of the words can be injected at either the head of the article or the tail, leading to the three variations of the injection attack. The head injection attack can be considered a control experiment, because transformer models are especially effective at capturing context at the start of long articles, and we are intentionally modifying that initial context with multiple attack candidates. The second naive approach is to perform **similarity switching** with a candidate set similar to SALSA, but without specifically identifying important words to switch in the article.

Finally, we incorporate both the candidate set generation strategy and the important token selection strategy into our SALSA algorithm which switches important tokens with similar ones from the candidate set.

Reproducibility. We fine-tuned two pretrained models available on huggingface, "distilbert-base-uncased"[1] and "roberta-base"[2]. For both, we added a classification head projecting the 768 dimensional output onto a 64 dimensional space, and using dropout probability of 0.4 before predicting the probabilities of the classification labels. We perform the fine-tuning process for 4 epochs and a batch size of 64. For tokenization purposes, we limit maximum length of a sequence to 512 as most articles fit within that size. We use Adam optimizer [9] with a learning rate of 5e−5 and an epsilon of 1e−8. All experiments are executed with a fixed random seed of 42. For the results of SALSA shown in Table 2, we have fixed the number of words to switch to 20 and the size of the candidate pool as 100. The values corresponding to this setting is highlighted in Table 3. Also, for all the attacks shown in Table 2, the candidate pool uses corresponds to the "Mixed POS" setting from Fig. 2, where no POS filtering is done. Code and results are available at https://github.com/iamshnoo/salsa.

4.2 Performance of SALSA (RQ1)

A consistent observation across all conducted experiments is the pronounced difference in the difficulty of executing fake attacks as compared to real attacks. This discrepancy is delineated in Table 2, where the ERI for real attacks invariably registers a higher percentage than that for fake attacks. A plausible explanation for

[1] https://huggingface.co/distilbert-base-uncased.
[2] https://huggingface.co/roberta-base.

this phenomenon may be grounded in the distinct semantic characteristics that differentiate fake news articles from genuine ones. Real news articles frequently contain key indicators, such as references to reputable sources like "Reuters", "Fox", "CNN", and they are often composed in a stylistically recognizable manner. Modern transformer architectures, trained on extensive data sets, are adept at discerning these subtleties, as is observed from high values of accuracy and other classification metrics on both the datasets that we analyzed. Consequently, deceiving the model into misclassifying a fake article as real presents a substantial challenge, depending on the selected attack method. Conversely, inducing a model to falsely label a real article as fake proves to be a relatively easier task.

An additional significant finding evident from Table 2 pertains to the contrasting performance of SALSA across two datasets: NELA-GT-2018 and TFG. In the case of NELA-GT-2018, SALSA outperforms other methods for both fake and real attacks, establishing itself as the most effective strategy. Conversely, for the TFG dataset, SALSA's performance in real attacks is comparable to the control setting of head injection, whereas its efficacy in fake attacks does not present a substantial improvement over other approaches. This can be attributed to the fact that the average article length for TFG is shorter than that for NELA-GT-2018, and consequently fake news articles are much easier to detect due to their semantic structure. Moreover, the classifier trained for TFG exhibits a remarkable 99% accuracy on the test set, in contrast to 91.6% for the NELA-GT-2018. Such a high degree of accuracy for TFG implies a robust generalization by the model, rendering it resilient to deception, particularly when restricted to implementing only subtle input perturbations.

4.3 Hyperparameter Study (RQ2)

To study the effect of varying the hyperparameters N and M, corresponding to the number of candidate tokens in the attack set, and the number of important tokens per article that we want to switch respectively, we perform an extensive evaluation in Table 3. We find that, choosing a large number of words (more than 30) for switching usually improves ERI for the same candidate set. However, we still limit to about 20 words to remain within our constraints of performing "subtle" perturbations. For the candidate set, larger is not necessarily better. In fact, we see many cases where having a set of 50 or 100 candidates performs better than having 200 candidates. This can be owed to the fact that the tokens in the candidate set were chosen after having been sorted by absolute salience score, so for a larger set we might end up having some less meaningful words in the set that do not affect the predictions as much as the top 50 would for instance. The final hyperparameters that we chose for our experiments were $M = 20, N = 100$, which gives subtle enough perturbations while being as optimal as possible in terms of performance.

4.4 Effect of Using Different POS for Attack Tokens (RQ2)

We perform another study that considers the effects of having different parts of speech in the candidate set. Figure 2 shows the results for real attack and fake

Table 3. Varying the number of words to switch and the size of the candidate pool shows that for the same candidate set size, switching more words leads to better attacks. Also, a smaller candidate pool with more relevant words performs better than larger sets, indicating the importance of choosing attack tokens carefully.

(*M*)	(*N*)	NELA-GT-2018 Dataset						TFG Dataset					
		Real Attack			Fake Attack			Real Attack			Fake Attack		
		ERI	ATR	ARR	ERI	ATR	ARR	ERI	ATR	ARR	ERI	ATR	ARR
5	25	33.17	39.37	75.89	20.38	25.27	57.07	55.06	55.58	91.43	4.08	4.96	27.91
5	50	32.19	38.41	74.76	18.16	22.95	56.10	55.06	55.60	88.57	5.30	6.17	30.23
5	100	33.13	39.34	75.80	23.17	28.00	60.49	47.53	48.02	85.71	4.88	5.71	32.56
5	150	31.87	38.13	74.11	22.60	27.35	61.03	34.05	34.44	85.71	5.28	6.12	32.56
5	200	31.73	38.00	73.82	23.17	27.97	60.92	34.19	34.58	85.71	5.09	5.96	30.23
10	25	39.37	45.83	79.66	19.45	24.37	55.67	66.38	67.04	85.71	4.29	5.18	27.91
10	50	38.16	44.64	78.34	18.33	23.24	54.60	67.30	67.91	91.43	7.27	8.11	34.88
10	100	39.48	46.01	79.19	24.79	29.61	62.31	55.77	56.36	82.86	5.94	6.77	34.88
10	150	37.55	44.04	77.59	23.50	28.43	59.53	38.02	38.48	80.00	7.30	8.14	34.88
10	200	37.02	43.63	75.99	24.68	29.45	62.85	38.13	38.55	85.71	6.95	7.79	34.88
20	25	49.31	56.15	86.06	15.90	21.30	45.93	78.37	79.14	82.86	5.06	5.98	25.58
20	50	47.35	54.19	84.27	15.07	20.63	42.93	82.01	82.77	88.57	9.78	10.59	39.53
20*	**100***	**51.01**	**57.92**	**87.19**	**26.17**	**31.18**	**61.24**	**67.55**	**68.21**	**85.71**	**7.51**	**8.36**	**34.88**
20	150	46.94	53.75	84.09	24.81	29.96	58.03	44.41	44.88	85.71	9.51	10.38	34.88
20	200	46.44	53.36	82.58	27.98	33.03	62.53	43.86	44.31	88.57	9.78	10.67	32.56
30	25	56.74	63.91	90.49	12.29	18.27	34.69	83.00	83.78	85.71	5.68	6.63	23.26
30	50	54.51	61.76	87.57	12.18	18.16	34.58	88.57	89.35	91.43	12.82	13.72	34.88
30	100	59.16	66.52	91.15	26.20	31.55	56.85	73.58	74.29	85.71	9.70	10.57	34.88
30	150	54.08	61.30	87.38	25.26	30.74	54.28	50.16	50.68	85.71	12.98	13.88	34.88
30	200	53.34	60.60	86.35	28.77	34.07	60.06	49.52	50.01	88.57	14.02	14.93	34.88
40	25	63.51	71.15	92.94	8.39	14.98	23.02	85.01	85.82	85.71	6.77	7.79	18.60
40	50	61.03	68.66	90.58	8.90	15.51	23.23	90.81	91.61	91.43	17.11	18.11	30.23
40	100	66.21	74.02	94.07	24.39	30.21	49.04	76.37	77.11	85.71	14.10	14.99	37.21
40	150	61.04	68.64	90.77	24.38	30.36	46.90	57.20	57.77	85.71	18.47	19.35	41.86
40	200	60.44	68.17	89.08	28.72	34.56	53.00	55.84	56.39	88.57	21.50	22.43	41.86

Table 4. Using different models for generating the inputs to SALSA via SHAP, and predicting the labels after attack, we find the larger model (RoBERTa) to be more susceptible to attacks generated using the smaller model.

	DistilBERT-attack						RoBERTa-attack					
	Real Attack			Fake Attack			Real Attack			Fake Attack		
	ERI	ATR	ARR	ERI	ATR	ARR	ERI	ATR	ARR	ERI	ATR	ARR
DistilBERT-victim	51.01	57.92	87.19	26.17	31.18	61.24	*33.32*	*40.03*	*71.37*	*11.15*	*15.73*	*51.82*
RoBERTa-victim	*71.66*	*78.44*	*95.62*	*6.04*	*12.96*	*11.46*	16.25	18.55	90.45	3.60	5.32	80.11

attack for NELA-GT-2018. The default setting that we explained in the results previously and also for our hyperparameter study contains candidate tokens with all types of parts of speech mixed in. But if we filter this set and only include candidate tokens that are verbs, we observe a higher attack ERI for real articles from NELA-GT-2018. But for fake attack, using only verbs does not perform as well as using only nouns or using only adverbs and adjectives for example. This leads us to conclude that different parts of speech might be more relevant for real attack and fake attack, and this setting might also vary across datasets.

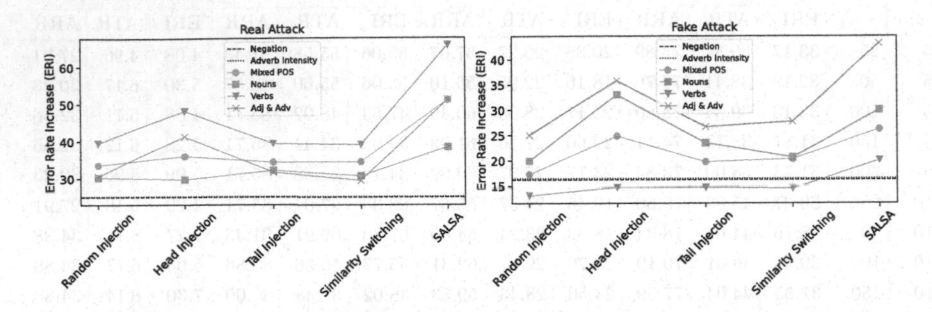

Fig. 2. Keeping only one part of speech as attack candidates has varying performances across different types of attacks.

4.5 Attack Transferability (RQ3)

While our results clearly show that SALSA works well under different hyperparameter settings for both fake and real attacks across datasets, the model m_1 that we used for generating SHAP outputs for candidate tokens and important words to switch, was the same as the model m_2 which we used for predictions before and after the attack, i.e. $m_1 = m_2$.

For attack transferability, we consider 2 distinct models m_1 (DistilBERT) and m_2 (RoBERTa), such that one of the models has much fewer parameters than the other. We would be using one of the models ("attacker model") to generate the inputs required by SALSA and then use the other model ("victim model") to predict labels on the perturbed inputs. Comparing this result with the basic setting of using the same model for both tasks, we get Table 4.

When we use the smaller model m_1 for predictions after generating SALSA inputs using the bigger model m_2, it is observed that all the three attack metrics fall for both real and fake attack compared to the default case of using m_1 for both predictions and SALSA. However, when we use the bigger model m_2 for predictions after using m_1 for SALSA, we notice an increase in the metrics, with a very notable increase in the ERI for real attack. Based on these experiments, our understanding is that bigger models are more susceptible to attacks generated using SALSA and a smaller model, but the reverse is not necessarily true.

5 Discussion

The proposed adversarial attack strategy represents an important step in identifying attacks and defending against them. Since our approach is interpretable, it would enable an easier understanding of adversarial inputs, which cause the model to flip predictions. Also, it would encourage the development of methods of defending against such subtle and interpretable attacks. Further, our results on attack transferability, which shows the susceptibility of larger models for words generated from smaller ones, can be a starting point to audit the vulnerabilities of large language models that have increasing usage over time.

6 Conclusion and Future Work

In this research, we conducted a thorough examination of adversarial attack strategies on fake news detection, with a specific focus on the inherently interpretable SALSA algorithm. Our findings underscored the nuanced challenges in executing fake attacks compared to real ones and highlighted the contrasting performance of SALSA across different datasets and semantic structures. In subsequent research, we aim to broaden the comparative framework by incorporating alternative interpretability methods such as LIME and Integrated Gradients.

A Appendix

See Fig. 3.

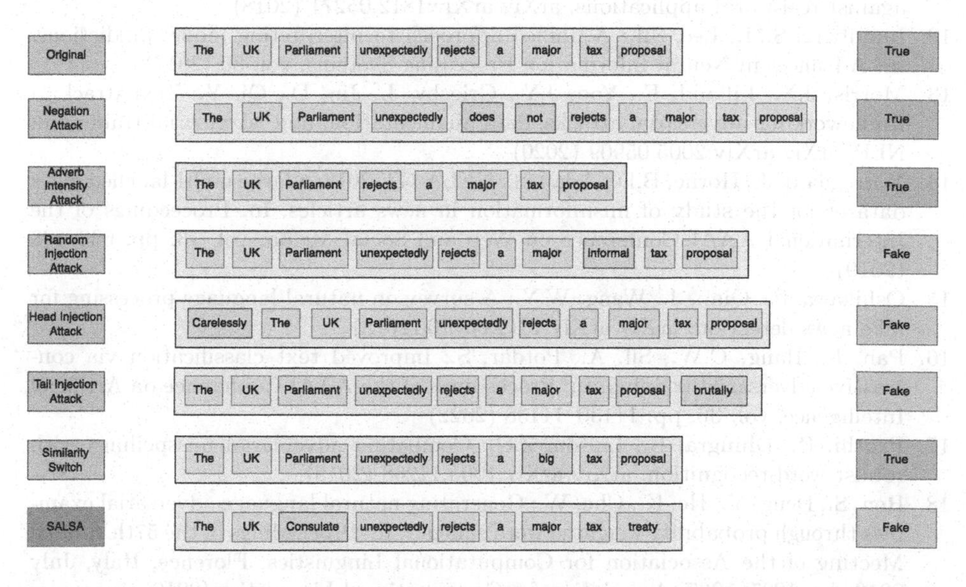

Fig. 3. Comparison of attack strategies on an example article with $y = 1$ (real attack)

References

1. https://huggingface.co/datasets/GonzaloA/fake_news
2. Ali, H., et al.: All your fake detector are belong to us: evaluating adversarial robustness of fake-news detectors under black-box settings. IEEE Access **9**, 81678–81692 (2021)
3. Chang, G., Gao, H., Yao, Z., Xiong, H.: TextGuise: adaptive adversarial example attacks on text classification model. Neurocomputing **529**, 190–203 (2023)
4. Ebrahimi, J., Rao, A., Lowd, D., Dou, D.: HotFlip: white-box adversarial examples for text classification. arXiv arXiv:1712.06751 (2017)
5. Flores, L.J.Y., Hao, Y.: An adversarial benchmark for fake news detection models. arXiv arXiv:2201.00912 (2022)
6. Gao, J., Lanchantin, J., Soffa, M.L., Qi, Y.: Black-box generation of adversarial text sequences to evade deep learning classifiers. In: 2018 IEEE Security and Privacy Workshops (SPW), pp. 50–56. IEEE (2018)
7. Ghaffari Laleh, N., et al.: Adversarial attacks and adversarial robustness in computational pathology. Nat. Commun. **13**(1), 5711 (2022)
8. Horne, B.D., Nørregaard, J., Adali, S.: Robust fake news detection over time and attack. ACM Trans. Intell. Syst. Technol. (TIST) **11**(1), 1–23 (2019)
9. Kingma, D.P., Ba, J.: Adam: a method for stochastic optimization. arXiv arXiv:1412.6980 (2014)
10. Koenders, C., Filla, J., Schneider, N., Woloszyn, V.: How vulnerable are automatic fake news detection methods to adversarial attacks? arXiv arXiv:2107.07970 (2021)
11. Li, J., Ji, S., Du, T., Li, B., Wang, T.: TextBugger: generating adversarial text against real-world applications. arXiv arXiv:1812.05271 (2018)
12. Lundberg, S.M., Lee, S.I.: A unified approach to interpreting model predictions. In: Advances in Neural Information Processing Systems, vol. 30 (2017)
13. Morris, J.X., Lifland, E., Yoo, J.Y., Grigsby, J., Jin, D., Qi, Y.: TextAttack: a framework for adversarial attacks, data augmentation, and adversarial training in NLP. arXiv arXiv:2005.05909 (2020)
14. Nørregaard, J., Horne, B.D., Adalı, S.: NELA-GT-2018: a large multi-labelled news dataset for the study of misinformation in news articles. In: Proceedings of the International AAAI Conference on Web and Social Media, vol. 13, pp. 630–638 (2019)
15. Oshikawa, R., Qian, J., Wang, W.Y.: A survey on natural language processing for fake news detection. arXiv arXiv:1811.00770 (2018)
16. Pan, L., Hang, C.W., Sil, A., Potdar, S.: Improved text classification via contrastive adversarial training. In: Proceedings of the AAAI Conference on Artificial Intelligence, vol. 36, pp. 11130–11138 (2022)
17. Pruthi, D., Dhingra, B., Lipton, Z.C.: Combating adversarial misspellings with robust word recognition. arXiv arXiv:1905.11268 (2019)
18. Ren, S., Deng, Y., He, K., Che, W.: Generating natural language adversarial examples through probability weighted word saliency. In: Proceedings of the 57th Annual Meeting of the Association for Computational Linguistics, Florence, Italy, July 2019, pp. 1085–1097. Association for Computational Linguistics (2019)
19. Sanh, V., Debut, L., Chaumond, J., Wolf, T.: DistilBERT, a distilled version of BERT: smaller, faster, cheaper and lighter. CoRR abs/1910.01108 (2019)
20. Shu, K., Cui, L., Wang, S., Lee, D., Liu, H.: dEFEND: explainable fake news detection. In: Proceedings of the 25th ACM SIGKDD International Conference on Knowledge Discovery & Data Mining, pp. 395–405 (2019)

21. Shu, K., Sliva, A., Wang, S., Tang, J., Liu, H.: Fake news detection on social media: a data mining perspective. ACM SIGKDD Explor. Newsl. **19**(1), 22–36 (2017)
22. Shu, K., Wang, S., Liu, H.: Beyond news contents: the role of social context for fake news detection. In: Proceedings of the Twelfth ACM International Conference on Web Search and Data Mining, pp. 312–320 (2019)
23. Simoncini, W., Spanakis, G.: SeqAttack: on adversarial attacks for named entity recognition. In: Proceedings of the 2021 Conference on Empirical Methods in Natural Language Processing: System Demonstrations, pp. 308–318 (2021)
24. Wolf, T., et al.: Transformers: State-of-the-Art Natural Language Processing, October 2020, pp. 38–45 (2020)
25. Xu, K., et al.: Structured adversarial attack: towards general implementation and better interpretability. arXiv arXiv:1808.01664 (2018)
26. Zeng, G., et al.: OpenAttack: an open-source textual adversarial attack toolkit. arXiv arXiv:2009.09191 (2020)
27. Zhang, X., Ghorbani, A.A.: An overview of online fake news: characterization, detection, and discussion. Inf. Process. Manage. **57**(2), 102025 (2020)
28. Zhou, X., Zafarani, R.: A survey of fake news: fundamental theories, detection methods, and opportunities. ACM Comput. Surv. (CSUR) **53**(5), 1–40 (2020)
29. Zhou, Z., Guan, H., Bhat, M.M., Hsu, J.: Fake news detection via NLP is vulnerable to adversarial attacks. arXiv arXiv:1901.09657 (2019)

Federated Conversational Recommender Systems

Allen Lin[1(✉)], Jianling Wang[1], Ziwei Zhu[2], and James Caverlee[1]

[1] Texas A&M University, College Station, TX, USA
al001@tamu.edu

[2] George Mason University, Fairfax, VA, USA

Abstract. Conversational Recommender Systems (CRSs) have become increasingly popular as a powerful tool for providing personalized recommendation experiences. By directly engaging with users in a conversational manner to learn their current and fine-grained preferences, a CRS can quickly derive recommendations that are relevant and justifiable. However, existing CRSs typically rely on a centralized training and deployment process, which involves collecting and storing *explicitly-communicated user preferences* in a centralized repository. These fine-grained user preferences are completely *human-interpretable* and can easily be used to infer sensitive information (e.g., financial status, political stands, and health information) about the user, if leaked or breached. To address the user privacy concerns in CRS, we first define a set of privacy protection guidelines for preserving user privacy then propose a novel federated CRS framework that effectively reduces the risk of exposing user privacy. Through extensive experiments, we show that the proposed framework not only satisfies these user privacy protection guidelines, but also achieves competitive recommendation performance comparing to the state-of-the-art non-private conversational recommendation approach.

1 Introduction

Conversational recommender systems (CRSs) have gained significantly increasing attention in the research community for their wide range of applications and promising functionality, spanning from streaming services [16] to E-Commerce [9]. These systems rely heavily on the elicited user preferences, which are typically conveyed in a human-interpretable manner, to make highly personalized recommendations. For example, a user might come to the CRS asking for an *affordable* restaurant that only serves *vegetarian* food in *Austin, Texas.* While these communicated preferences help the CRS to make personalized recommendations, such information, if leaked or breached, could easily be exploited to infer sensitive information (e.g., financial status, dietary restriction, and location) about the user [29,30]. Despite the success of interactive preference elicitation, ***addressing user privacy concerns in CRSs*** remains unexplored.

Current CRSs [9,16,25,26,35] are trained and deployed in a *centralized* manner, meaning that the communicated, human-interpretable, preferences information of all users are *fully accessible* to the service provider. Such a centralized

N. Goharian et al. (Eds.): ECIR 2024, LNCS 14612, pp. 50–65, 2024.
https://doi.org/10.1007/978-3-031-56069-9_4

recommendation framework raises serious concerns of user privacy leakage or unintended data breaches [12]. While many strategies [1,20,41] have been proposed to address the privacy concerns in a *static* recommendation setting, these approaches mainly focus on privatizing users' raw item-interaction history (e.g., in the form of a real-valued matrix), which is not directly applicable to protecting the user preferences that are communicated during the *interactive preference elicitation* process in a conversational recommendation setting.

Recognizing the research gap, we first define a set of guidelines for preserving user privacy in a conversational recommendation setting – **user information locality**, **user contribution anonymity**, and **interaction and recommendation locality** – then propose a novel federated CRS framework that satisfies these guidelines. Specifically, the framework starts with building a predictive model to form an initial estimation of users' historical interests. To satisfy **user information locality**, instead of uploading their personal interaction history to the server, users only upload gradient updates needed to tune the predictive model. Before uploading, these gradient updates are perturbed with a privatization mechanism to ensure user-level local differential privacy, satisfying **user contribution anonymity**. After obtaining the initial estimation of users' historical interests, a reinforcement learning (RL) based policy agent is deployed to initiate a personalized interactive preference elicitation process with each user. To account for the *environment heterogeneity*[1] caused by each user having unique needs and preferences, a local user embedding projection layer is integrated to further personalize the recommendation experience. This policy agent is trained in a similar fashion as the predictive model with each user uploading the perturbed gradient updates instead of the communicated preferences. Upon completion of the tuning process, a copy of the policy agent is sent to every user to infer recommendations locally, satisfying **interaction and recommendation locality**. In summary, the main contributions of this work are as follows:

- To the best of our knowledge, this is the first work to comprehensively study user privacy concerns under the conversational recommendation setting and introduce a set of guidelines to be satisfied.
- We propose the FedCRS framework with unique design and privatization mechanisms that provides personalized recommendation experiences to users while removing the need of centralized collection of any personal data.
- We show that the proposed FedCRS framework is able to achieve competitive recommendation performance when compared to the state-of-the-art conversational recommendation approach, while providing strong user privacy protection.

[1] A commonly encountered issue in federated reinforcement learning that hinders a uniform policy to deliver optimal interaction experience to every user [20].

2 Preliminaries

2.1 Conversational Recommender System

A CRS generally starts with a *historical interests estimation* stage where it utilizes a user's personal interaction history to establish an initial estimation of the user. However, since historical item-interactions might fail to represent a user's current interests [13,17,18], the system initiates a personalized *interactive preference elicitation* stage, in which the system directly engages with the user in a conversational manner to inquire about the user's current and fine-grained preferences. In many recent CRSs [9,16,25,26,35,40], this elicitation stage is governed by a RL based policy agent. Formally, let $P = (p_0, p_1, ..., p_m)$ denote a set of m domain-specific attributes that describe an item in the itemset. Based on the system's current understanding of the user, the policy agent decides whether to ask more questions or make a recommendation. If the policy agent thinks not enough preference evidence has been collected, it will pick one or more attribute(s) p from the set of unasked attributes to prompt the user. If the policy agent decides enough information has been collected, the CRS then ranks all the candidate items and recommends the top k ranked items to the user. This process repeats until either when the user accepts the recommendation or when a pre-defined number of turns has being reached by the system.

2.2 Federated Learning

Federated Learning [21,42] is a privacy-preserving technique that leverages the private data of users to collaboratively train a central model without the need to collect them. In the federated learning paradigm, the user's data never has to leave the client. Instead, clients train a local model on their private data and share model updates with the server. These updates are then aggregated to update the central model. Since model updates usually contain much less information than the raw user data, the risks of privacy leakage is effectively reduced [1,6,41]. In this work, we develop a Federated Learning framework for training and deploying CRSs to better protect users' privacy.

2.3 Local Differential Privacy

Differential privacy [10,11] guarantees that a model's results are negligibly affected by the participation of any individual client [24]. Differential privacy was introduced under a centralized setting where a trusted aggregator collects raw user data and enforces some privatization mechanism. However, the assumption that the aggregator can be trusted can often be unrealistic [3,8]. Thus, the notion of *local differential privacy* is proposed. Formally, let U be a set of clients, and each u has a private value v to be sent to an untrusted aggregator. Local differential privacy can guarantee that the leakage of private information for each user is strictly bounded via applying a randomized algorithm M to the private value v and sending the perturbed value $M(v)$ to the aggregator. For the privacy guarantee to hold, the randomized algorithm M needs to satisfy ϵ-local differential privacy such that: $Pr[M(v) = y] \leq e^{\epsilon} Pr[M(v') = y]$. Here, v and v' are two arbitrary input private values, and $y \in range(M)$. The ϵ, (≥ 0), is the privacy budget which controls the utility and privacy trade-off. When $\epsilon = 0$, we achieve perfect privacy preservation but zero utility, while for $\epsilon = \infty$ we would achieve zero privacy preservation but perfect utility. In this work, we use a carefully calibrated local differential privacy technique in conjunction with a federated learning paradigm to significantly reduce the risk of user privacy leakage.

3 Proposed Framework

We introduce a novel **Fed**erated learning framework for preserving user privacy in **C**onversational **R**ecommender **S**ystems – FedCRS. Specifically, it is designed to tackle three key challenges: (i) how to privatize the personal historical item-interactions during the *historical user interests estimation* stage; (ii) how to privatize the explicitly communicated user preferences during the *interactive preference elicitation* stage; and (iii) how to privatize the interactive conversational recommendation process during the *deployment* stage.

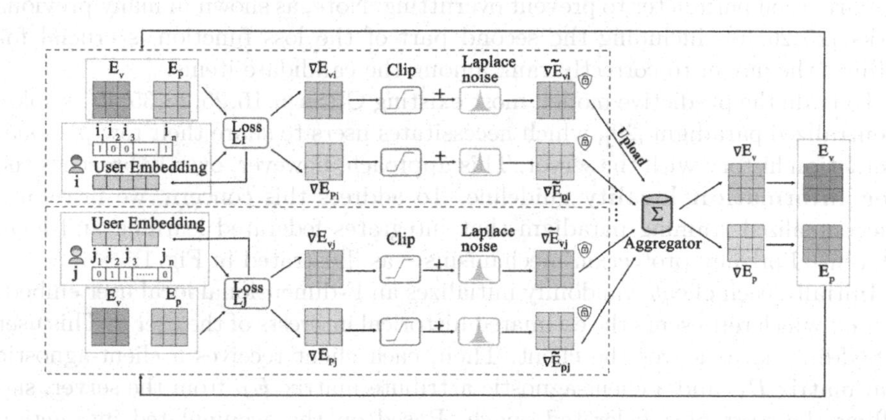

Fig. 1. Overall flow of decentralized historical interests estimation

3.1 Decentralized Historical Interests Estimation

For a CRS to provide a personalized recommendation experience, the system must start by establishing an initial understanding of the user based on his or her historical item-interactions. Thus, the goal of the *Decentralized Historical Interests Estimation* stage is to build a predictive model to learn a set of latent embeddings for all users, items, and attributes – denoted respectively as $\boldsymbol{E}_U$, $\boldsymbol{E}_V$, $\boldsymbol{E}_P$ – in a decentralized way to protect users' privacy. $\boldsymbol{E}_U$ denotes the set of user embeddings that encodes the estimated historical interests of each user, while $\boldsymbol{E}_V$ and $\boldsymbol{E}_P$ represent two sets of embeddings that respectively encode the latent characteristics or properties of each item or attribute. Following [25,26,40], we choose the factorization machine (FM) [33] as the predictive model.[2] Given user u, the user's explicitly stated preferences P_u, and an item v, u's predicted interest in v is computed as:

$$y(u, v, P_u) = \mathbf{e}_u^T \mathbf{e}_v + \sum_{p_i \in P_u} \mathbf{e}_v^T \mathbf{e}_{p_i}$$

where $\mathbf{e}_u$ and $\mathbf{e}_v$ respectively denotes the embedding of user u and item v, and $\mathbf{e}_{p_i}$ denotes the embedding for attribute $p_i \in P_u$. To train the model, we adopt the dual-loss pairwise Bayesian Personalized Ranking (BPR) objective function:

$$\begin{aligned} L = & \sum_{(u,v,v') \in D_1} -ln\sigma(y(u, v, P_u) - y(u, v', P_u)) \\ & + \sum_{(u,v,v') \in D_2} -ln\sigma(y(u, v, P_u) - y(u, v', P_u)) + \lambda_\theta ||\theta||^2 \end{aligned} \quad (1)$$

where D_1 denotes the set of pairwise instances where each pair is composed of an interacted item and a randomly chosen uninteracted item. And D_2 consists of instance pairs such that both the interacted and the uninteracted items have attributes matching P_u. σ denotes the sigmoid function and λ_θ denotes the regularization parameter to prevent overfitting. Note, as shown in many previous works [25,26,40], including the second part of the loss function is crucial for training the model to correctly rank among the candidate items.

To train the predictive model, most existing CRSs [9,16,25,26,35,40] employ a centralized paradigm [29], which necessitates users to share their raw personal interaction history with the server. This approach, however, overtly breaches the **user information locality** guideline. To address this concern, we introduce a decentralized training paradigm that integrates federated learning and local differential privacy protection mechanisms – as illustrated in Fig. 1.

Initially, each client[3] randomly initializes an F-dimensional local user embedding $\boldsymbol{e}_u$ which represents the estimated historical interests of the user u. This user embedding never leaves the client. Then, each client receives a client-agnostic item matrix $\boldsymbol{E}_V$ and a client-agnostic attribute matrix $\boldsymbol{E}_P$ from the server, signalling the start of a federated epoch. Based on the accumulated interaction history, the local user embedding $\boldsymbol{e}_u$, and the received $\boldsymbol{E}_V$ and $\boldsymbol{E}_P$, each client – with respect to Eq. (1) – computes (1) a local gradient update $\nabla \boldsymbol{e}_u$ for the local user embedding $\boldsymbol{e}_u$, which is, again, kept strictly locally; (2) a local gradient update $\nabla \boldsymbol{E}_{V_u}$ for the item matrix $\boldsymbol{E}_V$; and (3) a local gradient update $\nabla \boldsymbol{E}_{P_u}$ for the attribute matrix $\boldsymbol{E}_P$. The computed local user embedding gradient update $\nabla \boldsymbol{e}_u$ is directly used to *locally* update the current local user embedding by setting $\boldsymbol{e}_u = \boldsymbol{e}_u - \eta_1 \cdot \nabla \boldsymbol{e}_u$, where η_1 is the learning rate.

Meanwhile, the computed local gradient update for the item and attribute matrices – $\nabla \boldsymbol{E}_{V_u}$ and $\nabla \boldsymbol{E}_{P_u}$ – are uploaded onto at the server. To prevent

[2] Note that the choice of predictive model is flexible; we choose FM due to its widely demonstrated success in CRSs.

[3] Following [32], we regard each user device that participates in the training of the model as a *client*.

the risk of reverse-engineering [12], we further apply a local differential privacy mechanism to the local gradients via a randomized algorithm, M, defined as:

$$M(\nabla \boldsymbol{E}) = clip(\nabla \boldsymbol{E}, \delta) + Laplace(0, \lambda), \tag{2}$$

where $\nabla \boldsymbol{E}$ denotes a general gradient update matrix, in our case either $\nabla \boldsymbol{E}_{P_u}$ or $\nabla \boldsymbol{E}_{V_u}$, and δ denotes the clipping scale applied to the local gradients; 0 and λ respectively denotes the location and the scale parameter of the Laplace distribution used to drawn a Laplace noise. The scale parameter λ controls the strength of the Laplace noise such that larger λ brings a larger privacy budget but a lower utility of the perturbed gradients. The function $clip(\nabla \boldsymbol{E}, \delta)$ limits the value of the gradients in the scale of δ. It is used to help avoid potential gradient explosion and to control the degree of sensitivity of the local gradients (more details in Sect. 3.4). After performing the clipping and the randomization operations, each client uploads its privatized local gradients – $\nabla \widetilde{\boldsymbol{E}}_{V_u}$ and $\nabla \widetilde{\boldsymbol{E}}_{P_u}$ – to the server where all the local gradients are then aggregated by averaging. The aggregated gradients are then used to update the client-agnostic item and attribute matrices $\boldsymbol{E}_V$ and $\boldsymbol{E}_P$ stored on the server:

$$\boldsymbol{E}_V = \boldsymbol{E}_V - \eta_2 \cdot \frac{1}{|U|} \sum_{u \in U} \nabla \widetilde{\boldsymbol{E}}_{V_u} \tag{3}$$

$$\boldsymbol{E}_P = \boldsymbol{E}_P - \eta_3 \cdot \frac{1}{|U|} \sum_{u \in U} \nabla \widetilde{\boldsymbol{E}}_{P_u}, \tag{4}$$

where η_2 and η_3 respectively denote the learning rate used for optimizing the item and the attribute matrix. The updated client-agnostic item and attribute matrices $\boldsymbol{E}_V$ and $\boldsymbol{E}_P$ are then distributed back to each client for further tuning.

3.2 Decentralized Interactive Preference Elicitation

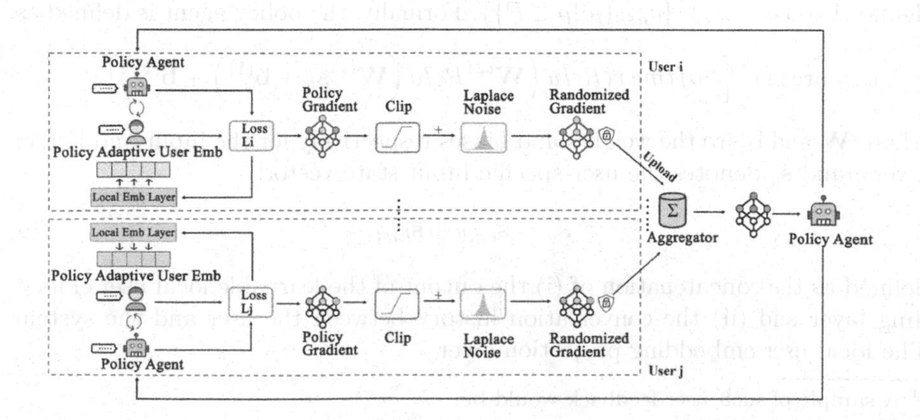

Fig. 2. Overall flow of decentralized interactive preference elicitation

Once the system acquires an initial estimation of each user's historical interests, it enters the personalized *interactive preference elicitation* stage to elicit users' current needs and preferences. To accomplish this, many existing CRSs [9,16,25,26,40] employ a RL based policy agent to customize the system's interactions (e.g. the attributes that the system inquires about) to deliver a personalized recommendation experience. This policy agent is typically trained through a trial-and-error process, in which it directly interacts with different users to learn *an optimal interaction strategy* that maximizes cumulative rewards. To obtain these rewards, many existing CRSs adopt the widely-accepted paradigm which requires users to share their raw feedback[4] with the server. However, this approach introduces two challenges. First, storing human-interpretable conversational feedback, which often contain explicit preferences of users, in a central repository dramatically increases the risk of unauthorized access and data breaches [23], violating the **interaction and recommendation locality** guideline. Second, since each user has distinct needs and preferences, a uniform interaction strategy can only provide sub-optimal recommendation experience [20].

To address these challenges, we propose a personalized federated RL paradigm that privatizes the training process of the policy agent, while accommodating environment heterogeneity caused by each user having unique preferences. In our proposed paradigm (see Fig. 2) all interactions between the policy agent and the user are confined to each user's device(s), while the policy agent is collaboratively updated using aggregated contributions from various users. To address the environment heterogeneity that hinders a uniform policy from delivering optimal recommendation experiences to every user, we employ a learnable local user embedding projection layer to locally finetune the policy agent to adapt to each user's distinct profile. This learnable local user embedding layer is directly tuned at each user's device(s) and never shared with the server.

To start, the server initializes the policy agent that will be trained collaboratively. To show the efficacy of the user-local embedding layer, we set the policy agent to a simple multi-layer perceptron neural network that maps the input state vector, $\mathbf{s}_u$, into an action, a, in the action space, A. Following [25], we define A to be $\{a_{rec} \cup \{a_{ask}(p)|p \in P\}\}$. Formally, the policy agent is defined as:

$$a = \arg\max \left[Softmax(Relu\left(\mathbf{W}^{(2)} Relu\left(\mathbf{W}^{(1)}\mathbf{s}_u + \mathbf{b}^{(1)}\right) + \mathbf{b}^{(2)}\right)\right], \quad (5)$$

where $\mathbf{W}$ and $\mathbf{b}$ are the weights and biases respectively for the input and hidden layers; and $\mathbf{s}_u$ denotes the user-specific input state vector:

$$\mathbf{s}_u = \mathbf{s}_{emb} \oplus \mathbf{s}_{hist}, \quad (6)$$

defined as the concatenation of (i) the output of the learnable local user embedding layer and (ii) the conversation history between the user and the system. The local user embedding projection layer

[4] A sample of such user feedback would be:
System: are you looking for more *country* music?
User: Not really, I used to like *country* music but now I am more into *jazz*.

$$\mathbf{s}_{emb} = Tanh\left(\mathbf{W}^{(u)}\mathbf{e}_u + \mathbf{b}_u\right) \tag{7}$$

takes in the initial user embedding, $\mathbf{e}_u$ (see Sect. 3.1), as the input and outputs a policy-adaptive embedding, $\mathbf{s}_{emb}$, that better suits the current interaction strategy. The conversation history, $\mathbf{s}_{hist}$, is represented by a one-hot vector with the same dimension as the total number of attribute, $|P|$, where the i-th dimension denotes whether the user has expressed preference for the attribute p_i.

Once the policy agent has been initialized on the server, its parameters θ are then shared with each client, signaling the start of a federated epoch. Each client interacts *locally* with the client-agnostic policy agent to collect N sequences of interactions between the client and the agent, in the form of state-action-reward tuples. Each sequence is known as a sampled trajectory, defined as $\tau = [(\mathbf{s}_{u_1}, a_1, r_{u_1}), ..., (\mathbf{s}_{u_t}, a_t, r_{u_t})]$. In a sampled trajectory, $\mathbf{s}_{u_i}$ denotes the user-specific state vector at the i-th time – as specified in Eq. (6); a_i denotes the system's action choose by the policy agent with $\mathbf{s}_{u_i}$ as the input – as specified in Eq. (5); and r_i denotes the numerical reflection or score of the user's feedback on the appropriateness of the chosen action, a_i. Based on the sampled trajectories, the policy agent is optimized locally with the objective to maximize the overall expected returns, defined as:

$$J(\theta)_u = \frac{1}{N}\sum_{i=1}^{N}\sum_{t}\gamma^t r(s_{i,t}, a_{i,t}),$$

where θ denotes the parameters of the policy agent shared from the server; γ^t denotes the discount parameter emphasizing conversation efficiency. Following [36], the local policy gradient of the objective function is defined as:

$$\nabla J(\theta)_u = \frac{1}{N}\sum_{i=1}^{N}\left[\left(\sum_{t}\nabla_\theta log\pi_\theta(a_{i,t}|s_{i,t})\right)\left(\sum_{t} r(s_{i,t}, a_{i,t})\right)\right] \tag{8}$$

Before sending local policy gradient updates, $\nabla J(\theta)$, back to the server, we apply the randomization algorithm M, introduced in Eq. (2), to perturb $\nabla J(\theta)$ to guarantee local differential privacy, for an enhanced user privacy protection:

$$\nabla \widetilde{J}_(\theta)_u = clip\left(\nabla J(\theta), \delta\right) + Laplace\left(0, \lambda\right), \tag{9}$$

where δ denotes the clipping scale applied to the local gradients; and 0 and λ respectively denotes the location and the scale parameter of the Laplace distribution used to drawn a Laplace noise. Once the clients have uploaded their local policy gradients, the server aggregates all the collected gradients by averaging:

$$\nabla J(\theta) = \frac{1}{|U|}\sum_{u\in U}\nabla \widetilde{J}_(\theta)_u$$

We use the averaged gradients to update the current policy agent:

$$\theta = \theta - \alpha \cdot \nabla J(\theta),$$

where θ denotes the parameters of the policy agent; α denotes the learning rate. It is important to note that, the parameters – $\mathbf{W}^{(u)}$ and $\mathbf{b}_u$ – of each user's local user embedding layer, specified in Eq. (7), are never shared with the server and are only updated locally based on the client's sampled trajectories. This approach not only enhances personalization of the policy agent but also provides an additional layer of user privacy protection.

3.3 Decentralized Deployment

Once the training of FedCRS has been completed, a copy of the predictive model and the policy agent is distributed to each client, allowing FedCRS to interact with the user and generate recommendations on each user's own devices - privatizing the interactive conversational recommendation process.

3.4 Privacy Analysis

We prove the privacy guarantee of FedCRS. Formally, we show that the randomized algorithm M – used in Eqs. (2) and (9) – guarantees ϵ, local differential privacy. Let $\mathbf{x}$ and $\mathbf{y}$ be any neighboring local gradient updates of dimension k, uploaded by two different clients. Let $P_{\mathbf{x}}(z)$ and $P_{\mathbf{y}}(z)$ be the probability density functions of $M(\mathbf{x})$ and $M(\mathbf{y})$ evaluated at an arbitrary point $z \in \mathbb{R}^k$. We show that the ratio of $P_{\mathbf{x}}(z)$ to $P_{\mathbf{y}}(z)$ is bounded by $exp(\epsilon)$:

$$\begin{aligned}\frac{P_{\mathbf{x}}(z)}{P_{\mathbf{y}}(z)} &= \frac{\prod_{i=1}^{k} exp(-\frac{\epsilon|\mathbf{x}_i - z_i|}{\Delta})}{\prod_{i=1}^{k} exp(-\frac{\epsilon|\mathbf{y}_i - z_i|}{\Delta})} = \prod_{i=1}^{k} exp(-\frac{\epsilon(|\mathbf{x}_i - z_i| - |\mathbf{y}_i - z_i|)}{\Delta}) \\ &\leq \prod_{i=1}^{k} exp(-\frac{\epsilon|\mathbf{x}_i - \mathbf{y}_i|}{\Delta}) = exp(\frac{\epsilon\sum_{i=1}^{k}|\mathbf{x}_i - \mathbf{y}_i|}{\Delta}) = exp(\frac{\epsilon||\mathbf{x} - \mathbf{y}||_1}{\Delta}) \\ &\leq exp(\epsilon),\end{aligned}$$

This proves that the randomized algorithm M guarantees local differential privacy, such that an **adversary cannot confidently distinguish** an individual user's local gradient updates from the local gradient updates of another user [39], satisfying **user contribution anonymity**. The Δ denotes the l_1 sensitivity between any two neighboring local gradient updates $\mathbf{x}$ and $\mathbf{y}$, defined as $\Delta = max_{\mathbf{x},\mathbf{y}}||\mathbf{x} - \mathbf{y}||_1$. The ϵ denotes the privacy budget of the randomized algorithm M and is defined as $\epsilon = \frac{\Delta}{\lambda}$, where λ is the scale parameter to draw the Laplace noise. Since a clipping function is applied in the randomized algorithm M - Eqs. (2) and (9), the l_1 sensitivity, Δ, is upper bounded by 2δ, with δ being the clipping scale. Thus, we have shown that the privacy budget, ϵ, is upper bounded by $\frac{2\delta}{\lambda}$.

4 Experiments

We conduct experiments to answer three key research questions: **RQ1**. How does recommendation performance of FedCRS compare to existing non-private

CRS approaches? **RQ2**. How does the estimated user historical interests and the personalized policy agent respectively contribute to the overall recommendation performance? **RQ3**. How does the privacy budget of the randomized algorithm affect the *recommendation performance* of FedCRS?

4.1 Experiment Setup

Dataset. We evaluate the proposed framework on two benchmark datasets – Yelp [35] and LastFM [25] – that are widely adopted [9,16,25,26,35] to evaluate CRSs. The two datasets are summarized in Table 1.

Table 1. Dataset statistics

Dataset	#users	#items	#interactions	#attributes
LastFM	1,801	7,432	76,693	33
Yelp	27,675	70,311	1,368,606	590

Baselines. To validate FedCRS, we compare its performance with the following **non-private** CRSs: **Max Entropy (MaxEnt)** adopts a rule-based policy agent that always asks the attribute with the highest entropy. **CRM** [35] uses a belief tracker to record a user's conveyed preference, and a policy agent to decide how to interact. **EAR** [25] builds a predictive model to estimate a user's historical preference and considers the feedback from the user. **SCPR** [26] leverages a knowledge graph to reduce attribute search space and uses Deep Q-Learning [37] to learn a policy agent. **UNICORN** [9] applies a dynamic-weighted-graph based RL approach to integrate the conversation and the recommendation components. **FPAN** [40] utilizes a user-item attribute graph to enhance the embeddings learned by the predictive model. **CRIF** [16] is the current state-of-the-art CRS. It explicitly infers a user's implicit preference conveyed during the conversation, and adopts inverse RL to learn a more flexible policy agent.

Evaluation Metrics. We use two widely adopted [9,16,25,26,35,40] metrics – success rate at turn t (SR@t) and average turn (AT) of conversations – for evaluation. SR@t is the accumulative ratio of successful conversational recommendation by the turn t, and AT is the average number of turns for all sessions.

User Simulator for Conversational Recommendation. Since conversational recommendation is a dynamic process, we follow [9,16,25,26,35,40] to create a user simulator to enable training and evaluation of a CRS. We simulate a conversation session for each observed interaction between users and items. Specifically, given an observed user-item interaction (u, v), we treat the v as the ground truth item and consider its attributes P_v as the oracle set of attributes preferred by the user in this conversation session. At the beginning, we randomly choose an attribute from the oracle set as the user's initial request to the CRS (e.g., Hi, can you find a *country* music?). Then the conversation session goes in the loop of the "model asks - user responses" process as introduced in Sect. 2.1. The session ends when either the CRS has recommended the groundtruth item v or a pre-defined number of allowed interactions has been reached. Code and Data will be released soon.

4.2 Recommendation Performance (RQ1)

As shown Table 2, it is surprising that FedCRS is able to outperform most of the baseline non-private CRSs on both LastFM and Yelp, given its primary intention is to preserve user privacy under the conversational recommendation setting. We conduct additional experiments to understand what contributes to this phenomenon. In Fig. 3, we compare the effectiveness of both the predictive model and the RL policy agent adopted respectively by FedCRS and CRIF.

Table 2. Overall Recommendation Performance Comparison, with SR@15 ↑ and AT ↓used as evaluation metrics.

		MaxEnt	CRM	EAR	SCPR	UNICORN	FPAN	CRIF	**FedCRS**
LastFM	SR@15	0.283	0.325	0.429	0.465	0.535	0.631	**0.866**	<u>0.831</u>
	AT	13.91	13.75	12.88	12.86	11.82	10.16	**9.17**	<u>9.54</u>
Yelp	SR@15	0.921	0.923	0.957	0.973	<u>0.984</u>	0.979	**0.986**	0.974
	AT	6.59	6.25	5.74	5.67	5.33	5.08	**4.35**	<u>4.92</u>

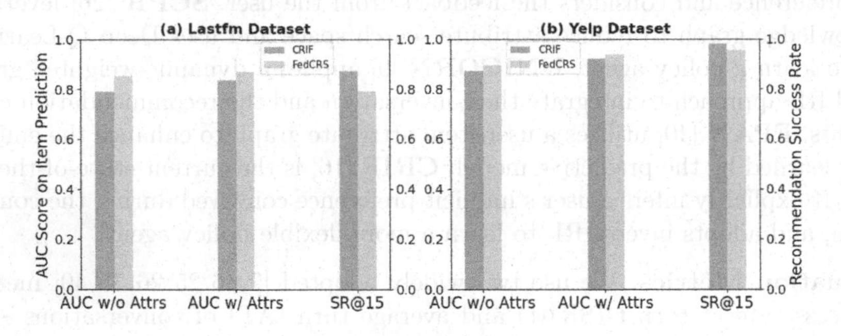

Fig. 3. Effectiveness of learned embeddings and adopted policy agent

To compare the effectiveness of predictive models, we report their respective AUC scores on item prediction with and without considering each item's associated attributes. The former task measures the effectiveness of the learned user and item embeddings, while the latter emphasizes on the learned attribute embeddings. As observed, FedCRS achieves higher AUC scores on both tasks than CRIF. This is primarily due to the *unweighted* aggregation of local matrices gradients, specified in Eqs. (3) and (4). The unweighted aggregation enforces the objective function to learn a globally optimized solution by simultaneously attending to all users, rather than disproportionately focusing on users with more item interactions [5,28]. To evaluate the effectiveness of the learned policy agents, we deploy the *same* predictive model for both FedCRS and CRIF and exclusively compare their adopted policy agents. As observed in Fig. 3, while using the same embeddings, the policy agent proposed in CRIF is able to slightly outperform FedCRS; however, the gap is relatively small.

4.3 Ablation Studies (RQ2)

We conduct ablation studies to further assess the contribution of each design component. For the *Decentralized Historical Interests Estimation* component, we measure its contribution by replacing the embeddings learned by the predictive model with (1) randomly initialized embeddings; and (2) CRIF's learned embeddings. As observed in Table 3, replacing the learned embeddings with randomly initialized embeddings causes dramatic drop in recommendation performance. In addition, replacing FedCRS's embeddings with CRIF's embedding also incurs a decrease in recommendation performance. For the *Decentralized Interactive Preference Elicitation* component, the uniqueness lies in the added learnable local user embedding layer that counters the environment heterogeneity [20]. As observed in Table 3, removing the learnable local user embedding layer causes significant decrease in recommendation performance.

Table 3. Ablation Studies with SR@15 $\uparrow$ and AT $\downarrow$ as metrics.

	Lastfm		Yelp	
	SR@15	AT	SR@15	AT
w/ Random Embeddings	0.251	13.96	0.327	12.05
w/ CRIF Embeddings	0.784	10.14	0.951	5.73
w/o Local Embedding Layer	0.722	11.07	0.919	6.14
FedCRS	**0.831**	**9.54**	**0.974**	**4.92**

4.4 The Privacy Budget Trade-Off (RQ3)

As introduced in Sect. 2.3, the privacy budget, ϵ, controls the trade-off between privacy preserved and the utility of the perturbed local gradients. In this section, we study the trade-off between privacy budget and recommendation performance in FedCRS. As shown in Table 3, the recommendation performance of FedCRS is closely related to the quality of embeddings learned by the predictive model during the *historical user interest estimation* stage and the effectiveness of the policy agent trained during the *interactive preference elicitation* stage. Therefore, we respectively investigate how different privacy budgets affects the effectiveness of the trained predictive model and policy agent. We use AUC scores as the metric to measure the effectiveness of the predictive model. As observed in Fig. 4(a), large privacy budgets enable FedCRS to train a more effective predictive model.

A similar phenomenon is observed in Fig. 4(b), where the recommendation success rate of the policy agent increases with the privacy budget. These observations align with our intuition that although smaller privacy budgets provide stronger privacy protection, the increased noise level (large λ in M) limits the utility of the uploaded local gradients. Note, both Fig. 4(a) and (b) exhibit a diminishing increase in slope for privacy budgets greater than 0.5. We select 0.5 as our privacy budget, as it reaches the optimal trade-off.

5 Related Work

5.1 Conversational Recommender System

CRSs have emerged as an increasingly popular tool to personalize a user's recommendation experience [9,16,40]. While early works on CRSs primarily rely on template conversation components [15,19], recent works have explored the possibilities of more effective conversational strategies with the help of RL. For example, EAR [25] proposes a novel three-staged framework that further strengthens the interaction between the recommender component and the conversation component. While sharing the same recommender component, SCPR [26] integrates knowledge graphs to improve the reasoning ability of the conversation component. Further utilizing the informativeness of knowledge graphs, UNICORN [9] proposes a dynamic weighted graph based reinforcement approach. Furthermore, CRIF [16] introduces the adaption of inverse reinforcement learning to learn a more flexible conversation policy. Although CRSs has shown promising capabilities, most of them centrally collect each user's complete *interaction* and *conversation* during training and inference, posing significant privacy concerns.

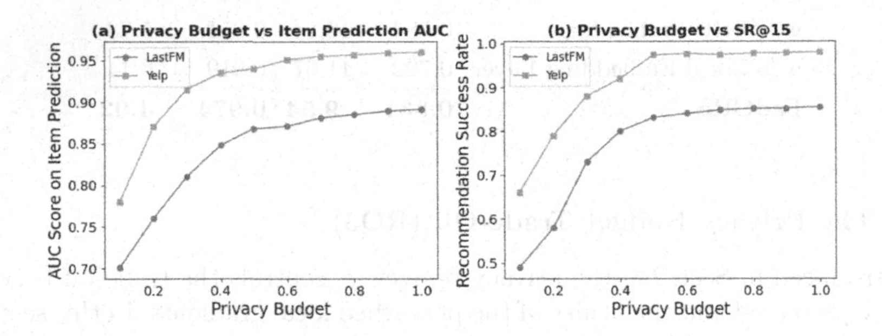

Fig. 4. Privacy budgets on effectiveness of the predictive model or policy agent

5.2 Privacy-Preserving Recommendation

Studies [12,32] indicate that even seemingly innocuous data, like movie ratings, can reveal sensitive user information. Research in this area typically divides based on trust in the recommender. When the recommender is assumed to be *trusted*, attacks aim to infer private information of the user from the released recommendation model or the recommendation results [12]. To protect the released model, [7] proposes to split the learned user-item embeddings into local and global segments. For protecting the recommendation results, [4,34] propose different means to perturb the results to satisfy differential privacy. When the recommender is assumed to be *distrusted*, existing solutions can be divided into two categories: private data collection [27,31] and distributed learning [2,22]. Data collection methods privatize the raw interaction data of the user before sharing it with the recommender [14,27,30], while distributed learning approaches distribute the training process such that only the gradient updates are uploaded to the server [32,38]. While the above approaches preserves user privacy in a static recommendation setting, none of them can be directly applied to privatize the explicitly communicated user preferences in CRSs.

6 Conclusion

In this work, we present, to the best of our knowledge, the first comprehensive study of user privacy concerns in CRSs and introduce a set of guidelines and a novel framework, FedCRS, aimed at protecting user privacy within the conversational recommendation setting. Through extensive experiments, we demonstrate that the proposed FedCRS not only satisfies all user privacy protection guidelines, but also provides strong recommendation performance.

References

1. Ammad-Ud-Din, M., et al.: Federated collaborative filtering for privacy-preserving personalized recommendation system. arXiv preprint arXiv:1901.09888 (2019)
2. Anelli, V.W., Deldjoo, Y., Di Noia, T., Ferrara, A., Narducci, F.: How to put users in control of their data in federated top-n recommendation with learning to rank. In: Proceedings of the 36th Annual ACM Symposium on Applied Computing, pp. 1359–1362 (2021)
3. Arachchige, P.C.M., Bertok, P., Khalil, I., Liu, D., Camtepe, S., Atiquzzaman, M.: Local differential privacy for deep learning. IEEE Internet Things J. **7**(7), 5827–5842 (2019)
4. Berlioz, A., Friedman, A., Kaafar, M.A., Boreli, R., Berkovsky, S.: Applying differential privacy to matrix factorization. In: Proceedings of the 9th ACM Conference on Recommender Systems, pp. 107–114 (2015)
5. Beutel, A., Chi, E.H., Cheng, Z., Pham, H., Anderson, J.: Beyond globally optimal: focused learning for improved recommendations. In: TheWebConf (2017)
6. Chai, D., Wang, L., Chen, K., Yang, Q.: Secure federated matrix factorization. IEEE Intell. Syst. **36**(5), 11–20 (2020)
7. Chen, C., Liu, Z., Zhao, P., Zhou, J., Li, X.: Privacy preserving point-of-interest recommendation using decentralized matrix factorization. In: Proceedings of the AAAI Conference on Artificial Intelligence, vol. 32 (2018)
8. Cormode, G., Jha, S., Kulkarni, T., Li, N., Srivastava, D., Wang, T.: Privacy at scale: local differential privacy in practice. In: Proceedings of the 2018 International Conference on Management of Data, pp. 1655–1658 (2018)

9. Deng, Y., Li, Y., Sun, F., Ding, B., Lam, W.: Unified conversational recommendation policy learning via graph-based reinforcement learning. In: Proceedings of the 44th International ACM SIGIR Conference on Research and Development in Information Retrieval, pp. 1431–1441 (2021)
10. Dwork, C.: Differential privacy. In: Bugliesi, M., Preneel, B., Sassone, V., Wegener, I. (eds.) ICALP 2006. LNCS, vol. 4052, pp. 1–12. Springer, Heidelberg (2006). https://doi.org/10.1007/11787006_1
11. Dwork, C.: Differential privacy: a survey of results. In: Agrawal, M., Du, D., Duan, Z., Li, A. (eds.) TAMC 2008. LNCS, vol. 4978, pp. 1–19. Springer, Heidelberg (2008). https://doi.org/10.1007/978-3-540-79228-4_1
12. Gao, C., Huang, C., Lin, D., Jin, D., Li, Y.: DPLCF: differentially private local collaborative filtering. In: Proceedings of the 43rd International ACM SIGIR Conference on Research and Development in Information Retrieval, pp. 961–970 (2020)
13. Gao, C., Lei, W., He, X., de Rijke, M., Chua, T.S.: Advances and challenges in conversational recommender systems: a survey. AI Open **2**, 100–126 (2021)
14. Gemulla, R., Nijkamp, E., Haas, P.J., Sismanis, Y.: Large-scale matrix factorization with distributed stochastic gradient descent. In: Proceedings of the 17th ACM SIGKDD International Conference on Knowledge Discovery and Data Mining, pp. 69–77 (2011)
15. Graus, M.P., Willemsen, M.C.: Improving the user experience during cold start through choice-based preference elicitation. In: Proceedings of the 9th ACM Conference on Recommender Systems, pp. 273–276 (2015)
16. Hu, C., Huang, S., Zhang, Y., Liu, Y.: Learning to infer user implicit preference in conversational recommendation. In: Proceedings of the 45th International ACM SIGIR Conference on Research and Development in Information Retrieval, pp. 256–266 (2022)
17. Iovine, A., Narducci, F., Semeraro, G.: Conversational recommender systems and natural language: a study through the ConveRSE framework. Decis. Support Syst. **131**, 113250 (2020)
18. Jannach, D., Manzoor, A., Cai, W., Chen, L.: A survey on conversational recommender systems. ACM Comput. Surv. (CSUR) **54**(5), 1–36 (2021)
19. Jiang, H., Qi, X., Sun, H.: Choice-based recommender systems: a unified approach to achieving relevancy and diversity. Oper. Res. **62**(5), 973–993 (2014)
20. Jin, H., Peng, Y., Yang, W., Wang, S., Zhang, Z.: Federated reinforcement learning with environment heterogeneity. In: International Conference on Artificial Intelligence and Statistics, pp. 18–37. PMLR (2022)
21. Kairouz, P., et al.: Advances and open problems in federated learning. Found. Trends® Mach. Learn. **14**(1–2), 1–210 (2021)
22. Kalloori, S., Klingler, S.: Horizontal cross-silo federated recommender systems. In: Proceedings of the 15th ACM Conference on Recommender Systems, pp. 680–684 (2021)
23. Lam, S.K.T., Frankowski, D., Riedl, J.: Do you trust your recommendations? An exploration of security and privacy issues in recommender systems. In: Müller, G. (ed.) ETRICS 2006. LNCS, vol. 3995, pp. 14–29. Springer, Heidelberg (2006). https://doi.org/10.1007/11766155_2
24. Lee, J., Clifton, C.: How much is enough? Choosing ε for differential privacy. In: Lai, X., Zhou, J., Li, H. (eds.) ISC 2011. LNCS, vol. 7001, pp. 325–340. Springer, Heidelberg (2011). https://doi.org/10.1007/978-3-642-24861-0_22
25. Lei, W., et al.: Estimation-action-reflection: towards deep interaction between conversational and recommender systems. In: WSDM (2020)

26. Lei, W., et al.: Interactive path reasoning on graph for conversational recommendation. In: KDD (2020)
27. Li, C., Palanisamy, B., Joshi, J.: Differentially private trajectory analysis for points-of-interest recommendation. In: 2017 IEEE International Congress on Big Data (BigData Congress), pp. 49–56. IEEE (2017)
28. Lin, A., Wang, J., Zhu, Z., Caverlee, J.: Quantifying and mitigating popularity bias in conversational recommender systems. In: Proceedings of the 31st ACM International Conference on Information & Knowledge Management, pp. 1238–1247 (2022)
29. Massa, P., Avesani, P.: Trust metrics in recommender systems. In: Golbeck, J. (eds.) Computing with Social Trust. Human-Computer Interaction Series. Springer, London (2009). https://doi.org/10.1007/978-1-84800-356-9_10
30. Minto, L., Haller, M., Livshits, B., Haddadi, H.: Stronger privacy for federated collaborative filtering with implicit feedback. In: Proceedings of the 15th ACM Conference on Recommender Systems, pp. 342–350 (2021)
31. Polat, H., Du, W.: SVD-based collaborative filtering with privacy. In: Proceedings of the 2005 ACM Symposium on Applied Computing, pp. 791–795 (2005)
32. Qi, T., Wu, F., Wu, C., Huang, Y., Xie, X.: Privacy-preserving news recommendation model learning. arXiv preprint arXiv:2003.09592 (2020)
33. Rendle, S.: Factorization machines. In: 2010 IEEE International Conference on Data Mining, pp. 995–1000 (2010)
34. Riboni, D., Bettini, C.: Private context-aware recommendation of points of interest: an initial investigation. In: 2012 IEEE International Conference on Pervasive Computing and Communications Workshops, pp. 584–589. IEEE (2012)
35. Sun, Y., Zhang, Y.: Conversational recommender system. In: SIGIR (2018)
36. Sutton, R.S., McAllester, D., Singh, S., Mansour, Y.: Policy gradient methods for reinforcement learning with function approximation. In: Advances in Neural Information Processing Systems, vol. 12 (1999)
37. Van Hasselt, H., Guez, A., Silver, D.: Deep reinforcement learning with double q-learning. In: Proceedings of the AAAI Conference on Artificial Intelligence, vol. 30 (2016)
38. Wu, C., Wu, F., Cao, Y., Huang, Y., Xie, X.: FedGNN: federated graph neural network for privacy-preserving recommendation. arXiv preprint arXiv:2102.04925 (2021)
39. Xiong, S., Sarwate, A.D., Mandayam, N.B.: Randomized requantization with local differential privacy. In: 2016 IEEE International Conference on Acoustics, Speech and Signal Processing (ICASSP), pp. 2189–2193. IEEE (2016)
40. Xu, K., Yang, J., Xu, J., Gao, S., Guo, J., Wen, J.R.: Adapting user preference to online feedback in multi-round conversational recommendation. In: Proceedings of the 14th ACM International Conference on Web Search and Data Mining, pp. 364–372 (2021)
41. Yang, L., Tan, B., Zheng, V.W., Chen, K., Yang, Q.: Federated recommendation systems. In: Yang, Q., Fan, L., Yu, H. (eds.) Federated Learning. LNCS (LNAI), vol. 12500, pp. 225–239. Springer, Cham (2020). https://doi.org/10.1007/978-3-030-63076-8_16
42. Yang, Q., Liu, Y., Chen, T., Tong, Y.: Federated machine learning: Concept and applications. ACM Trans. Intell. Syst. Technol. (TIST) **10**(2), 1–19 (2019)

FakeClaim: A Multiple Platform-Driven Dataset for Identification of Fake News on 2023 Israel-Hamas War

Gautam Kishore Shahi[1](✉), Amit Kumar Jaiswal[2], and Thomas Mandl[3]

[1] University of Duisburg-Essen, Duisburg, Germany
gautam.shahi@uni-due.de
[2] University of Surrey, Guildford, UK
a.jaiswal@surrey.ac.uk
[3] University of Hildesheim, Hildesheim, Germany
mandl@uni-hildesheim.de

Abstract. We contribute the first publicly available dataset of factual claims from different platforms and fake YouTube videos on the 2023 Israel-Hamas war for automatic fake YouTube video classification. The FakeClaim data is collected from 60 fact-checking organizations in 30 languages and enriched with metadata from the fact-checking organizations curated by trained journalists specialized in fact-checking. Further, we classify fake videos within the subset of YouTube videos using textual information and user comments. We used a pre-trained model to classify each video with different feature combinations. Our best-performing fine-tuned language model, Universal Sentence Encoder (USE), achieves a Macro F1 of 87%, which shows that the trained model can be helpful for debunking fake videos using the comments from the user discussion.

Keywords: Claim Extraction · YouTube Video classification · User Engagement · Polarisation · 2023 Israel-Hamas war

1 Introduction

Social media are a source of much unreliable information. During conflicts, the percentage of misinformation often increases [1,2] and, for example, false statements or misleading visual content are shared over different platforms. Fact-checking by experts has proven to help debunk misinformation and support a rational discourse. Debunking of claims propagated on social media platforms such as X[1] (formerly Twitter), YouTube, and Facebook [3] has increased in the last decade.

The fake news discourse changes frequently and follows current topics, like COVID-19, the Russia-Ukraine War, the 2023 Turkey-Syria earthquakes, and the 2023 Israel-Hamas war. Each time fake news spreads, it also makes use of new technology and patterns. The emergence of Generative Artificial Intelligence (GAI) led to fake news consisting of voice clips and altered media, which is

[1] We refer X as Twitter for clarity.

N. Goharian et al. (Eds.): ECIR 2024, LNCS 14612, pp. 66–74, 2024.
https://doi.org/10.1007/978-3-031-56069-9_5

hard to debunk. In contrast, ChatGPT is not able to detect fake claims without training data. For scientific analysis of the information space regarding conflicts from diverse angles, as well as for improving technology for supporting fact-checking, datasets are essential. We applied the AMUSED framework [3] to collect a dataset for fake claims on the Israel-Hamas conflict, which broke out in early October 2023. This dataset will allow research to analyze the information space regarding the war from diverse angles.

Contributions. We introduce the first fact-checked collection of claims related to the 2023 Israel-Hamas war called FakeClaim. It comprises 1,499 claims collected from 1,370 fact-checked articles published by 60 fact-checking organizations in 30 languages with background information about evidence pages to verify or reject the claims. We perform content extraction for the fact-checked articles, such as the origin of the claim on social media or unreliable news sources, and extract them as sample data as shown in Table 1. We demonstrate the utility of the dataset by forming a YouTube video classification by training state-of-the-art prediction models and find that user engagement, comments on the video, evidence pages and metadata significantly contribute to model performance. Finally, we benchmark our data with a fine-tuned version of the pre-trained embedding model, namely, Universal sentence encoder [4] that jointly ranks evidence pages and performs veracity prediction. The best-performing model achieves a Macro F1 of 87%, showing that combining these features can be used for video classification in future work.

Table 1. An example of a claim debunked by a fact-checking website and YouTube Video Link extraction from fact-checked articles (Claim data if extraction is possible like YouTube).

Feature	Value
ClaimId	factly_6
Claim	Visuals of Israelis protesting against Prime Minister Benjamin Netanyahu for war against Palestine
Label	False
Claim URL	https://tinyurl.com/25vm9h2e
Article Title	March 2023 Video Shared as Recent Israelis Protest Against Netanyahu Amid Ongoing Israel-Hamas Conflict
Article text	Amid the ongoing Israel-Hamas conflict that began in October 2023, a video depicting a massive crowd .... (refer to above link)
Published date	October 22, 2023
Language	English
Claim Source	YouTube
Link to Claim	https://www.youtube.com/watch?v=XDJP5Ow3ri
Claim(video) Data	Details extracted from claim(video)*

2 Related Work

Several datasets are available for recent conflicts and fake news from multiple social media platforms [3]. The analysis of public discussions on controversial issues has led to much research. A dataset for the Syrian war was assembled to support machine learning for fake news detection. The authors used the documentation of an NGO as ground truth [5].

For the Russian invasion of Ukraine, a dataset from Twitter was collected using related hashtags. An analysis showed that the activities of government sponsored actors were high at the beginning of the conflict [6]. A similar dataset for Facebook and Twitter shows how few actors contribute spread misinformation very intensely [7]. Another dataset was collected from Reddit and distinguished between military and conflict-related posts [8]. A study using topic modeling for a dataset on the Ukraine war has shown that identifying false news is unfeasible using unsupervised methods [9].

Technology for detecting false news articles is often based on supervised machine learning. In the shared task, CheckThat! which was organized in 2022, the classification of the veracity of text content proved to be hard for systems, and the best F1 measure obtained was 0.33 for a four-class problem [10,11]. Previously, a YouTube video classification by textual information was performed using BERT [12] without considering the evidence and user engagement [13]. Fake news classification has also been carried out for news articles using social user engagement [14].

There is a lack of datasets as combinations of claims, fake posts (like videos) and background evidence. Till now, Fake video detection has not yet been performed using a combination of claims, textual information, and user engagement. We define the task in Sect. 3.2 and present the implementation of our system and the results in Sect. 4.

3 Methodology

In this section, we explain the data collection process, formally define the fake news detection task for posts (videos) and explain the problem formation.

3.1 Data Collection

For data collection, we selected a list of fact-checking organizations publishing content on the 2023 Israel-Hamas conflict, such as Snopes, BoomLive. The AMUSED framework [3] for annotation was applied. It allows the extraction of the claimed source from either a social media post or from an online news source and maps the content to the label or verdict of the fact-checked article used in studying Twitter [15], YouTube [13].

First, we collect fact-checked articles from multiple fact-checking organizations. Starting from each collected article, we extract data available and translate it in case it is not written in English. Next, the content of the articles was filtered

Table 2. Distribution of claim debunked from different social media platforms (* some fact-checked article debunks multiple claims from different platforms)

Claim Source	Number of Claims
Facebook	579
Twitter	183
YouTube	389
Tiktok	186
Instagram	90
Telegram	72
Total(*)	1,499

with words related to the Israel-Hamas war using keywords like *Israel, Hamas.* We fetched social media links from filtered, fact-checked articles like YouTube videos and Instagram posts. Overall, we collected 1,370 fact-checking articles from 62 Fact-checking organizations in 30 languages. These fact-checked articles debunk 1,499 claims from different social media platforms (Table 2). Due to the availability of the research API and the possibility of extracting comments as an expression of user engagement, we perform a classification of the subset of YouTube videos. We obtained 404 fake YouTube videos, of which 388 still remain online on YouTube (checked on 29-10-2023). We randomly filtered 610 videos from YouTube using a keyword search related to the Israel-Hamas war to obtain videos with most likely correct information as another class.

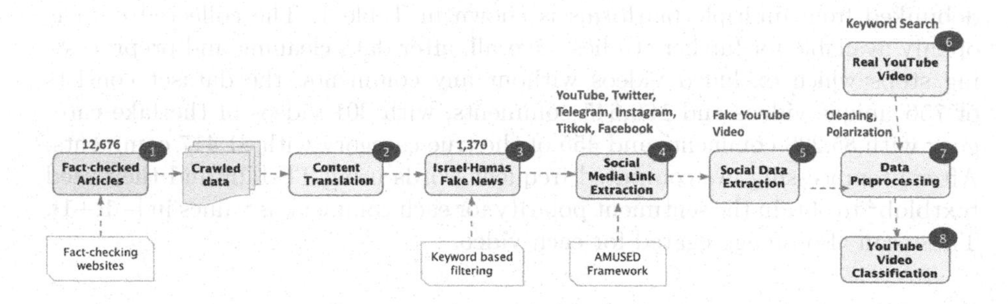

Fig. 1. The research framework used for extracting the dataset and classification problem.

3.2 Problem Formulation

Let D be a fake news detection dataset containing N samples. In the social media setting, we define the dataset as

$$\mathcal{D} = \{\mathcal{V}, \mathcal{U}, \mathcal{C}, \mathcal{F}\}$$

where $\mathcal{V} = \{v_1, v_2, v_3, ..., v_n\}$ is a set of textual descriptions of **YouTube videos**, $\mathcal{U} = \{u_1, u_2, u_3, ..., u_n\}$ is a set of related **social users** engaged in conversation related to the corresponding YouTube videos. $\mathcal{C}$ represent **user's comments** on videos, in which $c \in \mathcal{C}$ and $\mathcal{F}$ depicts the claim and provides correct background information to debunk the claim which is extracted from **fact-checked article**,$f \in \mathcal{F}$, R represents the set of **social audience engagements and background truth** as the composition of $\mathcal{V}$,$\mathcal{U}$,$\mathcal{C}$ in which $r \in R$ is defined as a quadruplet $\{(v, u, c, f)|v \in V, u \in U, c \in C, f \in F\}$ (i.e. user u has given c comments on YouTube video V and f was background truth provided by fact-checkers).

In line with most existing studies, we treat fake news detection on social media given the current scenario of YouTube framed as a binary classification problem (such as [13,16]). Specifically, V is split into a training set V_{train} and test set V_{test}. Each video v, which belongs to V_{train}, is associated with a ground truth label y of 1 if V is fake and 0 otherwise. We formulate the problem as follows:

Problem definition (Fake News Detection on YouTube Video): Given a YouTube video dataset D = V, U, C, F and ground-truth training labels y_{train}, the goal is to obtain a classifier f that, given test videos V_{train}, is able to predict the corresponding veracity labels y_{test}.

4 Experiments and Results

Data Prepossessing: After collection, articles are preprocessed by removing special characters, URLs, and emoji and applying a state-of-the-art classification model. The research framework is explained in Fig. 1. An overview of claims debunked from multiple platforms is shown in Table 1. The collected data is openly available for further studies. Overall, after data cleaning and preprocessing steps which excluded videos without any comments, the dataset consists of 756 unique videos and 166,645 comments, with 301 videos of the fake category with 85,228 comments and 455 of the true category with 81,417 comments. After preprocessing, we computed frequent words using TF-IDF and then used textblob[2] to obtain the sentiment polarity for each comment as values in [−1,+1]. These can also be aggregated for each video.

4.1 Experiment

Given our task of identifying fake claims, we utilize pre-trained word embedding models, where the multitude of textual comments spanned among YouTube videos entail diverse opinions, and so the social behaviors tend to upsurge conflict. We employ the polarity values for each video's comments as an implicit feature that distinguishes the spectrum of opinions reflecting online social interaction. We split our data by stratifying the binary class (fake/real claims) in the proportion of 70% for the train set and 30% for the test set and used precision (P), recall (R), and F1-score (F1) for evaluation.

[2] https://textblob.readthedocs.io/en/dev/quickstart.html.

Baseline Models: Given our collected corpus of fact claims data, including comments and metadata, our task tends to word-/sentence-level classification, and so we utilize pre-trained word embeddings and fine-tune the base model, namely, GNews-Swivel-20D [17]. The following baselines were implemented:

GN-Swivel-20D: The approach uses Swivel embeddings [17], which result from training on the extensive Google News 130 GB corpus. The GN-Swivel-20 dimensional model transforms words or phrases into 20-dimensional vectors, thereby adeptly encapsulating the semantic connotations and contextual interconnections that underlie the vocabulary, as discerned from an extensive textual corpus.

Universal Sentence Encoder (USE): We utilize the pre-trained Universal Sentence Encoder [4] tailored to transform sentences and text paragraphs into high-dimensional vector representations. It captures encapsulates the semantic essence and contextual information of complete sentences.

RoBERTa : RoBERTa is an encoder-only transformer model with 125 million parameters which underwent training on an extensive dataset comprising 160 GB of text corpus [18]. Our selection of this model is motivated by its common usage in classification and regression tasks, a practice shared with other encoder-only models such as BERT [12,19,20]. We employ the pre-trained $\text{RoBERTa}_{\text{base}}$ model[3] for comparison.

Fine-tuning Settings: Initially, we set the learning rate of both GN-Swivel-20D and USE models to be $1e{-}3$ with Adam as an optimizer and early stopping for monitoring the minimum validation loss and fine-tune the first layer for over ten epochs. Also, the model parameters are slightly smaller ($\sim$421k) in comparison to other variants of pre-trained embedding models. For fine-tuning the Universal Sentence Encoder model, we set the embedding size to 512, and then fine-tune the first layer. The model has $\sim$257 million parameters. We employ the pre-trained RoBERTa model without fine-tuning due to limited availability of resources.

Analysis: We outline how our results reported in Table 3 can reflect on mitigating factors for reducing fake claims in social discourse platforms. In Table 3, 'Video' refers to the video title. For this approach, we fed our model with concatenated pairs of video titles spanning comments and fact claims. We found out that the addition of fact-checked articles boosts the precision and F1 scores. On the contrary, we score the polarisation for each fact-checked article, where most articles and comments are highly skewed toward the words 'Israel,' 'Hamas,' and 'Gaza'. However, due to the skewness of the distribution of the sentiments

[3] https://huggingface.co/roberta-base.

for the comments tending to be high, our model performance fluctuates in the range of 72%–74%. Real videos are used without any fact-checked articles. Our current results are obtained without any additional polarisation features. This shows that the proliferation of users' engagement during discourse is an important factor in identifying fake claims.

Table 3. Results with pre-trained word embedding models for classification of Fake claims. Model performance is based on the F1 score, best models are in bold.

Model	Class & Features											
	fake (0)						real (1)					
	Video+ Comments+ Claims			Video+ Comments			Video+ Comments+ Claims			Video+ Comments		
	Scores											
	P	R	F1	P	R	F1	P	R	F1	P	R	F1
GN-Swivel-20D	0.831	0.814	0.822	0.748	0.772	0.76	0.822	0.828	0.825	0.768	0.77	0.769
Universal Sentence Encoder	0.874	0.867	**0.87**	0.803	0.798	**0.80**	0.868	0.882	**0.875**	0.796	0.813	**0.804**
RoBERTa	0.69	0.75	0.72	0.66	0.72	0.69	0.64	0.69	0.664	0.70	0.69	0.689

5 Conclusion and Future Work

In this study, we present a timely analysis of the 2023 Israel-Hamas war, providing a FakeClaim dataset that refers to the original fake posts and further scrapped YouTube videos and comments for the analysis. We formulated a classification problem for fake video descriptions by combining text information from videos, comments, claims and evidence and found that combining different features improved the F1 score. The study can be useful in fighting fake videos and help to mitigate fake news.

The limitations include the issue that the true class was not manually checked or fact-checked; in future work, we will try to also obtain labels for correct videos. Another limitation of the study is that collecting and extracting claims from fact-checked articles is difficult due to the lack of common standards for publishing fact-checked articles. It required continuous monitoring of the different websites and their structure to gather data. Finding around 1,499 claims in 3 weeks of conflict shows that the volume of claims is high, and an automated approach can help to debunk fake posts. The claims are spread over multiple platforms like YouTube, Twitter and Facebook. In future work, we intend to collect, explore and investigate FakeClaim data given the diversity of online comments from social platforms such as TikTok.

6 FAIR Data and Ethical Considerations

We have conducted all experiments on a macro level following strict data access, storage, and auditing procedures for the sake of accountability. Users personal information (like author name, profile picture) is neither used nor stored for classification in this paper The FakeClaim dataset, together with the ML/DL models generated in this study is available publicly for further research following the policy for sharing data. The dataset is available on Github (https://github.com/Gautamshahi/FakeClaim).

References

1. Lovelace, A.G.: Tomorrow's wars and the media. The US Army War College Quarterly: Parameters **52**(2), 117–134 (2022)
2. Shahi, G.K., Nandini, D.: FakeCovid–a multilingual cross-domain fact check news dataset for COVID-19. arXiv preprint arXiv:2006.11343 (2020)
3. Shahi, G.K., Majchrzak, T.A.: AMUSED: an annotation framework of multimodal social media data. In: Sanfilippo, F., Granmo, O.C., Yayilgan, S.Y., Bajwa, I.S. (eds.) Intelligent Technologies and Applications. INTAP 2021. CCIS, vol. 1616. Springer, Cham (2022). https://doi.org/10.1007/978-3-031-10525-8_23
4. Cer, D., et al.: Universal sentence encoder for English. In: Proceedings of the Conference on Empirical Methods in Natural Language Processing, EMNLP: System Demonstrations, Brussels, Belgium, October 31–November 4, pp. 169–174. Association for Computational Linguistics (2018)
5. Fatima, K. Salem, A. Feel, R.A. Elbassuoni, S. Jaber, M., Farah, M.: FA-KES: a fake news dataset around the Syrian war. In: Proceedings of the Thirteenth International Conference on Web and Social Media, ICWSM 2019, Munich, Germany, June 11–14, 2019, pp. 573–582. AAAI Press (2019)
6. Chen, E., Ferrara, E.: Tweets in time of conflict: a public dataset tracking the Twitter discourse on the war between Ukraine and Russia. In: Proceedings of the Seventeenth International AAAI Conference on Web and Social Media, ICWSM 2023, June 5–8, 2023, Limassol, Cyprus, pp. 1006–1013. AAAI Press (2023)
7. Pierri, F., Luceri, L., Jindal, N., Ferrara, E.: Propaganda and misinformation on Facebook and Twitter during the Russian Invasion of Ukraine. In: Proceedings of the 15th ACM Web Science Conference, WebSci, Austin, TX, USA, 30 April 2023–1 May, pp. 65–74. ACM (2023)
8. Zhu, Y., Haq, E., Lee, L., Tyson, G., Hui, P.: A Reddit dataset for the Russo-Ukrainian conflict in 2022. arXiv preprint arXiv:2206.05107 (2022)
9. Shin, Y., Sojdehei, Y., Zheng, L., Blanchard, B.: Content-based unsupervised fake news detection on Ukraine-Russia war. SMU Data Sci. Rev. **7**(1), 3 (2023)
10. Köhler, J., et al.: Overview of the CLEF-2022 checkthat! lab: task 3 on fake news detection. In: Proceedings of the Working Notes of CLEF 2022 - Conference and Labs of the Evaluation Forum, Bologna, Italy, Sept. 5th–8th, pp. 404–421. CEUR-WS.org (2022)
11. Shahi, G.K., Struß, J.M., Mandl, T.: Overview of the CLEF-2021 checkthat! lab: task 3 on fake news detection. In: Proceedings of the Working Notes of CLEF - Conference and Labs of the Evaluation Forum, Bucharest, Romania, September 21–24th, volume 2936 of CEUR Workshop Proceedings, pp. 406–423. CEUR-WS.org, (2021)
12. Devlin, J., Chang, M., Lee, K., Toutanova, K.: BERT: pre-training of deep bidirectional transformers for language understanding. In: Proceedings of the 2019 Conference of the North American Chapter of the Association for Computational Linguistics: Human Language Technologies, Volume 1 (Long and Short Papers), pp. 4171–4186 (2019)

13. Röchert, D., Shahi, G.K., Neubaum, G., Ross, B., Stieglitz, S.: The networked context of COVID-19 misinformation: Informational homogeneity on Youtube at the beginning of the pandemic. Online Soc. Netw. Media **26**, 100164 (2021)
14. Wu, J., Hooi, B.: Decor: degree-corrected social graph refinement for fake news detection. In: Proceedings of the 29th ACM SIGKDD Conference on Knowledge Discovery and Data Mining, pp. 2582–2593 (2023)
15. Shahi, G.K., Dirkson, A., Majchrzak, T.: An exploratory study of COVID-19 misinformation on Twitter. Online Soc. Netw. Media **22**, 100104 (2021)
16. Ajao, O., Bhowmik, D., Zargari, Z.: Fake news identification on Twitter with hybrid CNN and RNN models. In: Proceedings of the 9th International Conference on Social Media and Society, SMSociety 2018, Copenhagen, Denmark, July 18–20, 2018, pp. 226–230. ACM (2018)
17. Shazeer, N., Doherty, R., Evans, C., Waterson, C.: Swivel: improving embeddings by noticing what's missing. arXiv preprint arXiv:1602.02215 (2016)
18. Liu, Y., et al.: Roberta: a robustly optimized BERT pretraining approach. arXiv preprint arXiv:1907.11692 (2019)
19. Bommasani, R., et al.: On the opportunities and risks of foundation models. arXiv preprint arXiv:2108.07258 (2021)
20. Jaiswal, A., Liu, H.: Lightweight adaptation of neural language models via subspace embedding. In: Proceedings of the 32nd ACM International Conference on Information and Knowledge Management, pp. 3968–3972 (2023)

Countering Mainstream Bias via End-to-End Adaptive Local Learning

Jinhao Pan[1], Ziwei Zhu[2](✉), Jianling Wang[1], Allen Lin[1], and James Caverlee[1](✉)

[1] Texas A&M University, College Station, TX, USA
caverlee@tamu.edu
[2] George Mason University, Fairfax, VA, USA
zzhu20@gmu.edu

Abstract. Collaborative filtering (CF) based recommendations suffer from mainstream bias – where mainstream users are favored over niche users, leading to poor recommendation quality for many long-tail users. In this paper, we identify two root causes of this mainstream bias: (i) discrepancy modeling, whereby CF algorithms focus on modeling mainstream users while neglecting niche users with unique preferences; and (ii) unsynchronized learning, where niche users require more training epochs than mainstream users to reach peak performance. Targeting these causes, we propose a novel end-To-end Adaptive Local Learning (TALL) framework to provide high-quality recommendations to both mainstream and niche users. TALL uses a loss-driven Mixture-of-Experts module to adaptively ensemble experts to provide customized local models for different users. Further, it contains an adaptive weight module to synchronize the learning paces of different users by dynamically adjusting weights in the loss. Extensive experiments demonstrate the state-of-the-art performance of the proposed model. Code and data are provided at https://github.com/JP-25/end-To-end-Adaptive-Local-Leanring-TALL-.

Keywords: Recommender Systems · Collaborative Filtering · Mainstream Bias · Local Learning · Mixture-of-Experts

1 Introduction

The detrimental effects of algorithmic bias in collaborative filtering (CF) recommendations have been widely acknowledged [10,11,33,39]. Among these different types of recommendation bias, an especially critical one is **Mainstream Bias** (also called "grey-sheep" problem) [4,18,24,39], which refers to the phenomenon that *a CF-based algorithm delivers recommendations of higher utility to users with mainstream interests at the cost of poor recommendation performance for users with niche or minority interests.* For example, in a social media platform, mainstream users with interests in prevalent social topics will receive recommendations of high accuracy, while the system struggles to provide precise recommendations for niche users who focus on less common, yet equally important, topics. This makes the platform deliver unfair services to users with distinct interests and ultimately deteriorates the long-term prosperity of the platform.

N. Goharian et al. (Eds.): ECIR 2024, LNCS 14612, pp. 75–89, 2024.
https://doi.org/10.1007/978-3-031-56069-9_6

In this work, we identify two core root causes of such a mainstream bias: the discrepancy modeling problem and the unsynchronized learning problem.

Discrepancy Modeling Problem: CF algorithms estimate user preferences based on other users with similar tastes. So, the data from users with different preferences cannot help (or even play a negative role) in predicting recommendations for a target user. This issue functions bidirectionally – it impacts niche users, who differ from the majority, and mainstream users affected by the data from niche users. While prior studies [12,39] have used heuristic-based local learning to craft customized models for different user types (e.g., mainstream vs. niche), their efficacy is often bound by the quality of the underlying heuristics. This underscores the need for *adaptive approaches that learn to generate locally customized models for different user types in an end-to-end fashion.*

Unsynchronized Learning Problem: Another factor contributing to mainstream bias is the different learning paces between mainstream and niche users. Intuitively, mainstream users, with abundant training signals, tend to reach optimal learning faster. For instance, in Fig. 1, the high-mainstream subgroup peaks at around 60 epochs, whereas the low-mainstream group takes nearly 300 epochs. Training a model for all users without considering these learning pace disparities often results in models that cater primarily to mainstream users, sidelining niche users from reaching their optimal utility. Addressing this requires a method to *synchronize the learning process across users, irrespective of their mainstreamness.*

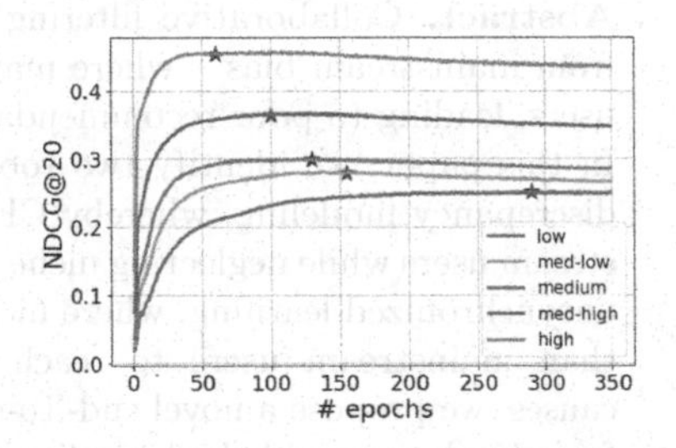

Fig. 1. Validation *NDCG@20* during training for users of varying mainstream levels. The star marks the epoch when reaching the peak performance of the subgroup.

To address these problems, we propose an end-**T**o-end **A**daptive **L**ocal **L**earning (TALL) framework. To tackle the discrepancy modeling problem affecting both niche and mainstream users, we devise a loss-driven Mixture-of-Experts structure as the backbone. This structure achieves local learning via an end-to-end neural network and adaptively assembles expert models using a loss-driven gate model, offering tailored local models for different users. Further, we develop an adaptive weight module to dynamically adjust the learning paces by weights in the loss, ensuring optimal learning for all user types. With these two complementary and adaptive modules, TALL can effectively **promote the utility for niche users while preserving or even elevating the utility for mainstream users**, leading to a significant debiasing effect based on the **Rawlsian Max-Min fairness principle** [29].

In sum, our contributions are: (1) we propose a loss-driven Mixture-of-Experts structure to tackle the discrepancy modeling problem, highlighted by an adaptive loss-driven gate module for customized local models; (2) we introduce an adaptive weight module to synchronize learning paces, augmented by a loss change and a gap mechanism for better debiasing; and (3) Extensive experiments demonstrate TALL's superior debiasing capabilities compared to leading alternatives, enhancing utility for niche users by 6.1% over the best baseline with equal model complexity. Data and code are available at https://github.com/JP-25/end-To-end-Adaptive-Local-Leanring-TALL-.

2 Preliminaries

Problem Formalization. Given a user set $\mathcal{U} = \{1, 2, \ldots, N\}$ consisting of N users and an item set $\mathcal{I} = \{1, 2, \ldots, M\}$ consisting of M items, we have the implicit feedback from users to items as the set $\mathcal{O} = \{(u, i)\}$, where $u \in \mathcal{U}$ refers to a user and $i \in \mathcal{I}$ refers to an item. This feedback set serves as the training data for training a recommendation model that generates recommendations for users. Each user u is represented by a binary vector of length M, denoted as $\mathbf{O}_u \in \{0, 1\}^M$. During the evaluation, the model provides a ranked list of K recommended items for every user. Various ranking evaluation metrics can be employed, such as NDCG@K and Recall@K [26].

Mainstream Bias. Typically, a recommender system is evaluated by averaging the utility over all users (such as NDCG@K), which essentially conceals the performance differences across different types of users. Previous work [39] formalizes the mainstream bias as *the recommendation performance difference across users of different levels of mainstreamness.* In this work, we follow the problem setting of [39] to measure the mainstream level for a user by calculating the average similarity of the user to all others: the more similar the user is to the majority, the more mainstream she is.

Debiasing Goal. In terms of the goal of debiasing, while addressing the issues of discrepancy modeling and unsynchronized learning benefits all user types, it is inappropriate to expect equalized utility across all users, which possibly encourages decreasing the utility for mainstream groups. Hence, in this work, we follow the **Rawlsian Max-Min fairness** principle of distribute justice [29]. To achieve fairness, this principle aims to maximize the minimum utility of individuals or groups, ensuring that no one is underserved. So, to counter the mainstream bias, we aim to **improve the recommendation utility of niche users while preserving or even enhancing the performance of mainstream users.**

3 End-to-End Adaptive Local Learning

To debias, we propose the end-**T**o-end **A**daptive **L**ocal **L**earning (**TALL**) framework, shown in Fig. 2. To address the *discrepancy modeling* problem, this framework integrates a **loss-driven Mixture-of-Experts** module to adaptively provide customized models for different users by an end-to-end learning procedure. To address the *unsynchronized learning* problem, the framework involves an **adaptive weight** module to synchronize the learning paces of different users by adaptively adjusting weights in the objective function.

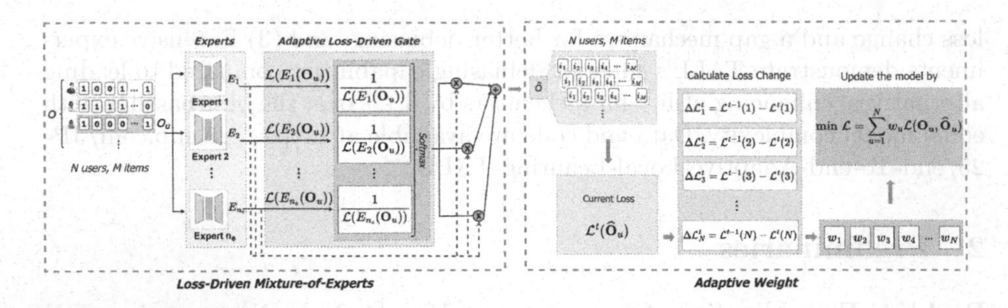

Fig. 2. The proposed End-to-End Adaptive Local Learning (TALL) framework.

3.1 Loss-Driven Mixture-of-Experts

Mixture-of-Experts. To address the discrepancy modeling problem, prior works propose the local learning method [12,39] to provide a customized local model trained by a small collection of local data for each user. However, how to curate such a local dataset and how to build a local model are completely hand-crafted in these algorithms, which significantly limits performance. To overcome this weakness, we adopt a Mixture-of-Experts (MoE) [14] structure as the backbone of our proposed framework to implement end-to-end local learning.

In detail, the MoE has two main components: gate and expert models. Specifically, an MoE comprises multiple expert models, each of which is trained to work for a specific task. For the recommendation problem, we adopt the MultVAE [26] as the expert model, and each MultVAE-based expert model in the MoE is responsible to process certain types of users. Then, with these expert models, for a target user, we rely on the gate model to distribute gate values to different expert models and generate the prediction for the user by weighted averaging the outputs of expert models based on gate values. Concretely, given an input user u with feedback record $\mathbf{O}_u$, the output of the MoE is $\widehat{\mathbf{O}}_u = \sum_{k=1}^{n_e} G_k(\mathbf{O}_u)E_k(\mathbf{O}_u)$, where n_e is the number of expert models, $G_k(\mathbf{O}_u)$ and $E_k(\mathbf{O}_u)$ are the k-th value of the gate model output and the output of the k-th expert model, and we also have the constraint $\sum_{k=1}^{n_e} G_k(\mathbf{O}_u) = 1$. The expert and gate models are trained together by the data in an end-to-end fashion.

Adaptive Loss-Driven Gate. However, the regular gate model, a free-to-learn feed-forward neural network (i.e., a multilayer perception) within a standard MoE, is also susceptible to various biases, including mainstream bias. The gate model is trained by data with more mainstream users and thus focuses more on how to assign gate values to improve utility for mainstream users while overlooking niche users. This results in an inability of the regular gate model to reasonably assign values to different users, especially niche users. Therefore, such a free-to-learn gate model cannot address the discrepancy modeling problem. A more precise and unbiased gate mechanism is needed.

A key principle of the gate model is that when we have a set of expert models, the gate model should assign high values for expert models that are effective for the target user, and low values for expert models less effective for the target user.

In this regard, the loss function serves as a high-quality indicator. Specifically, a high loss value in an expert model for a target user means that this expert model is not helpful for delivering prediction, and thus, it should not contribute to the aggregation of the final customized local model, receiving a small gate value. Conversely, a low loss indicates an effective expert model deserving a higher gate value. Based on this intuition, we propose the adaptive loss-driven gate module:

$$G_k(\mathbf{O}_u) = e^{(\mathcal{L}(E_k(\mathbf{O}_u)))^{-1}} \Big/ \sum_{t=1}^{n_e} e^{(\mathcal{L}(E_t(\mathbf{O}_u)))^{-1}},$$

where $\mathcal{L}(E_k(\mathbf{O}_u)))$ is the loss function of the k-th expert model for user u, and we adopt the multinomial cross-entropy loss function from [26].

This mechanism complies with the principle of allocating high gate values to more effective expert models and low values to irrelevant expert models. In practice, we can use the loss function calculated on training data or independent validation data for the proposed adaptive loss-driven gate mechanism. Due to that loss on previously unseen validation data is a more precise signal of model performance, in this work, we adopt the loss function on validation data to calculate gate values. By this adaptive loss-driven gate mechanism, we can automatically and adaptively assign gate values to expert models, based on which we generate effective customized models.

3.2 Synchronized Learning via Adaptive Weight

After addressing the discrepancy modeling problem, another core cause of the mainstream bias is the unsynchronized learning problem – the learning difficulties vary for different users, and users reach the performance peak during training at different speeds (refer to Fig. 1 for an example). Thus, a method to synchronize the learning paces of users is desired. We devise an adaptive weight approach to achieve learning synchronization so that the dilemma of performance trade-off between mainstream users and niche users can be overcome.

Adaptive Weight. The fundamental motivation of the proposed method lies in linking the learning status of a user to the loss function of the user at the current epoch. A high loss for a user signifies ineffective learning by the model, necessitating more epochs for accurate predictions. Conversely, a low loss indicates successful modeling, requiring less or even no further training. Moreover, we can use a weight in the loss function to control the learning pace for the user: a small weight induces slow updating, and a large weight incurs fast updating. Hence, we propose to synchronize different users by applying weights to the objective function based on losses users get currently – a user with a high loss should receive a large weight, and vice versa. We aim to achieve this intuition by solving the following optimization problem:

$$\max_{w} \sum_{u=1}^{N} w_u \mathcal{L}(\mathbf{O}_u, \widehat{\mathbf{O}}_u) - \alpha \|w\|_2^2, \quad \text{s.t.} \sum_{u=1}^{N} w_u = N, w \geq 0, \tag{1}$$

where we aim to maximize the weighted sum of losses of users, in which the solution will assign high weights to users with large losses; and the regularization $\alpha\|w\|_2^2$ is to control the skewness of the weight distribution – a larger α leads to a more even distribution. One extreme case is when $\alpha \to +\infty$, $w_u = 1$ for all users, and the other extreme case is when $\alpha = 0$, $w_u = N$ for user u with the largest loss. To solve this optimization problem, a closed-form solution by Lagrange Multipliers with Karush-Kuhn-Tucker (KKT) conditions can be derived [8]:

$$w_u^* = \max\left(\left(\mathcal{L}(\mathbf{O}_u, \widehat{\mathbf{O}}_u) - \lambda\right) / 2\alpha, 0\right), \quad \lambda = \left(\sum_{v=1}^{N} \mathcal{L}(\mathbf{O}_v, \widehat{\mathbf{O}}_v) - 2\alpha N\right) / N. \quad (2)$$

The proof for this solution can be found in [8]. Finally, with the computed weights, we can insert them into the original learning objective function of the framework and train the model by minimizing $\sum_{u=1}^{N} w_u \mathcal{L}(\mathbf{O}_u, \widehat{\mathbf{O}}_u)$. Similar to the proposed adaptive loss-driven gate model in Sect. 3.1, here we also rely on losses calculated by validation data to compute high-quality weights.

Loss Change Mechanism and Gap Mechanism. With the proposed adaptive weight approach, we take a solid step toward learning synchronization. However, two critical issues remain unaddressed. First, the scale of the loss function is innately different across different users. Usually, mainstream users possess a lower loss value than niche users because the algorithm can achieve better utility for mainstream users who are easier to model. Due to this *scale diversity problem*, computing weights by exact values of the loss can lead to the undesired situation that mainstream users get overly low weights and niche users get overly high weights, disturbing the learning process. On the other hand, the loss is not always stable, especially at the early stage of training. This *unstable loss problem* can deteriorate the efficacy of the problem adaptive weight method too.

To tackle the scale diversity problem, instead of directly using the loss, we propose to use loss change across epochs as the indicator for computing weights. If we observe the change of loss in recent epochs for a user is significant and the loss is in a decreasing trend, then we conclude that this user is in a fast learning stage and needs more epochs to converge. And we want to assign a high weight to this user. Thus, after each epoch, we will record the loss change $\Delta\mathcal{L}_u^t = \mathcal{L}^{t-1}(u) - \mathcal{L}^t(u)$. Furthermore, to make the indicator more robust to counter the scale diversity problem, in our experiment, we average recent L loss change values, denoted as $\overline{\Delta\mathcal{L}}_u^t$, to replace the loss in Eq. 1. We can then derive the solution:

$$w_u^* = \max\left(\left(\overline{\Delta\mathcal{L}}_u^t - \lambda\right) / 2\alpha, 0\right), \; \lambda = \left(\sum_{v=1}^{N} \overline{\Delta\mathcal{L}}_v^t - 2\alpha N\right) / N. \quad (3)$$

At last, since the losses are excessively unstable at the initial stage of training and the proposed adaptive weight module heavily relies on the stability of the loss value, the proposed method cannot perform well at the initial stage of training. Hence, we propose to have a gap at the beginning for our adaptive weight method. That is, we do not apply the proposed adaptive weight method to the framework at the first T epochs. Since at the early stage of the training, all users will be at a fast learning status (consider the first 50 epochs in Fig. 1), as there is no demand for learning synchronization. After T epochs of ordinary training, when the learning procedure is more stable and the loss is more reliable, we apply the adaptive weight method to synchronize the learning for different users. And this time synchronization is desired and plays an important role. The gap window T is a hyper-parameter and needs to be predefined.

4 Debiasing Experiments

In this section, we present a comprehensive set of experiments to highlight the strong debiasing performance of the proposed method, validate the effectiveness of various model components, assess the impact of the proposed adaptive weight module, and examine the impact of hyper-parameters.

Table 1. Data statistics.

	#users	#items	density
ML1M	6,040	3,706	4.46%
Yelp	20,001	7,643	0.32%
CDs & Vinyl	12,023	8,050	0.32%

4.1 Experimental Setup

Data and Metric. Table 1 summarizes statistics of datasets used in this paper. We use three public datasets for the experiments: **ML1M** [19], **Yelp** [1], and **Amazon CDs and Vinyl** [28]. For each dataset, we consider the ratings or reviews as positive feedback from users to items. Then, following the same evaluation scheme from [39], for each dataset, we uniformly randomly divide it into training, validation, and testing sets in the ratio of 70%, 10%, and 20%. Next, we calculate the mainstream scores of users. Specifically, for a user u, the mainstream score is calculated as $MS_u = \sum_{v \in \mathcal{U} \setminus u} Sim(\mathbf{O}_u, \mathbf{O}_v)/(N-1)$, where $Sim(\mathbf{O}_u, \mathbf{O}_v)$ is the user-user similarity between users u and v. The similarity is computed by Jaccard similarity between the implicit feedback record $\mathbf{O}_u$ and $\mathbf{O}_v$. Then, we sort users based on calculated mainstream scores in non-descending order and divide them into five subgroups with equal sizes. In the result, we denote the first 20% of users with the lowest mainstream scores as users of 'low' mainstream level, the subgroup of 20%-40% users as 'med-low' mainstream level, and so on for 40%-60% ('medium'), 60%-80% ('med-high'), and 80%-100% ('high') users. Last, we report and compare the average NDCG@20 for each subgroup to show the mainstream bias. We do not divide users by cutting at specific mainstream scores because this could lead to groups with extremely small numbers of users, deteriorating the reliability of reported utility and bias evaluation.

Considering the Rawlsian Max-Min fairness principle [29], **the goal of debiasing is to promote the average NDCG@20 for subgroups with low mainstream scores while preserving or even improving the utility for subgroups with high mainstream scores at the same time. Hence, we also anticipate an increase in the overall NDCG@20 of the model.**

Table 2. Comparing TALL with SOTA debiasing baselines on 3 datasets.

	ML1M						Yelp						CDs & Vinyl					
	NDCG @20	Subgroups of mainstream levels					NDCG @20	Subgroups of mainstream levels					NDCG @20	Subgroups of mainstream levels				
		L	ML	M	MH	H		L	ML	M	MH	H		L	ML	M	MH	H
MultVAE	.3260	.2354	.2764	.2986	.3652	.4546	.0877	.0686	.0710	.0733	.0901	.1355	.1367	.1100	.1316	.1366	.1457	.1596
WL	.3278	.2448	.2801	.2970	.3639	.4532	.0870	.0700	.0708	.0720	.0888	.1332	.1361	.1133	.1334	.1361	.1441	.1534
EnLFT	.3341	.2586	.2875	.3025	.3661	.4556	.0887	.0697	.0715	.0740	.0915	.1369	.1453	.1230	.1387	.1423	.1561	.1666
LOCA	.3308	.2551	.2780	.2972	.3622	.4617	.0942	.0723	.0758	.0764	.0970	.1494	.1573	.1341	.1510	.1556	.1665	.1795
LFT	.3416	.2707	**.2918**	.3072	.3727	.4657	.0927	.0740	.0738	.0768	.0956	.1432	.1557	.1343	.1481	.1515	.1678	.1770
TALL	**.3456**	**.2746**	.2903	**.3112**	**.3784**	**.4734**	**.0992**	**.0772**	**.0803**	**.0826**	**.1056**	**.1505**	**.1700**	**.1392**	**.1599**	**.1652**	**.1844**	**.2013**
$\Delta_{MultVAE}$(%)	6.01	16.65	5.03	4.22	3.61	4.14	13.11	12.54	13.1	12.69	17.2	11.07	24.36	26.54	21.5	20.94	26.56	26.13
Δ_{LFT}(%)	1.17	1.44	-0.51	1.30	1.53	1.65	7.01	4.32	8.81	7.55	10.46	5.10	9.18	3.65	7.97	9.04	9.89	13.73
Δ_{LOCA}(%)	4.47	7.64	4.42	4.71	4.47	2.53	5.31	6.78	5.94	8.12	8.87	0.74	8.07	3.80	5.89	6.17	10.75	12.14

L: low, ML: med-low, M: medium, MH: med-high, H: high

Baselines. In the experiments, we compare the proposed TALL with MultVAE and four state-of-art debiasing methods: (1) **MultVAE** [26] is the widely used vanilla recommendation model without debiasing. (2) **WL** [39] is a global method designed to assign more weights to niche users in the training loss of MultVAE. (3) **LOCA** [12] is a local learning model that trains multiple anchor models corresponding to identified anchor users and aggregates the outputs from anchor models based on the similarity between the target user and anchor users. (4) **LFT** [39] is the SOTA local learning model that first trains a global model with all data and then fine-tunes a customized local model for each user using their local data for the target user. (5) **EnLFT** [39] is the ensembled version of LFT, which is similar to LOCA but trains the anchor models by the approach of LFT.

To fairly compare the performance of different baselines and our proposed model, for all four debiasing baselines and our proposed TALL, we adopt the MultVAE as the base model (or the expert model in TALL). LOCA, EnLFT, and TALL have the same complexity with a fixed number of MultVAE in them. And owing to the end-to-end training paradigm, TALL takes less training time than other local learning baselines. Last, LFT has the largest complexity, which has an independent MultVAE for each user.

Reproducibility. All models are implemented in PyTorch and optimized by the Adam algorithm [20]. For the baseline MultVAE and the MultVAE component in other models, we set one hidden layer of size 100. And we maintain the number of local models at 100 for LOCA, EnLFT, and TALL for all datasets to ensure a fair comparison. All other hyper-parameters are grid searched by the validation sets. *All code and data can be found at* https://github.com/JP-25/end-To-end-Adaptive-Local-Leanring-TALL-.

4.2 Debiasing Performance

First, we conduct a comparative analysis to show the effectiveness of the proposed TALL. In Table 2, we evaluate the overall NDCG@20 and average NDCG@20 for five user subgroups with varying mainstream levels for all methods and datasets. The best results of each metric and subgroup for all datasets are marked in bold, and the improvement rate of the proposed TALL over the best baseline MultVAE, LOCA, and LFT is exhibited as well. The user subgroups are categorized based on their mainstream scores.

TALL vs. MultVAE & WL. First, we can observe that the utilities for users of all five subgroups are greatly promoted by our proposed TALL compared to the widely used model MultVAE. Moreover, we can see that although the global debiasing method WL can alleviate the mainstream bias to a certain degree compared to MultVAE, our proposed TALL can produce higher NDCG@20 for all five user subgroups than WL, depicting that the proposed TALL exhibits a more outstanding debiasing ability over WL.

TALL vs. EnLFT & LOCA. Hence, we next have a fair comparison between models of the same complexity and compare directly across local learning methods. From Table 2, we observe that LOCA and EnLFT are more effective in mitigating the mainstream bias than WL, as they remarkably enhance utility across all five groups on all datasets. Meanwhile, our TALL significantly outperforms LOCA and EnLFT across all user groups and datasets. The improvement is especially prominent for niche users: TALL improves NDCG@20 of the 'low' user group by 6.07% on average over LOCA and 10% over EnLFT. This shows that with the same model complexity, the proposed end-to-end adaptive local learning model is more effective than heuristic-based local learning models.

TALL vs. LFT. Last, we compare TALL with the state-of-the-art local learning baseline LFT, which is heavily computationally intense and time-consuming. But due to its special design that every user gets their own customized model trained by their local data, LFT can effectively address the discrepancy modeling problem. From Table 2, we observe that, for most of the time, LFT achieves the best performance for niche users among all baselines. And LFT can perform especially effectively for dense datasets (i.e., ML1M). In fact, given that the model complexity of LFT is much higher than TALL, it is unfair to compare them only based on recommendation accuracy (i.e., NDCG). For example, in the ML1M dataset, our TALL contains 100 expert models (MultVAE), while LFT trains 6,040 (#users) models separately, which is over 60 times larger than TALL. Although it is not a fair comparison, we can still observe in Table 2 that TALL can outperform LFT in most cases. Especially for the two sparse datasets Yelp and CDs&Vinyl, TALL produces significantly higher utilities for all types of users. This demonstrates the efficacy and necessity of an end-to-end local learning method compared to a heuristic-based one.

In sum, from Table 2, we see that for all datasets, the proposed TALL produces the greatest NDCG@20 improvement for each subgroup of different mainstream levels and leads to the state-of-the-art overall model performance. TALL can outperform baselines with lower and the same model complexity, and it can even outperform the baseline model that is way more complex than it.

4.3 Ablation Study

Next, we aim to investigate the effectiveness of different components in the proposed framework, including the proposed adaptive loss-driven gate module, the

Table 3. Ablation study on the adaptive loss-driven gate.

	NDCG@20	Subgroups of different mainstream levels				
		L	ML	M	MH	H
MultVAE	0.3260	0.2354	0.2764	0.2986	0.3652	0.4546
MoE	0.3230	0.2513	0.2668	0.2895	0.3519	0.4553
LMoE	**0.3401**	**0.2714**	**0.2893**	**0.3066**	**0.3736**	**0.4594**

L: low, ML: med-low, M: medium, MH: med-high, H: high

Table 4. Ablation study on the adaptive weight module, gap mechanism, and loss change mechanism.

	NDCG@20	Subgroups of different mainstream levels				
		L	ML	M	MH	H
LMoE	0.3401	0.2714	0.2893	0.3066	0.3736	0.4594
LMoE + LC	0.3371	0.2566	0.2835	0.3075	0.3724	0.4654
LMoE + gap + L	0.3396	0.2681	0.2878	0.3065	0.3741	0.4616
TALL (LMoE + gap + LC)	**0.3456**	**0.2746**	**0.2903**	**0.3112**	**0.3784**	**0.4734**

L: low, ML: med-low, M: medium, MH: med-high, H: high

adaptive weight module, the gap mechanism in the adaptive weight module, and the loss change mechanism in the adaptive weight module.

Adaptive Loss-Driven Gate. To verify the effectiveness of the proposed adaptive loss-driven gate module in the proposed TALL, we compare the MoE component with the adaptive loss-driven gate (denoted as LMoE, which is the TALL model without the adaptive weight module) to a conventional MoE (denoted as MoE) with the standard multilayer perceptron (MLP) as the gate learning from the dataset. By comparing LMoE and MoE, we can justify the effect of the proposed adaptive loss-driven gate. The results are present in Table 3, where we also include the result of MultVAE as a baseline. From the table, we can observe that MoE produces better performance for niche users compared to MultVAE but worse results for other users. This is caused by the MLP-based gate model in MoE, which cannot precisely allocate gate values to expert models to ensemble a strong customized model for different users. Conversely, we can see that even without the adaptive weight module, the LMoE can deliver greatly higher utilities for all types of users compared to MultVAE and MoE, showing the strong capability of the proposed adaptive loss-driven gate module in distributing gate values across expert models. And this result also demonstrates the efficacy of the proposed LMoE module in TALL in terms of addressing the discrepancy modeling problem.

Adaptive Weight. To address the unsynchronized learning problem, we develop the adaptive weight module to dynamically adjust the learning paces of different users. To verify the effectiveness of the proposed adaptive weight, we compare the complete TALL algorithm (with both the loss-driven MoE module and the adaptive weight module) to the loss-driven MoE module (LMoE). The results on ML1M are shown in Table 4, from which we can see that TALL outperforms LMoE for all types of users, manifesting the effectiveness of the proposed adaptive weight method. Furthermore, we explore the effectiveness of two special mechanisms, the gap mechanism and the loss change mechanism within the adaptive weight module, in the subsequent sections.

Gap Mechanism. To avoid the unstable loss problem at the initial training stage, we propose to have a gap for the adaptive weight method, i.e., we wait for a certain number of epochs at the initial training stage until the loss is stable and then apply the adaptive weight method. To verify the effectiveness of such a gap strategy, we compare the complete TALL (with a full version of the adaptive weight module including both the gap mechanism and the loss change mechanism) to a variation of TALL with the adaptive weight module without the gap mechanism. The comparison is presented in Table 4 as LMoE+LC vs. TALL. We can observe that the gap mechanism does have a significant influence on the model performance that the model with the gap mechanism (TALL) delivers better utilities for all types of users than the model without the gap mechanism (LMoE+LC).

Loss Change Mechanism. Last, we aim to verify the effectiveness of the proposed loss change mechanism in the adaptive weight module. The goal of loss change is to counter the scale diversity problem when applying the adaptive weight module. Here, we compare the complete TALL (with a full version of the adaptive weight module including both the gap mechanism and the loss change mechanism) to a variation of TALL (LMoE+gap+L) with the adaptive weight module that uses original loss as introduced in Sect. 3.2. From the comparison result shown in Table 4, we see that the model using loss change (TALL) performs better than the model using original loss for calculating weights (LMoE+gap+L). TALL outperforms LMoE+gap+L for all types of users.

In sum, by a series of comparative analyses, we show that the proposed adaptive loss-driven gate module, adaptive weight module, the gap mechanism in adaptive weight, and the loss change mechanism in adaptive weight are effective and play imperative roles in the proposed TALL framework.

4.4 Effect of the Adaptive Weight Module

Last, we turn our attention to investigating the effect of the adaptive weight module, studying how it synchronizes the learning paces of different users. We run TALL on the ML1M dataset and present the average weights for the five subgroups with the gap window ($\#gap = 40$) in Fig. 3. It can be observed that the adaptive weight module assigns

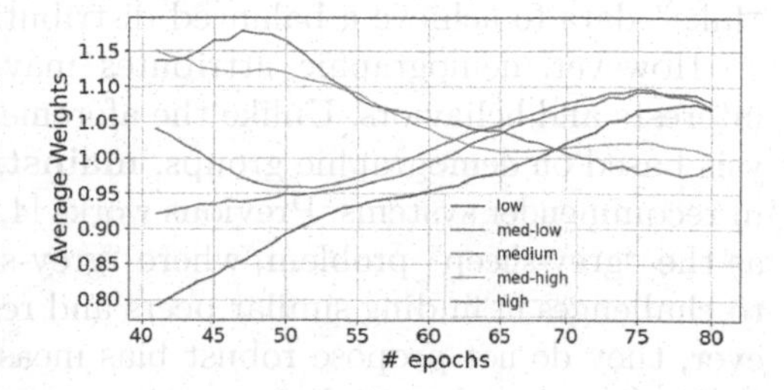

Fig. 3. Weights assignment across different types of users.

weights dynamically to different types of users to synchronize their learning paces. Initially, mainstream users receive higher weights because they are easier to learn and have a higher upper bound of performance than niche users. Then, when mainstream users reach the peak, the model switches the attention to niche users who are more difficult to learn, gradually increasing the weights for 'low', 'med-low', and 'medium' users until the end of the training procedure. However, 'med-high' and 'high' users, approaching converged, need a slower learning pace to avoid overfitting, leading to a decrease in the weights. Figure 3 illuminates the effectiveness and dynamic nature of the proposed adaptive weight module in synchronizing the learning procedures for different types of users.

4.5 Hyper-Parameter Study

Additionally, we have also conducted a comprehensive hyper-parameter study investigating the impacts of three hyper-parameters in TALL: (1) the gap window in the adaptive weight module; (2) α in the adaptive weight module; and (3) the number of experts. The complete results are in https://github.com/JP-25/end-To-end-Adaptive-Local-Leanring-TALL-/blob/main/Hyperparameter_Study.pdf.

5 Related Work

Fairness and bias issues in recommender systems have attracted increasing attention recently. Popularity bias [2,3,9,34,37,38], exposure bias [5,21,30,31], and item fairness [6,7,11,17,27,35] exemplify significant item-side biases. Besides prior works mainly focusing on the item perspective, several research studies have explored user biases, analyzing utility differences among diverse user groups based on user demographic attributes, like age or gender [10,15,16,25,32,33,36]. For instance, Ekstrand et al. [15] empirically investigated multiple recommendation models and demonstrated utility differences across user demographic groups. Schedl et al. [32] examined music preference differences among user age groups, revealing variations in recommendation performance. To address these issues, Fu et al. [16] proposed to leverage rich information from knowledge graphs, Li et al. [25] developed a re-ranking to narrow the utility gap between different user groups, and Chen et al. [10] implemented data augmentation by generating "fake" data to achieve a balanced distribution.

However, demographic attributes may not comprehensively capture user interests and behaviors. Unlike the aforementioned works focusing on bias analysis based on demographic groups, **mainstream bias** poses a critical challenge in recommender systems. Previous works [4,18,24] acknowledge mainstream bias as the "grey-sheep" problem, where "grey-sheep users" with niche interests lead to challenges in finding similar peers and result in poor recommendations. However, they do not propose robust bias measurements and debiasing methods. A more aligned study with better mainstream bias evaluations to this paper is [39], which also addresses mainstream bias and enhances utility for niche users using global and local methods. Prior existing local methods [12,13,22,23,39] and global methods [39] can mitigate the bias to some degree by improving the utility for niche users. The recently proposed Local Fine Tuning (LFT) [39] and local collaborative autoencoder (LOCA) [12] produce state-of-the-art performance by employing multiple multinomial variational autoencoders (Mult-VAE) [26] as base models and generating customized local models to capture special patterns of different types of user. Nonetheless, prior methods have a key limitation: their reliance on heuristics impacts performance, necessitating meticulous hyper-parameters tuning by practitioners. Thus, the performance of these prior heuristic-based local learning methods is limited. This work targets the mainstream bias problem by proposing an end-to-end adaptive local learning framework to automatically and adaptively learn customized local models for different users, overcoming the limitations of heuristic-based methods to mitigate mainstream bias.

6 Conclusion

In this study, we aim to address the mainstream bias in recommender systems that niche users who possess special and minority interests receive overly low utility from recommendation models. We identify two root causes of this bias: the discrepancy modeling problem and the unsynchronized learning problem. Toward debiasing, we devise an end-to-end adaptive local learning framework: we first propose a loss-driven Mixture-of-Experts module to counteract the discrepancy modeling problem, and then we develop an adaptive weight module to fight against the unsynchronized learning problem. Extensive experiments show the outstanding performance of our proposed method on both niche and mainstream users and overall performance compared to SOTA alternatives.

Acknowledgements. This research was funded in part by 4-VA, a collaborative partnership for advancing the Commonwealth of Virginia.

References

1. Yelp dataset (2021). https://www.yelp.com/dataset
2. Abdollahpouri, H., Burke, R., Mobasher, B.: Controlling popularity bias in learning-to-rank recommendation. In: Proceedings of the Eleventh ACM Conference on Recommender Systems, pp. 42–46 (2017)
3. Abdollahpouri, H., Burke, R., Mobasher, B.: Managing popularity bias in recommender systems with personalized re-ranking. In: The Thirty-second International Flairs Conference (2019)
4. Alabdulrahman, R., Viktor, H.: Catering for unique tastes: targeting grey-sheep users recommender systems through one-class machine learning. Expert Syst. Appl. **166**, 114061 (2021)
5. Ben-Porat, O., Torkan, R.: Learning with exposure constraints in recommendation systems. In: Proceedings of the ACM Web Conference 2023, pp. 3456–3466 (2023)

6. Beutel, A., et al.: Fairness in recommendation ranking through pairwise comparisons. In: Proceedings of the 25th ACM SIGKDD International Conference on Knowledge Discovery & Data Mining, pp. 2212–2220 (2019)
7. Cai, W., Feng, F., Wang, Q., Yang, T., Liu, Z., Xu, C.: A causal view for item-level effect of recommendation on user preference. In: Proceedings of the Sixteenth ACM International Conference on Web Search and Data Mining, pp. 240–248 (2023)
8. Chai, J., Wang, X.: Fairness with adaptive weights. In: International Conference on Machine Learning, pp. 2853–2866. PMLR (2022)
9. Chen, J., Wu, J., Chen, J., Xin, X., Li, Y., He, X.: How graph convolutions amplify popularity bias for recommendation? arXiv preprint arXiv:2305.14886 (2023)
10. Chen, L., et al.: Improving recommendation fairness via data augmentation. arXiv preprint arXiv:2302.06333 (2023)
11. Chen, X., et al.: Fairly adaptive negative sampling for recommendations. arXiv preprint arXiv:2302.08266 (2023)
12. Choi, M., Jeong, Y., Lee, J., Lee, J.: Local collaborative autoencoders. In: Proceedings of the 14th ACM International Conference on Web Search and Data Mining, pp. 734–742 (2021)
13. Christakopoulou, E., Karypis, G.: Local latent space models for top-n recommendation. In: Proceedings of the 24th ACM SIGKDD International Conference on Knowledge Discovery & Data Mining, pp. 1235–1243 (2018)
14. Eigen, D., Ranzato, M., Sutskever, I.: Learning factored representations in a deep mixture of experts. arXiv preprint arXiv:1312.4314 (2013)
15. Ekstrand, M.D., et al.: All the cool kids, how do they fit in?: popularity and demographic biases in recommender evaluation and effectiveness. In: Conference on Fairness, Accountability and Transparency, pp. 172–186. PMLR (2018)
16. Fu, Z., et al.: Fairness-aware explainable recommendation over knowledge graphs. In: Proceedings of the 43rd International ACM SIGIR Conference on Research and Development in Information Retrieval, pp. 69–78 (2020)
17. Geyik, S.C., Ambler, S., Kenthapadi, K.: Fairness-aware ranking in search & recommendation systems with application to linkedin talent search. In: Proceedings of the 25th ACM SIGKDD International Conference on Knowledge Discovery & Data Mining, pp. 2221–2231 (2019)
18. Gras, B., Brun, A., Boyer, A.: Can matrix factorization improve the accuracy of recommendations provided to grey sheep users? In: 13th International Conference on Web Information Systems and Technologies (WEBIST), pp. 88–96 (2017)
19. Harper, F.M., Konstan, J.A.: The movielens datasets: history and context. ACM Trans. Interactive Intell. Syst. (TIIS) **5**(4), 1–19 (2015)
20. Kingma, D.P., Ba, J.: Adam: a method for stochastic optimization. arXiv preprint arXiv:1412.6980 (2014)
21. Lee, J.w., Park, S., Lee, J.: Dual unbiased recommender learning for implicit feedback. In: Proceedings of the 44th International ACM SIGIR Conference on Research and Development in Information Retrieval, pp. 1647–1651 (2021)
22. Lee, J., Kim, S., Lebanon, G., Singer, Y.: Local low-rank matrix approximation. In: International Conference on Machine Learning, pp. 82–90. PMLR (2013)
23. Lee, J., Kim, S., Lebanon, G., Singer, Y., Bengio, S.: Llorma: local low-rank matrix approximation (2016)
24. Li, R.Z., Urbano, J., Hanjalic, A.: Leave no user behind: towards improving the utility of recommender systems for non-mainstream users. In: Proceedings of the 14th ACM International Conference on Web Search and Data Mining, pp. 103–111 (2021)

25. Li, Y., Chen, H., Fu, Z., Ge, Y., Zhang, Y.: User-oriented fairness in recommendation. In: Proceedings of the Web Conference 2021, pp. 624–632 (2021)
26. Liang, D., Krishnan, R.G., Hoffman, M.D., Jebara, T.: Variational autoencoders for collaborative filtering. In: Proceedings of the 2018 World Wide Web Conference, pp. 689–698 (2018)
27. Liu, W., Burke, R.: Personalizing fairness-aware re-ranking. arXiv preprint arXiv:1809.02921 (2018)
28. Ni, J., Li, J., McAuley, J.: Justifying recommendations using distantly-labeled reviews and fine-grained aspects. In: Proceedings of the 2019 conference on empirical methods in natural language processing and the 9th International Joint Conference on Natural Language Processing (EMNLP-IJCNLP), pp. 188–197 (2019)
29. Rawls, J.: Justice as fairness: A restatement. Harvard University Press (2001)
30. Saito, Y.: Unbiased pairwise learning from biased implicit feedback. In: Proceedings of the 2020 ACM SIGIR on International Conference on Theory of Information Retrieval, pp. 5–12 (2020)
31. Saito, Y., Yaginuma, S., Nishino, Y., Sakata, H., Nakata, K.: Unbiased recommender learning from missing-not-at-random implicit feedback. In: Proceedings of the 13th International Conference on Web Search and Data Mining, pp. 501–509 (2020)
32. Schedl, M., Bauer, C.: Online music listening culture of kids and adolescents: Listening analysis and music recommendation tailored to the young. arXiv preprint arXiv:1912.11564 (2019)
33. Wang, Y., Ma, W., Zhang, M., Liu, Y., Ma, S.: A survey on the fairness of recommender systems. ACM Trans. Inform. Syst. **41**(3), 1–43 (2023)
34. Wei, T., et al.: Model-agnostic counterfactual reasoning for eliminating popularity bias in recommender system. arXiv preprint arXiv:2010.15363 (2020)
35. Yao, S., Huang, B.: Beyond parity: fairness objectives for collaborative filtering. arXiv preprint arXiv:1705.08804 (2017)
36. Ying, Y., Zhuang, F., Zhu, Y., Wang, D., Zheng, H.: Camus: attribute-aware counterfactual augmentation for minority users in recommendation. In: Proceedings of the ACM Web Conference 2023, pp. 1396–1404 (2023)
37. Zhang, Y., et al.: Causal intervention for leveraging popularity bias in recommendation. arXiv preprint arXiv:2105.06067 (2021)
38. Zhang, Y., Cheng, D.Z., Yao, T., Yi, X., Hong, L., Chi, E.H.: A model of two tales: dual transfer learning framework for improved long-tail item recommendation. In: Proceedings of the Web Conference 2021, pp. 2220–2231 (2021)
39. Zhu, Z., Caverlee, J.: Fighting mainstream bias in recommender systems via local fine tuning. In: Proceedings of the Fifteenth ACM International Conference on Web Search and Data Mining, pp. 1497–1506 (2022)

Towards Optimizing Ranking in Grid-Layout for Provider-Side Fairness

Amifa Raj[1](✉) and Michael D. Ekstrand[2]

[1] Microsoft, Redmond, USA
amifaraj@u.boisestate.edu
[2] Department of Information Science, Drexel University, Philadelphia, USA
mdekstrand@drexel.edu

Abstract. Information access systems, such as search engines and recommender systems, order and position results based on their estimated relevance. These results are then evaluated for a range of concerns, including provider-side fairness: whether exposure to users is fairly distributed among items and the people who created them. Several fairness-aware ranking and re-ranking techniques have been proposed to ensure fair exposure for providers, but this work focuses almost exclusively on linear layouts in which items are displayed in single ranked list. Many widely-used systems use other layouts, such as the grid views common in streaming platforms, image search, and other applications. Providing fair exposure to providers in such layouts is not well-studied. We seek to fill this gap by providing a grid-aware re-ranking algorithm to optimize layouts for provider-side fairness by adapting existing re-ranking techniques to grid-aware browsing models, and an analysis of the effect of grid-specific factors such as device size on the resulting fairness optimization. Our work provides a starting point and identifies open gaps in ensuring provider-side fairness in grid-based layouts.

1 Introduction

Information access systems (IAS)—search engines, recommender systems, and similar—provide utility to their users, by retrieving relevant results, but also to the *providers* of the items (authors, artists, etc.) by exposing them to users who may read, purchase, or otherwise consume their creations. Providers receive both economic and reputational benefit from this exposure; however, a system may not always fairly allocate exposure to items, as some items may receive less exposure than others of similar relevance [26,38]. This *disparate exposure* [38] may lead to unfair outcomes for item providers on either individual or group bases. Item providers or producers are often associated with sensitive or *protected* group attributes such as race, gender, religion, age, and other demographic attributes and items with similar relevance may receive unfair exposure in ranking based

This paper reports work supported by the National Science Foundation under Grant 17-51278 and primarily conducted while the authors were at Boise State University.

N. Goharian et al. (Eds.): ECIR 2024, LNCS 14612, pp. 90–105, 2024.
https://doi.org/10.1007/978-3-031-56069-9_7

on their group membership [24,47]. Provider-side fairness in ranking seeks to correct this imbalance and ensure the fair allocation of exposure across item producers and providers in system results. Several re-ranking techniques have been proposed to improve result fairness [25,27,33,34,38], and the TREC Fair Ranking track [32,49] provided multiple tasks for which participants optimized their systems for both fairness and relevance. However, these efforts are limited to linear ranked lists, while many production IAS display results in grid-based layouts. The problem of optimizing ranking in grid layouts for fairness has received limited attention so far. Chen *et al.* [46] proposed one of the few grid-based re-ranking techniques, but did not consider fairness; re-ranking technique suitable for optimizing provider-side group fairness in grid layouts are still unknown.

Moreover, fair ranking metric scores vary depending on user browsing behavior and user browsing behavior varies across ranking layouts [37,48,52]. Hence, for the same set of ranked items, user attention can vary for the items depending on how they are displayed to users, resulting in different degrees of fair exposure to their providers. There is limited research on user browsing behavior in grid layouts, but the browsing models that are available have not yet been incorporated into ranking or layout strategies for fairness. Raj & Ekstrand [52] showed that a ranking that is optimized for fairness in linear layout may not preserve its fairness when rearranged into grid layout. Moreover, the geometry of grid layouts change depending on the user's device, as the number of columns changes with screen size. Therefore, fair grid layouts need to consider both suitable browsing models and the specific layout geometry to boe used. Our work helps fill this gap by providing the first re-ranking technique to optimize provider-side group fairness in grid layouts.

We adapt a commonly used re-ranking techniques from linear layouts and modify it for grid layout by incorporating grid-aware browsing models. Since designing fairness-aware re-ranking techniques for ranking in grid layouts depends on ranking design, user browsing behavior, and column size or user device, we study the impact of column sizes and browsing models on our method.

Our experimentals addrss the following research questions:

- **RQ1.** Does incorporating grid-aware browsing models to existing re-ranking technique improve fairness for results in grid layouts?
- **RQ2.** Does a ranking in a grid layout optimized for fairness on one device remain fair for other devices?
- **RQ3.** How can we optimize ranking in grid layouts for various screen sizes?

Our simple and re-configurable re-ranking approach for grid layout advances provider-side group fairness in IAS beyond simple linear ranked interfaces. As more specific user attention models are developed in the future for ranking in grid layouts, they can be plugged into our proposed method to provide more accurate fairness optimization. Our analyses also provide an initial guidance for practitioners to design more fine-tuned re-ranking approaches for grid layouts that may consider item metadata, tasks, and domain, and open several future research directions towards fairness concerns in additional layouts and variants.

2 Background and Related Work

This section provides background on grid-layout suitable user browsing models and fairness-aware re-ranking techniques in IAS.

2.1 User Browsing Models

Users do not provide equal attention to every position in ranked results [26], item at the lower position of a ranked list will not receive similar attention as the item at the top position. Since user attention varies across positions in ranked results, the position weight for each position in ranking depends on how users browse the displayed ranked results.

There are several user browsing models to demonstrate user browsing behavior in linear ranked lists. *Cascade* [11] and *geometric* [12] are two popularly used user browsing models to infer the probability of user visiting an item in a particular position in ranking. These models differ in their underlying components and parameter settings but can be cast as different configurations of the same model. In both geometric and cascade models, user attention or position weight decays exponentially with ranking positions but in cascade browsing model, user selection probability is a function of item relevance.

To implement grid layout-aware evaluation metrics, it is important to understand how users provide attention to items in grid layout or how user attention changes across items when they are displayed in a grid layout. Tatler [9] observed the *central fixation* tendency where users provide more attention at the center of the page but Djamasbi *et al.* [17] and Zhao *et al.* [22] found that users usually show an *F-shaped* reading pattern by focusing on the results located at the top left-hand side. The viewing pattern is dependent on task, content, and complexity of the web pages [8]. Xie *et al.* [23] showed various user browsing behaviors in grid layout in e-commerce search results and they showed that users show *row-skipping*, *slower-decay*, and *middle-bias* while browsing items in grid layout. Users often skip rows while browsing ranked results in a grid layout and they tend to show higher attention to the middle position of columns in ranking. Moreover, user attention decays slowly across items in grid layout than linear list. Raj & Ekstrand [52] provided modified versions of geometric and cascade browsing models incorporating grid layout-suitable *row-skipping* and *slower-decay* user browsing behaviors and they implemented fair ranking metrics in grid layouts by incorporating grid-aware browsing models.

2.2 Re-ranking Techniques

Several re-ranking and *learning to rank* (LTR) approaches have been proposed to optimize ranking for utility [3,5–7,13,14,16,27,34,35]. LTR methods learn to rank based on scoring functions which is used to determine an optimized ranking; individual items, list, or pair of ranked items are considered to measure loss function against ideal ranking. Depending on the design of the loss function, the LTR approaches are categorized into pairwise, point-wise, and list-wise approaches [30,35]. Pairwise approaches are often based on the change in ranking quality with the swap of each pair of items in ranking [15]. In RankNet [16] and LambdaMART [5], ranking quality is optimized by predicting an optimal ordering for each pair of items in ranked list before generating the final ranking. In point-wise optimization approaches, the ranking model is trained to minimize loss function determined from each individual item score [7]. In this approach, each of the item in candidate set is scored independently based on the target quality. Unlike previous two approaches, list-wise approaches consider the entire ranked list and the ranking function is trained on the entire list based on the minimization of the loss function [6,10,21].

Fairness-aware Re-ranking Techniques. Fairness optimization in ranking often involves trade-offs between utility and fairness score [19,28,31,38], where fairness-aware LTR and re-ranking approaches aim to improve fairness with minimum utility loss. Approaches to improve the fairness of algorithms, including IAS rankings, are often categorized into *pre-processing*, *in-processing*, and *post-processing* (typically re-ranking) techniques [44]. In pre-processing approaches, the potential bias in datasets or training labels are investigated in order to identify and mitigate bias in ranking [41,45]. In in-processing approaches, the IAS algorithms or models are adjusted to optimize for fairness or a combination of fairness and utility in the training phase [31,42]. Post-processing approaches take already-ranked results and reorder them to improve or optimize a fairness objective [24]. Constraint optimization approaches have also been proposed to re-rank results [28,34,38]; the optimization constraints often include both user satisfaction metrics and fair ranking metrics to preserve a balance between fairness and utility.

Various fair ranking metrics are used to measure fairness in ranking and to determine the target fairness score. Provider-side fairness in ranking is often measured by the discrepancy in between the expected exposure and the exposure providers receive from ranking [26,28,38,48]. Hence, fairness-aware re-ranking techniques consider the optimization of the fairness score derived from the fairness metrics. Liu *et al.* [34] proposed a personalized fairness-aware re-ranking algorithm for micro-lending recommendations where each item from the initial ranking will be assigned to a position in the displayed ranking based on the optimization or maximization of personalization and group fairness. Singh & Joachims [28] and Diaz *et al.* [38] considered exposure of provider-side in ranking in their fairness-aware ranking optimization techniques. However, all the approaches discussed above are proposed and implemented in linear ranked results when items are displayed in single-column list.

In this work, we modify a pairwise swap re-ranking technique to optimize ranking in grid layout for provider-side group fairness.

Table 1. Summary of notation.

$d \in D$	document or item
$q \in Q$	request (user or context)
L	ranked results of N items from D
$L(\mathrm{i})$	the item in position i of linear (1-column) layout
$L^{-1}(d)$	rank of item d in linear layout
$L(\mathrm{k}, \cdot)$	items in kth row in grid layout
$L(\mathrm{k}, \mathrm{c})$	items in row k and column c in grid layout
$y(d\|q)$	relevance of d to q
E_i	event: user examines the item at position i
S_i	event: user selects the item at position i
A_i	event: user abandons the process after examining the item at position i
K_k	event: user skipping the kth row

3 Problem Formulation and Proposed Approach

In this work, we consider a recommender system that recommends n items $d_1, d_2, \ldots, d_n \in D$ in response to information requests from users $q_1, q_2, \ldots, q_m \in Q$ based on their relevance to the request $y(d|q)$ and presents the results in a wrapped grid layout L (notation summarized in Table 1). Items are associated with producers or providers who in turn can be associated with demographic attributes identifying them with one or more of g groups. We model group membership of documents with group alignment vector $\mathcal{G}(d) \in [0,1]^g$ (s.t. $\|\mathcal{G}(d)\|_1 = 1$) forming a distribution over groups; this allows for mixed, partial, or uncertain membership in an arbitrary number of groups.

Wrapped grid layouts—in which a single ranking is laid out in a grid by filling each row before wrapping to the next—are not the only grid-based layout [52]; many systems such as streaming video platforms use *multi-list* grid layouts where each row is a separate list of recommendations, possibly produced by a different algorithm. We focus on wrapped layouts in this paper because there has not yet been sufficient research on user browsing behavior in multi-list layouts to produce the browsing models needed for fairness-aware re-ranking.

The ranking will be optimized for provider-side group fairness while preserving a balance between utility and fairness. The purpose of this work is to provide a preliminary approach to develop fairness-aware re-ranking techniques for ranking in grid layout, so we focus on fairness optimization in a single-ranking setting, leaving fair grid layouts in stochastic settings for future work. We use the same grid-layout suitable browsing models to measure utility and fairness for consistency.

3.1 Layout Objective

Our layouts strive to provide both fairness and utility; both are measured with a browsing model that accounts for grid-specific browsing behaviors instead of the simple linear models typically used in fairness and utility metrics.

User Browsing Model. In this work, we use the modified versions of geometric browsing models incorporating grid layout-suitable *row-skipping* and *slower-decay* user browsing behaviors provided by Raj & Ekstrand [52]. We use the generalized and configurable framework [51] of user browsing models that adapts *row-skipping* and *slower-decay* user behavior in *geometric* browsing model to measure position weight in grid layout. For a given ranking in grid layout, the visiting probability of item d in geometric-based row-skipping model is:

$$P_{RS(\text{geometric})}[V_d] = \left[\prod_{k=0}^{\text{row}(d)} (1-\gamma) \prod_{i \in L(\text{k},\cdot)} (1-\psi) + \prod_{k=0}^{\text{row}(d)} \gamma \right] \prod_{i \in \text{row}(d)} (1-\psi) \quad (1)$$

and the geometric visiting probability of item d with slower decay is:

$$P_{SD(\text{geometric})}[V_d] = \min(\beta^{\text{row}(d)} \prod_{i=[0,L^{-1}(d)]} (1-\psi), 1) \quad (2)$$

Target Fairness. To measure provide-side group fairness in single ranking layout, we follow recommendations from the comprehensive analysis of fair ranking metrics in [48] and use AWRF. Sapiezynski *et al.* [36] proposed attention-weighted rank fairness or AWRF which measures the difference between group exposure and configurable target distribution $\hat{\mathbf{p}}$ which represents the ideal exposure distribution over groups. Attention vector and the group alignment matrix is used to derive group exposure $\boldsymbol{\epsilon}_L$ ($\boldsymbol{\epsilon}_\text{L} = \mathcal{G}(\text{L})^\text{T}\mathbf{a}_\text{L}$) by aggregating the attention given to items of each group in proportion to their group membership as represented by the alignment vector. Since our distribution difference function in bounded by $[0, 1]$, we invert it so that AWRF $= 1$ at maximal fairness to be more directly comparable to the effectiveness metrics:

$$\text{AWRF}(L) = 1 - \Delta(\boldsymbol{\epsilon}_L, \hat{\mathbf{p}}) \quad (3)$$

Target Utility. To measure utility in ranking, we consider an effectiveness metric that consider items position weights in measurement. Moffat & Zobel [12] proposed *rank-biased precision* (RBP) which combined a geometric browsing model with binary relevance to measure the overall effectiveness of a ranking in a manner similar to nDCG, but with a re-configurable browsing model. The source of relevance can be the actual relevance judgement which generates RBP or system estimated relevance which generates $\hat{\text{RBP}}$. For a given ranking L, the rank-biased precision metric score is

$$\hat{\text{RBP}} = \psi \sum_{i=[0,L^{-1}(d)]} y(L(i)|q)\dot{(}1-\psi)^{i-1} \quad (4)$$

where $y(L(i)|q)$ is the systems estimated relevance score for the item in position i and the stopping probability ψ is decaying exponentially with ranking position. This metric can be adapted to measure $\hat{\text{RBP}}$ in grid layout by incorporating grid layout suitable browsing behavior. Thus, we modify the attention model used in this metric by considering geometric-based row-skipping model (Eq. 1) and geometric-based slower-decay (Eq. 2).

3.2 Re-ranking Algorithm

Pairwise swapping re-ranking is a commonly used post-processing approach that we adapt to optimize ranking in grid layout for provider-side group fairness. For a given initial ranking L, we optimize the ranking by considering alternative ranking position for each pair of ranked items and finally generate a fairness-aware ranked result L'. Starting from the top of the list, for each position i, we consider each potential swap with positions $j > i$, items swap their position and temporarily generate a new ranking $L_{i \longleftrightarrow j}$ keeping all the other items at the same place. Then we measure the lift in fairness as $\Delta\text{AWRF}(L, L_{i \longleftrightarrow j})$ and the loss in utility as $\Delta\text{RBP}(L, L_{i \longleftrightarrow j})$.

$$\Delta\text{RBP}(L, L_{i \longleftrightarrow j}) = \text{RBP}(L_{i \longleftrightarrow j}) - \text{RBP}(L) \tag{5}$$

$$\Delta\text{AWRF}(L, L_{i \longleftrightarrow j}) = \text{AWRF}(L_{i \longleftrightarrow j}) - \text{AWRF}(L) \tag{6}$$

Thus for each of the position i, we select the best swap by solving the maximization of lift function, $F(i \longleftrightarrow j | i, j \in \{1, ..., N\}, i < j)$:

$$F(i \longleftrightarrow j) = \arg \max_{j \in i, \ldots, N} \{\Lambda \Delta\text{AWRF}(L, L_{i \longleftrightarrow j}) \cdot (1 - \Lambda)\,(1 - \Delta\text{RBP}(L, L_{i \longleftrightarrow j}))\} \tag{7}$$

Algorithm 1 shows the formal algorithm for optimizing grid-ranking for provider-side group fairness. In each iteration, the item in position i is temporarily swapped with items that are in higher position than i and for each respective swap, it measures the AWRF improvement and inverse RBP loss. The swap that gives the maximum lift in fairness score with minimum utility loss is selected to generate a new ranking. Λ is used as a configurable balancing factor between fairness and utility.

4 Experimental Setup

We now present an experiment with real-world IAS dataset to observe whether and how the provider-side group fairness improves with our modified re-ranking techniques, addressing the research questions laid out in the introduction.

4.1 Data and Algorithms

In this work, we use *GoodReads* [29] book dataset integrated with the PIReT Book Data Tools [40] to obtain author metadata. This data records interactions from 870K users with 1.1M books. Consistent with the prior research using this data set [40,43], we used LensKit [39] to generate 1000 personalized book recommendations for 5000 test users with four implicit-feedback collaborative filtering (CF) algorithms as configured by Ekstrand & Kluver [40]: user-based

Algorithm 1. Fairness-Aware Re-ranking for Grid Ranking

Require: initial ranking L, user q, estimated relevance score $y(L|u)$, balancing factor Λ
Ensure: Re-ranked L'
1: **procedure** RE-RANK(L)
2: $L' \leftarrow L$
3: measure AWRF(L)
4: measure RBP(L)
5: **for** $i \in 1, ..., N$ **do**
6: **for** $j \in i, ..., N$ **do**
7: swap items in position i and j to generate $L_{i \longleftrightarrow j}$
8: measure ΔRBP($L, L_{i \longleftrightarrow j}$)
9: measure ΔAWRF($L, L_{i \longleftrightarrow j}$)
10: **end for**
11: $i' = \arg\max_{j \in i,...,N} \{\Lambda \Delta \mathrm{AWRF}(L, L_{i \longleftrightarrow j}) \cdot (1 - \Lambda)(1 - \Delta RBP(L, L_{i \longleftrightarrow j}))\}$
12: **if** $i' \neq i$ **then**
13: $L^{'} \leftarrow L_{i \longleftrightarrow i'}$
14: AWRF(L) $\leftarrow$ AWRF($L_{i \longleftrightarrow i'}$)
15: RBP(L) $\leftarrow$ RBP($L_{i \longleftrightarrow i'}$)
16: **end if**
17: **end for**
18: **return** L'
19: **end procedure**

CF (UU [2]), item-based CF (II [4]), matrix factorization (WRLS [18]), and Bayesian Personalized Ranking (BPR [15]). The fairness goal is to be fair to the book authors of different genders;[1] the data contains 177K books by women and 283K books by men, with other books having unknown author gender.

4.2 Experiment Design

We optimize provider-side group fairness in grid layout using the modified re-ranking technique considering two types of user browsing models. We also observe the affect of column sizes on fairness optimization in grid layout.

RQ1. *Improvement of Fairness in Grid Layout.* To observe the group fairness score improvement for provide-side fairness in grid layout,

- We implement the fair ranking metric AWRF to measure fairness in single ranking. We use distribution of male and female authors in book dataset to compute target distribution $\hat{\mathbf{p}}$. We compare the improvement of AWRF score in the re-ranked grid ranking where 1 is the highest score of fairness.
- To measure utility, we implement effectiveness metric RBP.

[1] Due to limitations of the underlying data set [40], we are only able to consider binary gender. We understand the potential harm of misrepresentation of gender in research [50]; our methods in this paper are extensible to non-binary gender or other attributes when suitable data is available.

- Both AWRF and RBP are implemented with grid-layout suitable browsing models, *row-skipping* and *slower-decay* with column size 5.
- We use 0.5 as the default value of the fairness-utility balancing parameter Λ.

RQ2. *Consistency of Optimized Fairness Across Devices.* Based on user devices, column size of grid layout changes. For example, *Goodreads* shows book recommendations in grid layout and the column size changes across devices; books are displayed in 5 columns on laptop, 2 columns on phone, and 9 columns on iPad. Hence, the system can display the same set of items in various column sizes depending on user device. Re-ranking the items by taking device size into consideration can help to preserve fairness across devices because optimizing the ranked results in grid layout for a particular device may not remain fair for other devices.

- We observe if and how the optimized fairness score from a re-ranked grid layout of column size n changes in other columns sizes.
- We optimize the grid-based ranked results with column size of 5 and use that fairness-aware re-ranked results to measure provider-side group fairness by changing column size to 2, 3, 4, 7, and 9.

RQ3. *Preserve Fairness Across Devices.* Since item exposure varies across column sizes in grid layout which affect the fairness score for provider groups, we want to preserve provider-side fairness across devices. With that goal,

- We implement the grid-aware re-ranking technique for multiple column sizes to maintain group fairness across user devices and observe the change in fairness optimization with the change of column sizes.
- We implement the grid-aware re-ranking algorithm for grid ranked results with common columns sizes of 2, 3, 4, 5, 7, and 9.

To observe the impact of browsing models on fairness optimization in grid layout, we implement both group fairness metric and effectiveness metric incorporating grid-layout suitable *row-skipping* and *slower-decay* browsing models with their default parameter settings.

4.3 Results

This section provide the results from our experiments.

RQ1. *Does incorporating grid-aware browsing models to existing re-ranking technique improve fairness for ranked results in grid layout?*

Figure 1(a) shows that the AWRF score increases in all the recommendation algorithms for both *row-skipping* and *slower-decay* browsing models. We do *paired t-test* [20] to observe the significance of this fairness improvement and find that for the algorithms in both browsing models, the AWRF score improvement is statistically significant with $p_{val} < 10^{-20}$. We round up the p-values at $\alpha = 0.05$ with Benjamini-Hochberg correction [1]. In both browsing models, the fairness score varies across recommendation algorithms during both pre and post-optimization showing the same patterns. For all the recommendation algorithms, the fairness scores improves significantly for ranking in grid layout when we consider grid-layout suitable browsing models. Figure 1(b) shows the RBP score and Fig. 1(c) shows the RBP $*$ AWRF score differences in between pre and post-optimization. For the *slower-decay* browsing model, the combined score improves in all the algorithms and the utility score improves after re-ranking. The improvement does not hold for *row-skipping* browsing model. This observation emphasizes the importance of using grid-aware re-ranking technique while optimizing ranked results displayed in grid layout.

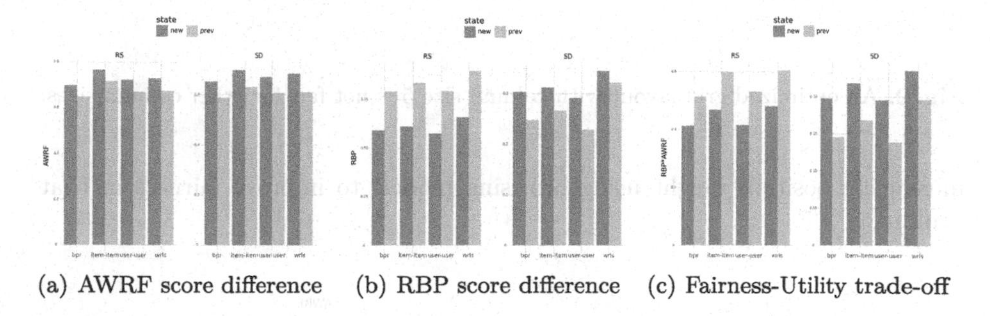

(a) AWRF score difference (b) RBP score difference (c) Fairness-Utility trade-off

Fig. 1. Pre and post-optimization fairness and utility scores in grid layout with column size 5

RQ2. *Does a ranking in grid layout optimized for fairness in a device remain fair for other devices?*

RQ2 shows the impact of column sizes on fairness optimization in grid layout. Figure 2 shows how fairness score for an optimized ranking changes with column sizes. A fairness-aware re-ranked 5-column grid layout does not remain fair when the column size is different and this pattern is true for all the algorithms. The pattern is more notable in *row-skipping* browsing model for all the algorithms. This result implies the need of considering appropriate column size to preserve fairness for the same set of ranked items across devices.

RQ3. *How can we optimize ranking in grid layout for various screen sizes?*

Figure 3 shows the improvement in fairness scores after optimizing ranking in the grid layout for various column sizes and the result shows a consistency in fairness improvement across column sizes. For all the considered column sizes, AWRF score improves significantly in all the recommendation algorithms ($p_{val} < 0.0001$ rounded at $\alpha = 0.05$ with Benjamini-Hochberg correction) after optimizing ranking in grid layout using grid-aware browsing models. By looking at Fig. 3, we can see that fairness score varies with the change of column sizes and this pattern remains consistent even after optimization in all the algorithms for both browsing models. This result shows that, fairness optimization of a given grid layout of column size n should consider the same column size while

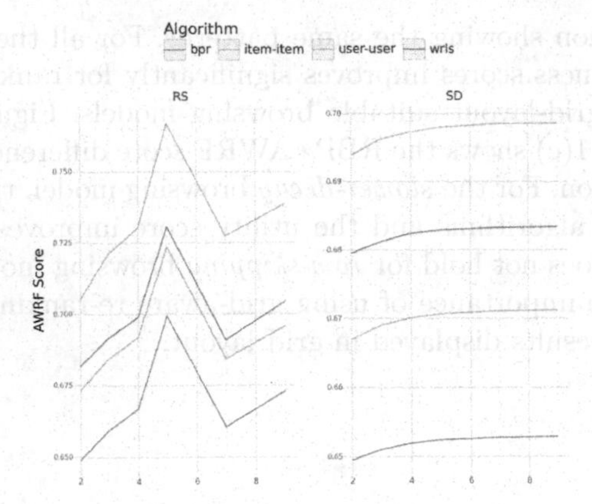

Fig. 2. An optimized grid layout with column size 5 is not fair for other column sizes.

measuring position weight using browsing models to improve fairness in that ranking.

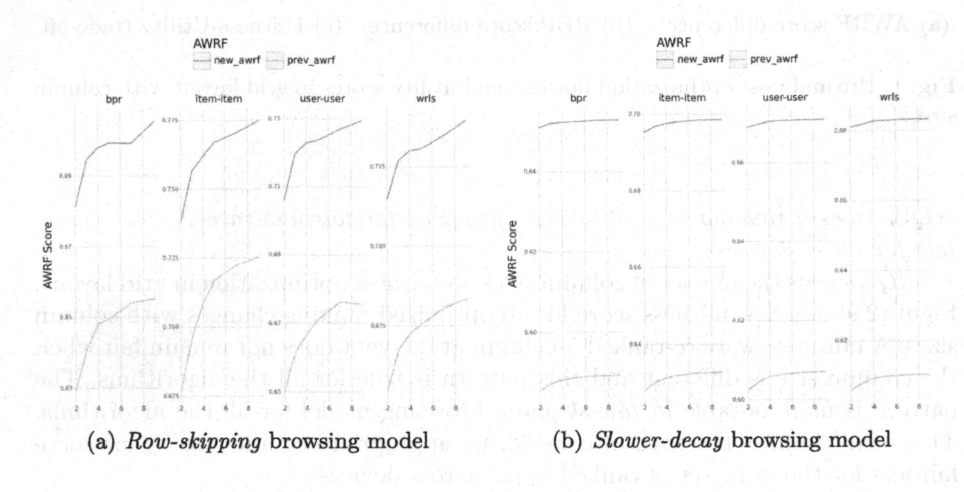

(a) *Row-skipping* browsing model (b) *Slower-decay* browsing model

Fig. 3. Improvement in fairness across column sizes in grid-aware browsing models

Discussion. Through our experiments we provide insights on the impact of device sizes and browsing models on fairness optimization in grid layout. We have made following observations from our analysis.

- It is possible to improve fairness in grid layout if we can make re-ranking techniques grid-aware by incorporating grid-layout suitable browsing models. However, the improvement in fairness score can vary depending on user browsing models. This observation highlight the importance of considering suitable

browsing models while measuring and optimizing group fairness in grid layout. Understanding how users browse grid layout and identifying various browsing tendencies can help to develop more accurate fairness optimization technique for grid layout.
- Device size is an important factor in improving fairness in grid layout. Optimizing provider-side group fairness in ranking in grid layout with a particular column size will not remain fair with the change of column sizes. Hence, a ranked result which is optimized for fairness while displaying in phone will not remain fair while displaying in a laptop. Therefore, to preserve fairness across devices, a retrieved results displayed in a particular device needs to be re-ranked considering the appropriate column size while displaying in another device.
- The consistency of fairness score across column sizes varies based on browsing models. For *row-skipping* browsing model, the fairness score varies notably across column sizes but for *slower-decay*, the fairness scores are more consistent across column sizes. This observation emphasizes the need of selecting suitable browsing model and column size while optimizing ranking in grid layout for provider-side group fairness.

5 Conclusion

In this paper, we work towards filling a gap in the research area of provider-side group fairness in ranking in IAS by studying fairness improvement in grid layout. We modify a widely used fairness-aware re-ranking technique to make it grid-aware by incorporating grid-layout suitable user browsing models. We implement the modified grid-aware re-ranking technique in real-world IAS dataset to observe the fairness improvement in ranking in grid layout. Our analysis shows that device size and user browsing models are crucial factors in designing fairness-aware re-ranking technique to optimize provider-side group fairness in grid layout in IAS.

This work opens up several potential research directions in improving provider-side fairness in grid layout. Our work shows the importance of using accurate user browsing models in fairness optimization for grid layout. User browsing behavior in ranking in grid layout has not received much attention yet, hence, further research work on understanding user browsing behavior in grid layout will help ensuring fairness in grid layout with minimum utility loss.

Moreover, in this work, we do not consider multi-list grid layout where items are displayed in multiple categories. Re-ranking technique designed for wrapped grid-layout may not work for multi-list grid layout because in multi-list grid, each rows represents different genre or categories. Moreover, same item can appear in multiple rows. Hence, future work is needed to optimize multi-list grid ranking for fairness by considering unique features and suitable user browsing models for multi-list ranking.

We believe this work will provide researcher and practitioners an guideline on what to expect while designing an optimization technique for fairness in grid layout and what factors to consider carefully.

References

1. Benjamini, Y., Hochberg, Y.: Controlling the false discovery rate: a practical and powerful approach to multiple testing. J. Roy. Stat. Soc.: Ser. B (Methodol.) **57**, 289–300 (1995)
2. Herlocker, J.L., Konstan, J.A., Borchers, A., Riedl, J.: An algorithmic framework for performing collaborative filtering. In: Proceedings of the 22nd Annual International ACM SIGIR Conference on Research and Development in Information Retrieval, pp. 230–237. Association for Computing Machinery, New York (1999). isbn: 1581130961
3. Friedman, J.H.: Greedy function approximation: a gradient boosting machine. Ann. Stat. 1189–1232 (2001)
4. Deshpande, M., Karypis, G.: Item-based Top-n recommendation algorithms. ACM Trans. Inf. Syst. (TOIS) **22**, 143–177 (2004)
5. Burges, C., et al.: Learning to rank using gradient descent. In: Proceedings of the 22nd International Conference on Machine Learning, pp. 89–96 (2005)
6. Cao, Z., Qin, T., Liu, T.-Y., Tsai, M.-F., Li, H.: Learning to rank: from pairwise approach to listwise approach. In: Proceedings of the 24th International Conference on Machine Learning, pp. 129–136 (2007)
7. Li, P., Wu, Q., Burges, C.: Mcrank: learning to rank using multiple classification and gradient boosting. Adv. Neural Inf. Process. Syst. **20** (2007)
8. Shrestha, S., Lenz, K.: Eye gaze patterns while searching vs. browsing a website. Usability News **9**, 1–9 (2007)
9. Tatler, B.W.: The central fixation bias in scene viewing: selecting an optimal viewing position independently of motor biases and image feature distributions. J. Vis. **7**, 4–4 (2007)
10. Xu, J., Li, H.: Adarank: a boosting algorithm for information retrieval. In: Proceedings of the 30th Annual International ACM SIGIR Conference on Research and Development in Information Retrieval, pp. 391–398 (2007)
11. Craswell, N., Zoeter, O., Taylor, M., Ramsey, B.: An experimental comparison of click position-bias models. In: Proceedings of the 2008 International Conference on Web Search and Data Mining, pp. 87–94 (2008)
12. Moffat, A., Zobel, J.: Rank-biased precision for measurement of retrieval effectiveness. ACM Trans. Inf. Syst. (TOIS) **27**, 1–27 (2008)
13. Taylor, M., Guiver, J., Robertson, S., Minka, T.: Softrank: optimizing nonsmooth rank metrics. In: Proceedings of the 2008 International Conference on Web Search and Data Mining, pp. 77–86 (2008)
14. Xia, F., Liu, T.-Y., Wang, J., Zhang, W., Li, H.: Listwise approach to learning to rank: theory and algorithm. In: Proceedings of the 25th International Conference on Machine Learning, pp. 1192–1199 (2008)
15. Rendle, S., Freudenthaler, C., Gantner, Z., Schmidt-Thieme, L.: BPR: bayesian personalized ranking from implicit feedback. In: Proceedings of the Twenty-Fifth Conference on Uncertainty in Artificial Intelligence, pp. 452–461. AUAI Press, Montreal, Quebec, Canada (2009), isbn: 9780974903958
16. Burges, C.J.: From ranknet to lambdarank to lambdamart: an overview. Learning **11**, 81 (2010)
17. Djamasbi, S., Siegel, M., Tullis, T.: Visual hierarchy and viewing behavior: an eye tracking study. In: International Conference on Human-Computer Interaction, pp. 331–340 (2011)

18. Takács, G., Pilászy, I., Tikk, D.: Applications of the conjugate gradient method for implicit feedback collaborative filtering. In: Proceedings of the Fifth ACM Conference on Recommender Systems, pp. 297–300. Association for Computing Machinery, Chicago (2011). isbn: 9781450306836
19. Dwork, C., Hardt, M., Pitassi, T., Reingold, O., Zemel, R.: Fairness through awareness. In: Proceedings of the 3rd Innovations in Theoretical Computer Science Conference, pp. 214–226 (2012)
20. Hsu, H., Lachenbruch, P.A.: Paired t test. Wiley StatsRef: statistics reference online (2014)
21. Lan, Y., Zhu, Y., Guo, J., Niu, S., Cheng, X.: Position-aware ListMLE: a sequential learning process for ranking. In: UAI, pp. 449–458 (2014)
22. Zhao, Q., Chang, S., Harper, F.M., Konstan, J.A.: Gaze prediction for recommender systems. In: Proceedings of the 10th ACM Conference on Recommender Systems, pp. 131–138 (2016)
23. Xie, X., et al. Investigating examination behavior of image search users. In Proceedings of the 40th International ACM Sigir Conference on Research and Development in Information Retrieval, pp. 275–284 (2017)
24. Yang, K., Stoyanovich, J.: Measuring fairness in ranked outputs. In: Proceedings of the 29th International Conference on Scientific and Statistical Database Management, pp. 1–6 (2017)
25. Zehlike, M., et al.: FA*IR: a fair top-k ranking algorithm. In: Proceedings of the 2017 ACM on Conference on Information and Knowledge Management, 1569–1578. Association for Computing Machinery, Singapore (2017). isbn: 9781450349185. https://doi.org/10.1145/3132847.3132938
26. Biega, A. J., Gummadi, K.P., Weikum, G.: Equity of attention: amortizing individual fairness in rankings. In: Proceedings of the 41st International ACM SIGIR Conference on Research and Development in Information Retrieval, pp. 405–414 (2018)
27. Ekstrand, M.D., Tian, M., Kazi, M.R.I., Mehrpouyan, H., Kluver, D.: Exploring author gender in book rating and recommendation. In: Proceedings of the 12th ACM Conference on Recommender Systems, pp. 242–250 (2018)
28. Singh, A., Joachims, T.: Fairness of exposure in rankings. In: Proceedings of the 24th ACM SIGKDD International Conference on Knowledge Discovery & Data Mining, pp. 2219–2228. Association for Computing Machinery, London (2018). isbn: 9781450355520. https://doi.org/10.1145/3219819.3220088
29. Wan, M., McAuley, J.: Item recommendation on monotonic behavior chains. In: Proceedings of the 12th ACM Conference on Recommender Systems, pp. 86–94. Association for Computing Machinery, Vancouver (2018). isbn: 9781450359016. https://doi.org/10.1145/3240323.3240369
30. Wu, L., Hsieh, C.-J., Sharpnack, J.: SQL-RANK: a listwise approach to collaborative ranking. In: International Conference on Machine Learning, pp. 5315–5324 (2018)
31. Beutel, A., et al.: Fairness in recommendation ranking through pairwise comparisons. In: Proceedings of the 25th ACM SIGKDD International Conference on Knowledge Discovery & Data Mining, pp. 2212–2220 (2019)
32. Biega, A.J., Diaz, F., Ekstrand, M.D., Kohlmeier, S.: Overview of the TREC 2019 fair ranking track. In: The Twenty-Eighth Text REtrieval Conference (TREC 2019) Proceedings (2019)

33. Geyik, S.C., Ambler, S., Kenthapadi, K.: Fairness-aware ranking in search & recommendation systems with application to linkedin talent search. In: Proceedings of the 25th ACM Sigkdd International Conference on Knowledge Discovery & Data Mining, pp. 2221–2231 (2019)
34. Liu, W., Guo, J., Sonboli, N., Burke, R., Zhang, S.: Personalized fairnessaware re-ranking for microlending. In: Proceedings of the 13th ACM Conference on Recommender Systems, pp. 467–471 (2019)
35. Pei, C., et al.: Personalized re-ranking for recommendation. In: Proceedings of the 13th ACM Conference on Recommender Systems, pp. 3–11 (2019)
36. Sapiezynski, P., Zeng, W., E Robertson, R., Mislove, A., Wilson, C.: Quantifying the impact of user attention on fair group representation in ranked lists. In: Companion Proceedings of The 2019 World Wide Web Conference, pp. 553–562. Association for Computing Machinery, San Francisco (2019). isbn: 9781450366755. https://doi.org/10.1145/3308560.3317595
37. Xie, X., et al.: Grid-based evaluation metrics for web image search. In: The World Wide Web Conference, pp. 2103–2114 (2019)
38. Diaz, F., Mitra, B., Ekstrand, M.D., Biega, A.J., Carterette, B.: Evaluating stochastic rankings with expected exposure. In: Proceedings of the 29th ACM International Conference on Information & Knowledge Management, pp. 275–284. Association for Computing Machinery, Virtual Event (2020). isbn: 9781450368599. https://doi.org/10.1145/3340531.3411962
39. Ekstrand, M.D.: LensKit for python: next-generation software for recommender systems experiments. In: Proceedings of the 29th ACM International Conference on Information & Knowledge Management, pp. 2999–3006. Association for Computing Machinery, Virtual Event (2020). isbn: 9781450368599. https://doi.org/10.1145/3340531.3412778
40. Ekstrand, M.D., Kluver, D.: Exploring author gender in book rating and recommendation. User Model. User-Adapt. Interact. (2020). https://md.ekstrandom.net/pubs/bag-extended
41. Jiang, H., Nachum, O.: Identifying and correcting label bias in machine learning. In: International Conference on Artificial Intelligence and Statistics, pp. 702–712 (2020)
42. Narasimhan, H., Cotter, A., Gupta, M.R., Wang, S.: Pairwise fairness for ranking and regression. In: AAA, vol. I, pp. 5248–5255 (2020)
43. Raj, A., Wood, C., Montoly, A., Ekstrand, M.D.: Comparing fair ranking metrics (2020). arXiv preprint arXiv:2009.01311
44. Pitoura, E., Stefanidis, K., Koutrika, G.: Fairness in rankings and recommendations: an overview. VLDB J. 1–28 (2021)
45. Sonoda, R.: A Pre-processing Method for Fairness in Ranking. arXiv preprint arXiv:2110.15503 (2021)
46. Chen, S., et al.: Reinforcement Re-ranking with 2D Grid-based Recommendation Panels. arXiv preprint arXiv:2204.04954 (2022)
47. Ekstrand, M.D., Das, A., Burke, R., Diaz, F., et al.: Fairness in information access systems. Found. Trends® Inf. Retr. **16**, 1–177 (2022)
48. Raj, A., Ekstrand, M.D.: Measuring fairness in ranked results: an analytical and empirical comparison. In: Proceedings of the 45th International ACM SIGIR Conference on Research and Development in Information Retrieval, pp. 726–736 (2022)
49. Ekstrand, M.D., McDonald, G., Raj, A., Johnson, I.: Overview of the TREC 2022 Fair Ranking Track. arXiv preprint arXiv:2302.05558 (2023)

50. Pinney, C., Raj, A., Hanna, A., Ekstrand, M.D.: Much ado about gender: current practices and future recommendations for appropriate gender-aware information access. arXiv preprint arXiv:2301.04780 (2023)
51. Raj, A., Ekstrand, M.: Unified browsing models for linear and grid layouts. arXiv preprint arXiv:2310.12524 (2023)
52. Raj, A., Ekstrand, M.D.: Towards measuring fairness in grid layout in recommender systems. arXiv preprint arXiv:2309.10271 (2023)

MedSumm: A Multimodal Approach to Summarizing Code-Mixed Hindi-English Clinical Queries

Akash Ghosh[1(✉)], Arkadeep Acharya[1], Prince Jha[1], Sriparna Saha[1], Aniket Gaudgaul[1], Rajdeep Majumdar[1], Aman Chadha[2,3], Raghav Jain[1], Setu Sinha[4], and Shivani Agarwal[4]

[1] Department of Computer Science And Engineering, Indian Institute of Technology Patna, Patna, India
akashghosh.ag90@gmail.com
[2] Stanford University, Stanford, USA
[3] Amazon GenAI, Seattle, USA
[4] Indira Gandhi Institute of Medical Sciences, Patna, India

Abstract. In the healthcare domain, summarizing medical questions posed by patients is critical for improving doctor-patient interactions and medical decision-making. Although medical data has grown in complexity and quantity, the current body of research in this domain has primarily concentrated on text-based methods, overlooking the integration of visual cues. Also prior works in the area of medical question summarisation have been limited to the English language. This work introduces the task of multimodal medical question summarization for codemixed input in a low-resource setting. To address this gap, we introduce the Multimodal Medical Codemixed Question Summarization (*MMCQS*) dataset, which combines Hindi-English codemixed medical queries with visual aids. This integration enriches the representation of a patient's medical condition, providing a more comprehensive perspective. We also propose a framework named *MedSumm* that leverages the power of LLMs and VLMs for this task. By utilizing our *MMCQS* dataset, we demonstrate the value of integrating visual information from images to improve the creation of medically detailed summaries. This multimodal strategy not only improves healthcare decision-making but also promotes a deeper comprehension of patient queries, paving the way for future exploration in personalized and responsive medical care. Our dataset, code, and pre-trained models will be made publicly available. https://github.com/ArkadeepAcharya/MedSumm-ECIR2024

Keywords: Mutimodal Summarization · LLM · VLM · Codemixing · Clinical Queries

1 Introduction

Disclaimer: The paper includes explicit medical imagery, necessary for an in-depth understanding of the subject matter.

A. Chadha—Work does not relate to position at Amazon.

N. Goharian et al. (Eds.): ECIR 2024, LNCS 14612, pp. 106–120, 2024.
https://doi.org/10.1007/978-3-031-56069-9_8

Recent surveys conducted by World Health Organsisation (WHO) reveals a drastic uneven doctor to population ratio, estimating a deficit of 12 million healthcare workers by 2030. This augmented with advancements in information and communication technologies (ICTs) has experienced surge in telehealth [23]. The COVID-19 pandemic has further accelerated the utilization of the internet for healthcare services, marking a remarkable surge in the past two decades and establishing a new norm in healthcare. In this context, one of the biggest challenges that doctors face is to quickly comprehend and understand the patient's query. To solve this problem, a medical question summarizer has emerged as a vital tool to distill information from consumer health questions, ensuring the provision of accurate and timely responses. Despite previous research efforts, the unexplored opportunity of combining textual information with visual data, such as images, has been largely ignored. Visual aids play a crucial role in medical question summarization (MQS) for several reasons. A considerable segment of the population isn't well-versed in medical terminology, making it difficult to precisely convey symptoms. Additionally, certain symptoms are inherently challenging to express using only text. Patients can sometimes get mixed up when it comes to similar symptoms, like trying to tell the difference between skin dryness and a skin rash. Using a combination of text and images in medical question summaries can improve accuracy and efficiency, giving a holistic picture of the patient's current medical status. This strategy acknowledges the intricate nature of patient inquiries, where visuals like symptom photos or medical reports can offer vital insights. By prioritizing image integration, researchers and healthcare professionals can address the changing demands of contemporary healthcare communication. By improving doctor-patient communication and understanding through multimodal summarization of health questions, this work has the potential to increase access to quality healthcare and promote health equity which is one of the SDG proposed by UNESCO[1].

Large Language Models (LLMs) [17] and Vision Language Models (VLMs) [33] have showcased impressive capabilities in generating human-like text and multimedia content. This capability has led to their application in the medical field, primarily for specialized tasks such as summarizing chest X-rays [26] and generating COVID-19 CT reports [21]. However, their use in summarizing medical questions that involve both text and images is a relatively unexplored area. Utilizing the zero-shot and few-shot learning abilities of these models [11] can be advantageous, particularly for tasks like multimodal medical question summarization, which often suffer from a lack of sufficient data.

Nonetheless, there are certain limitations to consider when employing LLMs and VLMs in this context. Generic LLMs and VLMs may not possess specialized knowledge in medical domains, potentially leading to summaries that lack important details such as symptoms, diagnostic tests, and medical intricacies. On the visual side, although VLMs have excelled in typical visual-linguistic tasks, medical imaging presents unique challenges. Models like SkinGPT4 [37], which fine-tune on skin disease images and clinical notes based on MiniGPT4 [38],

[1] https://en.unesco.org/sustainabledevelopmentgoals.

remain highly domain-specific. Medical images are inherently intricate, requiring a deep understanding of medical terminology and visual conventions, often necessitating input from expert medical professionals for accurate interpretation. This complexity, coupled with potential gaps in contextual comprehension, can result in models generating summaries that may be misleading or irrelevant.

Also in the 21st century where more than half of the population of the globe is multilingual, people generally switch between languages in conversation[2]. Though nowadays a lot of work is going on in this exciting area of codemixing, in the clinical domain there is a dearth of good datasets. This motivated us to study our proposed task in a codemixed context. We present a curated dataset tailored to this endeavor, ***MMCQS*** containing Hindi-English codemixed medical queries with their corresponding English summaries. This is the first dataset of its kind to validate this task. The dataset encompasses 3,015 medical questions along with their corresponding visual cues. Our approach is embedded within our novel architecture known as ***MedSumm***, a vision-language framework for medical question summarization task. This comprehensive framework operates using two primary inputs: the codemixed patient question and its associated visual cue. The methodology comprises four distinct components. They are employing pre-trained large language models to generate textual embeddings, utilizing vision encoders like ViT [12] to encode visual information, harnessing QLoRA [9], a low-rank adapter technique for efficient fine-tuning and employing a precise inference process to generate the corresponding symptom name and the synopsis. This careful, step-by-step structuring of ***MedSumm*** allows it to bridge the divide between general-purpose models and the niche requirements of medical question summarization, effectively integrating textual and visual data to create precise and context-aware medical summaries. A pipeline showing the application of MedSumm has be shown in Fig. 3. Our contributions can be summarized as follows:

- A novel task of **Multimodal Medical Question Summarization** for generating medically nuanced summaries.
- A novel dataset, ***MMCQS*** Dataset, to further research in this area.
- A novel framework, ***MedSumm*** that employs a combination of pre-trained language models, state-of-the-art vision encoders, and efficient fine-tuning methodologies like QLoRA [9] to seamlessly integrate visual and textual information for the final summary generation of multimodal clinical questions.
- A novel metric ***MMFCM*** to quantify how well the model captures the multimodal information in the generated summary.

2 Related Works

The following work has been relevant to following two research areas namely Medical Question Summarization and Multimodal Summarization.

[2] https://www.britishcouncil.org/voices-magazine/few-myths-about-speakers-multiple-languages.

Medical Question Summarization: In 2019, the field of Medical Question Summarization (MQS) emerged with the introduction of the MeQSum dataset, specifically designed for this purpose by Abacha et al. [1]. Initial MQS research utilized basic seq2seq models and pointer generator networks to generate summaries. In 2021, a competition centered around generating summaries in the medical domain was organized, as outlined in Abacha et al.'s [2] overview. Contestants leveraged various pre-trained models, including PEGASUS [34], ProphetNet [25], and BART [19]. Some innovative techniques, such as multi-task learning, were employed, using BART to jointly optimize question summarization and entailment tasks [22]. Another approach involved the use of reinforcement learning with question-aware semantic rewards, derived from two subtasks: question focus recognition (QFR) and question type identification (QTR) [31].

Multimodal Summarization: In order to ensure accurate diagnosis and guidance from medical professionals, it is crucial for us to communicate our medical symptoms effectively and efficiently. One way to enhance this communication is by supplementing textual descriptions with visual cues. Previous research has demonstrated the benefits of incorporating multimodal information in various medical tasks. For instance, Tiwari et al. [27] highlighted how multimodal information improves the performance of Disease Diagnosis Virtual Assistants. Delbrouck et al. [8] showed that integrating images leads to better summarization of radiology reports, while Gupta et al. [14] illustrated the advantages of incorporating videos in medical question-answering tasks. Kumar et al. [18] shows how multimodal information can help in summarizing news articles. The most recent work in this domain is done by Ghosh et al. [13] where they incorporated CLIP with LLMs to generate the final multimodal summaries. To the best of our knowledge, our work represents the first attempt to address the task of question summarization in the medical domain, particularly within a codemixed multimodal context.

3 MMCQS Dataset

3.1 Data Collection

Prior to this work, there was no multimodal codemixed question summarization dataset available in the healthcare domain with textual questions and corresponding medical images. We used the HealthCareMagic Dataset, derived from MedDialog data, with 226,395 samples, after removing 523 duplicates. We were led by medical doctors who were also the co-authors of the paper, identified 18 medical symptoms that are hard to convey through text, and divided into four groups: ENT, EYE, LIMB, and SKIN. The entire categorization is shown in Fig. 1. We selected symptoms and obtained images using the Bing Image Search API[3], which were verified by medical students. Our dataset features instances

[3] https://www.microsoft.com/en-us/bing/apis/bing-image-search-api.

mentioning body parts in questions and summaries. We used FlashText[4] for term matching and Textblob4[5] to correct misspellings, resulting in a final dataset of 3,015 samples for multimodal summarization.

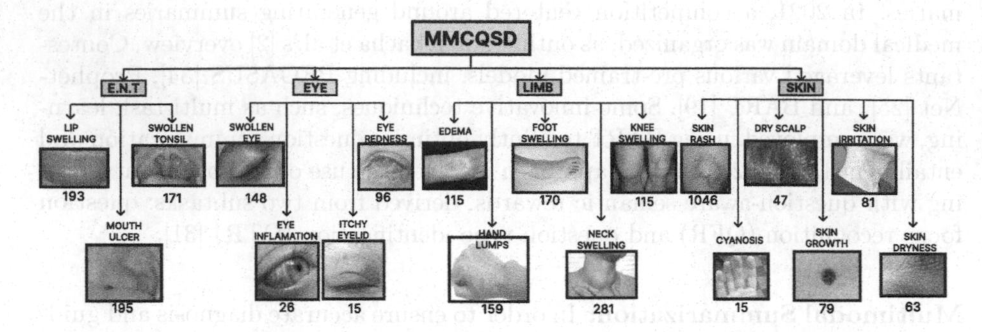

Fig. 1. Broad categorization of medical disorders in the MMCQS Dataset (MMCQSD). The number of data points corresponding to each category has been provided under each category in the above figure.

3.2 Data Annotation

We randomly selected 100 samples from the dataset and provided them to the medical experts, who are co-authors of this paper. They developed annotation guidelines for the annotation training process, divided into three categories: (A) Incorporation of Visual Cues, (B) Updating Golden Summaries, and (C) Hindi-English Codemixed Conversion.

A. Incorporation of Visual Cues:

The methodology involved medical experts annotating samples with visual cues. For instance, when a patient's description hints at a medical issue near their tonsils, the expert adds relevant statements and visual images to clarify the condition, such as adding ***Please see what happened to my tonsils in the image below***. This process integrates visual medical signs into the textual medical question

B. Updating the Golden Summaries:

The medical experts recognized a misalignment between conventional golden summaries and multimodal queries, leading them to update the summaries. They revised the questions, incorporated visual information, and adapted the golden summaries for the 100 samples. These 100 samples are used for annotation guidelines for the annotators.

C. Hindi-English Codemixed Annotation: For this task, around 100 samples were annotated by medical experts. We conducted experiments using

[4] https://pypi.org/project/flashtext/1.0/.

[5] https://textblob.readthedocs.io/en/dev/.

GPT-3.5 with few-shot prompting techniques on annotated samples due to its strong performance in generating codemixed text [32]. The goal was to generate a Hinglish (Hindi + English mixed) version of the English text. The generated Hinglish text had a code-mixing index [7] of 30.5, indicating a good mix of Hindi and English words. The specific prompt used is presented below. 80 samples were tested, with annotators rating code-mixing out of 5. After verification with the golden test examples, we get an average score of 3.2 out of 5 indicating the codemixed data is of reasonable quality. An instance of the Hindi-English Codemixed annotated data along with their corresponding English query and updated golder summaries has been shown in Fig. 2

Prompt used for codemixed text generation

You are a linguistic expert whose task is to convert the English passages into corresponding Hinglish codemixed ones.⟨Labelled Examples⟩: English: {text} Hinglish: {text}. Given the English passage: {text}, convert it into the corresponding Hinglish passage shown in the ⟨Labelled Examples⟩.

3.3 Annotation Training and Validation

To ensure high-quality annotation aligned with ethical guidelines, we engaged three postgraduate medical students proficient in Hindi and English. They received guidance and conducted sessions with medical experts to clarify task-related doubts. Breaks were provided every 45 min during the approximately 4-month, 90-training session annotation process[6].

In the validation phase, the dataset was divided into three parts, and annotators assessed fluency, adequacy, informativeness, and persuasiveness, maintaining inter-annotator agreement using the Cohen-Kappa coefficient. Annotators reviewed and discussed inaccuracies in each other's sets, leading to improved quality scores from the initial (fluency = 3.5, adequacy = 3.01, informativeness = 2.85, persuasiveness = 2.25) to the final phase (fluency = 4.8, adequacy = 4.7, informativeness = 4.1, persuasiveness = 4.45). The annotators achieved a kappa coefficient of 0.75, indicating annotation consistency.

[6] The medical students were compensated through gift vouchers and honorarium amount in lines with https://www.minimum-wage.org/international/india.

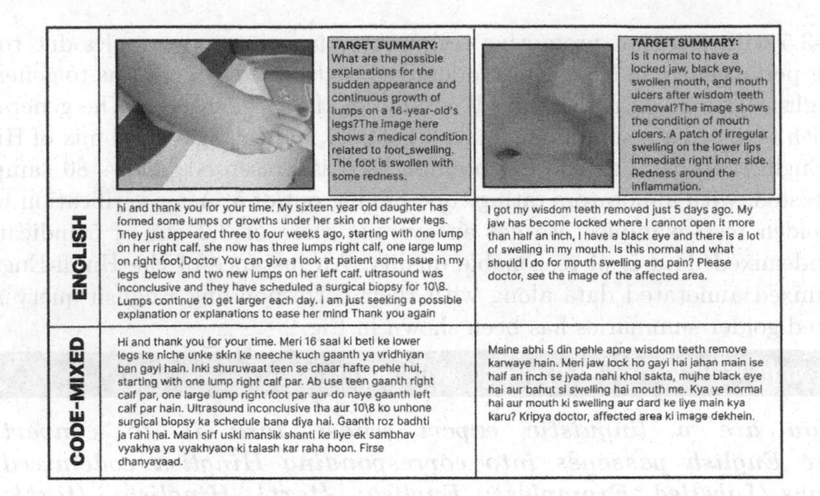

Fig. 2. Sample instance of the MMCQS Dataset with the corresponding English query and updated golden summary (target summary).

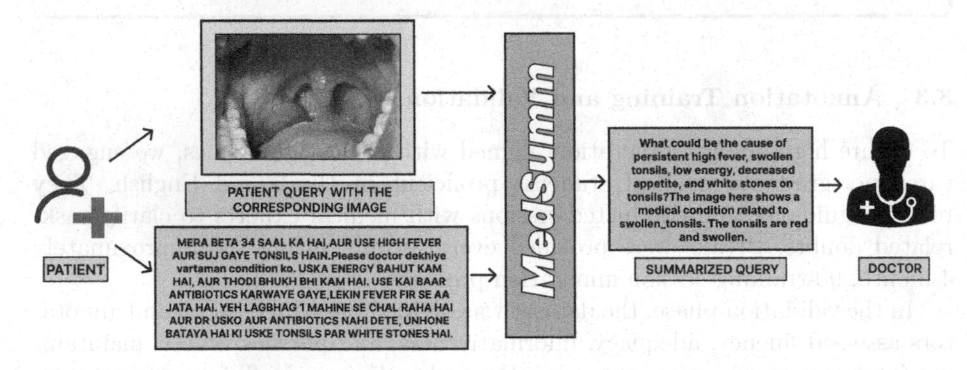

Fig. 3. Pipeline demonstrating the application of MedSumm for the summarization of multimodal patient queries in a code-mixed environment, aimed at facilitating expedited comprehension of patient queries for doctors and healthcare professionals.

4 Methodology

4.1 Problem Formulation

Each data point comprises a patient's textual inquiry, denoted as Q, along with an accompanying image, represented as I, illustrating the medical issue or concern the patient is seeking assistance within their text. The ultimate objective is to generate a natural language sequence Y represented as:

$$Y = \{S, T\} \tag{1}$$

where S represents the succinct summary that incorporates insights from both the textual and visual modalities and T represents a brief note about the medical condition associated with the visual cue. Figure 4 shows a diagrammatic representation of our proposed architecture MedSumm. MedSumm has 3 main stages namely: (i) **Question and Visual Symptom Representation**, (ii) **Adaptation Methods**, and (iii) **Inference**.

4.2 Textual and Visual Representation

We've utilized two distinct categories of data to represent a patient-clinician interaction: the patient's question in textual form and visual elements that the patient shared alongside the query.

Question Representation: In this context, each patient's query is presented as a text passage in which they explain their medical concerns to the doctor to receive relevant feedback. Recent advancements in language models, particularly decoder-based ones like LLaMA [28], GPT-3 [5], have demonstrated superior performance in encoding textual presentations compared to encoder-based models like BERT [10]. This technique involves employing Large Language Models (LLMs) for next-word prediction, from which we generate sentence embeddings by extracting the hidden vectors corresponding to the final word. Various LLMs, such as Vicuna [36] Llama2 [29], FLAN-T5 [6], Mistral-7B [16] and Zephyr-7B [30] , are utilized for this purpose. During this process, the text is tokenized into smaller units, and each token is ultimately transformed into a 4096-dimensional embedding. It is noteworthy that pre-trained models of these LLMs are employed in this task.

Visual Representation: We employ Vision-transformers ViT [12], for this task. These models take raw images as input and transform them into embeddings of size 768. Furthermore, we integrate a linear projection mechanism using a fully connected layer to map these 768-dimensional visual embeddings into a shared textual embedding space. This unified vision-language embedding is then fed into the decoder of the language model. Notably, only the linear projection layer is trainable in this process. This linear projection layer not only helps to extract meaningful information from the vision encoder but also helps in multimodal information fusion, thus producing richer information embeddings that can be effectively learned and finetuned.

4.3 Adaptation Methods

One of the biggest challenges of LLM-based techniques is fine-tuning them to domain-specific tasks as they are mostly bounded by resource constraints. There are several techniques that have been developed to overcome this problem namely prompt tunning, in-context learning, and low-rank adaptation techniques like LoRA [15] and QLoRA [9]. For this work, we are going with the latest and most parameter-efficient technique which is QLORA. QLoRA is a more memory-efficient version of LoRA providing 4-bit quantization to enable fine-tuning of larger LLMs using the same hardware constraints.

4.4 Inference

The inference module is an LLM that receives the attended multi-modal fusion representation vector, which combines the patient's textual query with visual cues. This module generates the corresponding symptom name and summary using a next-token prediction approach. In this method, the language model calculates probabilities for all tokens in its vocabulary, and these probabilities are used to predict the next predicted token. We use the following LLMs namely Llama-2, Mistral-7B, Vicuna, FLAN-T5 and Zephyr-7B for the final summary generation

5 Experiments and Results

To study the task of summarization of complex code-mixed medical queries we perform a meticulous fine-tuning and evaluation of both uni-modal and our multimodal model ***MedSumm*** on our ***MMCQS*** dataset. We have allocated 80% of the dataset for training, 5% for validation, and 15% for testing our fine-tuned models. We have leveraged the most recent and popular open source LLMs like Vicuna [36] Llama2 [29], FLAN-T5 [6], Mistral-7B [16] and Zephyr-7B [30] in both unimodal and multimodal (MedLLMSumm) setup. We utilized ROUGUE [20], BLEU [24], BERT score [35], METEOR [4] as automatic evaluation metrics[7]. For the purpose of human evaluation, we collaborate with a medical expert and a few medical postgraduate students. We have identified four distinct and medically nuanced metrics for this evaluation: clinical evaluation score, factual recall [3], hallucination rate [3], and our ***MMFCM*** metric.

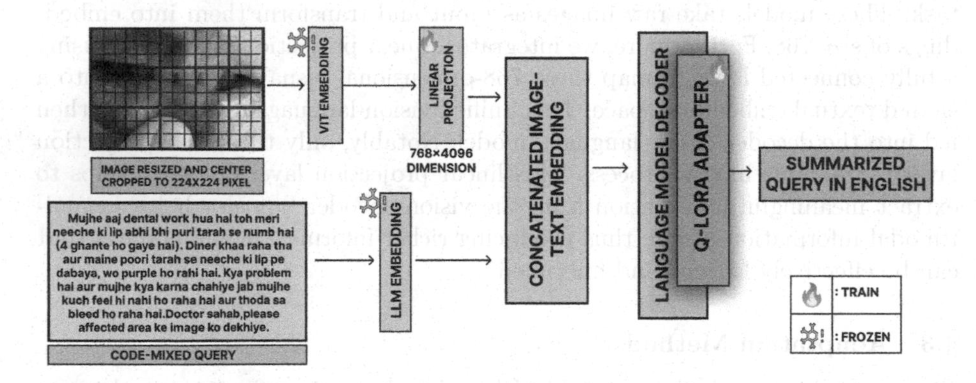

Fig. 4. Model structure of MedSumm showing the different stages of our proposed architecture. The frozen and trainable layers have also been indicated.

[7] To maintain uniformity in the results post-processing like removing extra spaces, repeated sentences are performed.

Automatic Evaluation: The automatic evaluation results in Table 1 reveal that **MedSumm** consistently outperforms all other LLMs across all defined metrics. This demonstrates the significant advantage of incorporating visual cues for generating more nuanced summaries. In unimodal settings, Mistral outperforms the rest, while Zephyr, as the most recent LLM, demonstrates balanced performance in both unimodal and multimodal scenarios. In the multimodal setting, LLAMA-2 takes the lead, closely followed by Vicuna in terms of performance. FLAN-T5's performance was the weakest across all metrics. This led us to the conclusion that decoder-only architectures perform better than encoder-decoder models like FLAN T5.

Human Evaluation: Existing automatic evaluation metrics can provide a misleading assessment of summary quality in the medical domain, where a single incorrect detail or omission of vital information can be perilous. A team of medical students, led by a doctor, conducted a human evaluation using a random 10% of the dataset to rate summaries from uni-modal and multi-modal models. Evaluation metrics include: (1) Clinical Evaluation Score for relevance, consistency, fluency, and coherence (rated 1 to 5), (2) Multi-modal Fact Capturing Metric (MMFCM) assessing the incorporation of medical facts and disorder identification, and (3) Medical Fact-Based Metrics, which employ Factual Recall and Omission Recall to measure the capture of medical facts compared to gold standard summaries. See Algorithm 1 for the detailed algorithm. Table 2 provides a comparison of both the uni-modal models and multi-modal models, underscoring the advantage of multimodal models over unimodal models for our task. This claim is also seconded by the improvement in the Factual Recall, Hallucination Rate, and our proposed MMFCM metrices.

Table 1. Performance of various *MedSumm* models and corresponding unimodal baselines, evaluated using automatic metrics with different LLMs.

	Model	ROUGE			BLEU				BERTScore	METEOR
		R1	R2	RL	B1	B2	B3	B4		
LLAMA-2	Unimodal(only texual query)	39.92	19.57	33.9	28.9	18.05	11.76	9.97	0.74	34.81
	MedSumm	**46.75**	**25.59**	**38.41**	**32.50**	**22.55**	**17.56**	**14.88**	**0.80**	**35.74**
Mistral	Unimodal(only texual query)	37.58	17.14	29.01	23.67	14.46	9.42	6.15	0.77	41.68
	MedSumm	**42.54**	**23.81**	**34.85**	**27.21**	**19.04**	**14.89**	**12.22**	**0.76**	**36.00**
Vicuna	Unimodal(only textual query)	38.64	21.55	32.9	24.12	15.88	11.09	9.13	0.75	32.25
	MedSumm	**45.09**	**26.11**	**37.05**	**27.58**	**19.98**	**16.44**	**14.33**	**0.80**	**32.10**
FLAN-T5	Unimodal(only textual query)	36.3	17.22	29.81	22.1	12.14	8.51	6.15	0.67	28.25
	MedSumm	**41.5**	**23.02**	**33.96**	**26.2**	**18.04**	**14.12**	**11.8**	**0.74**	**35.09**
Zephyr	Unimodal(only textual query)	36.67	17.54	30.12	22.93	14.01	8.97	5.80	0.70	35.32
	MedSumm	**44.55**	**25.37**	**34.97**	**27.05**	**19.48**	**15.84**	**13.63**	**0.77**	**33.37**

6 Qualitative Analysis

We conducted a thorough qualitative analysis of the summaries generated by different models in both unimodal and multimodal setting. We also performed

Algorithm 1. MMFCM Method

Require: $F_m = \{ fact_{m,1}, fact_{m,2} \cdots fact_{m,n-1}, fact_{m,n} \}$
\\Relevant Medical facts from golden summary of query 'm'.
$Sf_m = \{ Summfact_{m,1}, \ldots, Summfact_{m,n} \}$
\\Relevant Medical facts from the generated summary of query 'm'.
$\#CorrectFacts_m = |F_m \cap Sf_m|$
if (Correct Medical Disorder phrase $\in \{F_m \cap Sf_m\}$) **then**
  $\#CorrectFacts_m += 2$
else if (Partially correct disorder phrase $\in \{F_m \cap Sf_m\}$) **then**
  $\#CorrectFacts_m += 1$
else if (Incorrect disorder phrase $\in \{F_m \cap Sf_m\}$) **then**
  $\#CorrectFacts_m += -1$
else
  $\#CorrectFacts_m += 0$
end if
return MMFCM = $\tanh(\#CorrectFacts_m / |F_m|)$

Table 2. Human evaluation scores of various unimodal and multimodal models across different metrics.

Models	Clinical-EvalScore	Factual Recall	Hallucination Rate	MMFCM Score
LLAMA-2(U)	3.1	0.34	0.35	NA
LLAMA-2(M)	3.52	0.35	0.32	0.42
Mistral-7B(U)	3.41	0.36	0.29	NA
Mistral-7B(M)	3.43	0.36	0.3	0.41
Vicuna-7B(U)	3.32	0.33	0.33	NA
Vicuna-7B(M)	3.45	0.34	0.36	0.41
FLAN-T5(U)	2.8	0.28	0.31	NA
FLAN-T5(M)	3.1	0.31	0.33	0.34
Zephyr 7B β(U)	3.48	0.38	0.26	NA
Zephyr 7B β(M)	3.54	0.36	0.29	0.44
Annotated Summary	**4.1**	**0.88**	**0**	**0.87**

some case studies; one such instance is shown in Fig. 5. The analysis led to the following conclusion: (a) All models perform better in multimodal setting and is better at capturing the important visual information conveyed through the images and predicting the exact disorder phrase. (b) It was also observed that models like LLAMA-2 and Vicuna shows a tendency of hallucination (see Fig. 5) and sometimes generated facts which was completely out of context. The summaries generated by Mistral and Zephyr were more cohesive and showed lesser hallucinations. (c) FlanT5 showed the weakest results which made us conclude that decoder only LLMs perform better than LLMs with encoder-decoder architecture for our task.

CODE-MIXED QUESTION: Hello doctor,Do hafte se mujhe unbearable itching hai. Mujhe iske saath ek rash bhi tha jo plaque rash ka jaisa dikhta tha lekin ab wo fade ho gaya hai. Please doctor, dekhiye is vartaman condition ko. Mujhe super itching skin hai jo itch karne par bumps ban jata hai lekin thode time baad fade ho jata hai. Ye itching mujhe paagal kar rahi hai. Maine do doctors se mila hai aur unhe pata nahi kya ho raha hai, sirf ye pata hai ki ye contagious nahi hai kyunki mere husband aur bachhe ise nahi hai.

ENGLISH QUESTION: Hello doctor,For two weeks, I have been unbearably itchy. I had a rash to go along with it that looked more like little plaque rashes but those have faded now.Please doctor see the current condition below I am left with super itchy skin that bumps up when I itch but fades away after a while. The itching is driving me insane. I have seen two doctors and they have no clue what is going on, just that it is not contagious because my husband and kids do not have it.

TARGET SUMMARY: What could be causing the patient's unbearably itchy skin with plaque-like rashes that fade and are not contagious?The image shows the condition of skin rash. Round and swollen spots on the skin, red in color and multiple in the affected area.

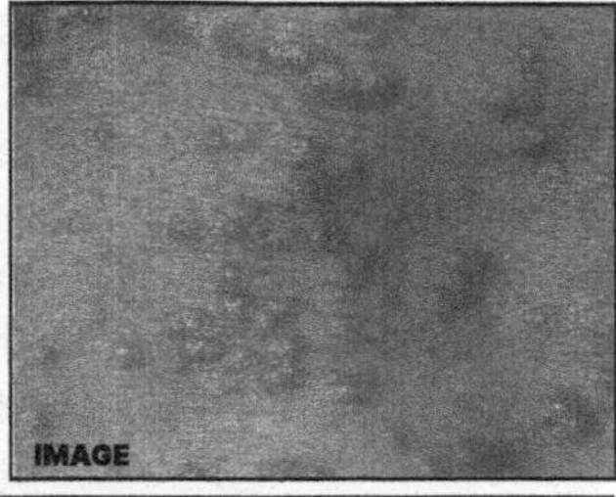

MODEL	UNIMODAL (USING ONLY TEXT QUERY)	MedSumm (IMAGE + TEXT QUERY)
LLAMA-2	40-year-old female with unbearable itching, rash, and bumps on skin. No contagiousness. Condition : Eczema	70-year-old male with unbearable itching, rash, and skin irritation. No contagion. The image shows the condition of skin rash. Round and swollen spots on the skin, red in color and multiple in the affected area. Little swelling and dryness on the skin.
MISTRAL-7B	What could be the cause of unbearable itching, a plaque rash, and a skin condition that causes bumps and fades away?	2 weeks of unbearable itching with a previous plaque rash, now fading, causing bumps and driving the patient mad. The image shows the condition of skin rash. Round and swollen spots on the skin, red in color and multiple in the affected area. Cluster of such spots. Little swelling and dryness on the spots. Possible skin irritation or allergy causing the rash. The image shows the condition of skin rash. Round and swollen spots on the skin, red in color and multiple in the affected area. Cluster of such spots.
VICUNA-7B	10 days of unbearable itching with a fading plaque rash, bumps that fade, and no contagion.	48-year-old female with unbearable itching, rash, and bumps on skin. The image shows the condition of skin rash. Round and swollen spots on the skin, red in color and multiple in the affected area. Round and swollen spots on the skin, red in color and multiple in the affected area. Condition- It could be a folliculitis or a boil. The exact diagnosis can be made only by a doctor after a physical examination. The affected area needs to be cleaned with soap and water and kept dry.
ZEPHYR-7B-BETA	What could be the cause of unbearable itching, a plaque rash, and skin bumps that fade away?	2 weeks of unbearable itching, rash that resembled plaque rash, bumps that fade, and driving me crazy. What could be the cause?The image shows the condition of skin rash. Round and swollen spots on the skin, red in color and multiple in the affected area. Condition- skin rash Skin rash on the leg's skin, red in color and multiple circular swelling on the skin. The circular swelling on the skin is red in color and swollen.
FLAN-T5	Describe the symptoms of a super itching rash on the skin, with a rash on the shins, and a bump on the shin.	What could be the cause of the itching and redness on the skin? The image shows the condition of skin rash. The skin is red and swollen with little or no rash.

Fig. 5. Samples summaries generated by the LLMs in unimodal and multimodal setting. The hallucinated phrases has been shown in red.

7 Risk Analysis

We must acknowledge several limitations in our approach. Initially, we confined our task to 18 symptoms conducive to image sharing. Therefore, introducing an image outside this scope may lead to the model generating potentially erroneous information in the summary. While our multimodal model shows promise, it is prudent to engage a medical expert for the ultimate verification, particularly in high-stakes scenarios. Our AI model serves as a tool, not a substitute for medical professionals.

8 Conclusion and Future Work

In this study, we delve into the impact of incorporating multi-modal cues, particularly visual information, on question summarization within the realm of healthcare. We present the MMCQS dataset the first of its kind dataset, comprising 3015 multimodal medical queries in Hindi-English codemixed language with golden summaries in English that merge visual and textual data. This novel collection fosters new assessment techniques in healthcare question summarization.

We also propose a model *MedSumm*, that incorporates the power of both LLM and Vision encoders to generate the final summary. In our future endeavors, we aspire to develop a Vision-Language model capable of extracting the intensity and duration details of symptoms and integrating them into the patient query's final summary generation. Furthermore, our expansion plans encompass incorporating medical videos and speeches data and addressing scenarios involving other low resources languages in Indian context.

Acknowledgements. Akash Ghosh and Sriparna Saha express their heartfelt gratitude to the SERB (Science and Engineering Research Board) POWER scheme(SPG/2021/003801) of the Department of Science and Engineering, Govt. of India, for providing the funding for carrying out this research

Ethical Considerations. In healthcare summarization, we prioritize ethical considerations, including safety, privacy, and bias. We took extensive measures with the MMCQS dataset, collaborating with medical professionals, obtaining IRB approval, and adhering to legal and ethical guidelines during data handling, image integration, and summary annotation. The dataset is based on the HealthcareMagic Dataset, and medical experts supervised the task. Identity protection was ensured for user privacy.

References

1. Abacha, A.B., Demner-Fushman, D.: On the summarization of consumer health questions. In: Proceedings of the 57th Annual Meeting of the Association for Computational Linguistics, pp. 2228–2234 (2019)
2. Abacha, A.B., M'rabet, Y., Zhang, Y., Shivade, C., Langlotz, C., Demner-Fushman, D.: Overview of the mediqa 2021 shared task on summarization in the medical domain. In: Proceedings of the 20th Workshop on Biomedical Language Processing, pp. 74–85 (2021)
3. Abacha, A.B., Yim, W.-W., Michalopoulos, G., Lin, T.: An investigation of evaluation metrics for automated medical note generation. arXiv preprint arXiv:2305.17364 (2023)
4. Banerjee, S., Lavie, A.: Meteor: an automatic metric for mt evaluation with improved correlation with human judgments. In: Proceedings of the ACL Workshop on Intrinsic and Extrinsic Evaluation Measures for Machine Translation and/or Summarization, pp. 65–72 (2005)
5. Brown, T., et al.: Language models are few-shot learners. Adv. Neural. Inf. Process. Syst. **33**, 1877–1901 (2020)
6. Chung, H.W., et al.: Scaling instruction-finetuned language models. arXiv preprint arXiv:2210.11416 (2022)
7. Das, A., Gambäck, B.: Identifying languages at the word level in code-mixed Indian social media text. arXiv preprint arXiv:2302.13971 (2014)
8. Delbrouck, J.-B., Zhang, C., Rubin, D.: QIAI at MEDIQA 2021: multimodal radiology report summarization. In: Proceedings of the 20th Workshop on Biomedical Language Processing, pp. 285–290. Association for Computational Linguistics (2021)
9. Dettmers, T., Pagnoni, A., Holtzman, A., Zettlemoyer, L.: Qlora: efficient finetuning of quantized llms. arXiv preprint arXiv:2305.14314 (2023)

10. Devlin, J., Chang, M.-W., Lee, K., Toutanova, K.: Bert: pre-training of deep bidirectional transformers for language understanding. arXiv preprint arXiv:1810.04805 (2018)
11. Dong, Q., et al.: A survey for in-context learning. *arXiv preprint* arXiv:2301.00234 (2022)
12. Dosovitskiy, A., et al. An image is worth 16×16 words: transformers for image recognition at scale. arXiv preprint arXiv:2010.11929 (2020)
13. Ghosh, A., Acharya, A., Jain, R., Saha, S., Chadha, A., Sinha, S.: Clipsyntel: clip and llm synergy for multimodal question summarization in healthcare. arXiv preprint arXiv:2312.11541 (2023)
14. Gupta, D., Attal, K., Demner-Fushman, D.: A dataset for medical instructional video classification and question answering (2022)
15. Hu, E.J., et al.: Lora: low-rank adaptation of large language models. *arXiv preprint* arXiv:2106.09685 (2021)
16. Jiang, A.Q., et al.: Mistral 7b. arXiv preprint arXiv:2310.06825 (2023)
17. Kojima, T., Gu, S., Reid, M., Matsuo, Y., Iwasawa, Y.: Large language models are zero-shot reasoners. arxiv (2023)
18. Kumar, R., Chakraborty, R., Tiwari, A., Saha, S., Saini, N.: Diving into a sea of opinions: multi-modal abstractive summarization with comment sensitivity. In: Proceedings of the 32nd ACM International Conference on Information and Knowledge Management, pp. 1117–1126 (2023)
19. Lewis, M., et al.: Bart: denoising sequence-to-sequence pre-training for natural language generation, translation, and comprehension. arXiv preprint arXiv:1910.13461 (2019)
20. Lin, C.-Y.: Rouge: a package for automatic evaluation of summaries. In: Text Summarization Branches Out, pp. 74–81 (2004)
21. Liu, G., et al.: Medical-vlbert: medical visual language bert for covid-19 ct report generation with alternate learning. IEEE Trans. Neural Netw. Learn. Syst. **32**(9), 3786–3797 (2021)
22. Mrini, K., Dernoncourt, F., Chang, W., Farcas, E., Nakashole, N.: Joint summarization-entailment optimization for consumer health question understanding. In: Proceedings of the Second Workshop on Natural Language Processing for Medical Conversations, pp. 58–65 (2021)
23. Nittari, G., et al.: Telemedicine practice: review of the current ethical and legal challenges. Telemed. e-Health **26**(12), 1427–1437 (2020)
24. Papineni, K., Roukos, S., Ward, T., Zhu, W.-J.: Bleu: a method for automatic evaluation of machine translation. In: Proceedings of the 40th Annual Meeting of the Association for Computational Linguistics, pp. 311–318 (2002)
25. Qi, W., et al.: Prophetnet: predicting future n-gram for sequence-to-sequence pre-training. arXiv preprint arXiv:2001.04063 (2020)
26. Thawkar, O., et al.: Xraygpt: chest radiographs summarization using medical vision-language models. arXiv preprint arXiv:2306.07971 (2023)
27. Tiwari, A., Manthena, M., Saha, S., Bhattacharyya, P., Dhar, M., Tiwari, S.: Dr. can see: towards a multi-modal disease diagnosis virtual assistant. In: Proceedings of the 31st ACM International Conference on Information & Knowledge Management, pp. 1935–1944 (2022)
28. Touvron, H., et al.: Llama: open and efficient foundation language models. arXiv preprint arXiv:2302.13971 (2023)
29. Touvron, H., et al. Llama 2: open foundation and fine-tuned chat models. *arXiv preprint* arXiv:2307.09288 (2023)

30. Tunstall, L., et al.: Zephyr: direct distillation of lm alignment (2023)
31. Yadav, S., Gupta, D., Abacha, A.B., Demner-Fushman, D.: Reinforcement learning for abstractive question summarization with question-aware semantic rewards. arXiv preprint arXiv:2107.00176 (2021)
32. Yong, Z.X., et al.: Prompting multilingual large language models to generate code-mixed texts: the case of south east Asian languages. In: Sixth Workshop on Computational Approaches to Linguistic Code-Switching (2023)
33. Zhang, J., Huang, J., Jin, S., Lu, S.: Vision-language models for vision tasks: a survey. arXiv preprint arXiv:2304.00685 (2023)
34. Zhang, J., Zhao, Y., Saleh, M., Liu, P.: Pegasus: pre-training with extracted gap-sentences for abstractive summarization. In: International Conference on Machine Learning, pp. 11328–11339. PMLR (2020)
35. Zhang, T., Kishore, V., Wu, F., Weinberger, K.Q., Artzi, Y.: Bertscore: evaluating text generation with bert. arXiv preprint arXiv:1904.09675 (2019)
36. Zheng, L., et al.: Judging llm-as-a-judge with mt-bench and chatbot arena. arXiv preprint arXiv:2306.05685 (2023)
37. Zhou, J., et al.: Skingpt-4: an interactive dermatology diagnostic system with visual large language model. arXiv preprint arXiv:2304.10691 (2023)
38. Zhu, D., Chen, J., Shen, X., Li, X., Elhoseiny, M.: Minigpt-4: enhancing vision-language understanding with advanced large language models. arXiv preprint arXiv:2304.10592 (2023)

Improving Exposure Allocation in Rankings by Query Generation

Thomas Jaenich(✉), Graham McDonald, and Iadh Ounis

University of Glasgow, Glasgow, UK
t.jaenich.1@research.gla.ac.uk,
{graham.mcdonald,iadh.ounis}@glasgow.ac.uk

Abstract. Deploying methods that incorporate generated queries in their retrieval process, such as Doc2Query, has been shown to be effective for retrieving the most relevant documents for a user's query. However, to the best of our knowledge, there has been no work yet on whether generated queries can also be used in the ranking process to achieve other objectives, such as ensuring a fair distribution of exposure in the ranking. Indeed, the amount of exposure that a document is likely to receive depends on the document's position in the ranking, with lower-ranked documents having a lower probability of being examined by the user. While the utility to users remains the main objective of an Information Retrieval (IR) system, an unfair exposure allocation can lead to lost opportunities and unfair economic impacts for particular societal groups. Therefore, in this work, we conduct a first investigation into whether generating relevant queries can help to fairly distribute the exposure over groups of documents in a ranking. In our work, we build on the effective Doc2Query methods to selectively generate relevant queries for underrepresented groups of documents and use their predicted relevance to the original query in order to re-rank the underexposed documents. Our experiments on the TREC 2022 Fair Ranking Track collection show that using generated queries consistently leads to a fairer allocation of exposure compared to a standard ranking while still maintaining utility.

1 Introduction

Information Retrieval (IR) systems are ubiquitously deployed to help users to navigate large amounts of information, enabling them to discover relevant products, services, or items that align with their specific information needs and preferences. The IR systems essentially display relevant information to users through rankings. Indeed, typically, relevant items, such as documents, are presented to the user in a ranking sorted by decreasing relevance, i.e., the most relevant document is ranked first. However, while useful for the information consumer, rankings that are ordered solely by the predicted relevance of documents can disadvantage particular societal groups, for example the information producers, due to how users interact with the produced rankings. Indeed, users typically analyse rankings starting at the top of the ranking, sequentially examining the

N. Goharian et al. (Eds.): ECIR 2024, LNCS 14612, pp. 121–129, 2024.
https://doi.org/10.1007/978-3-031-56069-9_9

documents in their ranked order. Moreover, the likelihood of examining an item is influenced by its position in the ranking. Consequently, lower-ranked documents receive less attention from the users since they are less likely to be examined than higher-ranked documents. Therefore, when documents from particular groups are systematically ranked lower by an IR system, the corresponding groups can suffer from lost opportunities and might even incur an unfair economic impact. For example, documents from authors that are ranked lower than other authors due to small differences in predicted relevance are more likely to go unnoticed by users. To avoid such issues, the exposure of documents to users in a ranking needs to be fairly distributed, e.g., equally distributed across all groups or according to a predefined target distribution that is deemed to be fair. In a standard IR system, exposure allocation depends on the underlying ranking model that creates the ranking. Increasingly, rankings are created using deep neural ranking models, such as BERT [5]. Such models use contextualised information to predict the relevance of documents to a user's query. More recently, sequence-to-sequence models such as T5 [18] have been deployed to generate relevant and representative content that can further increase the retrieval effectiveness. Specifically, generating relevant queries and appending them to documents has been shown to increase the effectiveness of a standard retrieval pipeline [8,20]. However, while generating queries has been shown to be effective in finding the most relevant documents, to the best of our knowledge, there has been no work investigating how generating such queries can be used to improve the allocation of exposure in a ranking. To address this gap, in this work, we propose a new re-ranking method that we call **Q**uery **S**imilarity **E**xposure **E**nhancement (QSEE). We introduce QSEE to study whether the generation of content is a suitable technique for averting societal problems that might emerge when deploying information retrieval (IR) systems. Our specific focus is on addressing the uneven distribution of attention towards underrepresented groups, thereby preventing the corresponding societal issues such as the loss of opportunity. QSEE can be deployed with three different policies, which use the similarity of a generated query to the original query that is issued by a user, in order to re-rank documents from underrepresented groups. Our experiments on the TREC 2022 Fair Ranking Track [7] test collection show that our proposed policies consistently lead to improvements in the fair allocation of exposure, as measured by Attention Weighted Ranked Fairness (AWRF), while still maintaining a comparable utility to the original ranking. Our best-performing strategy (Policy 3) improves the exposure allocation on all evaluated fairness categories by up to ~10% and increases the utility (nDCG) of a ranking on the majority (3/5) of the categories.

2 Related Work

Traditionally, the performance of an IR system is judged by how effective it is in estimating the relevance of a document for a user's query. In sparse ranking models such as BM25 [24] or DPH [1], the ability to predict the relevance of documents can be limited if there is no exact lexical match of the query tokens into the documents [12]. This is often referred to as the lexical mismatch problem. To alleviate the lexical mismatch problem, query expansion methods such as Rocchio [26] or Bo1 [2] have been introduced to reformulate the user's query to improve the lexical matching of the query to the documents. Moreover, the introduction of deep neural ranking models, which use contextualised information, such as BERT [5], allows to reduce the lexical mismatch. However, retrieval can still be insufficiently effective even when deep neural rankers are deployed [13]. Therefore, dense query reformulation methods using pseudo-relevance feedback such as ColBERT-PRF [30] have been introduced to improve retrieval effectiveness. Alternatively to expanding the user's query, augmenting the documents' contents with terms that are expected to be relevant and representative of the documents' contents has also been shown to be an effective approach for improving retrieval effectiveness [14]. Methods such as Doc2Query [19] and Doc2Query– [8] generate queries that are likely to be answered by a document, and append the generated queries to the original document's text. Appending such generated query terms can increase the chances of a relevant document being retrieved in response to a user's query. However, in addition to the traditional objective of providing the most relevant search results to a user, in this work we focus on the fair allocation of exposure to particular groups of documents. Distributing the exposure fairly in the search results has increasingly been recognised as an important objective for IR systems [6]. While there have been various prior work [3,9,10,16,17,27–29,31] into how a fair allocation of exposure in the search results can be achieved, in this paper we are specifically interested in whether generated additional content, such as queries, can be used to distribute exposure more fairly across groups of documents. To the best of our knowledge, our work is the first investigation into using generated queries to distribute exposure fairly.

3 Approach

In this work, we present a re-ranker that re-scores the relevance of documents from underexposed groups, using queries that are generated from the documents' text. Our re-ranking method, which we call Query Similarity Exposure Enhancement (QSEE), operates over groups of documents denoted as $g \in G$. The primary objective of QSEE is that all of the groups $g \in G$ should receive an equal amount of exposure to the user. In other words, the documents from these groups should be distributed across the ranking positions such that every group of documents has the same likelihood of being observed by a user. Standard re-ranking pipelines are prone to allocate exposure to groups unfairly due to, for example, unidentified biases in the ranking models or in the contents of the documents themselves [22,23]. With this in mind, for each document in an underexposed group, QSEE generates a set of queries that are representative of the document and then leverages the generated queries to better predict the relevance of the document to the user's query. We argue that by representing the document as a query, we can eliminate existing unidentified biases in the document's text to give the document a fairer chance of receiving an appropriate amount of exposure to

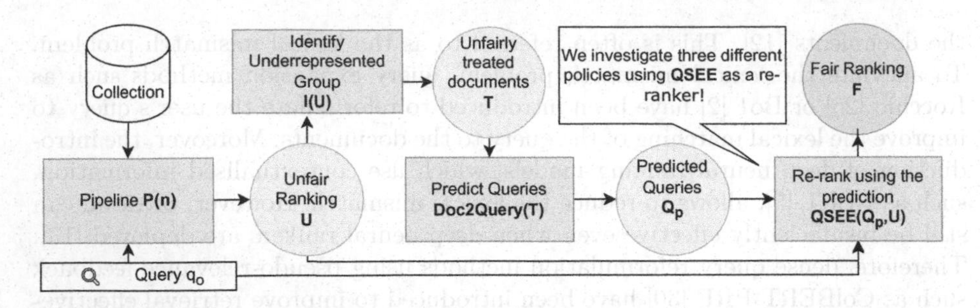

Fig. 1. The retrieval process integrating our QSEE approach and policies. Squares represent retrieval functions, while circles represent document sets and rankings.

the user. We propose three policies, which use the similarity of a generated query to the user's query as an additional relevance score, to re-score the document before re-ranking the documents, so as to improve the exposure allocation.

Re-ranking with Generated Queries: Figure 1 illustrates the retrieval and re-ranking process that integrates our proposed QSEE approach and policies. In Fig. 1, a standard retrieval pipeline $P(n)$ is deployed to find the documents that are the most relevant for a user's query q_o. Typically in $P(n)$, an inexpensive ranking model, such as BM25 [25], is deployed as a first-stage retrieval process, followed by a more effective but expensive neural ranking model, such as ELECTRA [4], to re-rank the documents that are retrieved in the first-stage retrieval. $P(n)$ returns a ranking U in which the documents are ranked by their predicted relevance score $r \in \mathbb{R}$. Once U is created, we can then calculate each group g's exposure in the ranking and identify the most underexposed group, T, i.e., the group that receives the least amount of exposure in the ranking. For each document in T, we generate a set of m generated queries, Q_p, that are representative of the document [18]. To do so, we use a sequence-to-sequence model, such as T5 [21]. In order to use the generated queries Q_p to re-score the documents in T, for each document in T, we first identify which of the document's generated queries, $q_p \in Q_p$, is the most relevant to the user query q_o. To do so, we use the same re-ranker that has been deployed in $P(n)$ to calculate a similarity score $s : q_o \times q_p$ for each $q_p \in Q_p$. For each document, we then select the generated query that is the most similar to the user's query (i.e., s_{max}). The similarity score s_{max} is then used in each of our policies to re-rank the documents in U in order to create a fairer ranking F based on new fair relevance scores r_f of the documents obtained by the policies.

Re-ranking Policies: In the following sections, we detail our proposed policies.

Policy 1: Replace Original Relevance Score with a Query Similarity Score. In this policy, to re-rank U, for each document in the underrepresented group T, we replace the document's relevance score r that was produced by $P(n)$ with the score of the document's generated query that is most similar to the user's query q_o, i.e., $r_f = s_{max}$, if $s_{max} > r$. Our intuition is as follows: If a generated query is more similar to the user's query than the document that the query was generated from, then the document may be more relevant to the user's query than was initially indicated by the score r produced by $P(n)$. To avoid penalising documents due to inaccurate query generation, we replace r only with s_{max} when the generated query's relevance score (s_{max}) is higher than the original score (r). However, replacing the entire score (r) can lead to overestimating relevance.

Policy 2: Mean of the Original Ranking and Query Similarity Scores. In this policy, if $s_{max} > r$, then we use the mean of the two scores as the document's final predicted relevance score, i.e.: $r_f = \frac{r+s_{max}}{2}$. By using the mean of the two scores, we aim to avoid overestimating the relevance of a document. However, scoring documents by the mean of the two scores may still be insufficient, since the total amount of exposure that is available in a ranking is finite. Therefore, increasing the exposure of an underrepresented group can decrease the exposure of one or more other groups. If the group that is the most underexposed in U is already receiving close to the group's ideal exposure then a document's final score, r_f, should be closer to the original relevance estimation, r, than to the similarity score s_{max} so that the group is not over-exposed.

Policy 3: Mixture model of Relevance Scores. For Policy 3, instead of calculating the highest query similarity scores for documents from only one underrepresented group, we generate queries for all underrepresented groups, i.e., for all documents from groups with less than the ideal amount of exposure. To obtain r_f, we incorporate a factor β that controls how much influence s_{max} has on the new relevance score r_f. β is obtained using the ratio of the target exposure exp_t to the actual exposure: $\beta = \frac{exp_t}{exp_a}$. We argue that if there is a big discrepancy between exp_a and exp_t, the original relevance estimation is less trustworthy and the prediction should obtain more weight when calculating r_f. Therefore we use β as a weight in $r_f = \frac{r+\beta * s_{max}}{2}$.

4 Experimental Setup

In this work, we aim to answer the following two research questions:

- **RQ1:** Can our QSEE policies distribute the available exposure to users more fairly among groups than a standard retrieval pipeline?
- **RQ2:** How do our re-ranking policies affect the relevance of the ranking?

Test Collection: We evaluate our policies using the 47 evaluation queries from the TREC 2022 Fair Ranking Track test collection [7]. The collection consists of English-language Wikipedia documents. Each document is linked to a single group within a fairness category. In the "Age of Topic" category, documents are grouped by the age of the document's subject (Unknown, Pre-1900s, 20th century, 21st century). The "Language" category groups documents by the number of different languages in Wikipedia the article appears in (5+ Languages, 2–5 Languages, Only English). "Age of Article" is based on the documents' creation dates (2001–2006, 2007–2011, 2012–2016, 2017–2022), and "Alphabetical" groups documents by the first letter (a-d, e-k, l-r, and s-z). "Popularity" is determined by the number of views in February 2022 (Low, Medium-Low, Medium-High, High). In the TREC 2022 Fair Ranking Track, submissions by participating groups were evaluated using intersectional fairness objectives, i.e., combinations of the categories. However, in this work, we evaluate performance on single categories. Evaluating policies on single categories can be more practical in real-world applications. Indeed, this allows for a simpler integration of targeted interventions without the overly complex challenges associated with intersectional evaluation.

Standard Retrieval: We use the PyTerrier IR Framework [15]. We index the documents using a Porter Stemmer and stop-word removal. As our baseline, we use a re-ranking pipeline that uses BM25 [25] for the initial retrieval and ELECTRA [4] as our re-ranker (i.e., BM25>>ELECTRA).

Query Generation: We use the PyTerrier implementation[1] of the T5 Doc-2Query model from Nogueira and Lin [18] to create 20 queries for every document.

Metrics: Following the TREC Fair Ranking Track, we use Attention Weighted Ranked Fairness (AWRF) as our fairness metric [7] and the Jensen-Shannon Distance (JSD) as a measure of how close the actual exposure, A, that a group receives in a ranking is to a desired target exposure T. We use the exposure drop-off from Discounted Cumulative Gain (DCG) [11], defined as: $\frac{1}{(log_2(n)+1)}$, as our user browsing model. We report nDCG as our measure of utility.

5 Results

To answer our two research questions, we evaluate our policies over rankings of size k=100. Table 1 presents the allocation of exposure over each of the categories, as measured by AWRF. In response to RQ1, we observe that applying Policy 1 results in a fairer exposure allocation than the baseline (BM25>>ELECTRA) in four out of the five categories. In two of these categories , Age of Article and Age of Topic, we observe statistically significant

[1] https://github.com/terrierteam/pyterrier_doc2query.

Table 1. Comparison of our policies to the baseline. We use * to indicate a significant difference to the standard re-ranking baseline in terms of AWRF.

Approaches	Alphabetical		Age of Article		Popularity		Age of Topic		Languages	
	AWRF	nDCG	AWRF	nDCG	AWRF	nDCG	AWRF	nDCG	AWRF	nDCG
Policy-1	0.7664	0.5762	0.7128*	0.5783	0.7125	0.5818	0.5744*	0.5739	0.7118	0.5749
Policy-2	0.8021*	0.5792	0.7072*	0.5787	0.7180*	0.5795	0.5595*	0.5769	0.7060*	0.5776
Policy-3	**0.8222***	**0.5863**	**0.7363***	0.5779	**0.7446***	**0.5940**	**0.5967***	**0.5811**	**0.7540***	0.5763
BM25>>ELECTRA	0.7756	0.5795	0.6790	**0.5795**	0.6931	0.5795	0.5478	0.5795	0.6843	**0.5795**

differences (*, t-test, p<0.05). Policy 2 consistently improves the exposure allocation in all of the categories with statistically significant differences compared to the baseline, showing a greater stability despite generally smaller improvements than Policy 1. Analysing the results for Policy 3, we can see that its AWRF is higher than the baseline, with statistically significant differences, and all other policies for all of the evaluated categories, resulting in improvements of up to ~10 % in AWRF. The results show that our policies can enhance the fairness of an IR system. However, the primary goal of any IR system is utility. Hence, in RQ2, we assess the relevance of the re-ranked search results using nDCG. For Policy 1, we can observe marginal drop-offs in nDCG on four out of five categories. However, by using the two one-sided t-tests (TOST), we observe a statistical equivalence in terms of utility. In the Popularity category, we can observe a small increase in nDCG. For Policy 2, we observe equivalence in nDCG to the baseline, with only slight decreases for all categories. Policy 3 decreases the nDCG slightly on two of the categories. However, in the other three categories, Policy 3 manages to increase nDCG. All of our policies show a statistical equivalence to the baseline in terms of nDCG. This indicates that all of our policies can be applied without significantly decreasing the utility of a ranking for a user (as per the TOST test). To further clarify how our policies operate, we provide an illustrative example for a re-ranked document in the Alphabetical for the query *Catholicism*, which consists of keywords such as "catholism papacy jesuit pope papal catholic bishop". In the original ranking, the group "a-d" (13 documents) is underrepresented. Therefore, our policy generates queries for the documents associated with the "a-d" group. After applying Policy 3, the document with the largest difference between the original ranking score and the generated ranking score is the document "Catholic Church in Germany". In the standard ranking, the document appears at rank 945 with an ELECTRA score of −6.56. However, we generate the query "who is the catholic bishop of Germany?", which yields a similarity score of −0.16 to the original query. After applying Policy 3, the document is ranked at position 33 with a score of −3.33.

6 Conclusions

In this paper, we have explored the potential of using generated queries to provide a fair exposure to different groups of documents in a ranking. We proposed QSEE, a new re-ranking method that uses generated queries for the documents from underrepresented groups to re-score the documents. Our experiments on the TREC 2022 Fair Ranking Track test collection showed that QSEE is able to significantly improve the fairness of exposure in a ranking without decreasing the ranking's utility. Our investigation into the relationship between query generation and fairness has shown promising results, indicating that generating queries for a fairer exposure allocation should be further explored in future research. Apart from the technical contribution, we view our work as a first step towards exploring how generated content can be used to address and overcome issues of unfairness.

References

1. Amati, G., Ambrosi, E., Bianchi, M., Gaibisso, C., Gambosi, G.: FUB, IASI-CNR and university of "Tor Vergata" at TREC 2008 blog track (2008)
2. Amati, G., Van Rijsbergen, C.J.: Probabilistic models of information retrieval based on measuring the divergence from randomness. ACM Trans. Inf. Syst. (TOIS) **20**(4), 357–389 (2002)
3. Biega, A.J., Gummadi, K.P., Weikum, G.: Equity of attention: amortizing individual fairness in rankings. In: Proceedings of SIGIR, pp. 405–414 (2018)
4. Clark, K., Luong, M.T., Le, Q.V., Manning, C.D.: ELECTRA: pre-training text encoders as discriminators rather than generators. In: Proceedings of ICLR (2019)
5. Devlin, J., Chang, M.W., Lee, K., Toutanova, K.: BERT: pre-training of deep bidirectional transformers for language understanding. In: Proceedings of NAACL (2019)
6. Ekstrand, M.D., Burke, R., Diaz, F.: Fairness and discrimination in retrieval and recommendation. In: Proceedings of SIGIR, pp. 1403–1404 (2019)
7. Ekstrand, M.D., McDonald, G., Raj, A., Johnson, I.: Overview of the TREC 2022 fair ranking track. In: Proceedings of TREC 2022 (2022)
8. Gospodinov, M., MacAvaney, S., Macdonald, C.: Doc2Query–: when less is more. In: Kamps, J., et al. (eds.) ECIR 2023. LNCS, vol. 13981, pp. 414–422. Springer, Cham (2023). https://doi.org/10.1007/978-3-031-28238-6_31
9. Heuss, M., Sarvi, F., de Rijke, M.: Fairness of exposure in light of incomplete exposure estimation. In: Proceedings of SIGIR, pp. 759–769 (2022)
10. Jaenich, T., McDonald, G., Ounis, I.: ColBERT-FairPRF: towards fair pseudo-relevance feedback in dense retrieval. In: Kamps, J., et al. (eds.) ECIR 2023. LNCS, vol. 13981, pp. 457–465. Springer, Cham (2023). https://doi.org/10.1007/978-3-031-28238-6_36
11. Järvelin, K., Kekäläinen, J.: Cumulated gain-based evaluation of IR techniques. ACM Trans. Inf. Syst. (TOIS) **20**(4), 422–446 (2002)
12. Krovetz, R., Croft, W.B.: Lexical ambiguity and information retrieval. ACM Trans. Inf. Syst. (TOIS) **10**(2), 115–141 (1992)
13. Lin, J., Ma, X., Mackenzie, J., Mallia, A.: On the separation of logical and physical ranking models for text retrieval applications. In: DESIRES, pp. 176–178 (2021)
14. MacAvaney, S., Nardini, F.M., Perego, R., Tonellotto, N., Goharian, N., Frieder, O.: Expansion via prediction of importance with contextualization. In: SIGIR, pp. 1573–1576 (2020)
15. Macdonald, C., Tonellotto, N.: Declarative experimentation in information retrieval using PyTerrier. In: Proceedings of ICTIR (2020)

16. McDonald, G., Macdonald, C., Ounis, I.: Search results diversification for effective fair ranking in academic search. Inf. Retrieval J. **25**(1), 1–26 (2022)
17. Morik, M., Singh, A., Hong, J., Joachims, T.: Controlling fairness and bias in dynamic learning-to-rank. In: Proceedings of SIGIR, pp. 429–438 (2020)
18. Nogueira, R., Jiang, Z., Pradeep, R., Lin, J.: Document ranking with a pretrained sequence-to-sequence model. In: Proceedings of EMNLP 2020, pp. 708–718 (2020)
19. Nogueira, R., Lin, J., Epistemic, A.: From doc2query to docTTTTTquery. Online preprint, vol. 6, 2 (2019)
20. Nogueira, R., Yang, W., Cho, K., Lin, J.: Multi-stage document ranking with BERT. arXiv preprint arXiv:1910.14424 (2019)
21. Raffel, C., et al.: Exploring the limits of transfer learning with a unified text-to-text transformer. J. Mach. Learn. Res. **21**(1), 5485–5551 (2020)
22. Rekabsaz, N., Kopeinik, S., Schedl, M.: Societal biases in retrieved contents: measurement framework and adversarial mitigation of BERT rankers. In: Proceedings of SIGIR, pp. 306–316 (2021)
23. Rekabsaz, N., Schedl, M.: Do neural ranking models intensify gender bias? In: Proceedings of SIGIR, pp. 2065–2068 (2020)
24. Robertson, S.E.: The probability ranking principle in IR. J. Doc. **33**(4), 294–304 (1977). https://doi.org/10.1108/eb026647
25. Robertson, S.E., et al.: Okapi at TREC-3. NIST Special Publication Sp, vol. 109 (1995)
26. Rocchio, Jr., J, J.: Relevance feedback in information retrieval. The SMART retrieval system: experiments in automatic document processing (1971)
27. Sarvi, F., Heuss, M., Aliannejadi, M., Schelter, S., de Rijke, M.: Understanding and mitigating the effect of outliers in fair ranking. In: Proceedings of WSDM, pp. 861–869 (2022)
28. Singh, A., Joachims, T.: Fairness of exposure in rankings. In: Proceedings of KDD (2018)
29. Usunier, N., Do, V., Dohmatob, E.: Fast online ranking with fairness of exposure. In: Proceedings of FACCT, pp. 2157–2167 (2022)
30. Wang, X., Macdonald, C., Tonellotto, N., Ounis, I.: ColBERT-PRF: semantic pseudo-relevance feedback for dense passage and document retrieval. ACM Trans. Web **17**(1), 1–39 (2023)
31. Zehlike, M., Castillo, C.: Reducing disparate exposure in ranking: a learning to rank approach. In: Proceedings of the Web Conference 2020, pp. 2849–2855 (2020)

The Open Web Index

Crawling and Indexing the Web for Public Use

Gijs Hendriksen[1](✉), Michael Dinzinger[2], Sheikh Mastura Farzana[3], Noor Afshan Fathima[4], Maik Fröbe[5], Sebastian Schmidt[6], Saber Zerhoudi[2], Michael Granitzer[2], Matthias Hagen[5], Djoerd Hiemstra[1], Martin Potthast[6,7], and Benno Stein[8]

[1] Radboud University, Nijmegen, The Netherlands
gijs.hendriksen@ru.nl
[2] University of Passau, Passau, Germany
[3] German Aerospace Center (DLR), Cologne, Germany
[4] CERN, Geneva, Switzerland
[5] Friedrich-Schiller-Universität Jena, Jena, Germany
[6] Leipzig University, Leipzig, Germany
[7] ScaDS.AI, Leipzig, Germany
[8] Bauhaus-Universität Weimar, Weimar, Germany

Abstract. Only few search engines index the Web at scale. Third parties who want to develop downstream applications based on web search fully depend on the terms and conditions of the few vendors. The public availability of the large-scale Common Crawl does not alleviate the situation, as it is often cheaper to crawl and index only a smaller collection focused on a downstream application scenario than to build and maintain an index for a general collection the size of the Common Crawl. Our goal is to improve this situation by developing the *Open Web Index*.

The Open Web Index is a publicly funded basic infrastructure from which downstream applications will be able to select and compile custom indexes in a simple and transparent way. Our goal is to establish the Open Web Index along with associated data products as a new open web information intermediary. In this paper, we present our first prototype for the Open Web Index and our plans for future developments. In addition to the conceptual and technical background, we discuss how the information retrieval community can benefit from and contribute to the Open Web Index—for example, by providing resources, by providing pre-processing components and pipelines, or by creating new kinds of vertical search engines and test collections.

1 Introduction

Web search is an important technology for accessing the information on the Web. However, operating a full-scale web search engine is far from trivial. Crawling, processing, and indexing the Web consumes a large amount of resources, even without factoring in the large volume of queries that a search engine might have to process. As a result, only a handful of large corporations have been able to develop and operate commercial search engines, and they currently dominate

N. Goharian et al. (Eds.): ECIR 2024, LNCS 14612, pp. 130–143, 2024.
https://doi.org/10.1007/978-3-031-56069-9_10

the search engine market. This is in contrast to the recent release of data sets and open source models for generative AI. While companies like OpenAI made rapid progress on large language models and commercialized their successes, withholding details about their training data, model infrastructures, and training methods, the open source community quickly caught up. Currently, dozens of open-source AI models rival the effectiveness of closed-source models [21,23]. Despite many efforts in previous years, this has not been the case with web search. The few alternative search engines that have emerged do not share their data, index, or other details.

We introduce the Open Web Index, the first collaborative and federated data structure for crawling, enriching, and indexing the Web, distributed across multiple European data centers. The Open Web Index is inspired by Lewandowski's idea of an "open web index" [13] and corresponding core principles [7]. However, we not only provide access via an API, but treat the entire index and associated data products as open data. Furthermore, we enrich the crawled web content with a variety of metadata that can in turn drive *vertical* search engines. By publishing the index itself, we enable a new landscape of search engines, where each vertical can target different audiences based on tailored ranking strategies that meet their respective values (e.g., sustainability or privacy). In addition, the Open Web Index enables the training of specialized language models on different subsets of the Web. Our goal is to gain traction in the information retrieval and open source communities, allowing interested parties to contribute to the Open Web Index. This may include new content analysis modules for the preprocessing pipeline and evaluation components for the open evaluation framework.

While the development of the index is part of the ongoing Open Web Search research project[1], the crawling, preprocessing, and indexing pipelines already run on two European data centers producing ca. 1TB of data per day and location. The current activities focus on three main areas: (1) conducting further research on the analysis of large web collections to expand the metadata provided in the index files, (2) implementation of the three major pipeline steps and onboarding of additional data centers, and (3) fostering the open source and research community around the Open Web Index.

2 Related Work

The idea of collecting data on a massive scale for various purposes is not new. Several projects in the past have engaged in such endeavors. Most notably, the Common Crawl project[2] stands out as a significant effort in this direction. The Common Crawl initiative collects data from the Web at large scale and makes it accessible to the public. Several derivative projects, such as C4 [20], the Pile [4], of Web Data Commons [18], have built upon the resources provided by Common Crawl, indicating its importance and far-reaching benefits in the community.

1 https://openwebsearch.eu/.
2 https://commoncrawl.org/.

Another related project, LAION,[3] takes a similar approach. The non-profit organization provides data sets, tools and models in order to strengthen open source machine learning research. Furthermore, Curlie,[4] which was previously known as the Open Directory Project (ODP) and DMOZ, offers a manually curated directory of the Web. By fully relying on the power of human editors, Curlie stands out due to its elaborate and qualitatively advanced approach to data categorization and authenticity.

However, our efforts are not only focused on data collection, but also on web search applications based on the crawled and enriched data. Aside from the few major search engines (among others, Google, Bing and Yandex), several alternative search engines have emerged over the years from which, in particular, DuckDuckGo, Ecosia, and Startpage.com became popular. They do not operate their own crawling and indexing infrastructure, but make use of a search API offered by Bing or Google. Other search engines, such as Mojeek and Qwant, have tried to present themselves as viable alternatives to the large commercial search engines by building their own index. Both providers are particularly committed to preserving their neutrality and user privacy. Although they have not been able to create and share their indexes with the public, their efforts underscore the need and necessity for more players in the search engine market. Yet, both Mojeek's[5] and Qwant's[6] index is only a fraction as large as that of the market leader Google. Qwant currently still relies on Bing's index to supplement its own index. We believe that a collaborative crawling and indexing effort can help make the Open Web Index a good alternative to the current gatekeepers' search indexes—both in terms of scope and quality.

Recent trends have highlighted the importance of personalized search experiences, where search engines strive to understand user intent and context to provide more tailored results. Additionally, the deployment of natural language understanding and conversational agents in search engines has transformed the way users interact with online information. However, the development of specialized search engines depends on the availability of specially curated data. The Open Web Index aims to support these needs and will offer search engine developers different types of curated indexes in different sizes.

In addition to the technical aspects, ethical considerations and responsible data handling are critical aspects that are increasingly becoming the focus of data collection initiatives. Recent regulations and discussions surrounding data privacy, such as the General Data Protection Regulation (GDPR) in Europe and the California Consumer Privacy Act (CCPA), reflect the global shift towards safeguarding individual privacy. In our initiative, we intend to respect all aspects of data privacy and ethics while crawling, storing and distributing web data.

---

[3] https://laion.ai/.
[4] https://curlie.org.
[5] https://blog.mojeek.com/2022/03/five-billion-pages.html.
[6] https://betterweb.qwant.com/en/2023/09/18/web-indexing-where-is-qwants-independence/.

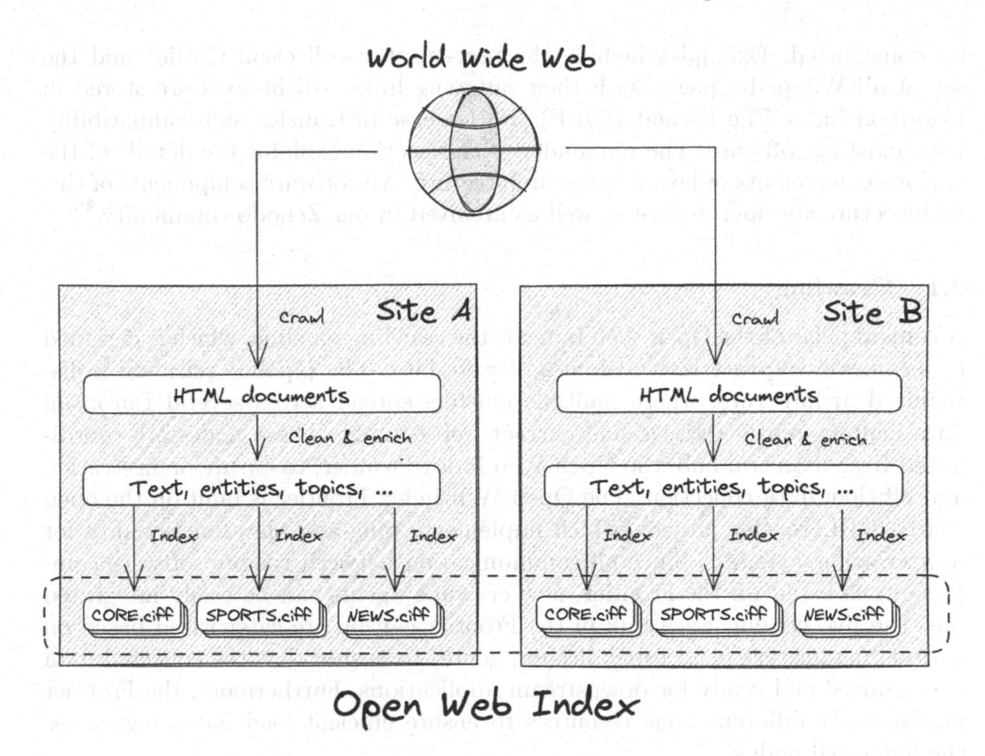

Fig. 1. Distributed pipeline architecture of the Open Web Index. Each data center is responsible for crawling, cleaning, enriching, and indexing its own subset of the web. Which documents are crawled by which data center is decided by the common frontier service. The definition and creation of index verticals and the choice of which enrichments to include are still open research questions.

3 Infrastructure of the Open Web Index

Figure 1 shows an overview of the Open Web Index architecture. Particularly noticeable are the improvements in accessing current search indexes compared to Lewandowski's [13] API-based proposal. Our federated data infrastructure crawls, enriches, and indexes web content in a distributed manner across multiple European data centers. The resulting indexes are divided into a set of predefined, possibly overlapping, verticals and are continuously updated over time. In addition to these vertical indexes, we also create "core indexes" that contain subsets of highly frequented or otherwise important websites. The underlying rationale is that a relatively small subset of the Web can already answer a large majority of queries. For example, Goel et al. [5] have shown that a subset of only 10,000 domains was responsible for about 80% of users' clicks in 2010, and, more recently, the creation of the ClueWeb22 corpus [19] confirms that a small number of domains still accounts for a large share of user clicks. In addition to the core index of highly frequented domains, other, use-case-specific indexes can be constructed. Examples include the user-curated collection Curlie[7] and the set of all Wikipedia pages with their outgoing links. All indexes are stored in Common Index File Format (CIFF) [15] for ease of transfer and compatibility with existing software. The remainder of this section explains the details of the various components or layers of our architecture. All software components of this architecture are open source as well as archived in our Zenodo community.[8][9]

3.1 Crawling

A central pillar of the Open Web Index is the crawling pipeline, which is designed to seamlessly capture large volumes of web data. The pipeline relies on a distributed architecture, where multiple clusters spread across several European data centers jointly retrieve and extract web content. These nodes are coordinated from a central hub, the Open Web Index Frontier, to ensure orchestration and efficient data collection. The Open Web Index Frontier is built on the open source URLFrontier project, which implements plug-and-play functionality for new crawling agents.[10] Since all communication between remote software components is based on the Frontier, new crawling agents can be easily integrated into the live system. Each link in the Frontier can be pre-categorized based on various parameters (e.g., topic, license, genre) to ensure that the collected data is organized and ready for downstream applications. Furthermore, the Frontier manages the different node resources to ensure efficient load balancing across the federated nodes.

To operationalize our crawling pipeline, we introduce a re-crawl mechanism specifically for Common Crawl dumps. Unlike Common Crawl, we crawl on a daily basis and save the results as WARC (Web ARChive) files.[11] This approach allows us to provide index delta files and supplemental data products at daily intervals, as opposed to monthly or bi-monthly data dumps. In addition, each link goes through a rigorous filtering process supported by an exclusion list mechanism. This list is constantly updated with malicious URLs, similar to platforms such as URLHaus.[12] This strategy ensures the highest quality of data.

By prioritizing certain top-level domains in our crawl exploration as well as using content negotiation headers and the IP locations of the various European data centers from which the crawls are conducted, we aim to collect web resources in many different languages. The goal is to increase the language diversity in the Open Web Index and thereby reduce the existing linguistic bias in downstream tasks towards high-resource languages such as English [12].

Our initial crawling runs across all our data centers resulted in a cumulative crawling rate of over 30 million web pages per day. Currently, our comprehensive crawling generates more than 1 terabyte of data daily.

[7] https://curlie.org.
[8] https://opencode.it4i.eu/openwebsearcheu-public/.
[9] https://zenodo.org/communities/owseu/.
[10] http://urlfrontier.net.
[11] https://iipc.github.io/warc-specifications/.
[12] https://urlhaus.abuse.ch/.

Crawling on Demand. Given the significant cost and resources required for crawling or scraping the Web, we introduce "crawling on demand". This allows authorized users to initiate a customized crawling process tailored to their specific needs. Users provide a curated list of seed URLs. Upon receiving this list, our system starts a dockerized crawl cluster tailored for the task at hand. Communicating with the Frontier service, the crawler detects whether a URL has already been crawled. If it has, its content is retrieved without unnecessary re-crawling, ensuring efficient resource usage. Users receive a daily update of WARC files corresponding to their seed URLs, ensuring access to the freshest and most relevant data. In addition, we are nearing completion of the addition of "index on demand" to this service. This service not only initiates the cleaning and enrichment of the crawled documents, but also delivers the data as index files for immediate integration into search engines.

Web Publisher Controls. As part of our crawling, we aim to improve the usage control facilities for web publishers. With the advent of generative AI applications, many content creators are looking for ways to better protect their publicly available web assets and opt out of their use for text and data mining or AI training. These opt-out signals are conveyed in various forms, including the robots exclusion protocol ("robots.txt") [11], meta tags such as "`noml`",[13] and emerging web standards such as the TDM Reservation Protocol.[14] We evaluate and incorporate these machine-readable signals in our crawlers, propagating publisher usage preferences to downstream users of the Open Web Index. Additionally, we adhere to established politeness policies, such as crawling intervals communicated by site managers through robots.txt.

3.2 Preprocessing

The preprocessing pipeline extracts various types of page-level metadata in addition to the cleaned text from the WARC files created during daily crawling activities. The extracted data is provided in Apache Parquet file format as part of the Open Web Index and can be used to enrich index files, as described in more detail in Subsect. 4.1.

Following the daily crawling activities, the preprocessing is executed as daily batch jobs at the different data centers storing the WARC files. First, an Apache Spark cluster is created on the respective HPC cluster using the Magpie script collection.[15] Then, the preprocessing job is submitted to the newly created Spark cluster and the extracted metadata is saved as Parquet files. Currently, the preprocessing pipeline extracts the plain text from the HTML code of each page, as well as various information from the WARC and HTTP headers and URL components. In addition, two types of metadata are created to enable partitioning of the index files: the language of the document and, if available, a label

[13] https://noml.info.
[14] https://www.w3.org/2022/tdmrep/.
[15] https://github.com/LLNL/magpie/.

for the domain based on the labels collected by the Curlie community [16]. We plan to incorporate more metadata into the preprocessing pipeline throughout the duration of the Open Web Search project.

Evaluation Benchmarks. Since the preprocessing pipeline is built on a modular architecture, it allows the integration of content analysis modules developed by third parties. This will help expand the amount of metadata extracted from crawls with the help of the open-source community. To ensure both sufficient quality and throughput of content analysis modules, new modules will be evaluated in a dedicated evaluation layer. This layer runs on the TIRA framework [2,3], which provides the means for evaluation as a service focused on information retrieval research. TIRA can host shared tasks on a given research problem and executes submitted software in sandbox machines without internet access to improve reproducibility.

Each new candidate module is evaluated against problem-specific benchmarking data, such as a set of labeled data for classification tasks. For tasks not previously included in the preprocessing pipeline, benchmarking data must be provided by the party developing the module. We also work on increasing the number of benchmarking datasets and submitted modules for a given task to support the development of high quality content analysis modules.

3.3 Indexing

The indexer takes the cleaned text from the preprocessing pipeline and converts it to a full-text index. The index is partitioned into a series of shards using various metadata values. Currently, each combination of top-level range, language, and Curlie topic is assigned to a different shard. However, determining the optimal metadata set for partitioning the index is still an open research question, and we will evaluate different approaches during the development of Open Web Index.

Each shard is a separate CIFF index [15] and can be easily downloaded along with the Parquet files containing the relevant metadata and clean text. A downstream search engine or end user can select any combination of these shards to create a custom vertical search application—public, commercial, or personal.

Similar to the preprocessing pipeline, indexing is performed as a daily Spark batch job that runs after all content for a day has been preprocessed. Magpie is used to provision the Spark cluster within an HPC allocation, after which the indexer is executed.

3.4 Challenges

While we have elaborately discussed the importance and necessity of an Open Web Index, there still exist crucial challenges regarding our proposed federated infrastructure. A major challenge to consider is the sustainability of the Open Web Index in the long term. The Open Web Search project is currently nurtured by a diverse team representing various institutes and countries. However, such a publicly funded project is inherently limited in both time and resources.

Within the project consortium, we are already discussing ways in which we can ensure the index's sustainability in the future. Identifying responsible parties or entities tasked with maintaining the index in the long run is an important point in these discussions. As part of the sustainability of the Open Web Index, we also wish to integrate the open source and open data communities into its development, and are discussing how this effort can be coordinated. The steps we take in these directions are crucial to guaranteeing the ongoing relevance and enduring presence of the Open Web Index beyond its initial phases.

Our federated infrastructure comes with additional challenges that any technical infrastructure has to tackle, such as security, hardware/software/service management, and agreements with end users through usage policies and service level agreements. These become especially difficult for a public project with limited resources, where no dedicated teams are in place to manage these issues. We are discussing and handling these challenges for our current infrastructure, and will consider ways in which they can be handled in the long term as well.

4 Usage of the Open Web Index

By publishing a comprehensive web index, we are supporting the information retrieval community in a very tangible way. The following section focuses on the data products provided as part of the Open Web Index. To further enhance the user experience, we also introduce an advanced concept to make parts of the index available to future developers of vertical search applications.

4.1 Data Products

The Open Web Index consists of a set of data products that can be used by downstream search engines or other data-intensive applications. Depending on ethical considerations and to the extent legally possible, we make this data available for public download. Below, we discuss the different types of data outputs we plan to generate, how they are created, and what uses they may have.

Index Files. The first and most important goal of the Open Web Index is to enable downstream (vertical) search engines. To achieve this, we periodically distribute the index files of all crawled and cleaned content and provide the inverted files in CIFF format [15]. Several search engines, including JASSv2, Lucene (and thus Anserini and Solr), PISA, OldDog, and Terrier, already support the import of CIFF indexes. This makes it easy to develop a vertical search engine based on the Open Web Index with mature software and minimal effort.

In recent years, the use of dense or sparse embeddings for ranking has become very important in the information retrieval community. Therefore, an interesting avenue for the Open Web Index is the inclusion of embedding-based indexes alongside the term-based CIFF index that we already offer. This option will further facilitate the creation of downstream search engines by eliminating the need to re-compute the embeddings every time, contributing to "Green IR" [22]. However, given the size of the web, we need to ensure that the embeddings are useful and can be computed efficiently before running comprehensive embedding models for the entire collection. In our future work on this problem, we will consider both dense [10] and sparse embeddings [1].

Clean Text. A search engine also needs access to the cleaned text of the indexed documents in order to present the search results correctly. In classical search engines with a results page containing "ten blue links", the cleaned text is used to generate informative snippets that are relevant to the query. In conversational web search systems that apply retrieval-augmented generation [14] to summarize search results in natural language, the document content is needed to generate the full-text response to a query. To support such use cases, we plan to provide the plain text (along with document-level metadata) in the form of Parquet files.

Another important opportunity arising from the availability of clean text is the ability to (pre-)train large open source language models. Currently, this is typically done with public datasets, of which Common Crawl (including derived datasets such as C4 [20] and the Pile [4]) are the largest. Our cleaned text could provide an alternative to the Common Crawl's WET (Web Extracted Text) files. In addition, our enriched content (described in more detail below) will allow us to select subsets of web content for training smaller, more focused language models. For example, filtering German web pages could prove useful for training German language models, and focusing on scientific data could result in a language model that can be used more effectively in the scientific domain.

Structured Information. The web is full of structured information that can be used to build knowledge graphs and support many downstream applications. Similar to Web Data Commons [18] (derived from Common Crawl), we plan to extract and share entities from Schema.org [8] from various semantic annotations embedded in web pages. Furthermore, the textual content of the page may contain additional entities that are not included in machine-readable markup, but are still useful for various use cases. For example, the text could contain names, dates, or location information. To extract such mentions, we include the REBL Batch Entity Linker [9,24] in our preprocessing pipeline. In addition, we implement and apply geoparsing tools that allow us to extract place names and place references and map them to physical geographic locations. Since there are currently no end-to-end geoparsers that can perform this task on the Web, we are using tools like Geoparsepy for inspiration [17].

Page-Level Metadata. For the development of vertical search engines, we would like to assign web documents to meaningful index domains. To achieve this, we will collect different types of page-level features that can be useful for index partitioning, such as language, topic, and genre.

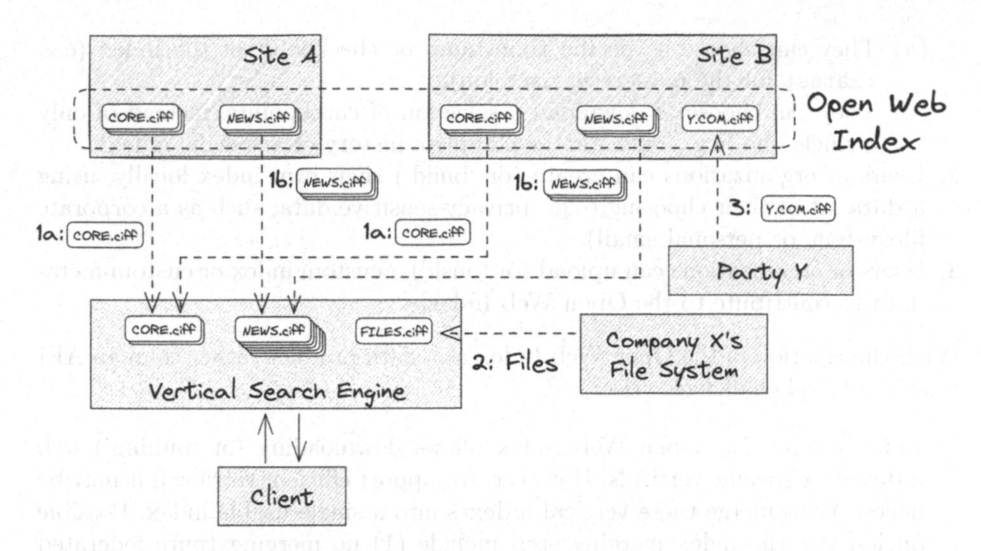

Fig. 2. Interaction between a search engine and the Open Web Index. Downstream search engines that rely on the Open Web Index can (1) retrieve prebuilt indexes, (2) create their own indexes, and (3) push indexes for sharing with others.

In addition, we will work on the development of an information nutrition label that includes benchmarks for the readability, factuality, and other aspects of any web document [6]. This will allow us to make judgments about the quality and trustworthiness of documents in the Open Web Index. In turn, this could enable search engine developers to create better ranking models, but more importantly, it allows end users to make an informed decision about which documents to access given a list of search results.

To empower developers and end users, we will develop classifiers to identify textual content that may be upsetting or otherwise disturbing to certain audiences. Adding such trigger warnings as page-level metadata allows search engine providers to display them alongside search results, empowering end users to avoid potentially harmful content. The necessary components in our pre-processing pipeline are built on existing research on trigger warning classification [25].

4.2 Interaction with the Open Web Index

Since the Open Web Index is intended to be used by a variety of search engines, it must also be easy for search engine developers to use the index (or parts of it) for their own purposes. To this end, we envision the Open Web Index becoming a (distributed) information system that can be used in a manner similar to the well-known Docker Hub. However, instead of container images, the Open Web Index contains prebuilt indexes that are immediately usable. The Open Web Index enables several downstream applications, which are depicted in Fig. 2:

1. Users or organizations can download (or 'pull') a specific, pre-built index.

   (a) They can choose a specific timestamp or checkpoint of the index (e.g. `:latest` for the most recent version).
   (b) They can choose to download a selection of checkpoints, instead of only a single one (e.g., `:all` for the complete history of a specific index).
2. Users or organizations can create (or 'build') their own index locally, using a data set of their choosing (e.g., privacy-sensitive data, such as a corporate filesystem, or personal email).
3. Users or organizations can upload (or 'push') a custom index or custom metadata to contribute to the Open Web Index.

With the creation of the Open Web Index as a data product rather than an API service, several challenges arise:

1. *Index merge:* The Open Web Index allows downloading (or 'pulling') web indexes for specific verticals. However, to support efficient retrieval, it may be necessary to merge these vertical indexes into a single usable index. Possible options for the index merging step include (1) no merging (pure federated search), (2) client-side merging, (3) server-side merging, and (4) a hybrid merging approach. Further research is needed to investigate the tradeoffs between efficiency and usability for each of these methods in order to make a decision on a method that is useful in practice.
2. *Freshness:* The Open Web Index and the indexes retrieved by downstream search engines need to be updated regularly to ensure that search results remain accurate. To accomplish this, we could (a) update an index incrementally by marking documents as obsolete and using a smaller, separate index for those updated documents, or (b) rebuild and replace the entire index from time to time. We will explore which method is best to ensure timeliness and also how we can extend these processes to also ensure the timeliness of downstream search engines. The federated structure of the Open Web Index also allows us to apply different "freshness" policies to specific sub-indexes, e.g., updating the news index more frequently compared to more static indexes.
3. *Index curation:* Eventually, the question arises how to ensure the quality of the contributed search indexes. One possible quality assurance model is to provide a small number of official and manually reviewed indexes. Similar to common practice in software repository platforms, users could also be given the option to star high-quality indexes provided by the community.

During the development of the Open Web Index, we continue to explore these issues and try to find solutions that make using the Open Web Index as easy and user-friendly as possible.

5 Conclusion

Inspired by recent advances in open-source AI models and Lewandowski's idea of an open web index, we introduce the Open Web Index. The main goal is to facilitate the development of search applications (e.g., to ground retrieval-augmented generation systems or to create vertical search engines for specific domains) without having to rely on the APIs of one of the few web-scale search engines typically operated by large corporations. We make the Open Web Index compatible with existing software by using the CIFF format and, in addition to the inverted files, we provide page-level metadata to support developers in customizing the selection of documents that meet the needs of their specific application. By using our 'crawling on demand' feature, developers can simply specify a list of URLs in order to receive the corresponding WARC files from either our existing crawls or from specially created new crawl clusters.

The Open Web Index is open for contributions from others. The IR community, for instance, can contribute by developing additional preprocessing components, by creating new kinds of vertical search engines and test collections, or by providing resources.

Acknowledgments. This work has received funding from the European Union's Horizon Europe research and innovation program under grant agreement No 101070014 (OpenWebSearch.EU, https://doi.org/10.3030/101070014).

References

1. Formal, T., Piwowarski, B., Clinchant, S.: SPLADE: Sparse Lexical and Expansion Model for First Stage Ranking, SIGIR 2021, pp. 2288-2292. Association for Computing Machinery, New York (2021), ISBN 9781450380379
2. Fröbe, M., et al.: The Information Retrieval Experiment Platform. In: Proceedings of the 46th International ACM SIGIR Conference on Research and Development in Information Retrieval (2023)
3. Fröbe, M., et al.: Continuous integration for reproducible shared tasks with TIRA.io. In: Advances in Information Retrieval. 45th European Conference on IR Research (ECIR 2023). LNCS. Springer (2023). https://doi.org/10.1007/978-3-031-28241-6_20
4. Gao, L., et al.: The Pile: An 800GB Dataset of Diverse Text for Language Modeling (Dec 2020)
5. Goel, S., Broder, A.Z., Gabrilovich, E., Pang, B.: Anatomy of the long tail: ordinary people with extraordinary tastes. In: Davison, B.D., Suel, T., Craswell, N., Liu, B. (eds.) Proceedings of the Third International Conference on Web Search and Web Data Mining, WSDM 2010, 4-6 February 2010, pp. 201–210. ACM, New York (2010)
6. Gollub, T., Potthast, M., Stein, B.: Shaping the Information Nutrition Label. In: Albakour, D., Corney, D., Gonzalo, J., Martinez, M., Poblete, B., Valochas, A. (eds.) 2nd International Workshop on Recent Trends in News Information Retrieval (NewsIR 2018) at ECIR. CEUR Workshop Proceedings, vol. 2079, pp. 9–11 (Mar 2018), ISSN 1613-0073
7. Granitzer, M., Voigt, S., et al.: Impact and Development of an Open Web Index for Open Web Search. J. Assoc. Inform. Sci. Technol. (2023)
8. Guha, R.V., Brickley, D., MacBeth, S.: Schema.org: evolution of structured data on the web: big data makes common schemas even more necessary. Queue **13**(9), 10–37 (2015), ISSN 1542-7730

9. Kamphuis, C., Hasibi, F., Lin, J., de Vries, A.P.: REBL: entity linking at scale. In: Alonso, O., Baeza-Yates, R., King, T.H., Silvello, G. (eds.) Proceedings of the Third International Conference on Design of Experimental Search & Information Retrieval Systems, San Jose, CA, USA, 30-31 August 2022. CEUR Workshop Proceedings, vol. 3480, pp. 68–75. CEUR-WS.org (2022)
10. Khattab, O., Zaharia, M.: ColBERT: efficient and effective passage search via contextualized late interaction over BERT. In: Proceedings of the 43rd International ACM SIGIR Conference on Research and Development in Information Retrieval, SIGIR 2020, pp. 39-48. Association for Computing Machinery, New York (2020), ISBN 9781450380164
11. Koster, M., Illyes, G., Zeller, H., Sassman, L.: RFC 9309 Robots Exclusion Protocol (2022)
12. Kreutzer, J., et al.: Quality at a Glance: An Audit of Web-Crawled Multilingual Datasets (2021)
13. Lewandowski, D.: The web is missing an essential part of infrastructure: an open web index. Commun. ACM **62**(4), 24 (2019)
14. Li, H., Su, Y., Cai, D., Wang, Y., Liu, L.: A Survey on Retrieval-Augmented Text Generation. arXiv preprint arXiv:2202.01110 (2022)
15. Lin, J., et al.: Supporting interoperability between open-source search engines with the common index file format. In: Proceedings of the 43rd International ACM SIGIR Conference on Research and Development in Information Retrieval, pp. 2149–2152 (2020)
16. Lugeon, S., Piccardi, T.: Curlie Dataset - Language-agnostic Website Embedding and Classification (Jan 2023). https://doi.org/10.6084/m9.figshare.19406693.v5, https://figshare.com/articles/dataset/Curlie_Dataset_-_Language-agnostic_Website_Embedding_and_Classification/19406693
17. Middleton, S.E., Kordopatis-Zilos, G., Papadopoulos, S., Kompatsiaris, Y.: Location extraction from social media: geoparsing, location disambiguation, and geotagging. ACM Trans. Inform. Syst. (TOIS) **36**(4), 1–27 (2018)
18. Mühleisen, H., Bizer, C.: Web data commons - extracting structured data from two large web corpora. In: Bizer, C., Heath, T., Berners-Lee, T., Hausenblas, M. (eds.) WWW 2012 Workshop on Linked Data on the Web, Lyon, France, 16 April 2012. CEUR Workshop Proceedings, vol. 937. CEUR-WS.org (2012)
19. Overwijk, A., Xiong, C., Liu, X., VandenBerg, C., Callan, J.: ClueWeb22: 10 Billion Web Documents with Visual and Semantic Information (Dec 2022)
20. Raffel, C., et al.: Exploring the limits of transfer learning with a unified text-to-text transformer. J. Mach. Learn. Res. **21**, 140:1–140:67 (2020)
21. Scao, T.L., et al.: BLOOM: A 176B-Parameter Open-Access Multilingual Language Model. CoRR arXiv: 2211.05100 (2022)
22. Scells, H., Zhuang, S., Zuccon, G.: Reduce, reuse, recycle: green information retrieval research. In: Amigó, E., Castells, P., Gonzalo, J., Carterette, B., Culpepper, J.S., Kazai, G. (eds.) SIGIR 2022: The 45th International ACM SIGIR Conference on Research and Development in Information Retrieval, Madrid, Spain, 11 - 15 July 2022, pp. 2825–2837. ACM (2022)
23. Touvron, H., et al.: LLaMA: Open and Efficient Foundation Language Models. CoRR arXiv: 2302.13971 (2023)
24. van Hulst, J.M., Hasibi, F., Dercksen, K., Balog, K., de Vries, A.P.: REL: an entity linker standing on the shoulders of giants. In: Proceedings of the 43rd International ACM SIGIR Conference on Research and Development in Information Retrieval, pp. 2197–2200. ACM, Virtual Event China (Jul 2020), ISBN 978-1-4503-8016-4

25. Wiegmann, M., Wolska, M., Schröder, C., Borchardt, O., Stein, B., Potthast, M.: Trigger warning assignment as a multi-label document classification problem. In: Rogers, A., Boyd-Graber, J., Okazaki, N. (eds.) Proceedings of the 61th Annual Meeting of the Association for Computational Linguistics (Volume 1: Long Papers), pp. 12113–12134. Association for Computational Linguistics, Toronto, Canada (Jul 2023)

A Conversational Robot for Children's Access to a Cultural Heritage Multimedia Archive

Thomas Beelen(✉), Roeland Ordelman, Khiet P. Truong, Vanessa Evers, and Theo Huibers

University of Twente, Enschede, The Netherlands
{t.h.j.beelen,roeland.ordelman}@utwente.nl

Abstract. In this paper we introduce a conversational robot designed to assist children in searching a museum's cultural heritage video archive. The robot employs a form of Spoken Conversational Search to facilitate the clarification of children's interest (their information need) in specific videos from the archive. Children are typically insufficiently supported in this process by common search technologies such as search-bar and keyboard, or one-shot voice interfaces. We present our approach, which leverages a knowledge-graph representation of the museum's video archive to facilitate conversational search interactions and suggest content based on the interaction, in order to study information-seeking conversations with children. We plan to use the robot test-bed to investigate the effectiveness of conversational designs over one-shot voice interactions for clarifying children's information needs in a museum context.

1 Introduction

New technologies are bringing opportunities for improved facilities for information retrieval in GLAM institutions (Galleries, Libraries, Archives, and Museums) that maintain such large cultural heritage collections, and wish to make them accessible to a wide audience [10]. One such technology, conversational agents, enables access to large and varied heritage collections in a way that better suits various user groups that are interested in these collections. Our goal is to investigate how multi-turn interactions can be supported (see [25]), focusing on the use case of children visiting an archive or museum. We wish to present them with a conversational robot that they can communicate with in an engaging manner to find interesting videos from the archive.

We focus on children because they are an important audience for museums, and because improving their access to archive materials improves their opportunity for learning [2]. Unfortunately, children are an under-served audience because search technology is often aimed at adults. To provide them with an engaging and natural conversational partner that could support conversational interactions, we use a robot.

Children tend to need more support when formulating their queries [19], and find it challenging to include all relevant context into one statement [21,24]. A

N. Goharian et al. (Eds.): ECIR 2024, LNCS 14612, pp. 144–151, 2024.
https://doi.org/10.1007/978-3-031-56069-9_11

conversational robot can help a child communicate their information need by leveraging multiple conversational turns so that the wants and needs of children can be discovered and addressed. To assess this idea, we develop the conversational robot and plan user studies. In this document, we show how the implementation of a conversational robot could help us to accelerate research and development of conversational search technology for children.

2 Background

2.1 Cultural Heritage Access

GLAM institutions aim to engage audiences with cultural heritage materials and media [3,13]. In this case study, we take the video archive from the Netherlands Institute for Sound & Vision (NISV) as a starting point for children to explore using conversational search. The NISV archive contains (among other things) a large collection of videos which are largely annotated with a thesaurus [15]. Currently, access to this content is limited to traditional, keyword-based search environments targeted at scholars, journalists, and APIs [12]. To make the content more accessible to children, we propose a conversational agent. Literature suggests an interactive strategy may work better for children [2], and an agent (or robot) can leverage Spoken Conversational Search [25].

2.2 Query Formulation

During a search, the information need that one has, should be formulated in a query that represents it well. It has been suggested [5] that children's limited abstract thinking abilities [16], limited knowledge base [11], and vocabulary [14] hamper them in creating rich enough queries that maximize the potential of finding interesting search results. Vanderschantz & Hinze [19] studied children's query formulation in a classroom search task. They found children's queries are often indeed too broad. Furthermore, children see query formulation as a major challenge in search [20]. Children also make more use of query-assistance tools such as *suggestions* in search engines, indicating that they may require more support than adults [20,22].

2.3 Search Paradigms

As described, children may benefit from more active support from search technology to formulate complete and specific queries. In the museum archive context, a more specific query would be able to narrow down the potential results more strongly. Unfortunately, many commercially available voice assistants (VAs) do not provide much support. They often use a *query-response* interaction paradigm. In this style of interaction the agent awaits a single search input from a user and gives results based on this initial statement. In this interaction the child is not supported with query formulation, which may explain why children often do not find what they were looking for [9]. Yarosh et al. [24] showed how children struggled more with spoken query reformulations (adapting the query after unsatisfactory results were returned) than adults, reaffirming that the query-response paradigm may not suit children.

3 System Design

Our goal is to leverage metadata in a knowledge graph to develop a conversational, multi-turn interaction that is able to engage children to provide a richer description of their search need. The metadata can be used to suggest additional terms that keep the conversation going, and narrow the search down to a set of results small enough to evaluate individual titles.

The conversational robot we built has a modular architecture, meaning there are distinct, rule-based, modules with different objectives. This design allows for a predictable system that does not require training data, as opposed to end-to-end architectures [8]. In this way, the system would also be applicable for other data sets and archives. Our design is based on the *Dialogue-State Architecture* described by Williams et al. [23]. We first describe the design rationale for the conversational policy of the robot, and then outline how we implemented it.

3.1 Conversational Design

We base our conversational design on Trippas et al. [18], who modelled the spoken conversational search process based on observations of human dyads of *seekers*, and *intermediaries* over an audio-only channel. The dyads were studied while working on a search task. The dyads' statements were annotated to create a model of the conversational search interaction. The model has three levels, the *discourse*, *task*, and *other* level. The *task* level contains all the search related functions. The *discourse* level includes the capabilities of discourse management, communication about system status, grounding, and navigation. A few miscellaneous utterances were grouped under the *other* level. The model provides an extensive description of all different actions that can be taken in search conversations. Since we are interested in helping children with creating specific queries in multiple turns, we begin implementing the actions deemed most relevant to the problem. We describe these below.

Task Level Capabilities. To help children formulate richer queries, we look at intermediary (or robot) actions of *search assistance* and *information requests.* The most common actions in human-human interactions were the *query refinement offer*, *query-rephrase*, and *enquiry for further information.* Since we know children struggle to put context to their query in a single statement [24], we want to provide a multi-turn interaction. We think this reduces complexity compared to having to put everything in one statement. To initiate a multi-turn interaction we select the action of *request for more details about information request.* We think this is a prompt that invites children to add to their description after the initial turn. Furthermore, children can benefit from query suggestions [20,22].

In the conversational model, these are named *query refinement offers.* In these actions, the system can make suggestions on keywords to add to the query to get to more specific results. We use these two actions as a starting point to develop an agent that can be used to explore search interactions with children. In Sect. 4 we describe in more detail how these actions are designed and implemented.

Discourse Level Capabilities. The capabilities (from [18]) are geared towards maintaining a smooth conversation. We start off designing the robot's discourse functions to an extent we anticipate will support a minimum viable conversation. In correspondence with the recommendations of [24], the robot states what it heard in case it did not understand what the child was searching for. This action is categorised as *feedback on what is happening* in Trippas et al.'s model. We combine this action with another *request for more details*-action from the task level to keep the conversation going.

4 Implementation

We chose the Furhat robot and SDK to develop our agent because it is suitable for natural conversations [1]. As stated, we follow the *Dialogue-State Architecture* by Williams et al. [23] which describes a set of interconnected modules. Below we discuss the implementation of the modules.

Speech Recognition and Keyword extraction. We used the default Google ASR that is supplied with the Furhat robot. Phrases coming from the ASR are processed to extract keywords by using KeyBERT [4]. KeyBERT was developed to select keywords that represent a document, but we found that it performs relatively well on the task of selecting potential keywords from a phrase. We built a server with KeyBERT that waits for input and returns any keywords to the main program. The threshold for keywords was set at *0.4.* The model we used is *paraphrase-multilingual-MiniLM-L12-v2.* The potential keywords are sent to the inflection module. Before running KeyBERT extraction, a few exceptions are removed from the input string that are redundant. Dutch, the original language).

Lemmatization. This next module that gets called changes the inflection of the potential keywords. Once words are plural, they can be checked against the thesaurus faster. Words are checked whether they are plural and inflected to plural if they were not. For this step different tools and techniques have been tested. Eventually, we decided to use a mostly handcrafted system that is rule based. In Dutch plural words generally end with *"-s"* or *"-en"*, if this is not the case, simpleNLG-NL [6] is used to inflect the word.

Identifying Known Terms. Once potential keywords are selected from input, and changed to plural, they will be checked one by one against the terms in the thesaurus. This is done by an API call with a SPARQL query. The query finds whether the term is known and has any watchable videos associated with it.

The query was developed and tested by using a Communica environment [17], and implemented in the robot afterwards. Matched terms return a GTAA [15], which is a unique identifier. The GTAA identifiers are stored in the current list of keywords.

Retrieval Module. The robot can find videos that are related to the currently stored topics using another SPARQL query. This query is built up dynamically to ensure it works for one or more keywords.

Result Presentation. A tablet is used to display retrieved videos. We built a graphical user interface using the Furhat GUI tools that is hosted on a computer. The GUI is opened on the tablet (as seen in Fig. 1) and waits until it receives a video link event from the main program.

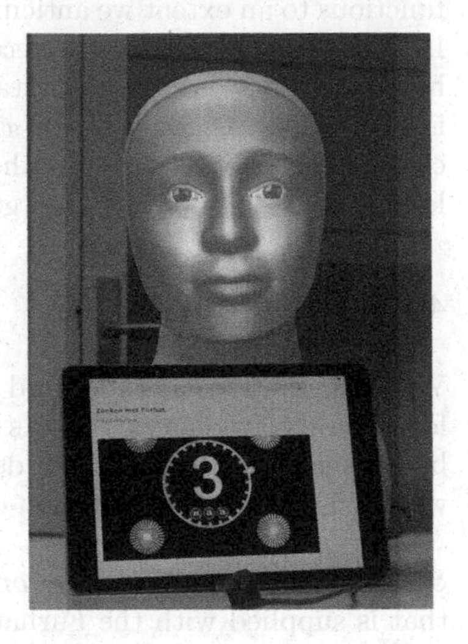

Fig. 1. The Furhat robot with tablet for displaying archive videos.

Suggestion Module. As described in 3.1, the robot will make *query refinement offers*; it will suggest additional keywords to the user. To ease query expansion, children can choose from provided keywords rather than having to come up with their own. This action is triggered in case a new keyword has not been found for more than one turn. The suggestion is based on the current list of keywords in the state. These keywords have an associated list of videos that match. The additional keywords that are attached to those videos are gathered in a set and sorted by frequency. The robot suggests three keywords, if none are picked by the child, it will present the next three. The way suggestions are constructed is also shown in Fig. 2.

State Tracker. The state tracking implementation keeps the currently captured keywords, the set of keywords of last turn, and the set of videos that the current keywords yield. Furthermore, data is kept about suggestions, such as a counter to time suggestions, previously suggested keywords, etc. The dialog manager can decide what to do based on the current state.

Dialog Management. The dialog manager determines the next action based on the current state. The tracking is based on a recursive function that asks input from the child to progress. This could be e.g. asking whether the child wants to follow a suggestion, asking for more topics to add to the search, or asking which of the results the child wants to see.

Logging. The robot logs interactions to JSON files for later analysis.

5 Evaluation

The robot we describe above has been used in a study (to be published), where children were asked to find videos using the robot. The forthcoming work focuses on the information need descriptions and user experience of the two conditions. From an implementation viewpoint, we also noted potential areas of improvement for further implementations. Firstly, the ASR module struggled to consistently, correctly interpret input (in correspondence with [7]). Secondly, problems occasionally arose with the simple rule-based systems for lemmatization and matching, especially when children employed synonyms. This, in turn, reduced the rate of successfully extracted search terms. This pipeline may need to be replaced by a more robust system to reduce breakdowns in the conversations. Finally, the suggestion module correctly provided children with possible additional keywords, and *state tracking* and *dialog management* worked adequately in keeping track of the conversation and deciding upon next actions.

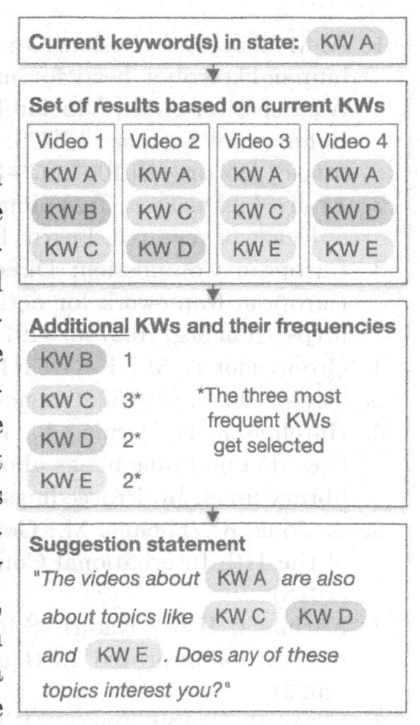

Fig. 2. An illustration of how the suggestion module works. *KW* represents a keyword.

6 Discussion and Future Work

As we outlined, children could benefit from support to create more specific queries during spoken search. We described the design of a robot that is aimed at providing support via multi-turn conversation with prompts for more keywords, as well as suggestions. We described how we implemented this design in a robot connected to a museum archive. This robot will be used for studies with children through which we can learn more about supporting children during search. Future work involves a study where children worked on search tasks with the robot, where we investigated whether conversation can aid children with their query formulations, as well as provide a good user experience. In this document we highlighted some early impressions from this study about the performance of the different robot modules. We also plan to run additional studies to optimize the design of suggestions, and to develop a novel approach for evaluating search conversations with children. By sharing this work, we contribute to the growing body of research on children's information retrieval.

References

1. Al Moubayed, S., Beskow, J., Skantze, G., Granström, B.: Furhat: a back-projected human-like robot head for multiparty human-machine interaction. In: Esposito, A., Esposito, A.M., Vinciarelli, A., Hoffmann, R., Müller, V.C. (eds.) Cognitive Behavioural Systems. LNCS, vol. 7403, pp. 114–130. Springer, Heidelberg (2012). https://doi.org/10.1007/978-3-642-34584-5_9
2. Andre, L., Durksen, T., Volman, M.L.: Museums as avenues of learning for children: a decade of research. Learn. Environ. Res. **20**(1), 47–76 (2017)
3. European Commission: Directorate-General for Education, Youth, S., Culture: European framework for action on cultural heritage. Publications Office (2019). https://doi.org/10.2766/949707
4. Grootendorst, M.: KeyBERT: minimal keyword extraction with BERT (2020). https://doi.org/10.5281/zenodo.4461265
5. Hutchinson, H., Druin, A., Bederson, B.B., Reuter, K., Rose, A., Weeks, A.C.: How do I find blue books about dogs? The errors and frustrations of young digital library users. In: Proceedings of HCII 2005, pp. 22–27. Citeseer (2005)
6. de Jong, R., Theune, M.: Going Dutch: creating simpleNLG-NL. In: Proceedings of the 11th International Conference on Natural Language Generation, pp. 73–78 (2018)
7. Kathania, H.K., Kadiri, S.R., Alku, P., Kurimo, M.: A formant modification method for improved ASR of children's speech. Speech Commun. **136**, 98–106 (2022)
8. Lison, P., Kennington, C.: Who's in charge? Roles and responsibilities of decision-making components in conversational robots. In: HRI 2023 workshop Human-Robot Conversational Interaction (2023). https://doi.org/10.48550/ARXIV.2303.08470, https://arxiv.org/abs/2303.08470
9. Lovato, S.B., Piper, A.M., Wartella, E.A.: Hey Google, do unicorns exist? Conversational agents as a path to answers to children's questions. In: Proceedings of the 18th ACM International Conference on Interaction Design and Children, pp. 301–313. IDC 2019. Association for Computing Machinery, Boise, ID, USA (2019). https://doi.org/10.1145/3311927.3323150
10. Machidon, O.M., Tavčar, A., Gams, M., Duguleană, M.: CulturalERICA: a conversational agent improving the exploration of European cultural heritage. J. Cultural Heritage **41**, 152–165 (2020)
11. Moore, P.A., St George, A.: Children as information seekers: the cognitive demands of books and library systems. Sch. Libr. Media Q. **19**(3), 161–68 (1991)
12. Ordelman, R., Melgar, L., Van Gorp, J., Noordegraaf, J.: Media suite: unlocking audiovisual archives for mixed Media scholarly research. In: Selected papers from the CLARIN Annual Conference, vol. 159, pp. 133–143 (2018)
13. Ordelman, R., et al.: Data stories in CLARIAH: developing a research infrastructure for storytelling with heritage and culture data. In: DARIAH Annual Event 2022 (2022)
14. van der Sluis, F., van Dijk, B.: A closer look at children's information retrieval usage. In: 33st Annual International ACM SIGIR Conference on Research and Development in Information Retrieval (SIGIR 2010) (2010)
15. Netherlands Institute for Sound and Vision: Gemeenschappelijke thesaurus audiovisuele archieven (30-10-2023). https://www.beeldengeluid.nl/kennis/kennisthemas/metadata/gemeenschappelijke-thesaurus-audiovisuele-archieven

16. Spavold, J.: The child as naïve user: a study of database use with young children. Int. J. Man-Mach. Stud. **32**(6), 603–625 (1990)
17. Taelman, R., Van Herwegen, J., Vander Sande, M., Verborgh, R.: Comunica: a modular sparql query engine for the web. In: Proceedings of the 17th International Semantic Web Conference (2018). https://comunica.github.io/Article-ISWC2018-Resource/
18. Trippas, J.R., Spina, D., Thomas, P., Sanderson, M., Joho, H., Cavedon, L.: Towards a model for spoken conversational search. Inform. Process. Manage. **57**(2), 102162 (2020)
19. Vanderschantz, N., Hinze, A.: Do internet search engines support children's search query construction: a visual analysis (2017)
20. Vanderschantz, N., Hinze, A.: Children's query formulation and search result exploration. Int. J. Digit. Libr. **22**(4), 385–410 (2021)
21. Vanderschantz, N., Hinze, A., Cunningham, S.J.: "Sometimes the internet reads the question wrong" children's search strategies & difficulties: "Sometimes the Internet reads the question wrong" children's search strategies & difficulties. Proceedings of the American Society for Information Science and Technology **51**(1), 1–10 (2014). https://doi.org/10.1002/meet.2014.14505101053, http://doi.wiley.com/10.1002/meet.2014.14505101053
22. Weber, I., Jaimes, A.: Who uses web search for what: and how. In: Proceedings of the Fourth ACM International Conference on Web Search and Data Mining, pp. 15–24. WSDM 2011, Association for Computing Machinery, New York (2011). https://doi.org/10.1145/1935826.1935839, https://dl.acm.org/doi/10.1145/1935826.1935839
23. Williams, J.D., Raux, A., Henderson, M.: The dialog state tracking challenge series: a review. Dialogue Discourse **7**(3), 4–33 (2016)
24. Yarosh, S., et al.: Children asking questions: speech interface reformulations and personification preferences. In: Proceedings of the 17th ACM Conference on Interaction Design and Children, pp. 300–312. IDC 2018. Association for Computing Machinery, Trondheim, Norway (2018). https://doi.org/10.1145/3202185.3202207
25. Zamani, H., Trippas, J.R., Dalton, J., Radlinski, F.: Conversational Information Seeking. arXiv preprint arXiv:2201.08808 (2022)

Towards Robust Expert Finding in Community Question Answering Platforms

Maddalena Amendola[1,2](✉), Andrea Passarella[2], and Raffaele Perego[3]

[1] University of Pisa, Pisa, Italy
maddalena.amendola@phd.unipi.it
[2] IIT-CNR, Pisa, Italy
[3] ISTI-CNR, Pisa, Italy

Abstract. This paper introduces TUEF, a topic-oriented user-interaction model for fair Expert Finding in Community Question Answering (CQA) platforms. The Expert Finding task in CQA platforms involves identifying proficient users capable of providing accurate answers to questions from the community. To this aim, TUEF improves the robustness and credibility of the CQA platform through a more precise Expert Finding component. The key idea of TUEF is to exploit diverse types of information, specifically, content and social information, to identify more precisely experts thus improving the robustness of the task. We assess TUEF through reproducible experiments conducted on a large-scale dataset from StackOverflow. The results consistently demonstrate that TUEF outperforms state-of-the-art competitors while promoting transparent expert identification.

Keywords: Expert Finding · Community Question & Answering

1 Introduction

The Expert Finding (EF) task in Community Question Answering (CQA) platforms aims to identify and recognize community users with a high level of expertise. Newly posted questions are forwarded to them, reducing the waiting time for answers. EF is crucial for enhancing the platform's overall quality, as it determines, by and large, the credibility and trustworthiness of the platform. Precise identification of experts is a key feature in guaranteeing high-quality answers and the credibility of the CQA platform. Given a question posted by a user of the online community, the EF task can be cast directly to the task of *computing a short, ranked list of community members, i.e., experts, that are likely to provide an accurate answer to the question.*

Numerous studies have suggested solutions for the EF task that rely solely on textual information. Nevertheless, CQA platforms provide information through which an implicit social network among users can be defined and exploited to identify experts. Exploiting multiple sources of information (such as content and

N. Goharian et al. (Eds.): ECIR 2024, LNCS 14612, pp. 152–168, 2024.
https://doi.org/10.1007/978-3-031-56069-9_12

social relationships) is also a way to make the EF task more robust, as algorithms can be built on complementary and largely orthogonal sources of information.

Based on these considerations, this paper proposes TUEF, a *Topic-oriented, User-interaction model for EF.* TUEF jointly leverages *content* and *social* information available in the CQA by defining a topic-based Multi-Layer graph (MLG) that represents CQA users' relationships based on their similarities in providing answers. TUEF stands out for its integrated approach, blending social network analysis with information retrieval techniques. One of its notable features is its transparency. Unlike deep neural network approaches that operate as black boxes, TUEF's approach ensures a transparent expert selection process: the method's graph exploration begins from key nodes, ensuring a straightforward selection process for candidate experts. Further enhancing its performance, TUEF employs LambdaMart [4], a decision tree-based learning-to-rank method. This not only refines the ranking of experts but also showcases the contribution of the various features to the decision process, allowing for insights into the importance of various factors in the ranking outcome. Figure 1 outlines the main building blocks of the proposed solution. TUEF first generates an MLG, where each layer corresponds to one of the main topics discussed in the community. In each layer, the nodes represent users actively participating in topic discussions, while edges model similarities and relationships among users under specific topics. At inference time, given a question q, the MLG G, and a ranking model r, TUEF first determines the main topics to which the question q belongs and the corresponding graph layers. Next, for each layer, it selects the candidate experts from two perspectives: i) *Network*, by identifying central users that may have considerable influence within the community; ii) *Content*, by identifying users who previously answered questions similar to q. In both cases, the graph is used to collect candidate experts through appropriate exploration policies. Then, TUEF extracts features based on text, tags, and graph relationships for the selected experts. Finally, it uses a learned, precision-oriented model r to score the candidates and rank them by expected relevance.

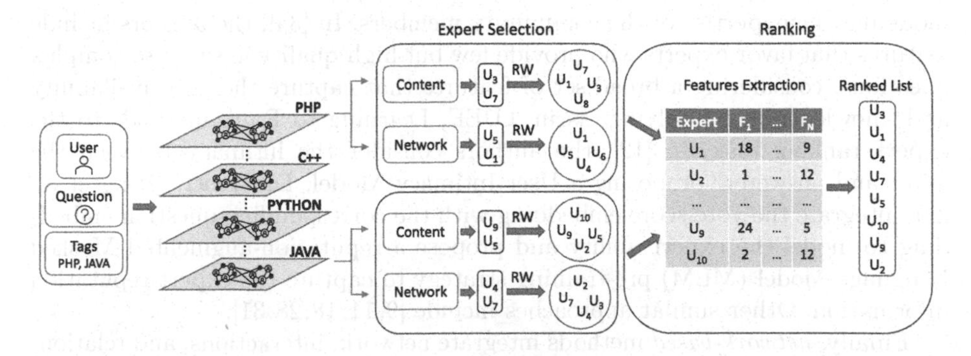

Fig. 1. TUEF approach representation highlighting the distinct components.

The different components of TUEF sketched in Fig. 1 are presented in detail in Sect. 3. We evaluate the performance of TUEF in terms of ranking quality metrics by comparing it with state-of-the-art proposals, as well as with reference benchmarks which only consider part of the information included in the TUEF model (Sects. 4 and 5). The reproducible experiments conducted are based on a large-scale dataset available from StackOverflow[1], the largest community within the Stack Exchange network. Before detailing the methodology adopted for TUEF, we discuss related work in Sect. 2, while in Sect. 6 we draw the main conclusion of our study.

The experimental results show that TUEF consistently outperforms the competitors and effectively exploits the heterogeneous sources of information modeling users and questions. Specifically, TUEF outperforms the best-performing competitor by 65.06% for Mean Reciprocal Rank and 112% for P@1.

2 Related Work

Research proposals in the field of the EF task for CQA platforms can be grouped into three broad groups: text-based, feature-based, and network-based methods. *Text-based* methods address the EF task by relying on the similarities between the current and previously answered questions. In [35], the authors propose a temporal context-aware representation learning model, which models temporal dynamics by multi-shift and multi-resolution settings. The model learns the expert representation in the context of a question's semantic and temporal information using a pre-trained deep bidirectional transformer [8]. In [13], the authors define a new approach based on the Recurrent Memory Reasoning Network (RMRN), composed of different reasoning memory cells that implement attention mechanisms to focus on different aspects of the question. Recently, Liu et al. [24] proposed a Non-sampling Expert Finding model that could learn accurate representations of questions and experts from whole training data. Other similar approaches include [6,7,23]

Feature-based methods methods are based on a set of hand-crafted features modeling the expertise of the community members. In [34], the authors include features that favor experts who provide few but high-quality answers to complex questions, considering a broad set of features that capture the user availability and knowledge and applying, as in TUEF, Learning to Rank methods to the expert ranking task. In [12], the authors consider the intimacy between the asker and answerer, proposing a User Intimacy Model. Differently, Peng et al. [29] integrate the vote score embedding with the corresponding question embedding to model the expert ability and propose a reputation-augmented Masked Language Model (MLM) pre-training strategy to capture the expert reputation information. Other similar approaches include [9,11,18,28,31].

Finally, *network-based* methods integrate network, interactions, and relationship information. Kundu et al. [17] define a framework that includes a text-based component, which estimates an expert knowledge of a topic, and a Competition

[1] https://stackoverflow.com.

Based Expertise Network [1] that exploits link analysis techniques [21,26]. Differently, Sun et al. [33] follow the idea that a user's expertise is language-agnostic. The authors build a competition graph containing users and question nodes, with the edges representing relationships among them, capable of encoding the hierarchical structure of questions. Other approaches in this class are presented in [19,20,22,32].

Position of our Work. TUEF can be considered a network-based, holistic model that jointly integrates text, community, and social information. It models user relationships under community topics with an MLG. It applies Information Retrieval and Social Network analysis techniques to identify the candidate experts to forward the new question. Finally, it uses a Learning-to-Rank model to precisely order these candidates.

3 TUEF Design

In typical CQA, users tag their questions to characterize them. In addition, it is possible to exploit users' behavior to derive implicit "social" relationships based on the similarity of users' behaviors. To effectively incorporate the valuable information derived from tags, content, and user relationships, TUEF employs an MLG (Sect. 3.1) representing the macro topics discussed in the CQA platform. Each graph layer models the users' relationships at the level of the specific macro topic by considering the users' similarities as captured by tags, questions, and answering behaviors. TUEF adopts an exploratory algorithm that comprehensively takes full advantage of the MLG structure to select a set of candidate experts, i.e., the users who have earned a reputation by consistently providing accepted answers to questions, for each macro topic the current question belongs to (Sect. 3.2). The exploratory algorithm jointly leverages *social* and *content* information, considering experts' centrality in the network and the users' expertise in the topics of the question, increasing the likelihood of selecting high-quality experts. The selection process is inherently transparent, guided by the graph structure that delineates user relationships. Finally, TUEF extracts features representing the identified experts and applies Learning-to-Rank (LtR) techniques (Sect. 3.3) to sort them according to their expertise and likelihood of answering the question. TUEF uses a decision tree-based LtR approach providing a high level of clarity on the significance of different factors influencing the ranking result [27].

3.1 User Interaction Model

Hereinafter, let U be the set of active users in the CQA platform considered, and Q the set of historical questions posted and answered by users in U.

Topic Identification. TUEF clusters the tags associated with the historical questions posted on the CQA platform by adopting the solution discussed in [15], which applies the k-means algorithm to a tag co-occurrence matrix M. Tags' co-occurrence patterns provide helpful information for clustering since tags often associated with the same questions are likely semantically related [14]. Moreover, the adopted methodology is based on the observation that CQA users often use very common and broad tags paired with more specific ones to ensure greater precision in topic identification. We can enhance the questions' categorization by considering the co-occurrences of discriminating tags with the broader ones.

Let $T = \{t \mid t \in tags(q),\ q \in Q\}$ be the set of all tags associated with questions in Q, where $tags(q)$ is the tags list of the question q, and F the set of the λ most frequent tags occurring in questions of Q representing the clustering features. The co-occurrences matrix $M^{|T|\times|F|}$ is constructed as follows:

$$m_{i,j} = |\{q \in Q \mid \{t_i, f_j\} \subseteq tags(q),\ t_i \in T,\ f_j \in F\}| \quad (1)$$

Element $m_{i,j}$ indicates the number of questions in Q where the i_{th} tag and the j_{th} feature co-occur. We obtained a normalized matrix $\hat{M}$ by normalizing M's rows to express the fraction of tag co-occurrences relative to each feature.

Given a value k, the k-means hyperparameter, the clustering algorithm returns k disjoint clusters of tags that represent the main domain areas of the community considered. As detailed in Sect. 4, we use the silhouette maximization criteria [30] to identify the proper value of k.

Multi-layer Graph. TUEF models users' relationships within each layer by representing users U as nodes and establishing a connection between two nodes if the corresponding users have a similar pattern of providing accepted answers to questions related to the specific layer. Formally, TUEF uses a MLG $G = [L_1, ..., L_k]$ where L_i represents the layer of G associated with the $i-th$ tag cluster. Each layer $L_i = (V_i, E_i)$ is an independent graph (i.e., there are no edges between nodes of different layers), where V_i is the set of nodes representing users in layer L_i and E_i is the set of edges representing users' relationships.

To focus on users who can consistently provide accurate answers for questions related to the specific layer, we implement a filtering mechanism that includes in V_i only users that provided at least ϵ accepted answers, i.e., $V_i \subseteq U$.

To model the users' knowledge, we build for each user $u \in V_i$ a topic vector b_u^i where the j_{th} position of b_u^i stores the number of accepted answers provided by u to questions associated with the j_{th} tag of L_i, normalized based on the total number of accepted answers provided by the user in all the layers[2].

After representing each user u of V_i with the topic vector b_u^i, we compute the cosine similarity between all pairs of vectors within the same layer. If the similarity between two users u_a and u_b exceeds a predetermined threshold value δ, we insert in E_i an edge (u_a, u_b) weighted by the cosine similarity value.

It is important to note that each question in TUEF is associated with a list of tags, where each tag is assigned to only one layer within the MLG G.

[2] When modeling the knowledge, we discard the λ high-frequent tags, used as features during the clustering phase, to focus on more discriminating ones.

Consequently, a question may belong to multiple layers within G. Furthermore, the users who answer these questions are represented in all the associated layers, implying that a user's expertise and social relationships span across all the layers to which the questions they answer belong. Considering the multidimensionality of user expertise across various layers, TUEF captures a complete representation of users' knowledge and interactions within the CQA platform.

3.2 Expert Selection

Given a new question q, the Expert Selection component is performed multiple times, one for each layer q belongs to. The selection process is divided into three phases: the *sorting* phase, which sorts the nodes in each layer; the *collection* phase, which selects an initial set of experts from the sorted lists; the *exploratory* phase, which expands the initial set of experts by exploring the graph.

Expert Identification. Experts in CQA platforms are identified effectively by considering the number of *accepted answers* they provided in the past. Another signal of user trust is given by the acceptance ratio r_u, i.e., the ratio between the number of accepted answers and the total number of answers a given user provides. As in [5], the set $C \subseteq U$ of *candidate experts* is selected by considering all the users having a number of accepted answers greater or equal to a specified threshold β. Finally, the set of *experts* $E \subseteq C$ includes the candidates whose acceptance ratio r_u is greater than the overall average $\overline{r}$. Note that each layer includes *experts* and *non-experts* nodes. The former is the target of the Expert Selection process; the latter serves as node transit to fully explore the MLG.

Sorting. To exploit content and social information, TUEF considers two complementary perspectives: the users' centrality within the network (*Network-based* perspective) and the relevance of the user to the newly posted query q based on her previously answered questions (*Content-based* perspective).

The Network-based approach uses the Betweenness centrality [10], which assesses the centrality of nodes within a graph by considering the shortest paths and representing the influence a node has over the flow of information in a graph. Central nodes will have a higher value of Betweenness centrality. The expert nodes v_j^i in the layer L_i are sorted considering their centrality score s_j^i.

In contrast, the Content-based approach sorts the layer's expert nodes based on the similarity to the query q of the questions answered in the past. To compute content-based similarity, TUEF rely on Information Retrieval techniques and pre-built indexes. Specifically, it instantiate two indexes: the *TextIndex* and the *TagIndex*, indexing the text and tags associated with historical questions, respectively. Given the new question with associated tags, the Content-based method uses a retrieval model (e.g., BM25) on both indexes, each returning a sorted list of questions along with the information about the experts who provided the accepted answer. For the *TagIndex*, the query is composed by concatenating the question's tags, while for the *TextIndex*, the query is the concatenation of the question's title and body. Finally, the two lists are merged by alternatively taking one item from each list while preserving the elements' original order. The final result is a query-dependent ordering of the nodes of each layer L_i.

Candidate Collection. Candidate collection involves selecting a subset of experts $D_i \subseteq V_i$ for each method in each layer L_i the question belongs to. On the one hand, the objective is to consider a set of candidates large enough to achieve a high probability of obtaining the correct answer from them, i.e., high recall. Conversely, we aim to consider a set of candidates potentially including only users who are experts on the specific question, i.e., high precision. To trade-off between precision and recall, we estimate the probability p of not receiving an answer from any user in D_i. TUEF starts collecting the experts from the sorted list computed in the sorting step. The probability p, initialized to 1, is incrementally updated as experts are included in D_i. Specifically, it is reduced as a function of μ_u corresponding to any new expert u added to D_i, which is the ratio between the number of accepted answers over the total number of answers provided by u. Moreover, to better model the topic-based expertise, μ_u is smoothed by considering the user's activity on the specific layer L_i. This smoothing is computed by multiplying μ_u by the ratio between the number of expert answers in the layer and the maximum number of answers provided in L_i among all users.

The value of μ_u estimates the probability that an expert will answer the question by modeling her capability and overall activity in the layer. The probability p is thus updated as $p = p \cdot (1 - \mu_u)$, until it becomes less than or equal to a threshold α. At this point, we start the Exploratory phase.

Exploratory Phase. During the Exploratory phase, TUEF explores the graph structure to gather additional experts by leveraging user relationships. The experts D_i from each layer L_i serve as the starting nodes for the exploration. For each node $v_i \in D_i$, TUEF conducts a fixed number of Random Walks and, at each step, randomly selects the next node to visit based on the probability distribution d computed considering the neighbors of the current node v_i and its links weights, representing the similarities with its neighbors. Specifically, given an expert v_i, we denote with $\mathcal{N}_i = \{v_{i1}, \ldots, v_{iN_i}\}$ the neighbours of v_i (where $N_i = |\mathcal{N}_i|$ is the total number of v_i's neighbours). The probability of visiting v_{ij} is given by the ratio between its weight w_{ij} and the sum of the neighbours' weights in N_i. This process is computed independently for each layer to which the question belongs. Additionally, within each layer, Random Walks are performed for both the Network-based and Content-based methods.

3.3 Ranking

To learn an effective ranking function from the training data we resort to *Learning to Rank* (LtR) [25]. LtR algorithms exploit a *ground-truth* to learn a scoring function σ mimicking the ideal ranking function hidden in the training examples. TUEF extracts a subset of the historical questions used to model the users' relationships in Sect. 3.1 as the training set for the LtR algorithm. Specifically, each query q in the training set is associated with a set of candidate experts $U = \{u_0, u_1, \ldots\}$. Each query-candidate pair (q, u_i) is in turn associated with a *relevance judgment* l_i establishing if u_i is an expert for query q or not. Query-candidate pairs (q, u_i) are represented by a vector of features x, able to describe the query, the expert, and their relationship. The LtR algorithm learns a function $\sigma(x)$ predicting a relevance score for the input feature vector x. Such function $\sigma(x)$ is finally used at inference time to compute the candidate experts' scores and rank them accordingly.

The features modeling the query and a candidate expert can be categorized into two groups: *static* and *query-dependent* features. The *static* features model the relative importance of each expert selected during the Expert Selection process, disregarding any relevance to the specific query. They include: (i) the **Reputation** of the expert; (ii) the total number of **Answers** and **AcceptedAnswers** provided by the expert; (iii) the **Ratio** between the Answers and AcceptedAnswers; (iv) **AvgActivity** and **StdActivity** features, which are the average and standard deviation derived from the time differences between consecutive answers provided by the expert, respectively. In contrast, the *query-dependent* features are derived from the Expert Selection process and concern the content and topics of the specific query. The *query-dependent* features modelling the query q and the expert u are the following: (i) **LayerCount**: the number of distinct graph layers in which the expert is selected during the Expert Selection process; (ii) **QueryKnowledge**: the ratio between the number of Answers and AcceptedAnswers provided by u in the layers relevant for q; (iii) **VisitCountContent** and **VisitCountNetwork**: the total number of times an expert is encountered in the Collection and Exploratory phases using the Content-based and Network-based techniques, respectively; (iv) **StepsContent** and **StepsNetwork**: the number of steps necessary to first discover u, either during the Collection or Exploratory phase; (v) **BetweennesPos** and **BetweennessScore**: the expert's rank in the list of users ordered by Betweenness score and the Betweenness score itself; (vi) **ScoreIndexTag** and **ScoreIndexText**: the sum of BM25 scores of historical questions answered by the experts in the IndexTag and IndexText, respectively; (vii) **FrequencyIndexTag** and **FrequencyIndexText**: the number of distinct questions answered by the expert returned by the respective indexes; (viii) **Eigenvector**, **PageRank** and **Closeness**: the scores of the Eigenvector, PageRank, and Closeness centrality measure, respectively; (ix) **Degree**: the node Degree of the expert node; and (x) **AvgWeights**: the average of the links' weights of the expert node. Note that features like Eigenvector, PageRank, Closeness, Degree, AvgWeights, are query-dependent and not static features because the TUEF graph layers on which they are computed depend on the tags associated with the query.

For each candidate expert selected in more than a layer, we consider the maximum of values for the Network-based features and the sum of values for the

Content-based, *LayerCount* and *QueryKnowledge* features. For *StepsContent*, *StepsNetwork*, *BetweennesPos* we consider the minimum value.

Table 1. Statistics of the StackOverflow dataset used.

	Questions	Answers	Users	**Tags**
Raw Dataset	877,180	1,155,189	618,659	36,759
Train Dataset	268,185	373,472	211,057	23,245
Test Dataset	17,002	21,879	18,181	6,380

4 Experimental Setup

We conducted experiments using a large-scale dataset from StackOverflow, the largest community within the Stack Exchange network. The StackOverflow data dump is publicly available[3] and contains more than 22 million questions posted by community members from 2008-07-31 up today.

Preprocessing. We selected the data from 2020-07-01 to 2020-12-31 to focus on a six-months time range. As in [28], we followed well-established cleaning practices to ensure data quality. First, we removed all questions and answers without a specified Id and OwnerUserId. We kept only questions with an AcceptedAnswerId and answers with a specified ParentId, corresponding to the Id of the question it relates to. Finally, we removed questions where the asker and the best answerer were the same user. We divide the questions using 80% for training/validation and the remaining 20% for the test set. Importantly, we preserve the temporal order of the questions. Statistics about the resulting dataset are reported in Table 1.

User Interaction Model. To identify the macro topics discussed in the StackOverflow community, we apply the clustering technique detailed in Sect. 3.1 to questions' tags of the training set. Considering $\lambda = 10$ most frequent tags, we obtain $k = 10$ clusters as the value of k maximising the Silhouette score. Each cluster represents a layer of the MLG where we define the interaction between users who gave at least $\epsilon = 3$ accepted answers. We represent the users as topic vectors, as reported in Sect. 3.1. Finally, when modeling relationships, we keep the edges with a similarity equal to or greater than $\delta = 0.5$.

[3] https://archive.org/details/stackexchange.

Expert Selection. We model the experts by selecting all users with at least $\beta = 20$ accepted answers and following the procedure explained in Sect. 3.2, resulting in 1,230 experts. During the expert selection process, the two indexes are queried for the 1,000 past questions most similar to the query. After sorting the nodes, the collection phase selects the initial set of experts D, reaching a probability $p = 0.001$ representing the probability of not getting an answer. Finally, we perform 5 Random Walks of 10 steps for each selected expert.

Ranking. We use the implementation of LambdaMART available in LightGBM [16]. We build the LtR training set by considering the last $50K$ queries in the training dataset used for building the graph. For each query, we consider all the candidate experts selected by TUEF. We remove from the set of queries those for which TUEF is not able to include, among the candidates, the actual expert who provided the accepted answer, and we extract, for each query and candidate expert pair, the features detailed in Sect. 3.3. The resulting LtR training dataset includes 43,001 queries with 182 candidate experts each, on average. We split the train set into train and validation sets using the 0.8/0.2 splitting criteria. We performed hyper-parameter tuning with MRR on the validation set by exploiting the HyperOpt library [3]. We optimized four learning parameters: `learning_rate` $\in [0.0001, 0.15]$, `num_leaves` $\in [50, 200]$, `n_estimators` $\in [50, 150]$, `max_depth` $\in [8, 15]$, and `min_data_in_leaf` $\in [150, 500]$.

4.1 Baselines and Competitors

We compare TUEF with state-of-the-art competitors and variants of the proposed solution exploiting each only a subset of components. By examining these different configurations, we can assess and quantify the impact of each component and gain insights into the effectiveness of combining social and content information in our approach for addressing the EF task for CQA platforms:

- **BC**: It uses the MLG and simply sorts the experts in the layers related to the new question based on their Betweenness centrality score.
- **BM25**: It uses the MLG and sorts the experts in the question's layers based on the BM25 score computed between the query and their previously answered questions. As in TUEF, the lists of the indexes are merged in alternating the elements.
- **TUEF$_{\text{NB}}$**: It uses the MLG and applies the Network-based method only. Candidates are ranked using a LtR model exploiting the static and the following query-dependent features: LayerCount, QueryKnowledge, VisitCountNetwork, StepsNetwork, BetweennesPos, BetweennessScore, Eigenvector, PageRank, Closeness, Degree, and AvgWeights.
- **TUEF$_{\text{CB}}$**: It uses the MLG and applies the Content-based method only. Candidates are ranked using a LtR model exploiting the static and the following query-dependent features: LayerCount, QueryKnowledge, VisitCountContent, StepsContent, ScoreIndexTag, ScoreIndexText, FrequencyIndexTag, FrequencyIndexTex, Degree, and AvgWeights.

- **TUEF$_{SL}$**: In contrast to TUEF, it represents users' relationships in a graph with a Single Layer. All other phases are unchanged.
- **TUEF$_{NoRW}$**: In contrast to TUEF, it skips the Exploratory phase and does not perform Random Walks to extend the set of candidate experts.
- [31]: It ranks the experts according the solution of [31] which uses a linear combination of features modeling experts, questions, and users' expertise.
- **NeRank** [22]: It models the CQA platform as a heterogeneous network to learn representation for question raisers and question answerers through a metapath-based algorithm. With the heterogeneous network, NeRank preserves the information about relationships while it models the question's content with a single-layer LSTM. Finally, a CNN associates a score to each expert for a given question, representing the probability of the expert providing the accepted answer.

Evaluation Metrics. We test the model using the first 5,000 queries of the test set, which comprise only questions for which the best answerer is an expert. We use Precision@1 (P@1), Normalized Discounted Cumulative Gain @3 (NDCG@3), Mean Reciprocal Rank (MRR), and Recall@100 (R@100) as our evaluation metrics and we compute statistical significance tests with the RanX Library [2]. The cutoffs considered are low as for the EF task it is essential to find the relevant results at the top of the ranked lists. We mark statistically-significant performance gain/loss with respect to the best baseline (TUEF$_{CB}$) with the symbols ▲ and ▼ (paired t-test with p-value < 0.05). For NeRank we report P@1, Hit@5 and MRR as these are the only metrics returned by the code. The source code of TUEF is publicly available[4].

Table 2. Comparison between TUEF and the baselines.

	P@1	NDCG@3	R@100	MRR
BC	0.020▼	0.033▼	0.076▼	0.033▼
BM25	0.234▼	0.355▼	0.808▼	0.369▼
TUEF$_{NB}$	0.066▼	0.087▼	0.213▼	0.0930▼
[31]	0.264▼	0.359▼	0.874▲	0.383▼
TUEF$_{SL}$	0.436▼	0.560▼	0.826▼	0.559▼
TUEF$_{CB}$	0.447	0.573	0.849	0.572
TUEF$_{NoRW}$	0.443	0.561▼	0.754▼	0.552▼
TUEF	**0.453**	**0.578▲**	**0.874▲**	**0.579▲**

[4] https://github.com/maddalena-amendola/TUEF.

5 Results and Discussion

We first compare TUEF with all baselines but NeRank, as the latter requires some additional operations to make the approaches comparable, as explained later on. Table 2 shows that TUEF consistently outperforms all baseline methods across all metrics considered. The best-performing competitor is $TUEF_{CB}$, which relies only on the Content-based method. The improvements of TUEF over $TUEF_{CB}$ are statistically significant for all the metrics reported but P@1 where the two methods have similar performances even if TUEF has a slightly higher P@1. These findings emphasize the substantial contribution of the Content-based method to the model's performance, with the combined utilization of the Network-based method yielding a slightly superior performance. When comparing TUEF to the $TUEF_{NoRW}$ baseline, which stands as the best-performing baseline in terms of P@1, TUEF exhibits a relative improvement of 15.91% for R@100. This result highlights the importance of the exploration of the MLG through Random Walks that identify candidate experts not retrieved with the other previous techniques. The P@1 value, however, suggests that the higher recall is not fully exploited by the TUEF ranking model, who seems not perfectly able to push the right candidate to the top position. Notably, the $TUEF_{NB}$ baseline, which solely employs the Network-based method, reports an R@100 of 0.213, merely a quarter of the R@100 achieved by the Content-based baseline. The third most effective baseline, $TUEF_{SL}$, is outperformed by TUEF by 3.9% for P@1 and 5.81% for R@100. This outcome highlights the advantages introduced by the TUEF MLG: the representation of experts under different macro topics enhance significantly the selection and the ranking of candidate experts. Furthermore, even $TUEF_{CB}$, the Content-based baseline, outperforms $TUEF_{SL}$, emphasizing once more the significance of capturing question similarities and user relationships across distinct macro topics. Instead, the method [31] shows comparable performance in terms of R@100, meaning that the linear combination of features successfully push the proper experts into the top-100 positions. However, the LtR techniques applied in TUEF allow to reach performance almost twice higher in terms of P@1. Using LtR techniques allows, in fact, to capture non-linear relations between the features, remarkably improving the ranking performance. Lastly, the results of the $TUEF_{BC}$ and $TUEF_{BM25}$ baselines highlight that centrality measure alone is not a strong feature to choose the right expert, while who answered in the past the most similar question is the right expert in approximately 23% of cases.

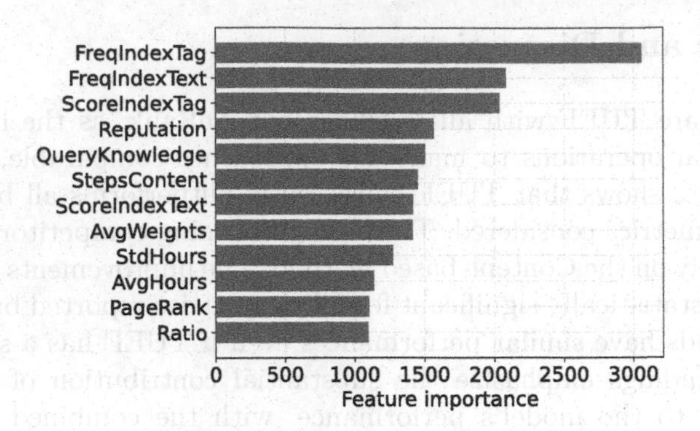

Fig. 2. The feature importance as determined by the Learning-to-Rank approach.

Feature Importance Analysis. Figure 2 highlights the twelve most influential factors in the TUEF algorithm's decision-making process, as determined by the LtR algorithm. Notably, *FreqIndexTag* and *FreqIndexText* stand out as the most significant features, emphasizing the importance of the frequency with which experts have addressed questions sourced from specific indexes. Concurrently, *ScoreIndexTag*, *Reputation*, *QueryKnowledge* and *ScoreIndexText* indicate the importance given to the quality, relevance, and reliability of expert responses. The presence of *StepsContent*, *AvgWeight*, and *PageRank* reveal the role of network-related features, emphasizing both the path-based exploration and the structural importance of experts within the network. Additionally, the inclusion of *StdHours* and *AvgHours* features accentuates the algorithm's recognition of temporal activity patterns.

Table 3. Comparison between TUEF and NeRank [22].

	StackOverflow			Unix			AskUbuntu		
	P@1	Hit@5	MRR	P@1	Hit@5	MRR	P@1	Hit@5	MRR
NeRank	0.34	0.708	0.498	0.568	0.882	0.698	0.499	0.89	0.676
TUEF	**0.721**	**0.951**	**0.822**	**0.648**	**0.933**	**0.769**	**0.61**	**0.91**	**0.74**

Comparison with NeRank. NeRank adopts a different experimental setting. Specifically, for each test query q, NeRank generates and ranks a set of 20 candidate experts that include the n experts that provided the answers to q, plus $20 - n$ other experts randomly selected from the top 10% most responsive users of the community. The apriori knowledge of the users answering the question from one side provides NeRank with a considerable advantage over the setting used for TUEF, but from the other side, it makes NeRank not suited for routing the query to the proper experts in the community. To fairly compare TUEF and NeRank, we applied the same experimental setting for TUEF. Specifically, for each query q, we consider, as in NeRank, the n experts that answered q plus $20 - n$ other experts randomly selected from the ones collected during the TUEF exploratory phase. To highlight the contribution of TUEF, we experiment with two more StackExchange communities: Unix and AskUbuntu. Given the high memory footprint and computational cost of NeRank, for all the tests we used only a sample of about 30,000 questions. Specifically, for StackOverflow, we consider the last 30,000 questions between those used in the previous experiments. Instead, for Unix and AskUbuntu, we consider the last 30,000 questions after extracting the data from January 2015 up to today. Given the small sample, a user is labeled as an expert if she has at least $\beta = 5$ accepted answers. Table 3 reports the results comparing the two models considering the metrics returned by NeRank. We can see that TUEF remarkably outperforms NeRank, with a relative improvement on StackOverflow of 65.06% for MRR and exceeding 112% for P@1. Moreover, the expert can be found in the top-5 ranked experts (Hit@5) in about 95.1% of the TUEF predictions. We can also notice an important improvement for the AskUbuntu community and an inferior one for the Unix community. This experimental methodology that uses the ground truth of the queried questions to identify the candidate experts remarkably facilitates the EF task as highlighted by the superior performance reported in Table 3 w.r.t. that in Table 2.

6 Conclusions

This work addressed the EF task in the context of CQA platforms. We contributed TUEF, a novel EF solution that combines content and social information available in CQA platforms. The empirical evaluation, performed on datasets from StackOverflow, demonstrates the superior performance of TUEF compared to existing methods. The improvements reach as high as 65.06% for MRR and 112% for P@1. By providing effective expert predictions, TUEF contributes to making CQA engaging, trusted and credible. In future work, we will investigate the interpretability and fairness of TUEF predictions.

Acknowledgements. This work was partially supported by: the H2020 SoBigData++ project (#871042); the CAMEO PRIN project (#2022ZLL7MW) funded by the MUR; the HEU EFRA project (#101093026) funded by the EC under the NextGeneration EU programme. A. Passarella's and R. Perego's work was partly funded under the PNRR - M4C2 - Investimento 1.3, PE00000013 - "FAIR" project. However, the views and opinions expressed are those of the authors only and do not necessarily reflect those of the EU or European Commission-EU. Neither the EU nor the granting authority can be held responsible for them.

References

1. Aslay, Ç., O'Hare, N., Aiello, L.M., Jaimes, A.: Competition-based networks for expert finding. In: Proceedings of the 36th International ACM SIGIR Conference on Research and Development in Information Retrieval, pp. 1033–1036 (2013)
2. Bassani, E.: ranx: a blazing-fast python library for ranking evaluation and comparison. In: Hagen, M., et al. (eds.) ECIR (2). Lecture Notes in Computer Science, vol. 13186, pp. 259–264. Springer, Heidelberg (2022). https://doi.org/10.1007/978-3-030-99739-7_30
3. Bergstra, J., Yamins, D., Cox, D.D.: Making a science of model search: Hyperparameter optimization in hundreds of dimensions for vision architectures. In: Proceedings of the 30th International Conference on International Conference on Machine Learning, ICML 2013, vol. 28, pp. I-115–I-123. JMLR.org (2013)
4. Burges, C.J.: From ranknet to lambdarank to lambdamart: an overview. Learning **11**(23–581), 81 (2010)
5. Nobari, A.D., Gharebagh, S.S., Neshati, M.: Skill translation models in expert finding. In: Proceedings of the 40th International ACM SIGIR Conference on Research and Development in Information Retrieval, pp. 1057–1060 (2017)
6. Dehghan, M., Abin, A.A.: Translations diversification for expert finding: a novel clustering-based approach. ACM Trans. Knowl. Disc. Data **13**(3), 32:1–32:20 (2019). https://doi.org/10.1145/3320489
7. Dehghan, M., Biabani, M., Abin, A.A.: Temporal expert profiling: with an application to t-shaped expert finding. Inf. Process. Manag. **56**(3), 1067–1079 (2019). https://doi.org/10.1016/j.ipm.2019.02.017
8. Devlin, J., Chang, M., Lee, K., Toutanova, K.: BERT: pre-training of deep bidirectional transformers for language understanding. In: Burstein, J., Doran, C., Solorio, T. (eds.) Proceedings of the 2019 Conference of the North American Chapter of the Association for Computational Linguistics: Human Language Technologies, NAACL-HLT 2019, Minneapolis, MN, USA, 2–7 June 2019, vol. 1 (Long and Short Papers), pp. 4171–4186. Association for Computational Linguistics (2019). https://doi.org/10.18653/v1/n19-1423
9. Faisal, M.S., Daud, A., Akram, A.U., Abbasi, R.A., Aljohani, N.R., Mehmood, I.: Expert ranking techniques for online rated forums. Comput. Hum. Behav. **100**, 168–176 (2019)
10. Freeman, L.C.: A set of measures of centrality based on betweenness. In: Sociometry, pp. 35–41 (1977)
11. Fu, C.: Tracking user-role evolution via topic modeling in community question answering. Inf. Process. Manag. **56**(6), 102075 (2019)
12. Fu, C.: User correlation model for question recommendation in community question answering. Appl. Intell. **50**, 634–645 (2020)
13. Fu, J., et al.: Recurrent memory reasoning network for expert finding in community question answering. In: Proceedings of the 13th International Conference on Web Search and Data Mining, pp. 187–195 (2020)
14. Giannakidou, E., Koutsonikola, V., Vakali, A., Kompatsiaris, Y.: Co-clustering tags and social data sources. In: 2008 The Ninth International Conference on Web-Age Information Management, pp. 317–324 (2008). https://doi.org/10.1109/WAIM.2008.61
15. Hoffa, F.: Making sense of the metadata: clustering 4,000 stack overflow tags with bigquery k-means (2019). https://stackoverflow.blog/2019/07/24/making-sense-of-the-metadata-clustering-4000-stack-overflow-tags-with-bigquery-k-means/

16. Ke, G., et al.: LightGBM: a highly efficient gradient boosting decision tree. In: Guyon, I., et al. (eds.) Advances in Neural Information Processing Systems, vol. 30. Curran Associates, Inc. (2017). https://proceedings.neurips.cc/paper_files/paper/2017/file/6449f44a102fde848669bdd9eb6b76fa-Paper.pdf
17. Kundu, D., Mandal, D.P.: Formulation of a hybrid expertise retrieval system in community question answering services. Appl. Intell. **49**, 463–477 (2019)
18. Kundu, D., Pal, R.K., Mandal, D.P.: Finding active experts for question routing in community question answering services. In: Deka, B., Maji, P., Mitra, S., Bhattacharyya, D., Bora, P., Pal, S. (eds.) Pattern Recognition and Machine Intelligence: 8th International Conference, PReMI 2019, Tezpur, India, 17–20 December 2019, Proceedings, Part II, pp. 320–327. Springer, Heidelberg (2019). https://doi.org/10.1007/978-3-030-34872-4_36
19. Kundu, D., Pal, R.K., Mandal, D.P.: Preference enhanced hybrid expertise retrieval system in community question answering services. Decis. Supp. Syst. **129**, 113164 (2020)
20. Le, L.T., Shah, C.: Retrieving people: identifying potential answerers in community question-answering. J. Am. Soc. Inf. Sci. **69**(10), 1246–1258 (2018)
21. Li, L., Shang, Y., Zhang, W.: Improvement of hits-based algorithms on web documents. In: Proceedings of the 11th International Conference on World Wide Web, pp. 527–535 (2002)
22. Li, Z., Jiang, J.Y., Sun, Y., Wang, W.: Personalized question routing via heterogeneous network embedding. In: Proceedings of the AAAI Conference on Artificial Intelligence, vol. 33, pp. 192–199 (2019)
23. Liang, S.: Unsupervised semantic generative adversarial networks for expert retrieval. In: Liu, L., et al. (eds.) The World Wide Web Conference, WWW 2019, San Francisco, CA, USA, 13–17 May 2019, pp. 1039–1050. ACM (2019). https://doi.org/10.1145/3308558.3313625
24. Liu, H., Lv, Z., Yang, Q., Xu, D., Peng, Q.: Efficient non-sampling expert finding. In: Proceedings of the 31st ACM International Conference on Information & Knowledge Management, pp. 4239–4243 (2022)
25. Liu, T.Y.: Learning to rank for information retrieval. Found. Trends Inf. Retr. **3**(3), 225–331 (2009). https://doi.org/10.1561/1500000016
26. Liu, X., Bollen, J., Nelson, M.L., Van de Sompel, H.: Co-authorship networks in the digital library research community. Inf. Process. Manag. **41**(6), 1462–1480 (2005)
27. Lucchese, C., Nardini, F.M., Orlando, S., Perego, R., Veneri, A.: Ilmart: interpretable ranking with constrained lambdamart. In: Proceedings of the 45th International ACM SIGIR Conference on Research and Development in Information Retrieval, SIGIR 2022, Pp. 2255-2259. Association for Computing Machinery, New York (2022). https://doi.org/10.1145/3477495.3531840
28. Mumtaz, S., Rodriguez, C., Benatallah, B.: Expert2vec: experts representation in community question answering for question routing. In: Giorgini, P., Weber, B. (eds.) Advanced Information Systems Engineering: 31st International Conference, CAiSE 2019, Rome, Italy, 3–7 June 2019, Proceedings, vol. 31, pp. 213–229. Springer, Heidelberg (2019). https://doi.org/10.1007/978-3-030-21290-2_14
29. Peng, Q., Liu, H.: Expertplm: pre-training expert representation for expert finding. In: Findings of the Association for Computational Linguistics: EMNLP 2022, pp. 1043–1052 (2022)
30. Rousseeuw, P.J.: Silhouettes: a graphical aid to the interpretation and validation of cluster analysis. J. Comput. Appl. Math. **20**, 53–65 (1987)

31. Roy, P.K., Singh, J.P., Nag, A.: Finding active expert users for question routing in community question answering sites. In: Machine Learning and Data Mining in Pattern Recognition: 14th International Conference, MLDM 2018, New York, NY, USA, 15–19 July 2018, Proceedings, Part II, vol. 14. pp. 440–451. Springer, Heidelberg (2018). https://doi.org/10.1007/978-3-319-96133-0_33
32. Sun, J., Bandyopadhyay, B., Bashizade, A., Liang, J., Sadayappan, P., Parthasarathy, S.: ATP: directed graph embedding with asymmetric transitivity preservation. In: The Thirty-Third AAAI Conference on Artificial Intelligence, AAAI 2019, The Thirty-First Innovative Applications of Artificial Intelligence Conference, IAAI 2019, The Ninth AAAI Symposium on Educational Advances in Artificial Intelligence, EAAI 2019, Honolulu, Hawaii, USA, 27 January–1 February 2019, pp. 265–272. AAAI Press (2019). https://doi.org/10.1609/aaai.v33i01.3301265
33. Sun, J., Moosavi, S., Ramnath, R., Parthasarathy, S.: QDEE: question difficulty and expertise estimation in community question answering sites. In: Proceedings of the Twelfth International Conference on Web and Social Media, ICWSM 2018, Stanford, California, USA, 25–28 June 2018, pp. 375–384. AAAI Press (2018). https://aaai.org/ocs/index.php/ICWSM/ICWSM18/paper/view/17854
34. Tondulkar, R., Dubey, M., Desarkar, M.S.: Get me the best: predicting best answerers in community question answering sites. In: Proceedings of the 12th ACM Conference on Recommender Systems, pp. 251–259 (2018)
35. Zhang, X., et al.: Temporal context-aware representation learning for question routing. In: Caverlee, J., Hu, X.B., Lalmas, M., Wang, W. (eds.) WSDM 2020: The Thirteenth ACM International Conference on Web Search and Data Mining, Houston, TX, USA, 3–7 February 2020, pp. 753–761. ACM (2020). https://doi.org/10.1145/3336191.3371847

Demo Papers

QuantPlorer: Exploration of Quantities in Text

Satya Almasian(✉), Alexander Kosnac, and Michael Gertz

Heidelberg University, Heidelberg, Germany
{almasian,gertz}@informatik.uni-heidelberg.de,
alexander.kosnac@stud.uni-heidelberg.de

Abstract. Quantities play an important role in documents of various domains such as finance, business, and medicine. Despite the role of quantities, only a limited number of works focus on their extraction from text and even less on creating respective user-friendly document exploration frameworks. In this work, we introduce QuantPlorer, an online quantity extractor and explorer. Through an intuitive web interface, QuantExplorer extracts quantities from unstructured text, enables users to interactively investigate and visualize quantities in text, and it supports filtering based on diverse features, i.e., value ranges, units, trends, and concepts. Furthermore, users can explore and visualize distributions of values for specific units and concepts. Our demonstration is available at https://quantplorer.ifi.uni-heidelberg.de/.

1 Introduction

Factual information is essentially conveyed through quantities, measurements, and numerical values. However, very few systems in the literature focus on quantity extraction [4,9,11], and even those have limited capabilities, require manual installation, and lack a user-friendly interface for annotation and exploration of quantities in documents. Quantitative information in existing systems is further narrowed down to metric units and values, ignoring noun-based quantities, e.g., "200 people" or how a value is changing, e.g., "DAX gained 2%", showing an upward trend. Moreover, the extraction of only units and values provides little information without further context. For example, consider the mention of "100 m", which could both be the height of a building or the distance traveled. In our prior work [2], we address the shortcomings of existing quantity extractors and devise a comprehensive system that can handle a variety of quantities from different domains and extract contextual information in terms of related concepts and changes. This demo builds an interactive exploration tool on top of this system. It offers a user-friendly interface for the annotation of multiple documents and accessible filtering based on units, specific ranges of values, related concepts, and changes. The system also provides an overview and summary of quantities contained in the text. Users can gain an overview of a document by looking at the distribution of values associated with a specific concept or the frequency of different units and concepts in the text. Annotations can be

N. Goharian et al. (Eds.): ECIR 2024, LNCS 14612, pp. 171–176, 2024.
https://doi.org/10.1007/978-3-031-56069-9_13

downloaded in JSON format for further processing, and for a large number of documents, manual interactions with the web interface are eliminated by using our programmatic API at https://quantplorer.ifi.uni-heidelberg.de/api/extract.

2 Related Work

In the following, we briefly outline related works, noting that to the best of our knowledge, there is no system supporting the rich functionalities of QuantPlorer.

Extractors: Quantulum3[1], Recognizers-Text [9], and Grobid-quantities [4] are open source packages that help detect and normalize scientific units and values. Illinois Quantifier [11] is the first extractor that extends the definition of quantities from unit and value to include changes. However, all these systems focus on specific types of quantities or specific domains, mainly ignore noun-based units, and do not extract concepts associated with units and values.

Interfaces: QFinder [1] supports retrieving documents based on user-specified numerical conditions and ranges of values for a given unit and set of keywords. For this, it uses a numerical index and Quantulum3 for extraction. SciHarvester [12] is similar to QFinder, but focuses mostly on agricultural documents. In addition to relevant documents, SciHarvester shows the distribution of values in the retrieved documents. Qsearch [5,6] and AnaSearch [10] are two other search-oriented systems with a focus on quantities related to named entities. The quantity extractors for these systems are Illinois Quantifier and Recognizers-Text, respectively. Moving from unstructured text to tables, QuTE [7,8] retrieves relevant cells in tables based on queries containing numerical conditions. None of the mentioned interfaces are purely focused on quantity extraction.

3 System Overview

We briefly review the Comprehensive Quantity Extractor (CQE) [2], and explain the core functionalities of our demo interface. For a more detailed explanation, refer to the paper. The backend is based on the output of the CQE package[2], where the units and values are enriched with contextualized information about associated concepts and how values change.

Quantity: A quantity is a tuple $\langle v, u, ch, cn \rangle$ with the following components:

1. *Value (v):* A real number or a range of values in a standardized format, e.g., "she earns between \$10 to \$20 an hour" has a range of $v = (10, 20)$, and "she earns \$1k" has a single value $v = 1000$.

[1] https://github.com/nielstron/quantulum3 DLA: 12.09.2023.
[2] https://github.com/vivkaz/CQE DLA: 12.09.2023.

2. *Unit (u):* The atomic unit of measure. It can be part of predefined physical units or currencies, e.g., "kilometer" or "dollar", or any noun phrase referring to the multitude of an object, e.g., "two apples". Predefined units have varied forms (e.g., "pounds", "GBP", "£"), causing ambiguity (e.g., "She weighs 50 pounds" indicates a unit of weight, not a currency). Normalizing units to a uniform representation is crucial in such cases.
3. *Change (ch):* The value's modifier defines changes, based on six categories: = (equal), ∼ (approximate), > (more than), < (less than), *up* (upward), and *down* (downward). E.g, in "DAX fell 2%", "fell" implies a downward trend.
4. *Concept (cn):* Nouns that are either a phenomenon measured by a quantity (e.g., "height of the Eiffel Tower") or an entity involved in a quantity-based action (e.g., "Google investing $200 million" with $cn = Google$). Concepts are not always present, as in "200 people were at the concert."

Extraction Pipeline: After preprocessing to remove unnecessary punctuations, the input text is tokenized into words using a quantity-aware tokenizer that improves subword splitting for quantities, e.g., "km/h" and "10.3" are kept as a single token. The extraction of different quantity components is performed using 61 rules based on POS tags and dependency parsing. Standardization of values and normalization of units is done with the help of value and unit dictionaries. The units dictionary is a set of 531 units, their surface forms, and symbols. The value dictionary contains information for converting values to real numbers, such as the list of suffixes of scales, e.g., "B: billion". Ambiguous unit surface forms are classified to the correct normalized form with a BERT [3] classifier.

Web Interface: The front-end is built with the Flask[3] web framework. Before passing an input text to CQE, documents are split into sentences using the SpaCy[4] library, and passed independently to the extractor. The design is done using HTML, JavaScript, and the Bootstrap 5[5] library is used for a responsive layout. Charts and histograms are generated using Chart.js[6].

4 Demonstration

In this section, we describe typical user workflows and highlighting the four main capabilities of our system.

Extraction: The entry point to the interface is shown in Fig. 1a. The user can explore the provided examples or add a new document by clicking on the "+"

[3] https://flask.palletsprojects.com/ DLA: 28.08.2023.
[4] https://spacy.io/ DLA: 28.08.2023.
[5] http://getbootstrap.com DLA: 28.08.2023.
[6] https://www.chartjs.org/ DLA: 28.08.2023.

symbol. The document is created either dragging a text file or directly typing in the editor. By clicking the "Extract" button, all quantity components are extracted and color-coded in the text for easy identification. The *Quantities* tab in the sidebar to the right in Fig. 1a demonstrates the extractions, where values are standardized and units are normalized. The download button in the bottom right corner can be used to download extraction results in JSON format.

Filtering: The *Filters* tab allows the user to view and download annotations for quantities that match on certain criteria. Users can specify ranges of a value, filter based on a specific concept and unit, or type a specific keyword in the filter textbox. Upon choosing a filter, the *Quantities* tab is automatically updated to reflect the selections, which appear in closable spans over the list of annotations. The download button will then export only the quantities matching the filters. In Fig. 1b, 9 out of 105 quantities contain the concept "Google" and the unit "dollar", where the values lie between 2 and 90 billion.

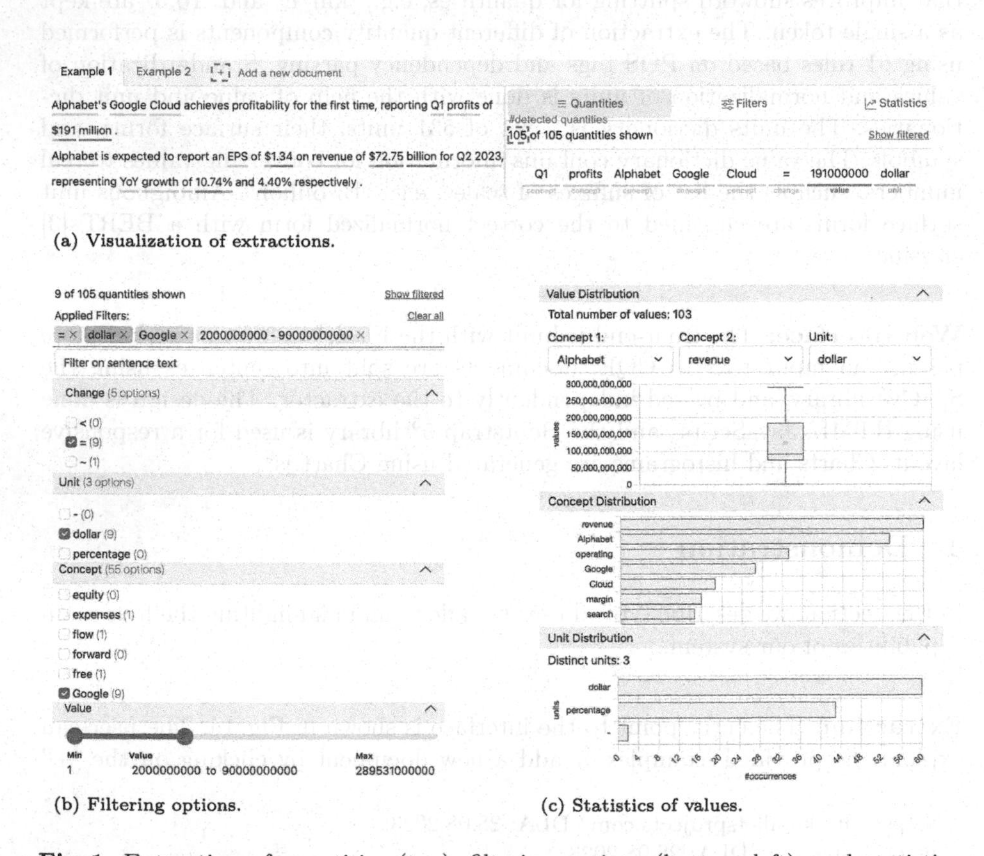

(a) Visualization of extractions.

(b) Filtering options.

(c) Statistics of values.

Fig. 1. Extraction of quantities (top), filtering options (bottom left), and statistics tabs for inspecting the distributions (bottom right).

Statistics: A summary of the document in terms of quantities is shown in the *Statistics* tab, where "Unit distribution" and "Concept Distributions" depict the frequency of different unit, and concept uni-grams in the given document. Figure 1c shows that the example document is dominated by the unit "percentage" and quantities mainly revolve around "Alphabet" and "revenue". The "Value Distribution" tab presents a more fine-grained analysis, where the user can choose a single or combination of two concepts with a specified unit to analyze the distribution of values. In the example in Fig. 1c, the user can see the distribution of "Alphabet revenue" in "dollars" to get a sense of ranges and scales of values.

API: To quickly obtain extractions for a collection of documents, without the need to install the CQE package or interact with the web interface, we additionally provide an API endpoint that returns extracted data in JSON format.

5 Conclusion and Future Work

We presented an interactive web interface along with a programmatic API designed for annotating and exploring quantities in text documents. We are currently extending the interface to explore and aggregate quantities across multiple documents and investigate methods for clustering of similar concepts.

References

1. Almasian, S., Bruseva, M., Gertz, M.: QFinder: a framework for quantity-centric ranking. In: SIGIR 2022: The 45th International ACM SIGIR Conference on Research and Development in Information Retrieval, Madrid, Spain, 11–15 July 2022, pp. 3272–3277. ACM (2022)
2. Almasian, S., Kazakova, V., Göldner, P., Gertz, M.: CQE: a comprehensive quantity extractor. In: Proceedings of the 2023 Conference on Empirical Methods in Natural Language Processing, EMNLP 2023, Singapore, 6–10 December 2023, pp. 12845–12859. Association for Computational Linguistics (2023)
3. Devlin, J., Chang, M., Lee, K., Toutanova, K.: BERT: pre-training of deep bidirectional transformers for language understanding. In: Proceedings of the 2019 Conference of the North American Chapter of the Association for Computational Linguistics: Human Language Technologies, NAACL-HLT 2019, Minneapolis, MN, USA, 2–7 June 2019, vol. 1 (Long and Short Papers), pp. 4171–4186. Association for Computational Linguistics (2019)
4. Foppiano, L., Romary, L., Ishii, M., Tanifuji, M.: Automatic identification and normalisation of physical measurements in scientific literature. In: Proceedings of the ACM Symposium on Document Engineering 2019, Berlin, Germany, 23–26 September 2019, pp. 24:1–24:4. ACM (2019)
5. Ho, V.T., Ibrahim, Y., Pal, K., Berberich, K., Weikum, G.: Qsearch: answering quantity queries from text. In: Ghidini, C., et al. The Semantic Web - ISWC - 18th International Semantic Web Conference, Proceedings, Part I. Lecture Notes in Computer Science, vol. 11778, pp. 237–257. Springer, Heidelberg (2019). https://doi.org/10.1007/978-3-030-30793-6_14

6. Ho, V.T., Pal, K., Kleer, N., Berberich, K., Weikum, G.: Entities with quantities: extraction, search, and ranking. In: WSDM 2020: The Thirteenth ACM International Conference on Web Search and Data Mining, Houston, TX, USA, 3–7 February 2020, pp. 833–836. ACM (2020)
7. Ho, V.T., Pal, K., Razniewski, S., Berberich, K., Weikum, G.: Extracting contextualized quantity facts from web tables. In: WWW 2021: The Web Conference 2021, Virtual Event/Ljubljana, Slovenia, 19–23 April 2021, pp. 4033–4042. ACM/IW3C2 (2021)
8. Ho, V.T., Pal, K., Weikum, G.: QuTE: answering quantity queries from web tables. In: SIGMOD 2021: International Conference on Management of Data, Virtual Event, China, 20–25 June 2021, pp. 2740–2744. ACM (2021)
9. Huang, W., Lin, Z., McConnell, C., Karlsson, B.F.: Recognizers-Text: recognition and resolution of numbers, units, and date/time entities expressed across multiple languages (2017)
10. Li, T., Fang, L., Lou, J., Li, Z., Zhang, D.: AnaSearch: extract, retrieve and visualize structured results from unstructured text for analytical queries. In: WSDM 2021, The Fourteenth ACM International Conference on Web Search and Data Mining, Virtual Event, Israel, 8–12 March 2021, pp. 906–909. ACM (2021)
11. Roy, S., Vieira, T., Roth, D.: Reasoning about quantities in natural language. Trans. Assoc. Comput. Linguistics **3**, 1–13 (2015)
12. Rybinski, M., et al.: SciHarvester: searching scientific documents for numerical values. In: Proceedings of the 46th International ACM SIGIR Conference on Research and Development in Information Retrieval, SIGIR 2023, Taipei, Taiwan, 23–27 July 2023, pp. 3135–3139. ACM (2023)

Interactive Document Summarization

Raoufdine Said and Adrien Guille(✉)

Université de Lyon, Lyon 2, ERIC UR 3083, Lyon, France
{raoufdine.said,adrien.guille}@univ-lyon2.fr

Abstract. With the advent of modern chatbots, automatic summarization is becoming common practice to quicken access to information. However the summaries they generate can be biased, unhelpful or untruthful. Hence, in sensitive scenarios, extractive summarization remains a more reliable approach. In this paper we present an original extractive method combining a GNN-based encoder and a RNN-based decoder, coupled with a user-friendly interface that allows for interactive summarization.

Keywords: Interactive Summarization · Extractive Summarization · Graph Neural Network · Recurrent Neural Network · User Interface

1 Introduction

One of the most common uses for modern chatbots is automatic summarization [1]. While the abstractive summaries generated by these tools can be relevant, the language models they're based on might be biased, *e.g.* politically biased [9] or gender biased [5], they can sometimes be untruthful because large language models are prone to hallucinate [6], or they can simply be unhelpful, as chatbots might fail to follow instructions correctly [8]. This limits their reliability and can make them ill-suited to summarize sensitive documents, like scientific articles, encyclopedic articles or press articles, among others, where information shouldn't be altered. A viable alternative under this scenario is extractive summarization, which consists in composing the summary from actual sentences picked from the original document. In this paper, we first present an original method for extractive summarization, by combining two existing methods. The implementation of the model is available here: https://github.com/baragouine/radsum. Next, we present an interface that allows users to interactively summarize documents, the code of which is available here: https://github.com/baragouine/radsum_app/. For a video demonstration, see https://youtu.be/vBenEaCIwkI.

2 Summarization Method

We combine two existing methods, namely (i) HeterSUMGraph [11], which we slightly modify and only use for the encoding part of our architecture, and (ii) SummaRuNNer [7], which we use for the decoding part only. Figure 1 illustrates the overall methodology. The rational behind combining these methods is to have a more expressive encoder while having an interpretable decoder.

N. Goharian et al. (Eds.): ECIR 2024, LNCS 14612, pp. 177–181, 2024.
https://doi.org/10.1007/978-3-031-56069-9_14

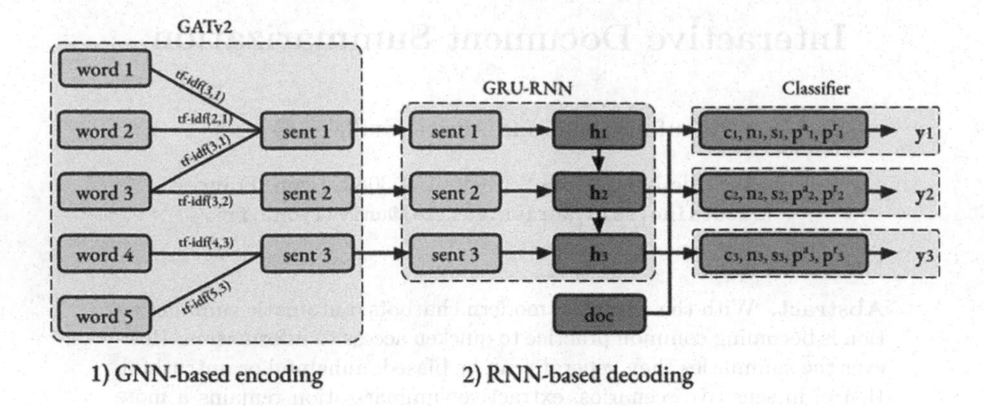

Fig. 1. Overall architecture of the implemented summarization method.

Document Encoding. We convert the input document to a graph as in HeterSUMGraph [11]. It is a bipartite graph with two kinds of vertices: word vertices and sentence vertices. More precisely there is one vertex per unique word in the document and one vertex per sentence in the document, connected according to the composition of the sentences. Edges are weighted with *tf-idf* scores measured at the sentence level, meaning *tf* is calculated as the number of times the word occurs in the sentence, while *idf* is defined as the inverse of the degree of the word vertex. Word embeddings are propagated with a 2-layer graph neural network to obtain sentence representations. Whereas HeterSUMGraph uses GAT layers [10], we implement more expressive GATv2 layers [2], with an attention mechanism that accounts for discretized tf-idf weights via edge embeddings.

Summary Decoding. We pass the sentence representations to a recurrent neural network based on the GRU cell [3] to further contextualize them and proceed similarly to SummaRuNNer in classifying sentences sequentially, following their order in the document. The probability to keep the ith sentence is calculated in terms of 5 scores:

- **content score**: a linear function of the representation of this sentence, $\mathbf{W}_c\mathbf{h}_i$;
- **salience score**: a bilinear function of the representation of this sentence and the representation of the whole document, $\mathbf{h}_i^\top \mathbf{W}_s \mathbf{d}$;
- **novelty score**: a bilinear function of the representation of sentence i and the representation of the document up to sentence $i-1$, $\mathbf{h}_i^\top \mathbf{W}_r \tanh(\mathbf{s}_i)$;
- **absolute position score**: a linear function of the embedding of the absolute position of the sentence, $\mathbf{W}_{ap}\mathbf{p}_i^a$;
- **relative position score**: a linear function of the embedding of the relative position of the sentence, $\mathbf{W}_{rp}\mathbf{p}_i^r$.

The probability to keep the ith sentence in the summary is calculated as follows:

$$\mathrm{p}(1|\mathbf{h}_i, \mathbf{s}_i, \mathbf{d}) = \sigma\left(\mathbf{W}_c\mathbf{h}_i + \mathbf{h}_i^\top \mathbf{W}_s \mathbf{d} + \mathbf{h}_i^\top \mathbf{W}_r \tanh(\mathbf{s}_i) + \mathbf{W}_{ap}\mathbf{p}_i^a + \mathbf{W}_{rp}\mathbf{p}_i^r\right), \tag{1}$$

where $\mathbf{d}$ is the representation of the whole document obtained by averaging all the hidden states $\{\mathbf{h}_1, \mathbf{h}_2, \ldots, \mathbf{h}_n\}$, and $\mathbf{s}_i$ is the representation of the document up to the previous sentence, a weighted average of $\{\mathbf{h}_1, \ldots, \mathbf{h}_{i-1}\}$:

$$\mathbf{s}_i = \sum_{j=1}^{i-1} \mathbf{h}_j \mathrm{p}(1|\mathbf{h}_j, \mathbf{s}_j, \mathbf{d}). \tag{2}$$

3 User Interface

Table 1. Performance on the NYT corpus. The gain over SummaRuNNer is given after the + sign.

	ROUGE-1	ROUGE-L
SummaRuNNer	45.3	34.65
HeterSUMGraph	46.76 +2.0%	35.21 +1.6%
RadSum	46.91 +2.4%	35.35 +2.0%

For the purpose of the demonstration, we train our method, RadSum, on the New York Times annotated corpus [4]. Its performance in terms of ROUGE-1 and ROUGE-L scores is reported in Table 1, along with the scores achieved by SummaRuNNer and HeterSUMGraph. Figure 2 shows the default interface of the proposed application. The left side of the windows allows to input the document to summarize and adjust general settings. The right side of the window shows the summary extracted from that document. For each sentence, the contribution of the 5 scores to the overall probability is depicted with a bar chart. Each sentence is colored according to the dominant score (scaled between 0 and 1 with the sigmoid function). When inputting a press article about the Nobel Prize in Physics awarded in October 2023, it produces a 4-sentence long summary that focuses on the laureates and the implications of their discovery by picking the 1st sentence, particularly for its position in the document, and sentences 4-6 mostly because of their content and salience. Figure 3 shows how one could leverage the app to tailor the summary according to its needs. Another 4-sentence long summary focusing on the discovery itself is obtained by selecting sentences 19, 28, 29 and 31 solely based on the salience score (using the filter button). It also shows how one can manually filter the sentences: here, sentence 31 has been removed from the summary, which resulted in the automatic addition of sentence 30 (the next most salient sentence) to keep the summary at a length of 4 sentences.

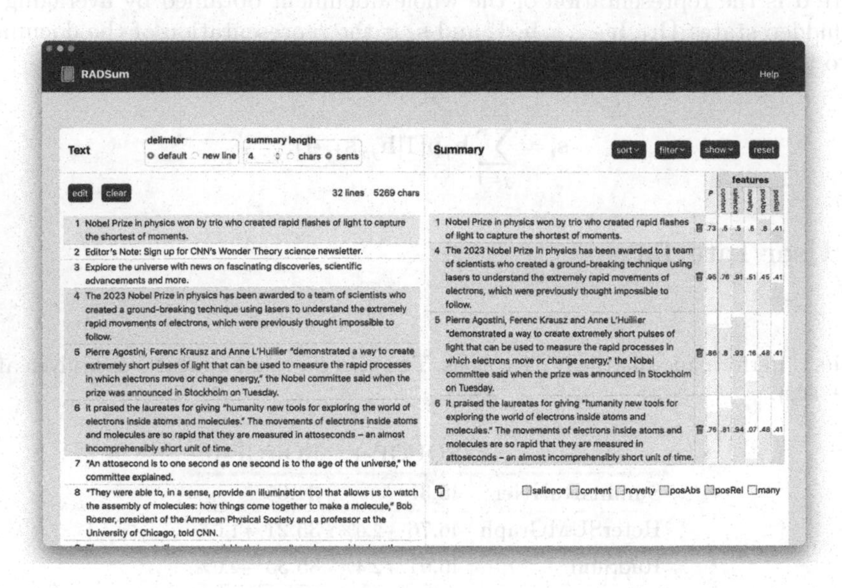

Fig. 2. Default interface. The P column (in yellow) corresponds to the probability to keep the sentence. (Color figure online)

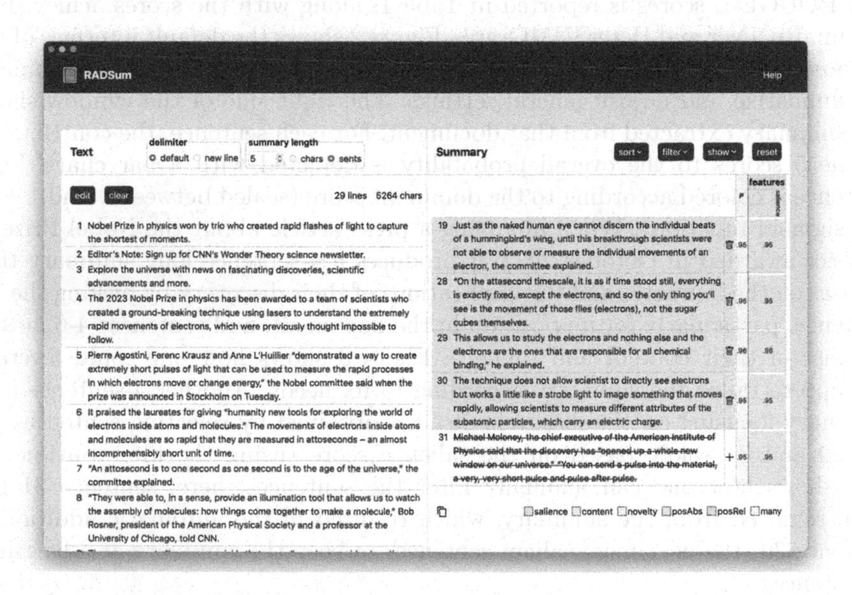

Fig. 3. Example of interaction: the user has selected salience as the only score and has removed sentence 31 from the summary.

References

1. Azaria, A., Azoulay, R., Reches, S.: Chatgpt is a remarkable tool – for experts (2023)
2. Brody, S., Alon, U., Yahav, E.: How attentive are graph attention networks? In: ICLR (2022)
3. Cho, K., et al.: Learning phrase representations using rnn encoder-decoder for statistical machine translation. In: EMNLP (2014)
4. Durrett, G., Berg-Kirkpatrick, T., Klein, D.: Learning-based single-document summarization with compression and anaphoricity constraints. In: ACL (2016)
5. Jentzsch, S., Turan, C.: Gender bias in BERT - measuring and analysing biases through sentiment rating in a realistic downstream classification task. In: GeBNLP Workshop @ ACL (2022)
6. Ji, Z., et al.: Survey of hallucination in natural language generation. ACM Comput. Surv. **55**(12), 1–38 (2023)
7. Nallapati, R., Zhai, F., Zhou, B.: Summarunner: a recurrent neural network based sequence model for extractive summarization of documents. In: AAAI (2017)
8. Ouyang, L., et al.: Training language models to follow instructions with human feedback. ArXiV Technical Report (2022)
9. Rozado, D.: The political biases of chatgpt. Soc. Sci. **12**(3), 148 (2023)
10. Veličković, P., Cucurull, G., Casanova, A., Romero, A., Liò, P., Bengio, Y.: Graph attention networks. In: ICLR (2018)
11. Wang, D., Liu, P., Zheng, Y., Qiu, X., Huang, X.: Heterogeneous graph neural networks for extractive document summarization. In: ACL (2021)

KnowFIRES: A Knowledge-Graph Framework for Interpreting Retrieved Entities from Search

Negar Arabzadeh(✉), Kiarash Golzadeh, Christopher Risi, Charles L. A. Clarke, and Jian Zhao

University of Waterloo, Waterloo, Canada
{narabzad,kgolzade,cjrisi,claclark,jianzhao}@uwaterloo.ca

Abstract. Entity retrieval is essential in information access domains where people search for specific entities, such as individuals, organizations, and places. While entity retrieval is an active research topic in Information Retrieval, it is necessary to explore the explainability and interpretability of them more extensively. KnowFIRES addresses this by offering a knowledge graph-based visual representation of entity retrieval results, focusing on contrasting different retrieval methods. KnowFIRES allows users to better understand these differences through the juxtaposition and superposition of retrieved sub-graphs. As part of our demo, we make KnowFIRES (Demo: http://knowfires.live, Source: https://github.com/kiarashgl/KnowFIRES) web interface and its source code publicly available (A demonstration of the tool: https://www.youtube.com/watch?v=9u-877ArNYE).

1 Introduction

Entity retrieval is crucial in information access domains where users search for specific items, such as individuals, organizations, and places, that are distinguishable by their characteristics, attributes, and connections to other entities [7,12,15]. Researchers estimate that over 40% of web search queries are for entities and major web search engines use extensive knowledge bases to respond to these requests [4,9,23,24,27,32,33,38]. While entity retrieval has been investigated extensively, little attention has been given to the interpretability and explainability of such systems [34]. Increased explainability of search results has been shown to [1,35]: 1) increase searcher's trust in the system and thus the searcher's satisfaction rate, 2) increase the probability of satisfying the information need behind the query, and 3) decrease the chance of spreading misinformation due to more informative reasoning. In this demo, we introduce an explainable entity retrieval system called the Knowledge-graph Framework for Interpreting Retrieved Entities from Search (KnowFIRES) that not only presents the entity names in a ranked list but also visualizes the entities through a knowledge graph representation. KnowFIRES focuses on highlighting similarities and

N. Arabzadeh, K. Golzadeh, and C. Risi—Equal Contributions.

N. Goharian et al. (Eds.): ECIR 2024, LNCS 14612, pp. 182–188, 2024.
https://doi.org/10.1007/978-3-031-56069-9_15

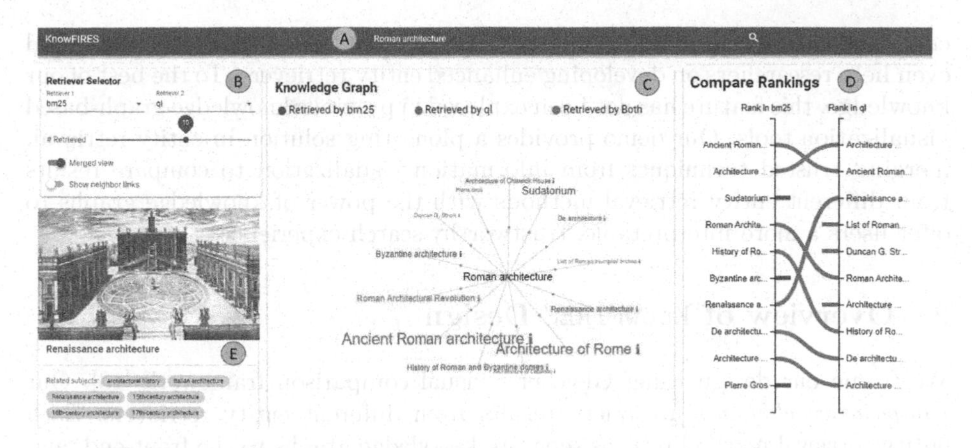

Fig. 1. *KnowFIRES* interface. A) Search Bar, B) Entity Retrieval Selection Panel, C) Knowledge Graph View, D) Entity Ranking Comparison View, and E) Related content.

differences of the retrieved entities on a per-retriever basis. By including information beyond the entity names, which reflects the relationship between them, we hope to allow searchers to gain additional insights through more explainable results.

Knowledge graph presentation of entities has been explored in domains such as biomedical and graph visualization [5,10,18,39,40]. However, to the best of our knowledge, `KnowFIRES` is the first visualization tool that combines *ad hoc* retrieval results with entity presentation in a unified framework and recreates the knowledge graph on top of these results. In `KnowFIRES`, we first retrieve relevant entities from a collection of entities and their metadata. Then, given the top-k most relevant entities, we build a knowledge graph and allow searchers to traverse this graph to better understand relationships between the entities, as well as their relationship to the original query. Furthermore, `KnowFIRES` offers searchers the option of accessing related content associated with each displayed entity, providing a more comprehensive understanding. More importantly, `KnowFIRES` allows easier comparison between different entity retrievers, by following the visual comparison conceptual framework proposed by Gleicher [14].

While the community has explored many different methods, there is no single retriever that can satisfy all possible use cases and applications [2,16]. Reasons include the variability of data sources and contextual variability. Additionally, performance trade-offs matter in real-world applications in terms of accuracy, speed, and resource requirements. Given these considerations, it is often necessary to use a combination of retriever methods that could complement each other for different purposes to satisfy the requirements of different use cases and applications [3,6,25]. To tackle this problem, in `KnowFIRES`, we visualize the degree of complementarity of a set of retrievers of interest and notably visualize the similarities and differences between their retrieved results. As such, `KnowFIRES` can be utilized for deciding the appropriate retriever on per application basis and even help researchers on developing enhanced entity retrievers. To the best of our knowledge, this feature has not been explored in previous knowledge graph-based visualization tools. Our demo provides a pioneering solution in entity retrieval, merging trusted techniques from information visualization to compare results from different entity retrieval methods with the power of knowledge graphs to offer users a more interpretable, trustworthy search experience.

2 Overview of KnowFIRES Design

We frame our design using Gleicher's visual comparison framework [13]. Our *comparative elements* are query results from different entity retrievers. Each entity retrieval method returns separate knowledge graphs to the front-end, and we contrast these results through our visualization. As for our *comparative challenges*, knowledge graphs' size and complexity are the greatest areas of concern. The visualization could become cluttered and uninterpretable for retrieved entities with a very dense set of relations. We address these challenges by incorporating various designs and making our graphs interactive, so the searcher can drag around the frame and zoom in on specific nodes. Our *comparative strategy* for dealing with the dense knowledge graph challenge is to trim our results with *subset selection*. The slider in our design enables the user to limit the number of retrieved entities from the entity retrievers. We typically expect a small number of entities to provide more than enough information. We include a toggle for visualizing one-hop connections from the retrieved entities. The one-hop linked entities add important contextual information, but allowing the users to remove them from the visualization was an important addition to enhance interpretability. Finally, our *comparative design*, *KnowFIRES* (Fig. 1), is a *superposition* or *juxtaposition* of knowledge graphs highlighting the similarities and differences between the results by colour-coding the knowledge graph's nodes.

2.1 Entity Retrieval Methods

We employ various entity retrieval method that are categorized into two groups: 1) *sparse retrievers*, which match exact keywords using inverted indexes, and 2) *dense retrievers*, which measure semantic similarity using dense vector representations [8,19,29,30]. More specifically, our sparse retrievers include **BM25** [30] and **QL** [36], with **BM25-PRF** and **QL-PRF** augmented by RM3 pseudo-relevance feedback. For dense retrievers, we use **SentenceBERT** [28,37] and **ColBERT** [20,31], each offering unique approaches to estimating relevance. This diversity enables us to explore the impact of sparse versus dense retrievers.

2.2 System Architecture

We employ DBpedia V2 [16,21], a vast knowledge base extracted from Wikimedia projects, as our entity corpus. While adaptable to other corpora, our demonstration uses the English subset of DBpedia version 2015-10 [11,16,17,26]. This subset mandates entities to have a title and abstract (rdfs:label and rdfs:comment predicates), excluding certain page types but including list pages. Our corpus comprises 4.6 million entities, each uniquely identified by URI. Entity retrievers were implemented in Python 3.8 using the pyserini library [22]. Data was served to the front-end via Flask and deployed on the backend server. The front-end communicates with the back-end through a REST API, enabling search queries, retriever selection, and result entity count specification, with responses provided in JSON format, including entity rank, score, relevant entities, and metadata. The front-end, implemented as a single-page *VueJS* application, utilizes the Vuetify framework for Material Design components like sliders, dropdowns, and grids. Visualizations rely on the *D3* library. For the knowledge graph, we employ the *Force-Graph* library, a D3 node-link diagram wrapper that integrates the D3-force library for node arrangement. The Entity Ranking Comparison View is custom-built with D3. During each search query, the front-end retrieves entities from each retriever through the backend API. JSON responses are parsed and stored to construct nodes and links in the Knowledge Graph view. In the final step, the code is production-ready, built using Webpack, and deployed on a Linux VPS using the *serve utility* . Finally, the code was built for production with Webpack and deployed on a Linux VPS using serve.

2.3 Visual Design

The Knowledge Graph View. The primary visual representation, as shown in Fig. 1-*C*. It displays the most relevant entities related to a query, with nodes color-coded based on each of the entity retrievers (green, blue, or magenta for common entities). The node size reflects the relevance of the entity to the query.
Superposition vs Juxtaposition. This feature offers two graph layouts. The superposition layout merges knowledge graphs from two retrieval methods, with color encodings helping differentiate between them. Alternatively, the juxtaposition layout shows two separate graphs, maintaining the color distinction.
Neighbour Links. It is possible to navigate through neighbor links of an entity, providing context about a particular node. This feature, includes grey links with descriptive text labels giving insight into the relationship between entities.
The Entity Ranking Comparison View. This interface, depicted in Fig. 1-*D*, enables users to compare entity rankings between retrieval methods. Entities from each method are shown side by side, with a magenta line connecting common entities, highlighting ranking disparities between the two methods.
The Additional Content Panel. Beyond the entity's name, we provide supplemental information, including a link to the entity's Wikipedia page, the entity type, related tags, and occasionally an image as shown in Fig. 1-*E*.

2.4 Interactivity

Our design prioritizes interactivity to empower users in exploring and comparing entity retrieval results. The entity retriever selection panel (Fig. 1-B) offers dropdowns to select entities for comparison, a slider to determine linked entity quantity and a toggle switch for juxtaposition or superposition knowledge graph views. Another toggle hides neighbor links for decluttering dense queries. In the Knowledge Graph View (Fig. 1-C), users can zoom using the scroll wheel, move nodes by click-and-drag, and click nodes for additional information in the *Related Content* panel. Clicking a node highlights its rank in the Compare Ranking panel (Fig. 1-D).

3 Concluding Remarks

In KnowFIRES, we visualize differences between entity retrievers within a knowledge graph, enhancing explainability and interpretability of retrieved results. Our demo showcases its efficiency and potential for broader applications. Future improvements include user-friendly visualizations, API integration, customization in graph traversal, and addition of metadata like external links.

References

1. Anand, A., Lyu, L., Idahl, M., Wang, Y., Wallat, J., Zhang, Z.: Explainable information retrieval: a survey. arXiv preprint arXiv:2211.02405 (2022)
2. Arabzadeh, N., Mitra, B., Bagheri, E.: Ms marco chameleons: challenging the ms marco leaderboard with extremely obstinate queries. In: Proceedings of the 30th ACM International Conference on Information and Knowledge Management, pp. 4426–4435 (2021)
3. Arabzadeh, N., Yan, X., Clarke, C.L.: Predicting efficiency/effectiveness trade-offs for dense vs. sparse retrieval strategy selection. In: Proceedings of the 30th ACM International Conference on Information and Knowledge Management, pp. 2862–2866 (2021)
4. Balog, K.: Entity retrieval (2018)
5. Barsky, A., Munzner, T., Gardy, J., Kincaid, R.: Cerebral: visualizing multiple experimental conditions on a graph with biological context. IEEE Trans. Visual. Comput. Graph. **14**(6), 1253–1260 (2008). https://doi.org/10.1109/TVCG.2008.117
6. Cormack, G.V., Clarke, C.L., Buettcher, S.: Reciprocal rank fusion outperforms condorcet and individual rank learning methods. In: Proceedings of the 32nd International ACM SIGIR Conference on Research and Development in Information Retrieval, pp. 758–759 (2009)
7. De Cao, N., Izacard, G., Riedel, S., Petroni, F.: Autoregressive entity retrieval. arXiv preprint arXiv:2010.00904 (2020)
8. Devlin, J., Chang, M., Lee, K., Toutanova, K.: BERT: pre-training of deep bidirectional transformers for language understanding. arXiv preprint arXiv:1810.04805 (2018)
9. Fetahu, B., Gadiraju, U., Dietze, S.: Improving entity retrieval on structured data. In: Arenas, M., et al. (eds.) The Semantic Web - ISWC 2015, pp. 474–491. Springer, Cham (2015). https://doi.org/10.1007/978-3-319-25007-6_28

10. Fujiwara, T., Zhao, J., Chen, F., Ma, K.L.: A visual analytics framework for contrastive network analysis. In: 2020 IEEE Conference on Visual Analytics Science and Technology (VAST), pp. 48–59. IEEE (2020). https://doi.org/10.1109/VAST50239.2020.00010
11. Gerritse, E.J., Hasibi, F., de Vries, A.P.: Graph-embedding empowered entity retrieval. In: Jose, J.M., et al.: (eds.) Advances in Information Retrieval: 42nd European Conference on IR Research, ECIR 2020, pp. 97–110. Springer, Cham (2020). https://doi.org/10.1007/978-3-030-45439-5_7
12. Gillick, D., et al.: Learning dense representations for entity retrieval. arXiv preprint arXiv:1909.10506 (2019)
13. Gleicher, M.: Considerations for visualizing comparison. IEEE Trans. Visual Comput. Graphics **24**(1), 413–423 (2017)
14. Gleicher, M., Albers, D., Walker, R., Jusufi, I., Hansen, C.D., Roberts, J.C.: Visual comparison for information visualization. Inf. Vis. **10**(4), 289–309 (2011)
15. Hasibi, F., Balog, K., Bratsberg, S.E.: Exploiting entity linking in queries for entity retrieval. In: Proceedings of the 2016 ACM International Conference on the Theory of Information Retrieval, pp. 209–218 (2016)
16. Hasibi, F., et al.: Dbpedia-entity v2: a test collection for entity search. In: Proceedings of the 40th International ACM SIGIR Conference on Research and Development in Information Retrieval, pp. 1265–1268 (2017)
17. Jafarzadeh, P., Amirmahani, Z., Ensan, F.: Learning to rank knowledge subgraph nodes for entity retrieval. In: Proceedings of the 45th International ACM SIGIR Conference on Research and Development in Information Retrieval, pp. 2519–2523 (2022)
18. Kang, H., Getoor, L., Shneiderman, B., Bilgic, M., Licamele, L.: Interactive entity resolution in relational data: a visual analytic tool and its evaluation. IEEE Trans. Visual Comput. Graph. **14**(5), 999–1014 (2008). https://doi.org/10.1109/TVCG.2008.55
19. Karpukhin, V., et al.: Dense passage retrieval for open-domain question answering. arXiv preprint arXiv:2004.04906 (2020)
20. Khattab, O., Zaharia, M.: Colbert: efficient and effective passage search via contextualized late interaction over bert. In: Proceedings of the 43rd International ACM SIGIR Conference on Research and Development in Information Retrieval, pp. 39–48 (2020)
21. Lehmann, J., et al.: DBpedia – a large-scale, multilingual knowledge base extracted from wikipedia. Semantic Web **6**(2), 167–195 (2015). https://doi.org/10.3233/SW-140134
22. Lin, J., Ma, X., Lin, S.C., Yang, J.H., Pradeep, R., Nogueira, R.: Pyserini: a python toolkit for reproducible information retrieval research with sparse and dense representations. In: Proceedings of the 44th International ACM SIGIR Conference on Research and Development in Information Retrieval, pp. 2356–2362 (2021)
23. Lin, X., Lam, W., Lai, K.P.: Entity retrieval in the knowledge graph with hierarchical entity type and content. In: Proceedings of the 2018 ACM SIGIR International Conference on Theory of Information Retrieval, pp. 211–214 (2018)
24. Macdonald, C., Ounis, I.: Voting for candidates: adapting data fusion techniques for an expert search task. In: Proceedings of the 15th ACM International Conference on Information and Knowledge Management, pp. 387–396 (2006)
25. McCabe, M.C., Chowdhury, A., Grossman, D.A., Frieder, O.: A unified environment for fusion of information retrieval approaches. In: Proceedings of the Eighth International Conference on Information and Knowledge Management, pp. 330–334 (1999)

26. Nikolaev, F., Kotov, A.: Joint word and entity embeddings for entity retrieval from a knowledge graph. In: Jose, J.M., et al. (eds.) Advances in Information Retrieval: 42nd European Conference on IR Research, ECIR 2020, pp. 141–155. Springer, Cham (2020). https://doi.org/10.1007/978-3-030-45439-5_10
27. Pound, J., Mika, P., Zaragoza, H.: Ad-hoc object retrieval in the web of data. In: Proceedings of the 19th International Conference on World Wide Web, pp. 771–780 (2010)
28. Reimers, N., Gurevych, I.: Sentence-bert: sentence embeddings using siamese bert-networks. arXiv preprint arXiv:1908.10084 (2019)
29. Robertson, S., Zaragoza, H., et al.: The probabilistic relevance framework: Bm25 and beyond. Found. Trends® Inf. Retriev. **3**(4), 333–389 (2009)
30. Robertson, S.E., Walker, S., Jones, S., Hancock-Beaulieu, M.M., Gatford, M., et al.: Okapi at trec-3. Nist Spec. Publ. SP **109**, 109 (1995)
31. Santhanam, K., Khattab, O., Saad-Falcon, J., Potts, C., Zaharia, M.: ColBERTv2: effective and efficient retrieval via lightweight late interaction. http://arxiv.org/abs/2112.01488
32. Sciavolino, C., Zhong, Z., Lee, J., Chen, D.: Simple entity-centric questions challenge dense retrievers. arXiv preprint arXiv:2109.08535 (2021)
33. Shehata, D., Arabzadeh, N., Clarke, C.L.A.: Early stage sparse retrieval with entity linking. arXiv:2208.04887 (2022)
34. Shehata, D., Arabzadeh, N., Clarke, C.L.A.: Early stage sparse retrieval with entity linking (2022)
35. Singh, J., Anand, A.: EXS: Explainable search using local model agnostic interpretability. In: Proceedings of the Twelfth ACM International Conference on Web Search and Data Mining, pp. 770–773 (2019)
36. Song, F., Croft, W.B.: A general language model for information retrieval. In: Proceedings of the Eighth International Conference on Information and Knowledge Management, pp. 316–321 (1999)
37. Thakur, N., Reimers, N., Daxenberger, J., Gurevych, I.: Augmented sbert: data augmentation method for improving bi-encoders for pairwise sentence scoring tasks (2021)
38. Wu, L., Petroni, F., Josifoski, M., Riedel, S., Zettlemoyer, L.: Scalable zero-shot entity linking with dense entity retrieval. arXiv preprint arXiv:1911.03814 (2019)
39. Zhao, J., Cao, N., Wen, Z., Song, Y., Lin, Y.R., Collins, C.: Fluxflow: visual analysis of anomalous information spreading on social media. IEEE Trans. Visual. Comput. Graph. **20**(12), 1773–1782 (2014). https://doi.org/10.1109/TVCG.2014.2346922
40. Zhao, J., Glueck, M., Breslav, S., Chevalier, F., Khan, A.: Annotation graphs: a graph-based visualization for meta-analysis of data based on user-authored annotations. IEEE Trans. Visual. Comput. Graph. **23**(1), 261–270 (2017). https://doi.org/10.1109/TVCG.2016.2598543

Physio: An LLM-Based Physiotherapy Advisor

Rúben Almeida[1(✉)], Hugo Sousa[1,2(✉)], Luís F. Cunha[1,2(✉)], Nuno Guimarães[1,2], Ricardo Campos[1,3,4], and Alípio Jorge[1,2]

[1] INESC TEC, Porto, Portugal
{ruben.f.almeida,hugo.o.sousa,luis.f.cunha,nuno.r.guimaraes, ricardo.campos,alipio.jorge}@inesctec.pt
[2] University of Porto, Porto, Portugal
[3] University of Beira Interior, Covilhã, Portugal
[4] Ci2 - Smart Cities Research Centre, Tomar, Portugal

Abstract. The capabilities of the most recent language models have increased the interest in integrating them into real-world applications. However, the fact that these models generate plausible, yet incorrect text poses a constraint when considering their use in several domains. Healthcare is a prime example of a domain where text-generative trustworthiness is a hard requirement to safeguard patient well-being. In this paper, we present Physio, a chat-based application for physical rehabilitation. Physio is capable of making an initial diagnosis while citing reliable health sources to support the information provided. Furthermore, drawing upon external knowledge databases, Physio can recommend rehabilitation exercises and over-the-counter medication for symptom relief. By combining these features, Physio can leverage the power of generative models for language processing while also conditioning its response on dependable and verifiable sources. A live demo of Physio is available at https://physio.inesctec.pt.

Keywords: Retrieval-augmented generation · Information extraction · Conversational health agents

1 Introduction

Although language models (LMs) have long been studied by the research community, they only reached mainstream attention with the release of the ChatGPT application by OpenAI [3]. This application granted the public access to a highly effective generative model, GPT-3.5 [15], that was capable of producing coherent conversations on various topics, a novelty at the time. This development naturally led to the emergence of numerous applications and discussions regarding the potential applications of generative models in various domains, such as law [16], education [17], and health [11,12]. However, these models also exhibited significant limitations that hindered their implementation in those domains. At the

R. Almeida, H. Sousa and L.F. Cunha—Equal contribution.

N. Goharian et al. (Eds.): ECIR 2024, LNCS 14612, pp. 189–193, 2024.
https://doi.org/10.1007/978-3-031-56069-9_16

top of that list is the hallucination problem [14], *i.e.*, their propensity to generate incorrect yet convincing answers. This limitation prompted increased research into grounding the text generated by these models on reliable sources, a task known as retrieval-augmented generation [13]. The general approach starts by retrieving documents relevant to the input query and subsequently using them to generate an answer. By doing so, one can link the generated texts to the original documents, thereby providing the user references where he/she can get more information supporting the generated answer [9,10]. This research gave rise to systems like BingChat [2] and Bard [1], search engines that combine the personalization of answers generated by LMs with the trustworthiness provided by the retrieval component of the system. This concept can be taken one step further to be applied to domain-specific applications by constraining the retrieval component to a specialized set of documents. This is the main idea behind the demo we present in this paper, Physio, a chat-based application tailored to help users in physical rehabilitation. Physio's answers are generated from an augmented prompt that compiles the user input with documents from a curated knowledge base specifically tailored to only contain reliable sources for physical rehabilitation. Furthermore, the sentences in the generated answer present references to those documents. Apart from that, Physio's answer can also contain exercise and over-the-counter medication recommendations whenever appropriate. The source code for Physio is open-source and available on GitHub[1].

2 Physio

To some extent, Physio can be thought of as an initial step toward an artificial intelligent physiatrist. To this end, we designed it to offer functionalities akin to those found in a physiatrist appointment, specifically: explaining/discussing the problem that the user is querying; and recommending exercises and medication when deemed appropriate. For the LM, Physio relies on OpenAI GPT-4 model which is the most effective LM at the time of development of this demo [14]. An example answer from Physio is illustrated in Fig. 1 for the prompt "I feel pain in my lower back. What can I do?" In the following sections, we explain how Physio's answer is built.

2.1 Knowledge-Base Construction

The construction of the knowledge for Physio commenced by scraping the Rehab Hero website [6]. This website boasts an extensive compilation of exercises, each accompanied by an instructional video demonstrating execution, and information on the specific physical conditions they address. After scraping this website, we compiled all the physical conditions that we found on the Rehab Hero pages and queried each of them in a set of 20 websites that provided more information

[1] https://github.com/liaad/physio.

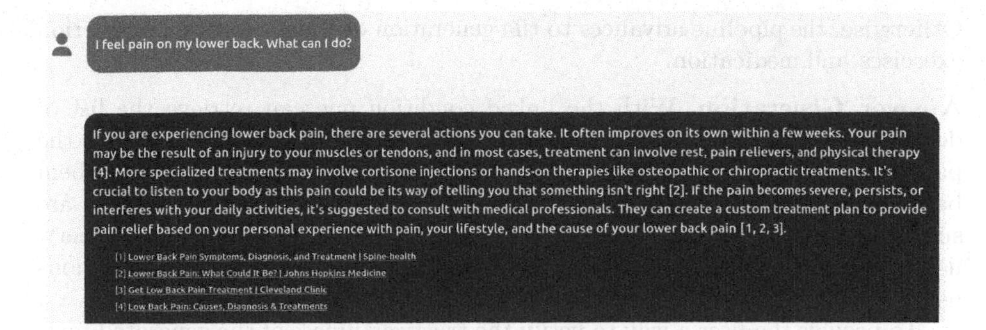

Fig. 1. Screenshot from Physio web demonstration. The user input is in the grey box, while the system answer is presented in the blue box. (Color figure online)

about the condition[2]. To ensure the reliability of our system the list of websites was curated and validated by a physiotherapist and includes sources like Mayo Clinic [4], NHS [7], and OrthoInfo [5]. For the medication-related aspects, we rely on the DrugBank database [18]. The information obtained was then indexed in a MongoDB database composed of three collections: `exercises`, `webpages`, and `medications`. The first two are indexed based on the associated condition, while the last is indexed by drug name.

2.2 Data Pipeline

When a user submits a query in Physio, the text undergoes processing through a data pipeline to generate a response. The initial step in this pipeline is to **verify if it is an English physiotherapy-related prompt**. This validation is achieved by using the LM and a predefined validation prompt template that assesses whether the user's input is related to physiotherapy and written in English. The validation prompt template conditions the LM's response to produce a boolean output (either "True" or "False") so that it can be programmatically interpreted. If the input is deemed invalid, the system provides a default response. Instead, if the input is validated, the application proceeds to the **condition identification** step, where it determines the condition in the user's question. This is accomplished by employing a few-shot template to instruct the LM to identify the condition. For instance, when presented with the input "I have sprained my ankle" the model should identify the condition as "ankle sprain."

Once the condition is identified, it is **linked to one of the entries in our database**. This linkage process first attempts an exact match, followed by a search in the list of aliases (for instance "lumbago" is in the list of aliases for "back pain"), and, as a last resort, employs substring matching. In case no match is found, the LM is prompted with the user query, and the answer is returned.

[2] More details about the list of websites selected can be found on the GitHub repository.

Otherwise, the pipeline advances to the generation of the answer and extraction exercises and medication.

Answer Generation. With the linked condition one can retrieve the list of documents related to that condition from the `webpages` collection. Among the pages available, we employ the BM25 retrieval model [8] to search and rank them based on their relevance to the user's input. The top five ranked documents are subsequently provided to the generative model, along with the user's input query, using a prompt template designed to instruct the model to answer user questions using the information contained in these pages.

To provide the user a way to verify the trustworthiness of the generated text, a list of references is incorporated by ranking the sentences from the original source pages in relation to the sentences in the generated text, again, using the BM25 ranking method. Note that determining the optimal number of references to include in the final answer is not trivial. In our application, we establish a heuristic to use as reference the top-N ranked sentence-document pairs, where N is the number of generated sentences.[3] However, the final answer may not necessarily contain the same number of references as sentences, as a generated sentence can be highly similar to multiple sentences within a given document.

Exercise and Medication Extraction. The linked condition is also used to fetch the exercises directly from the `exercises` collection, as they are indexed by condition. For the linked condition we randomly sample up to five exercises to be presented in the web interface.

The final element of the response pertains to medication recommendations. This is accomplished by instructing the LM to provide medication suggestions based on the user's query, the linked condition, and the generated answer. The prompt explicitly specifies that the response should be in the form of a JSON-parsable list of strings, where each string represents a medication. After parsing these strings, we conduct a search for the recommended drugs within the `medication` collection of our database, first by exact matching and subsequently with fuzzy matching.

The last task of the pipeline is to combine the three components of the answer, send them to the frontend, and cache the result in the database.

Acknowledgment. This work is financed by National Funds through the Fundação para a Ciência e a Tecnologia, within the project StorySense (DOI 10.54499/2022.09312.PTDC) and the Recovery and Resilience Plan within the scope of the Health From Portugal project.

Ethical Considerations. Given the sensitive nature of this domain, ethical considerations are paramount. As a result, we include a disclaimer on Physio's website, explicitly stating that it is a research demonstration, and we strongly advise users to consult with

[3] While this heuristic has been effective in practice, ongoing research is aimed at refining the reference selection process based on similarity scores.

a specialist before making any decisions regarding their health. Furthermore, we have limited medication recommendations to include only over-the-counter options.

References

1. Bard. https://bard.google.com/chat. Accessed 05 Oct 2023
2. Bing chat. https://www.microsoft.com/en-us/edge/features/bing-chat. Accessed 05 Oct 2023
3. Chatgpt. https://chat.openai.com/. Accessed 05 Oct 2023
4. Mayo clinic. https://www.mayoclinic.org/. Accessed 05 Oct 2023
5. Orthoinfo. https://orthoinfo.aaos.org/. Accessed 05 Oct 2023
6. Rehab hero. https://www.rehabhero.ca/. Accessed 05 Oct 2023
7. United kingdom national health service. https://www.nhs.uk/. Accessed 05 Oct 2023
8. Amati, G.: BM25, pp. 257–260. Springer US, Boston, MA (2009). https://doi.org/10.1007/978-0-387-39940-9_921
9. Gao, T., Yen, H., Yu, J., Chen, D.: Enabling large language models to generate text with citations (2023)
10. Huang, J., Chang, K.C.C.: Citation: A key to building responsible and accountable large language models (2023)
11. Kung, T.H., et al.: Performance of ch1ChatGPT on USMLE: potential for AI-assisted medical education using large language models. PLOS Digit. Health **2**(2), e0000198 (2023). https://doi.org/10.1371/journal.pdig.0000198
12. Levine, D.M., et al.: The Diagnostic and Triage Accuracy of the GPT-3 Artificial Intelligence Model. https://doi.org/10.1101/2023.01.30.23285067
13. Lewis, P., et al.: Retrieval-augmented generation for knowledge-intensive NLP tasks. In: Larochelle, H., Ranzato, M., Hadsell, R., Balcan, M., Lin, H. (eds.) Advances in Neural Information Processing Systems, vol. 33, pp. 9459–9474. Curran Associates, Inc. (2020). https://proceedings.neurips.cc/paper_files/paper/2020/file/6b493230205f780e1bc26945df7481e5-Paper.pdf
14. OpenAI: Gpt-4 technical report (2023)
15. Ouyang, L., et al.: Training language models to follow instructions with human feedback (2022)
16. Savelka, J.: Unlocking Practical Applications in Legal Domain: Evaluation of GPT for Zero-Shot Semantic Annotation of Legal Texts (2023). https://doi.org/10.1145/3594536.3595161,arXiv: 2305.04417
17. Savelka, J., Agarwal, A., Bogart, C., Song, Y., Sakr, M.: Can Generative Pre-trained Transformers (GPT) Pass Assessments in Higher Education Programming Courses? (2023). arXiv: 2303.09325
18. Wishart, D.S., et al.: DrugBank 5.0: a major update to the DrugBank database for 2018. Nucleic Acids Res. **46**(D1), D1074–D1082 (2018). https://doi.org/10.1093/nar/gkx1037

MathMex: Search Engine for Math Definitions

Shea Durgin, James Gore, and Behrooz Mansouri(✉)

University of Southern Maine, Portland, Maine, USA
{shea.durgin,james.gore,behrooz.mansouri}@maine.edu

Abstract. This paper introduces MathMex, an open-source search engine for math definitions. With MathMex, users can search for definitions of mathematical concepts extracted from a variety of data sources and types including text, images, and videos. Definitions are extracted using a fine-tuned SciBERT classifier, and the search is done with a fine-tuned Sentence-BERT model. MathMex interface provides means of issuing a text, formula, and combined queries and logging features.

Keywords: Math Search · Formula Search · Definition Extraction

1 Introduction

Math information retrieval refers to information retrieval where information needs are regarding math. When users issue math queries such as "Pythagorean theorem", they may be looking for its definition, examples, proof, or applications. This paper demonstrates MathMex,[1] named after Memex [2], the first hypothetical information retrieval system, carrying the idea of associative links in information. While this version primarily focuses on mathematical definitions, the ultimate goal of the MathMex project is to connect various pieces of information related to mathematical concepts, including definitions, proofs, and applications, in subsequent stages.

Our collection consists of math definitions from Wikipedia, community question-answering websites (both text and image), YouTube videos, and arXiv papers. To extract these definitions, a fine-tuned SciBERT [1] model is used to determine whether a sentence contains a definition. After extracting definitions, the semantic vector representations of definitions are extracted, using a fine-tuned Sentence-BERT [9] model. The vectors are then loaded in OpenSearch, where dense vector retrieval is performed by approximate k-NN search.

MathMex is the first math search engine, focusing on searches for specified information related to a mathematical concept. It is also the first search engine to provide a means of searching for various sources including text, images, and videos. Previous work on math search includes systems such as MathDeck [3]

S. Durgin and J. Gore—These authors contributed equally to this work

[1] https://www.mathmex.com.

N. Goharian et al. (Eds.): ECIR 2024, LNCS 14612, pp. 194–199, 2024.
https://doi.org/10.1007/978-3-031-56069-9_17

that focuses on different query editors for math formula query, and Approach0 [13] that provides means of searching over the Math Stack Exchange platform. Compared to these works, MathMex focuses on definitions and considers a wide variety of documents as the collection, indexing ~5.8M math definitions.

Fig. 1. MathMex Search Interface. The input query is "What is Pythagorean Theorem?"; the most relevant result from Wikipedia is returned.

2 MathMex Demo

This section introduces MathMex and its main features.

A. Collection. MathMex collection consists of the following sources:

- *ArXiv*: Papers from 2019 to July 2023. These papers are processed using their HTML format from ar5iv.[2]
- *Math CQ&A Websites*: Math Stack Exchange, MathOverFlow, and Mathematica Stack Exchange, using their July 2023 snapshot from the Internet Archive.
- *Wikipedia*: Math Wikipedia pages using NTCIR-12 [11] collection.
- *YouTube*: Math-related videos manually extracted from math courses.
- *Math-related Images*: Images from questions in CQ&A websites.

[2] https://ar5iv.labs.arxiv.org/.

B. Interface Features. MathMex utilizes various information sources. Users can filter the sources they wish to search through by employing the "Source" drop-down menu (see Fig. 1).

Using MathLive web component,[3] MathMex users can issue queries using text, LaTeX (for math), or both. This is achieved by switching the input mode of the search bar by pressing the tab or clicking the second button in the bar with T for text and $f(x)$ for math (see Fig. 2). Whenever the user pastes a text, the pasted characters will either be text or math based on the current mode.

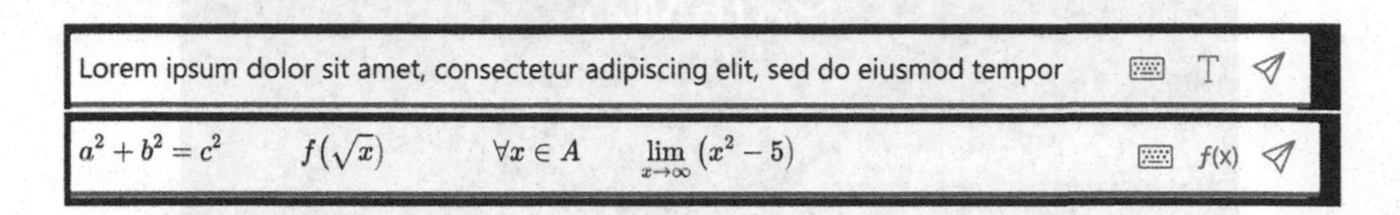

Fig. 2. MathMex Text and Formula Keyboards.

MathMex provides both text and math autocompletion features. For text mode, the suggested queries for autocompletion are determined after 3 characters are inserted (except for the first word that needs the first 5 characters). We use the bucket aggregation feature of OpenSearch on the MathMex collection to find the most common word that completes our last token. It first searches for the preceding word and if that does not return 5 unique results, it searches just using the incomplete token. The system then returns the top-5 ordered list of the potential completions of the current word of the user's query.

For the math mode, we rely on the MathLive library auto-complete feature. In this mode, after the user inserts '\', the suggested formula(e) will appear, with a live update after inserting the characters. For instance, when the user inserts '\si', queries such as sin, sigma, and sim are suggested for autocompletion.

C. Logging. Query logs are a valuable resource to study users' behavior through a search session [5]. MathMex offers explicit feedback buttons for each returned result (see Fig. 1). MathMex keeps a record of the user's activities by logging their feedback, issued query, returned definition, DateTime, and Session ID. A Session ID is assigned to a user after connecting to MathMex by assigning a UUID (universally unique ID). We plan to make these logs available to researchers every other month.

D. Indexing and Search. To extract definitions, we fine-tuned a SciBERT model [1] using data from the DEFT corpus [10], which is specifically designed for definition extraction from English text. One subtask involves determining whether a sentence contains a definition. For fine-tuning, we considered this as a binary classification task, determining whether a sentence contains a definition.

After fine-tuning the model, we utilized the spaCy[4] library to extract sentences from each document in the MathMex collection. For the text sources, each

[3] https://github.com/arnog/mathlive.
[4] https://spacy.io/.

sentence in the text was passed to the model. In the case of images, we focused on the question titles and implemented a filtering process using the CLIP [8] (Contrastive Language-Image Pre-training) model (*clip-ViT-L-14*). Images that fell below a 25% similarity threshold with their corresponding titles were removed. When it comes to YouTube videos, we considered their transcripts. Each sentence was passed through the fine-tuned SciBERT model, and if the classification result was positive with a score of ≥ 0.9, we categorized it as a definition and included it in the index.

MathMex indexing and search are done using a Sentence-BERT [9] model. We examined the bi-encoder model *'all-mpnet-base-v2'* search results, both with and without fine-tuning on ARQMath-3 test collection [7], and measured their effectiveness on the top-5 results. For fine-tuning, we used the data from ARQMAth-1 [12] and -2 [6]. ARQMath test collection provides a set of math questions, each accompanied by associated answers labeled with relevance scores 0, 1, 2, and 3.

To fine-tune the Sentence-BERT model, with ARQMath data, we considered all answers with a relevance score of 0 as non-relevant, and those with a score higher than 0, as relevant. For optimization, we utilized a cosine similarity loss function, with 10 epochs, and a batch size of 16. As presented in Table 1, the fine-tuned model provided better effectiveness across all measures. Therefore, the MathMex search model is fine-tuned using ARQMath-1, -2, and -3 test collections with a 95:5 training validation split, using the same parameters as used in our initial experiment.

Table 1. Model Results on ARQMath-3

Model	MRR@5	P@5	NDCG@5
all-mpnet-base-v2	0.77	0.62	0.41
Fine-Tuned all-mpnet-base-v2	**0.81**	**0.63**	**0.45**

For indexing, each definition is passed to the fine-tuned Sentence-BERT model, and its semantic vector representation is loaded in OpenSearch. For retrieval, the input query's vector is generated (using the same model). This vector is then compared to the definition vectors through an approximate k-NN search using *nmslib* and *Faiss* [4]. The user's chosen sources are individually searched, and the top-5 results from each source are then fused together to form the final search results. The results are fused by selecting the highest-ranked results from each source until the top-5 results are filled.

E. Search Results. MathMex provides the most relevant definition for an input query, with the next top-4 relevant results being presented, upon clicking on "Next Result" button. Users are provided with the relevance level of each retrieved instance: high, medium, or low. We evaluated the relevance levels using the cosine similarity score, where values greater than or equal to 80% are considered high, values between 60% and 80% are deemed medium, and values less than 60% are categorized as low. For each returned item, a link to the source is available, and for YouTube videos, users are directed to specific seconds of the video.

3 Conclusion

In this paper, we demonstrated MathMex, an open-source search engine for math definitions. While users can search over a text collection, MathMex stands out as the first solution to address the challenge of searching for mathematical content within image and video resources. Moreover, it offers the flexibility of accommodating input queries in both text and mathematical modalities, empowering users to express their specific information needs. For future work, we plan to extend MathMex's capabilities to encompass a broader spectrum of information related to mathematical concepts, including proofs and practical applications.

References

1. Beltagy, I., Lo, K., Cohan, A.: SciBERT: a pretrained language model for scientific text. In: Proceedings of the 2019 Conference on Empirical Methods in Natural Language Processing and the 9th International Joint Conference on Natural Language Processing (EMNLP-IJCNLP) (2019)
2. BUSH, V.: As we may think. Atlantic Monthly (1945)
3. Diaz, Y., Nishizawa, G., Mansouri, B., Davila, K., Zanibbi, R.: The mathdeck formula editor: interactive formula entry combining LaTeX, structure editing, and search. In: Extended Abstracts of the 2021 CHI Conference on Human Factors in Computing Systems
4. Johnson, J., Douze, M., Jegou, H.: Billion-scale similarity search with GPUs. IEEE Trans. Big Data **7**(3), 535–547 (2019)
5. Mansouri, B., Zanibbi, R., Oard, D.: Characterizing searches for mathematical concepts. In: 2019 ACM/IEEE Joint Conference on Digital Libraries (JCDL) (2019)
6. Mansouri, B., Zanibbi, R., Oard, D., Agarwal, A.: Overview of ARQMath-2 (2021): second CLEF lab on answer retrieval for questions on math. In: Experimental IR Meets Multilinguality, Multimodality, and Interaction: 12th International Conference of the CLEF Association, CLEF 2021 (2021)
7. Mansouri, B., Novotny, V., Agarwal, A., Oard, D., Zanibbi, R.: Overview of ARQMath-3 (2022): third CLEF lab on answer retrieval for questions on math. In: International Conference of the Cross-Language Evaluation Forum for European Languages (2022)
8. Radford, A., et al.:Others learning transferable visual models from natural language supervision.In: International Conference On Machine Learning (2021)
9. Reimers, N., Gurevych, I.: Sentence-BERT: sentence embeddings using siamese BERT-networks. In: Proceedings of the 2019 Conference on Empirical Methods in Natural Language Processing and the 9th International Joint Conference on Natural Language Processing (EMNLP-IJCNLP) (2019)
10. Spala, S., Miller, N., Yang, Y., Dernoncourt, F., Dockhorn, C.: DEFT: a corpus for definition extraction in free-and semi-structured text. In: Proceedings of the 13th Linguistic Annotation Workshop (2019)
11. Zanibbi, R., Aizawa, A., Kohlhase, M., Ounis, I., Topic, G., Davila, K.: NTCIR-12 MathIR task overview. In: NTCIR (2016)

12. Zanibbi, R., Oard, D., Agarwal, A., Mansouri, B.: Overview of ARQMath 2020: CLEF lab on answer retrieval for questions on math. In: Experimental IR Meets Multilinguality, Multimodality, and Interaction: 11th International Conference of the CLEF Association, CLEF 2020 (2020)
13. Zhong, W., Zanibbi, R.: Structural similarity search for formulas using leaf-root paths in operator subtrees. In: Advances in Information Retrieval: 41st European Conference on IR Research (2019)

XSearchKG: A Platform for Explainable Keyword Search over Knowledge Graphs

Leila Feddoul[1(✉)], Martin Birke[1], and Sirko Schindler[2]

[1] Heinz Nixdorf Chair for Distributed Information Systems, Friedrich Schiller University Jena, Jena, Germany
{leila.feddoul,martin.birke}@uni-jena.de

[2] Institute of Data Science, German Aerospace Center DLR, Jena, Germany
sirko.schindler@dlr.de

Abstract. One of the most user-friendly methods to search over knowledge graphs is the usage of keyword queries. They offer a simple text input that requires no technical or domain knowledge. Most existing approaches for keyword search over graph-shaped data rely on graph traversal algorithms to find connections between keywords. They mostly concentrate on achieving efficiency and effectiveness (accurate ranking), but ignore usability, visualization, and interactive result presentation. All of which offer better support to non-experienced users. Moreover, it is not sufficient to just show a raw list of results, but it is also important to explain why a specific result is proposed. This not only provides an abstract view of the capabilities and limitations of the search system, but also increases confidence and helps discover new interesting facts. We propose XSearchKG, a platform for explainable keyword search over knowledge graphs that extends our previously proposed graph traversal-based approach and complements it with an interactive user interface for results explanation and browsing.

Keywords: Explainability · Keyword Search · Knowledge Graph

1 Introduction

Keyword search is a familiar method for information retrieval that can also be adapted to enable access to knowledge graphs (KGs). Using this common search paradigm is intuitive and especially suitable for end users. It allows to take advantage of the graph nature of the data by exploring it and thus discovering prominent associations and deriving deductions. Moreover, it does not require any technical expertise (e.g., SPARQL [1]) or knowledge of the underlying data structure. Recently, various approaches were developed to solve the problem of keyword search over KGs [4,6,9,14,15,17]. Most of those systems aim at efficiently finding and ranking subgraph-shaped structures that connect the graph elements corresponding to the keywords. However, only very few systems propose a user interface for result presentation [2,8,10,13]. This reduces the suitability

N. Goharian et al. (Eds.): ECIR 2024, LNCS 14612, pp. 200–205, 2024.
https://doi.org/10.1007/978-3-031-56069-9_18

for end users. Furthermore, none of the systems aims at explaining why a specific item appears among the search results in a detailed way. We observe in some cases only the usage of basic features such as the highlighting of keyword mentions in the search result snippets [2,10]. Recently, the concept of explainability has gained importance especially in the field of artificial intelligence (AI), where the aim is to support understanding the predictions performed by machine learning models [5], in order to increase user trust and causality [11]. However, this is not only relevant to AI, but also other use cases such as recommendation systems or information retrieval in general. In the context of keyword search over KGs, we define the explainability of a system as its capability to answer the following questions: *Why is a specific result relevant to the query?* and *Why is a specific result more or less important compared to others?*.

In this demo paper, we present XSearchKG, a full-stack web application for explainable keyword search over KGs. The latter is based on our approach proposed in [6,7][1], where first, the off-line phase consists of summarizing the input KG and assigning query-independent costs to its edges. The second (runtime) phase receives user keywords and a target (types of entities to be retrieved) as input, finds the nodes corresponding to the keywords and target in the summary graph, explores the graph to find subgraphs connecting keywords and target, ranks the subgraph templates (schema-level), and generates SPARQL queries corresponding to the graphs. We extend the pipeline by combining the results generated by executing the top-k SPARQL queries and propose a ranking of retrieved entities based on the scores of their subgraph templates. The implemented features to support explainability are detailed in Sect. 2. These are mainly based on giving the user more transparency about the generated results leveraging the connections in the schema and instance levels of the graph. The proposed visualizations allow a better understanding of the results without overwhelming the user. Moreover, they enable the discovery of new insights by getting explanations for results that may appear not correct at first glance. The conducted user evaluation is shortly described in Sect. 3. The source code is publicly available on GitHub [3][2]. We will demonstrate the system using MONDIAL [12] as input KG. We will use queries of different complexity levels and explore some generated results. We will also go through the different explanation views and show possible user interactions (e.g., ignore specific connections).

2 XSearchKG

The application comprises two fundamental building blocks: (1) A Node.js Express server (backend) produces ranked results with essential information such as labels, descriptions, and subgraph bindings. To avoid regenerating identical data

[1] Note that the used code version uses the cosine similarity between the embeddings of the node labels as a semantic relatedness metric. The embeddings are generated using the Universal Sentence Encoder lite.

[2] https://github.com/fusion-jena/XSearchKG (a link to the demo video can be found in the README of the repository).

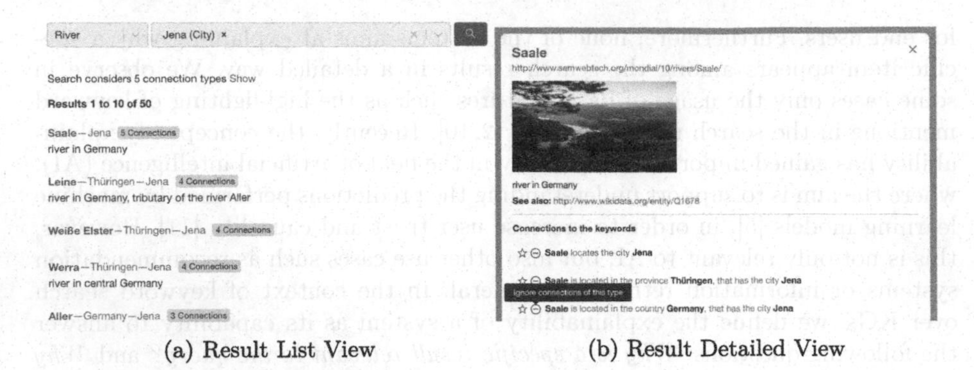

(a) Result List View (b) Result Detailed View

Fig. 1: Screenshots of some of the system views.

and improve performance, we utilize both standard HTTP cache headers and in-memory caches. The runtime phase of [6,7] was integrated into the backend component. (2) A React app (frontend) provides a user-friendly and responsive web interface. The application requires a running SPARQL endpoint and the files representing the summary graph generated at the first off-line phase of [6,7]. More details about the architecture are provided on GitHub. In the following, we give an overview about the different platform's features:

Input Suggestion. To perform a search, the user must enter a target type corresponding to a node on the summary graph and a set of keywords representing resources on the KG. Therefore, we assist with a suggestion service that provides appropriate candidates based on the labels allowing the user to scroll through the list and get inspiration.

Result List View. The result list (Fig. 1a) shows the individual results with a label and a text description. If the KG does not provide this information we try to retrieve it from semantically identical resources in other KGs[3]. Results are ranked based on the subgraph template scores, where entities with bindings from a larger number of templates will be higher ranked. The view also offers a first glimpse of how the result is related to the keywords by showing a binding of the subgraph with the highest ranking. As in all views, we display the subgraphs in the form of a tree, with the result being the root on the left side and the keywords being (typically) the leaf nodes on the right side. This provides a clear reading direction. Since we do not want to clutter the result list, we show a sparse view consisting of only the nodes of the subgraph omitting any edge labels together with the number of subgraph templates that have a binding for this result to show results' importance. For example, in Fig. 1a, the user can see that the river *Saale* is directly connected to the keyword *Jena* while the

[3] In MONDIAL, this was done by leveraging `owl:sameAs` links to Wikidata [16].

top subgraphs of the other results contain an intermediate node (*Thüringen* and *Germany*, respectively).

Result Detailed View. Clicking on an item in the result list opens an overlay with detailed information about the result (Fig. 1b). In addition to a description and an image, this includes a list of all subgraph templates that have a binding for this result. In this view, we provide meaningful edge labels to help understand the specific connection. If a subgraph template has multiple bindings, we will indicate this and allow the user to load more example bindings. We also offer two interaction options for each subgraph: (1) If a user considers a connection to be irrelevant, it can be ignored in the result generation by clicking on the minus icon, and (2) if a connection exactly matches the user's intention, it can be used as the only subgraph template by clicking on the star icon.

Connection Types View. Above the result list (Fig. 1a), we show the number of subgraph templates that are considered in the result generation. By clicking the *Show* link, the user gets a list of all subgraph templates. This representation is similar to the one in the detailed view, but instead of bindings, the node types are shown. We also offer the same types of interactions. By clicking on any of them, the user immediately sees the effect on the result set, allowing for experimentation and finding the subset of subgraph templates that best fits the intent.

3 Evaluation

We conducted a qualitative user evaluation in live sessions with 6 participants with heterogeneous backgrounds in terms of domain and technological knowledge. We observed, that the users considered the top search results to be mostly relevant and reasonably ordered. Further down in the result list, participants often questioned whether individual results were correct. With the subgraphs provided, they were able to understand the context and to decide whether or not to consider the result. By ignoring irrelevant subgraph templates in the result generation, participants were able to obtain result lists that better matched their intentions. In addition, the evaluation revealed a number of usability issues that we addressed by making mostly minor changes to the user interface, such as providing easier access to the interactions.

4 Conclusion and Future Work

We introduced XSearchKG, a platform that makes the process of keyword search over KGs more transparent to the end users and allows more interaction capabilities. We designed features with various levels of abstraction for result explanation, ranging from sparse to detailed views. The system has a modular client-server architecture that communicates with a triple store endpoint and uses different caching mechanisms. An initial user evaluation in live sessions revealed that the proposed interface is intuitive in most of the cases without prior knowledge about KGs or the used algorithm. This also helped improve some features by leveraging the feedback of the users. We plan to test the platform using larger KGs and conduct more extensive evaluations.

References

1. Sparql 1.1 query language: W3c recommendation 21 March 2013 (2013). https://www.w3.org/TR/sparql11-query/
2. Bhalotia, G., Hulgeri, A., Nakhe, C., Chakrabarti, S., Sudarshan, S.: Keyword searching and browsing in databases using BANKS. In: Proceedings of the 18th International Conference on Data Engineering, San Jose, CA, USA, February 26 - March 1, 2002, pp. 431–440. IEEE Computer Society (2002). https://doi.org/10.1109/ICDE.2002.994756
3. Birke, M., Feddoul, L.: fusion-jena/XSearchKG v1.0.0 (Oct 2023). https://doi.org/10.5281/zenodo.8430121
4. Dosso, D., Silvello, G.: Search text to retrieve graphs: a scalable RDF keyword-based search system. IEEE Access **8**, 14089–14111 (2020). https://doi.org/10.1109/ACCESS.2020.2966823
5. Došilović, F.K., Brčić, M., Hlupić, N.: Explainable artificial intelligence: a survey. In: 2018 41st International Convention on Information and Communication Technology, Electronics and Microelectronics (MIPRO), pp. 0210–0215 (2018). https://doi.org/10.23919/MIPRO.2018.8400040
6. Feddoul, L.: Semantics-driven Keyword Search over Knowledge Graphs. In: Proceedings of the Doctoral Consortium at ISWC 2020 co-located with 19th International Semantic Web Conference (ISWC 2020), Athens, Greece, November 3rd, 2020. CEUR Workshop Proceedings, vol. 2798, pp. 17–24. CEUR-WS.org (2020). https://ceur-ws.org/Vol-2798/paper3.pdf
7. Feddoul, L.: Top-k Query generator (Semantics-driven Keyword Search over Knowledge Graphs) (Oct 2023). https://doi.org/10.5281/zenodo.8414093
8. García, G., Izquierdo, Y., Menendez, E., Dartayre, F., Casanova, M.A.: RDF keyword-based query technology meets a real-world dataset. In: Proceedings of the 20th International Conference on Extending Database Technology, EDBT 2017, Venice, Italy, March 21–24, 2017, pp. 656–667. OpenProceedings.org (2017). https://doi.org/10.5441/002/edbt.2017.86
9. Ghanbarpour, A., Niknafs, K., Naderi, H.: Efficient keyword search over graph-structured data based on minimal covered r-cliques. Frontiers Inf. Technol. Electron. Eng. **21**(3), 448–464 (2020). https://doi.org/10.1631/FITEE.1800133
10. Kadilierakis, G., Nikas, C., Fafalios, P., Papadakos, P., Tzitzikas, Y.: Elas4RDF: Multi-perspective triple-centered keyword search over RDF using elasticsearch. In: Harth, A., et al. (eds.) ESWC 2020. LNCS, vol. 12124, pp. 122–128. Springer, Cham (2020). https://doi.org/10.1007/978-3-030-62327-2_21
11. Lipton, Z.C.: The mythos of model interpretability. Commun. ACM **61**(10), 36–43 (sep 2018). https://doi.org/10.1145/3233231
12. May, W.: Information extraction and integration with Florid: The Mondial case study. Tech. Rep. 131, Universität Freiburg, Institut für Informatik (1999). https://dbis.informatik.uni-goettingen.de/Mondial

13. Menendez, E.S., Casanova, M.A., Leme, L.A.P.P., Boughanem, M.: Novel node importance measures to improve keyword search over RDF graphs. In: Database and Expert Systems Applications - 30th International Conference, DEXA 2019, Linz, Austria, August 26–29, 2019, Proceedings, Part II. Lecture Notes in Computer Science, vol. 11707, pp. 143–158. Springer (2019). https://doi.org/10.1007/978-3-030-27618-8_11
14. Shi, Y., Cheng, G., Kharlamov, E.: Keyword search over knowledge graphs via static and dynamic hub labelings. In: WWW '20: The Web Conference 2020, Taipei, Taiwan, April 20–24, 2020, pp. 235–245. ACM / IW3C2 (2020). https://doi.org/10.1145/3366423.3380110
15. Shi, Y., Cheng, G., Tran, T., Kharlamov, E., Shen, Y.: Efficient computation of semantically cohesive subgraphs for keyword-based knowledge graph exploration. In: WWW '21: The Web Conference 2021, Virtual Event / Ljubljana, Slovenia, April 19–23, 2021, pp. 1410–1421. ACM / IW3C2 (2021). https://doi.org/10.1145/3442381.3449900
16. Vrandečić, D., Krötzsch, M.: Wikidata: a free collaborative knowledgebase. Commun. ACM **57**(10), 78–85 (2014). https://doi.org/10.1145/2629489
17. Zhang, Z., Yu, J.X., Wang, G., Yuan, Y., Chen, L.: Key-core: cohesive keyword subgraph exploration in large graphs. World Wide Web **25**(2), 831–856 (2022). https://doi.org/10.1007/s11280-021-00926-y

Result Assessment Tool: Software to Support Studies Based on Data from Search Engines

Sebastian Sünkler[1(✉)], Nurce Yagci[1], Sebastian Schultheiß[1], Sonja von Mach[1], and Dirk Lewandowski[1,2]

[1] Department Information, Media and Communication, Hamburg University of Applied Sciences, Finkenau 35, 22081 Hamburg, Germany
{sebastian.suenkler,nurce.yagci,sebastian.schultheiss, sonja.vonmach,dirk.lewandowski}@haw-hamburg.de, dirk.lewandowski@uni-due.de

[2] University of Duisburg-Essen, Campus Duisburg, 47057 Duisburg, Germany

Abstract. The Result Assessment Tool (RAT) is a software toolkit for conducting research with results from commercial search engines and other information retrieval (IR) systems. The software integrates modules for study design and management, automatic collection of search results via web scraping, and evaluation of search results in an assessment interface using different question types. RAT can be used for conducting a wide range of studies, including retrieval effectiveness studies, classification studies, and content analyses.

Keywords: search engine evaluation · web scraping · retrieval tests

1 Introduction

Conducting research using search engine data is challenging. Data from commercial search engines is not publicly available, and only a few search engines offer an API for accessing search results. In addition, recruiting jurors to evaluate search results is challenging, making it difficult to conduct studies on a large scale.

The information retrieval (IR) community has been developing software tools to conduct retrieval effectiveness studies for decades. However, the tools primarily consist of one-time use tools (e.g., [1–3]), prototypes that have not been developed further [4, 5], and software to be used with test collections instead of real-world data [6–8] or specific use cases [9, 10]. Therefore, we propose the Result Assessment Tool (RAT) as a sustainable solution that integrates all the necessary steps to conduct studies based on search engine data.

2 Significance of the Result Assessment Tool

On the one hand, the significance of the RAT derives from the need for a sustainable solution within the IR community to conduct large-scale studies based on search engine data. On the other hand, RAT is not limited to IR. In the field of health, for instance,

N. Goharian et al. (Eds.): ECIR 2024, LNCS 14612, pp. 206–211, 2024.
https://doi.org/10.1007/978-3-031-56069-9_19

health experts evaluated the quality (e.g., [11]) or manually coded the content (e.g., [12]) of health-related search results. In addition, researchers in the field of media and communication science classified search results e.g. based on content types and ideological biases (e.g., [13]). Since such studies usually rely on small data sets that are manually collected, evaluated, and analyzed, we see great potential for RAT, as its scalability can improve the studies by providing larger data sets and making jurors' work easier. Due to its modular construction, RAT possesses a high level of adaptability. Thus, its functionality can be expanded based on the needs of its users by adding new modules.

3 RAT Use Cases

Since evaluating the quality of information retrieval systems is an everyday use case when analyzing search engine data, the Result Assessment Tool was initially developed to conduct retrieval effectiveness studies. Even though quality is a multifaceted concept, the retrieval efficiency of search engines remains the foundation of all comprehensive quality evaluations. The retrieval effectiveness of Google and Bing was the subject of a study conducted with this early RAT version [14]. The jurors evaluated search results for 1,000 informational and navigational queries using a crowdsourcing strategy. Conducting classification studies is another use case. RAT supports both manual (i.e., by hand) and automatic (i.e., algorithmic) classifications of search results. The researchers or judges can perform the manual classification through the assessment interface. Automatic classifications can be accomplished by utilizing existing classifiers or by adding one's own classifiers. We refer to the study conducted with RAT by Hinz et al. [15] concerning automatic result classification. The authors analyzed whether candidates use search engine optimization (SEO) on their personal websites for the 2021 federal election[1].

RAT is also capable of facilitating content analysis based on search results. An example is the work by Haider et al. [17]. For the Swedish term for wind power (vindkraft), the query sampler extension [18] generated 252 queries, for which the RAT scraped 5,710 search results[2].

The software also supports domain analyses, such as comparisons between search engines and countries. Yagci et al. [19] analyzed the source diversity of Google and alternative search engines and the degree to which their root domains overlap. The top 10 search results from Google, Bing, DuckDuckGo, and MetaGer were scraped for 3,537 queries, resulting in 141,480 results.

4 Functionality of RAT

The Result Assessment Tool is an adaptable web-based software toolkit built with Python, the PostgreSQL database, and Selenium for web scraping. Researchers can use a web interface to design studies, while participants evaluate search results for predefined

[1] The SEO-classifier implementation is described at [16].

[2] We developed a script that generates search queries based on keyword suggestions generated by the Google Ads API: https://developers.google.com/google-ads/api/.

questions using the same interface. In addition to traditional IR studies, classification studies, and data analyses, qualitative content analyses are also possible due to the modular design of the toolkit. We develop the software based on the principles of user-centered design (UCD). The evaluation of the usability of the software is an integral part of the UCD process [20, 21], which is why continuous usability tests and heuristic evaluations are conducted. Figure 1 shows the software architecture of RAT with its applications and modules.

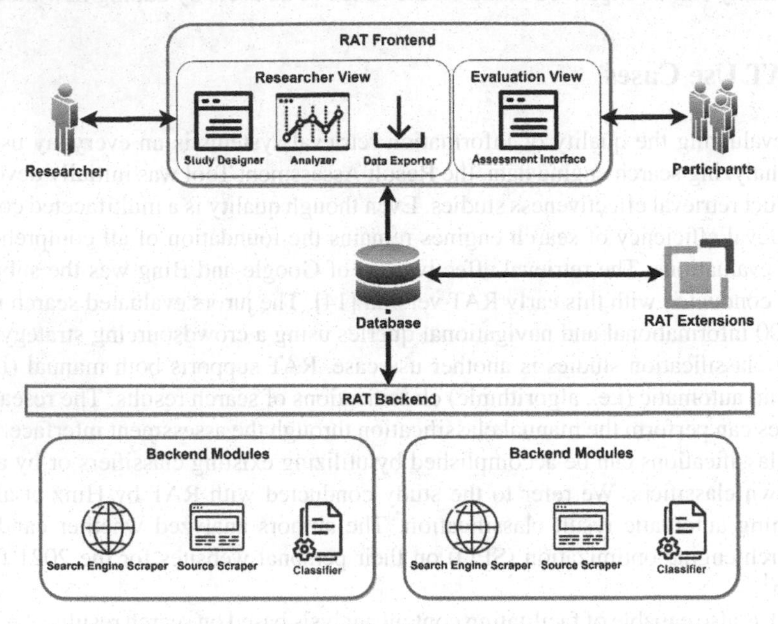

Fig. 1. Overview of the applications and their modules in RAT

The software consists of two applications that can be installed on separate computers and are linked through the database. This allows researchers to share the resources required for time-consuming and computationally intensive processes. The backend application provides scraping processes and classification tasks, while the frontend application provides a graphical interface for researchers to design studies and for study participants to evaluate search results. In addition, we provide an infrastructure for developers and researchers to create extensions for RAT that will be connected to RAT through the database.

The RAT Frontend application is a Flask GUI designed for researchers and study participants. It includes a Researcher View for designing studies and analyzing results and an Evaluation View for collecting participant assessments. The Study Designer is the basic module researchers use to define the study type, the assessment result type (search results and/or snippets from search result pages), and the type of access to the assessment interface. In the Researcher View, researchers can invite participants and define questions using Likert scales, open-ended questions, sliders, and multiple-choice questions. The Analyzer module computes and reports statistics about the study, including search queries, the expected number of results, and evaluation statistics. The Evaluation View in the RAT Frontend allows participants to register using a link provided by researchers. This approach enables anonymous access to participation and does not collect identifiable data. Answers are stored in a database, accessible using the Data Exporter to download the results.

The RAT Backend application processes inputs from the Researcher View in the RAT Frontend. The main module in the backend is the result scraper. During the scraping of results, metadata, source code, and a screenshot of the result are collected. The result scraper is based on Selenium, a test suite for web applications. A framework for automatically adding classifiers to analyze search results has also been implemented. The RAT Backend's architecture is based on a job management system using the Advanced Python Scheduler library (APScheduler). Jobs for all modules are created through inputs in the Researcher View at the RAT Frontend, and search results are collected by the Search Engine Scraper. Alternatively, lists of uniform resource locators (URLs) can be uploaded to be made available for assessment. Researchers can use scrapers we already provide (Google, Bing, DuckDuckGo) or add their search engine scrapers. The classification module allows automatic classifications based on the collected data. RAT provides the possibility of adding any classifier using templates and the database.

Availability of Software Demo, Source Code, and Research Data

To adhere to the Findability, Accessibility, Interoperability, and Reusability (FAIR) principles [22], we make the research data on the studies we conducted with the RAT available via the Open Science Framework (OSF)[3], provide a software demo[4], and provide the source code[5]. This is part of our sustainability strategy, which also serves as a marketing measure for building an international community of researchers and developers. Registering an account through the launch demo button is necessary for the software demo. The user is supported in creating the study within the tool.

Acknowledgments. This work is funded by the German Research Foundation (Deutsche Forschungsgemeinschaft (DFG); Grant No. 460676551).

References

1. Bar-Ilan, J., Levene, M.: A method to assess search engine results. Online Inf. Rev. **35**, 854–868 (2011). https://doi.org/10.1108/14684521111193166
2. Tawileh, W., Griesbaum, J., Mandl, T.: Evaluation of five web search engines in Arabic language. In: Atzmüller, M., Benz, D., Hotho, A., and Stumme, G. (eds.) Proceedings of LWA 2010, Kassel, Germany, pp. 1–8 (2010)
3. Trielli, D., Diakopoulos, N.: Partisan search behavior and Google results in the 2018 U.S. midterm elections. Inf. Commun. Soc. 1–17 (2020). https://doi.org/10.1080/1369118X.2020.1764605

[3] Repository for the research data generated with RAT: https://osf.io/t3hg9/

[4] The RAT software demo is available at https://rat-software.org/

[5] Repository for the source code: https://github.com/rat-software/

4. Lingnau, A., Ruthven, I., Landoni, M., van der Sluis, F.: Interactive search interfaces for young children - the PuppyIR approach. In: 2010 10th IEEE International Conference on Advanced Learning Technologies, pp. 389–390. IEEE (2010). https://doi.org/10.1109/ICALT.2010.111
5. Renaud, G., Azzopardi, L.: SCAMP. In: Proceedings of the 4th Information Interaction in Context Symposium on - IIIX 2012, pp. 286–289. ACM Press, New York (2012). https://doi.org/10.1145/2362724.2362776
6. Dussin, M., Ferro, N.: Design of a digital library system for large-scale evaluation campaigns. In: Christensen-Dalsgaard, B., Castelli, D., Ammitzbøll Jurik, B., Lippincott, J. (eds.) Research and Advanced Technology for Digital Libraries, pp. 400–401. Springer, Heidelberg (2008). https://doi.org/10.1007/978-3-540-87599-4_45
7. Koopman, B.: Semantic search as inference. ACM SIGIR Forum. (2014). https://doi.org/10.1145/2701583.2701601
8. Ogilvie, P., Callan, J.P.: Experiments using the lemur toolkit. In: Proceedings of Tenth Text REtrieval Conference, TREC 2001, Gaithersburg, MD, USA, 13–16 November 2001 (2001)
9. Digitalmethods: DMI Tools. https://wiki.digitalmethods.net/Dmi/ToolDatabase. Accessed 23 Feb 2023
10. Thelwall, M.: Introduction to webometrics: quantitative web research for the social sciences. Synth. Lect. Inf. Concepts Retr. Serv. (2009). https://doi.org/10.2200/s00176ed1v01y200903icr004
11. Janssen, S., Käsmann, L., Fahlbusch, F.B., Rades, D., Vordermark, D.: Side effects of radiotherapy in breast cancer patients: the Internet as an information source. Strahlenther. Onkol. Organ Dtsch. Rontgengesellschaft Al **194**, 136–142 (2018). https://doi.org/10.1007/s00066-017-1197-7
12. Rachul, C., Marcon, A.R., Collins, B., Caulfield, T.: COVID-19 and 'immune boosting' on the internet: a content analysis of Google search results. BMJ Open **10**, e040989 (2020). https://doi.org/10.1136/bmjopen-2020-040989
13. Ballatore, A.: Google chemtrails: a methodology to analyze topic representation in search engine results. First Monday **20** (2015)
14. Lewandowski, D.: Evaluating the retrieval effectiveness of web search engines using a representative query sample. J. Assoc. Inf. Sci. Technol. **66**, 1763–1775 (2015). https://doi.org/10.1002/asi.23304
15. Hinz, K., Sünkler, S., Lewandowski, D.: SEO im Wahlkampf. In: Korte, K.-R., Schiffers, M., von Schuckmann, A., Plümer, S. (eds.) Die Bundestagswahl 2021, pp. 1–28. Springer, Wiesbaden (2023). https://doi.org/10.1007/978-3-658-35758-0_19-1
16. Lewandowski, D., Sünkler, S., Yagci, N.: The influence of search engine optimization on Google's results: a multi-dimensional approach for detecting SEO. In: 13th ACM Web Science Conference 2021, WebSci 2021, 21–25 June 2021 Virtual Event UK (2021). https://doi.org/10.1145/3447535.3462479
17. Haider, J., Ekström, B., Wallin, E.T., Lorentzen, D.G., Rödl, M., Söderberg, N.: Tracing online information about wind power in Sweden: an exploratory quantitative study of broader trends (2023). https://doi.org/10.13140/RG.2.2.27914.13766
18. Schultheiß, S., Lewandowski, D., Von Mach, S., Yagci, N.: Query sampler: generating query sets for analyzing search engines using keyword research tools. PeerJ Comput. Sci. **9**, e1421 (2023). https://doi.org/10.7717/peerj-cs.1421
19. Yagci, N., Sünkler, S., Häußler, H., Lewandowski, D.: A comparison of source distribution and result overlap in web search engines. Proc. Assoc. Inf. Sci. Technol. **59**, 346–357 (2022). https://doi.org/10.1002/pra2.758
20. Abras, C., Maloney-krichmar, D., Preece, J.: User-centered design. In: Bainbridge, W. (ed.) Encyclopedia of Human-Computer Interaction, pp. 445–456. Sage Publications, Thousand Oaks (2004)

21. International Organization for Standardization: ISO 9241-210 (2019). https://www.iso.org/standard/77520.html. Accessed 10 Oct 2023
22. Wilkinson, M.D., et al.: The FAIR guiding principles for scientific data management and stewardship. Sci. Data. **3**, 160018 (2016). https://doi.org/10.1038/sdata.2016.18

eval-rationales: An End-to-End Toolkit to Explain and Evaluate Transformers-Based Models

Khalil Maachou, Jesús Lovón-Melgarejo(✉), Jose G. Moreno, and Lynda Tamine

Université Paul Sabatier, IRIT, UMR 5505 CNRS, Toulouse, France
khalil.maachou@univ-tlse3.fr, {jesus.lovon,jose.moreno,tamine}@irit.fr

Abstract. State-of-the-art (SOTA) transformer-based models in the domains of Natural Language Processing (NLP) and Information Retrieval (IR) are often characterized by their opacity in terms of decision-making processes. This limitation has given rise to various techniques for enhancing model interpretability and the emergence of evaluation benchmarks aimed at designing more transparent models. These techniques are primarily focused on developing interpretable models with the explicit aim of shedding light on the rationales behind their predictions. Concurrently, evaluation benchmarks seek to assess the quality of these rationales provided by the models. Despite the availability of numerous resources for using these techniques and benchmarks independently, their seamless integration remains a non-trivial task. In response to this challenge, this work introduces an end-to-end toolkit that integrates the most common techniques and evaluation approaches for interpretability. Our toolkit offers user-friendly resources facilitating fast and robust evaluations.

Keywords: Interpretable · Evaluation · Transformers

1 Introduction

The rapid evolution of neural models in Natural Language Processing (NLP) and Information Retrieval (IR) has yielded exceptional performance on various benchmarks [11,18]. However, the high performance achieved often obfuscates the inner workings of these models, reducing their interpretability [12]. Consequently, the field of eXplainable Artificial Intelligence (XAI) has garnered increased attention within the AI community, aiming to enhance approaches for elucidating the decision-making processes of these models [3,5].

Particularly, in the context of transformers-based models for NLP [10,14] and IR [2,8], different efforts have emerged towards explaining model predictions at different granularity levels. Most common approaches focus on *local* explanations, which seek to clarify the reasons behind a single input's prediction, while less common approaches for *global* explanations aim to explain predictions for general input. Moreover, while there are different approaches to understanding

N. Goharian et al. (Eds.): ECIR 2024, LNCS 14612, pp. 212–217, 2024.
https://doi.org/10.1007/978-3-031-56069-9_20

what makes a model interpretable, we specifically focus on the use of *local rationales*, following recent work [4,6,9]. These *rationales* are subsets of the input elements (words, phrases, or sentences) that explain a prediction [17].

The rationale-based XAI techniques are fundamentally focused on elucidating the "why" behind a model's specific output. A step forward in this research line seeks to evaluate the "quality" of the explanations found by these methods in order to compare them. However, as a novel field, there is little agreement on how this evaluation should be performed; consequently, comprehensive benchmarks such as ERASER have emerged [4].

Various programming frameworks and libraries have been developed to facilitate the implementation of these XAI techniques, including well-known tools such as LIME [13], Captum[1], ELI5[2], Shap [15], and others. Similarly, evaluation resources, like the ERASER benchmark, are readily accessible online. However, integrating the evaluation of XAI techniques with predefined benchmarks is a challenging task. For a regular user, the integration of both frameworks can be a limiting factor and a reason to avoid adopting benchmark evaluation as a good practice to evaluate their models. Moreover, as the number of freely available pre-trained models and datasets increases, using hubs such as HuggingFace has become an important source to explore and enhance new approaches. However, the available tools are limited to a fixed list of models and datasets, and integrating HuggingFace resources is not straightforward.

To address these challenges, we present *eval-rationales*, an end-to-end toolkit that integrates *local* XAI techniques from different libraries with the ERASER evaluation benchmark. Our toolkit can also integrate transformers-based models and datasets from the HuggingFace hub, thus facilitating the integration of state-of-the-art models for evaluation and exploration. Our toolkit integrates the main functions of the Captum, LIME, and ERASER libraries, and we abstract all the integration processes of the mentioned frameworks to compute the rationales prediction and evaluation. The user is only required to input a Transformers model, a Dataset, and an XAI technique. As a final output, the user obtains detailed metrics for their inputs, reflecting the quality of their model. Furthermore, we have empowered this toolkit with classical functions such as highlighted visualization for each input.

2 The *eval-rationales* Toolkit

The presented end-to-end toolkit provides a robust set of rationale-based metrics, enabling researchers and practitioners to evaluate the quality of explanations generated by transformers-based models on evaluation datasets. Specifically, our toolkit consists of two interconnected modules: i) the prediction rationales module, responsible for predicting rationales from model inputs using various XAI techniques, and ii) the metrics module, which computes the quality metrics based on these predicted rationales (see Fig. 1).

[1] https://captum.ai/.
[2] https://eli5.readthedocs.io/en/latest/.

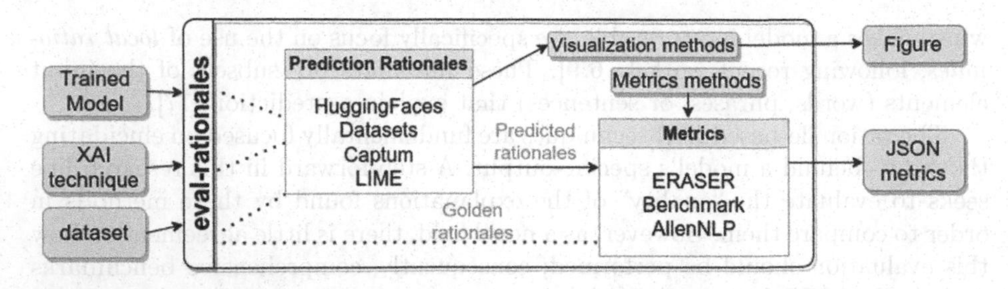

Fig. 1. The eval-explanation toolkit consists of two modules: prediction rationales, and metrics. The output is a file json file containing the computed metrics.

The prediction rationales module of our toolkit takes as input three elements: a model, an evaluation dataset, and an XAI technique. Our contribution is tailored to assess models primarily oriented toward sequence classification tasks, specifically focusing on binary classification models and datasets. To foster integration with the latest developments in NLP and IR, we have adapted our toolkit to support models based on the HuggingFace library, which can be seamlessly imported from the online repository or utilized from local storage. Regarding the XAI techniques, our toolkit includes attention-based methods [1,16], LIME [13], and gradient-based [7] approaches, mainly based on the Captum library. Additionally, we have included a Random method, which serves as a baseline, following the methodology proposed in previous work [4].

The metrics module leverages rationale-based metrics following the ERASER benchmark. Specifically, we compute two essential metrics: comprehensiveness and sufficiency. Comprehensiveness evaluates whether the predicted rationales include all the features necessary to make a prediction. Sufficiency assesses whether the extracted rationales contain enough signal to make an informed decision. Additionally, we provide the Area Over the Perturbation Curve (AOPC), originally proposed by ERASER, to further gauge the quality of explanations.

Furthermore, when the evaluation dataset includes golden rationales, we compute metrics at a token-level for accuracy, recall, and F1 score. These metrics are stored in a JSON file for easy access and manipulation by users.

In the following section, we demonstrate the utility and versatility of our evaluation toolkit through a case study involving two different datasets.

3 Case Study

Let us consider a hypothetical scenario where a user implements and trains a transformer-based model for enhanced interpretability using the Hugging Face library. This user's main objective is to assess the quality of the model's rationales predictions evaluated against the ERASER benchmark with Movie Rationales[3] and MIMIC IV[4] datasets. This evaluation process involves implementing

[3] https://huggingface.co/datasets/movie_rationales.

[4] https://physionet.org/content/mimiciv/0.4/.

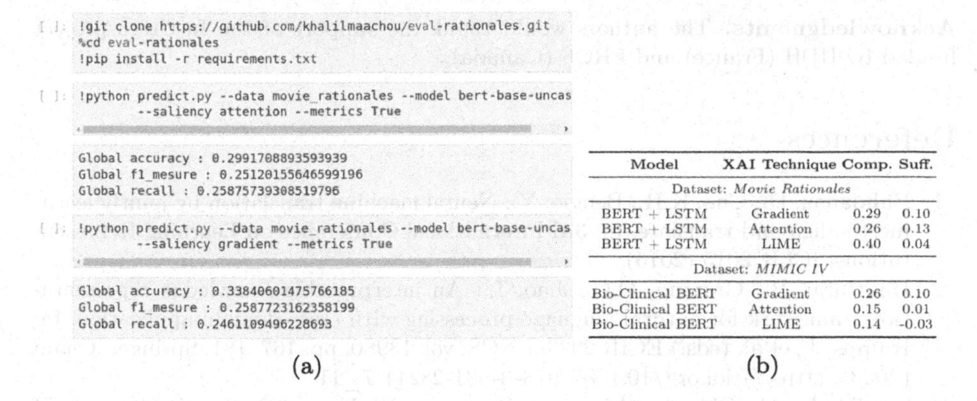
[ ]: !git clone https://github.com/khalilmaachou/eval-rationales.git
%cd eval-rationales
!pip install -r requirements.txt
[ ]: !python predict.py --data movie_rationales --model bert-base-uncas
--saliency attention --metrics True
Global accuracy : 0.2991708893593939
Global f1_mesure : 0.25120155646599196
Global recall : 0.25875739308519796
[ ]: !python predict.py --data movie_rationales --model bert-base-uncas
--saliency gradient --metrics True
Global accuracy : 0.33840601475766185
Global f1_mesure : 0.2587723162358199
Global recall : 0.2461109496228693
(a)

Model	XAI Technique	Comp.	Suff.
Dataset: *Movie Rationales*			
BERT + LSTM	Gradient	0.29	0.10
BERT + LSTM	Attention	0.26	0.13
BERT + LSTM	LIME	0.40	0.04
Dataset: *MIMIC IV*			
Bio-Clinical BERT	Gradient	0.26	0.10
Bio-Clinical BERT	Attention	0.15	0.01
Bio-Clinical BERT	LIME	0.14	-0.03

(b)

Fig. 2. (a) eval-rationales running on a fresh JupyterLab noteebok. This code includes installation and running two XAI techniques for a given model. (b) Sample results for Movie Rationales and MIMIC IV datasets.

Captum library code for rationale prediction and adapting the ERASER evaluation code, typically requiring significant coding effort.

In contrast, our toolkit provides a concise one-line command that seamlessly integrates the user's trained model into the end-to-end toolkit. By reducing the coding workload associated with rationale generation and evaluation, our toolkit empowers researchers to focus on refining the core model. It is developed on Python and tailored to the prevailing framework, Hugging Face, preferred by NLP and IR practitioners. Once installed, one can simply run the evaluation with a single-line command or use it as a library in a given script. Figure 2 presents an example of installation and usage of the toolkit, and a sample of our experiment results for this case study. The *eval-rationales* toolkit is freely available on GitHub[5], where detailed instructions are given and a video demo[6].

4 Conclusion and Future Work

This paper introduces an end-to-end toolkit to simplify the evaluation of predicted rationales generated by interpretable models. Our toolkit offers seamless integration with HuggingFace models, Datasets, and the ERASER benchmark, mitigating the challenges of integrating these distinct frameworks. This tool can benefit researchers and practitioners, significantly simplifying the model evaluation process. We plan to broaden the toolkit's adaptability with more XAI techniques tailored to textual-based transformer models. We also aim to integrate innovations and resources from the interpretability domain into ERASER benchmark and evaluation metrics.

[5] https://github.com/khalilmaachou/eval-rationales.
[6] https://youtu.be/3M1MJPhmMQE.

Acknowledgments. The authors would thank the support of the In-Utero project funded by HDH (France) and FRQS (Canada).

References

1. Bahdanau, D., Cho, K.H., Bengio, Y.: Neural machine translation by jointly learning to align and translate. In: 3rd International Conference on Learning Representations, ICLR 2015 (2015)
2. Bhattarai, B., Granmo, O.C., Jiao, L.: An interpretable knowledge representation framework for natural language processing with cross-domain application. In: Kamps, J., et al. (eds.) ECIR 2023. LNCS, vol. 13980, pp. 167–181. Springer, Cham (2023). https://doi.org/10.1007/978-3-031-28244-7_11
3. Danilevsky, M., Qian, K., Aharonov, R., Katsis, Y., Kawas, B., Sen, P.: A survey of the state of explainable AI for natural language processing. In: Proceedings of the 1st Conference of the Asia-Pacific Chapter of the Association for Computational Linguistics and the 10th International Joint Conference on Natural Language Processing, pp. 447–459 (2020)
4. DeYoung, J., et al.: ERASER: a benchmark to evaluate rationalized NLP models. In: Proceedings of the 58th Annual Meeting of the Association for Computational Linguistics, pp. 4443–4458. Association for Computational Linguistics (2020). https://doi.org/10.18653/v1/2020.acl-main.408
5. Dwivedi, R., et al.: Explainable AI (XAI): core ideas, techniques, and solutions. ACM Comput. Surv. **55**(9), 1–33 (2023)
6. Hayati, S.A., Kang, D., Ungar, L.: Does BERT learn as humans perceive? Understanding linguistic styles through lexica. In: Proceedings of the 2021 Conference on Empirical Methods in Natural Language Processing, pp. 6323–6331. Association for Computational Linguistics, Online and Punta Cana, Dominican Republic (2021). https://doi.org/10.18653/v1/2021.emnlp-main.510
7. Karlekar, S., Niu, T., Bansal, M.: Detecting linguistic characteristics of Alzheimer' dementia by interpreting neural models. In: Proceedings of the 2018 Conference of the North American Chapter of the Association for Computational Linguistics: Human Language Technologies, Volume 2 (Short Papers), pp. 701–707 (2018)
8. Lyu, L., Anand, A.: Listwise explanations for ranking models using multiple explainers. In: Kamps, J., et al. (eds.) ECIR 2023. LNCS, vol. 13980, pp. 653–668. Springer, Cham (2023). https://doi.org/10.1007/978-3-031-28244-7_41
9. Mathew, B., Saha, P., Yimam, S.M., Biemann, C., Goyal, P., Mukherjee, A.: Hatexplain: a benchmark dataset for explainable hate speech detection. In: Proceedings of the AAAI Conference on Artificial Intelligence, vol. 35, no. 17, pp. 14867–14875 (2021). https://doi.org/10.1609/aaai.v35i17.17745
10. Narang, S., Raffel, C., Lee, K., Roberts, A., Fiedel, N., Malkan, K.: Wt5?! training text-to-text models to explain their predictions. arXiv preprint arXiv:2004.14546 (2020)
11. Qiu, X., Sun, T., Xu, Y., Shao, Y., Dai, N., Huang, X.: Pre-trained models for natural language processing: a survey. SCIENCE CHINA Technol. Sci. **63**(10), 1872–1897 (2020)
12. Ras, G., Xie, N., Van Gerven, M., Doran, D.: Explainable deep learning: a field guide for the uninitiated. J. Artif. Intell. Res. **73**, 329–396 (2022)

13. Ribeiro, M.T., Singh, S., Guestrin, C.: "Why should I trust you?": explaining the predictions of any classifier. In: Proceedings of the 22nd ACM SIGKDD International Conference on Knowledge Discovery and Data Mining, San Francisco, CA, USA, 13–17 August 2016, pp. 1135–1144 (2016)
14. Ross, A., Marasović, A., Peters, M.E.: Explaining NLP models via minimal contrastive editing (MICE). In: Findings of the Association for Computational Linguistics: ACL-IJCNLP 2021, pp. 3840–3852 (2021)
15. Sundararajan, M., Najmi, A.: The many shapley values for model explanation. In: Proceedings of the 37th International Conference on Machine Learning, ICML 2020. JMLR.org (2020)
16. Vaswani, A., et al.: Attention is all you need. In: Advances in Neural Information Processing Systems, vol. 30 (2017)
17. Wiegreffe, S., Marasovic, A.: Teach me to explain: a review of datasets for explainable natural language processing. In: Thirty-Fifth Conference on Neural Information Processing Systems Datasets and Benchmarks Track (Round 1) (2021)
18. Yates, A., Nogueira, R., Lin, J.: Pretrained transformers for text ranking: BERT and beyond. In: Kondrak, G., Bontcheva, K., Gillick, D. (eds.) Proceedings of the 2021 Conference of the North American Chapter of the Association for Computational Linguistics: Human Language Technologies: Tutorials, pp. 1–4. Association for Computational Linguistics, Online (2021). https://doi.org/10.18653/v1/2021.naacl-tutorials.1

SELMA: A Semantic Local Code Search Platform

Anja Reusch(✉), Guilherme C. Lopes, Wilhelm Pertsch, Hannes Ueck, Julius Gonsior, and Wolfgang Lehner

Dresden Database Systems Group, Technische Universität Dresden, Dresden, Germany
{anja.reusch,guilherme.lopes,wilhelm.pertsch,hannes.ueck, julius.gonsior,wolfgang.lehner}@tu-dresden.de

Abstract. Searching for the right code snippet is cumbersome and not a trivial task. Online platforms such as Github.com or searchcode.com provide tools to search, but they are limited to publicly available and internet-hosted code. However, during the development of research prototypes or confidential tools, it is preferable to store source code locally. Consequently, the use of external code search tools becomes impractical. Here, we present SELMA (Code and Videos: https://anreu.github.io/selma): a local code search platform that enables term-based and semantic retrieval of source code. SELMA searches code and comments, annotates undocumented code to enable term-based search in natural language, and trains neural models for code retrieval.

Keywords: Code Retrieval · Transformer Models

1 Introduction

Software development plays a key role in many aspects of modern life. In this context, platforms like StackOverflow.com play a huge role in the daily work of software developers and researchers. According to a recent survey[1], at least 80% of them visit Stack Overflow a few times a week, 50% even daily. This fact points to the importance of code search for software development. Especially, when new developers join existing projects, they undergo a steep learning curve, since internal code is often not as easily searchable as well-known libraries. In order to enable the search in these internal projects, online tools can usually not be applied since the project and its code need to be kept confidential.

Therefore, we present SELMA, a semantic code search platform that runs within a local environment. Semantic code search refers to the task of using natural language to formulate a query in order to search in documents written in source code. At its core, it consists of two code retrieval models, one traditional BM25 method and a Transformer-Encoder [14]. Transformer-Encoder

[1] https://survey.stackoverflow.co/2022/.

N. Goharian et al. (Eds.): ECIR 2024, LNCS 14612, pp. 218–222, 2024.
https://doi.org/10.1007/978-3-031-56069-9_21

models, of which BERT [2] is the most prominent approach, have already demonstrated great success in modeling natural languages and were therefore incorporated into current retrieval systems (for a survey, see [15]). Apart from natural language, Transformer-Encoders were also trained for source code understanding [4,5,7,9,11]. Also, models using Transformer-Decoders for code were developed, for which GitHub Co-Pilot[2] is one example. These models are trained to generate code given a prompt in natural language. We, however, want to rely on code that is actually existent in the code base, and not on code that is only based on some code snippets the model was trained on. In addition, a recent study by Nguyen et al. [12] found that the generated code is written correctly in only half of the evaluated cases. Therefore, we decided to combine CodeBERT [4], a Transformer-Encoder trained for source code understanding instead of generation, with ColBERT [8] to enable fast retrieval times.

SELMA also includes code expansion, where documentation is generated for undocumented code snippets to enable natural language queries also for BM25. Different configurations of retrieval mechanisms and indexes can be applied depending on the code base and hardware setup. A preview of the platform can be seen in Fig. 1. In the following, we will present a walk-through of the system from the perspective of a user and provide an overview of the different components SELMA is built of.

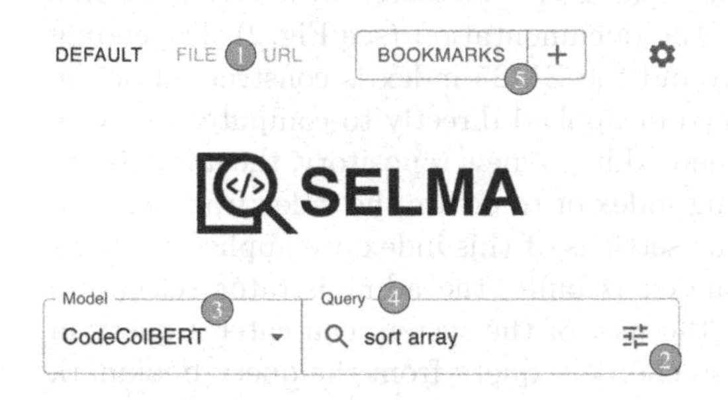

Fig. 1. Screenshot of SELMA.

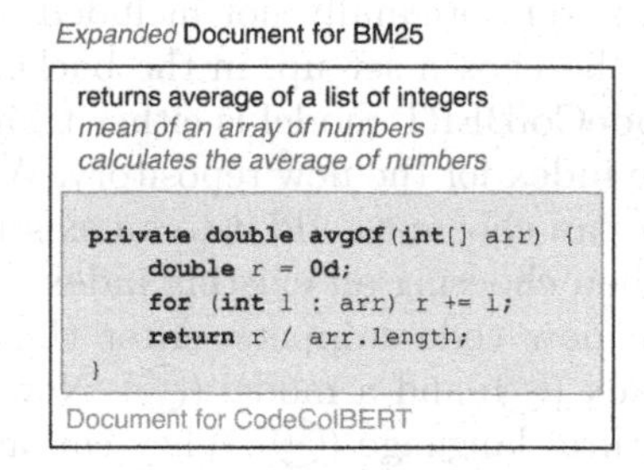

Fig. 2. Code expansion.

2 System Overview

In this section, we present the main features of SELMA. Figure 1 and 3 show screenshots of our platform.

The system first needs to be set up for a new code base. During the set up process, a Git repository URL is entered, which will then be cloned automatically (1). The user who sets up the system - usually an administrator - can decide which of the two retrieval methods are used to build the index: the term-based BM25 and the Transformer-Encoder-based CodeColBERT. The BM25 index is

[2] https://github.com/features/copilot.

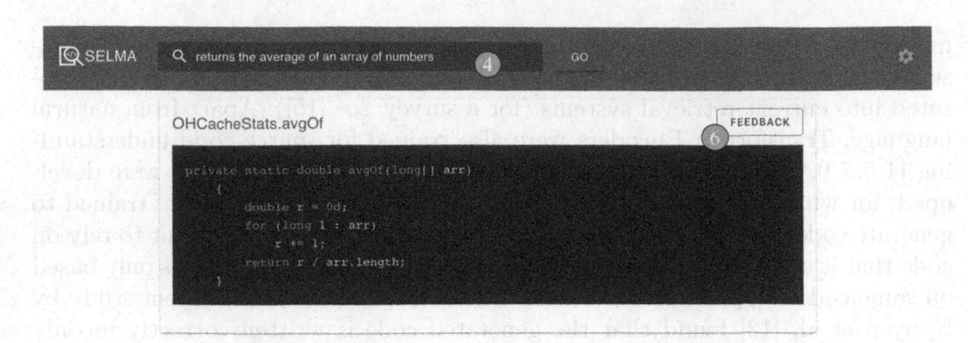

Fig. 3. Result preview for the query "returns the average of an array of numbers".

built using the code snippet and its documentation as an index-able document, while CodeColBERT is applied directly on the code snippets. Due to CodeColBERT's high code understanding capabilities, this index delivers the best results. It requires, however, a GPU to run, which is why we also offer the BM25 index as a second option. To provide semantic search even when using the BM25 index, the administrator can choose to use SELMA's code expansion feature, which runs two Transformer models that generate documentation. The system benefits from the generated documentation since it expands the document by additional terms that were originally not included in the documentation (see Fig. 2). Depending on the chosen set up, in the background the BM25 index is constructed or the CodeColBERT model is either trained or applied directly to compute and store the index for the new repository. When adding a new repository, the administrator can choose to add it to an existing index or to create the index from scratch. When choosing an existing index, the settings of this index are applied to index the new code snippets. After the index is built, the administrator selects an index (2) and a model (3). Now, the user of the system can enter a query in natural language (4). They can also choose a query from the query bookmark menu (5), which stores frequently executed queries. A screenshot of the results page can be found in Fig. 3. For each result returned by the selected model, the user can provide feedback on the relevance of the code snippet (6).

Pre-Processing. After cloning a code repository, we use the tool Treesitter[3] to parse the code, split it into method level granularity and extract existing documentation. Tokenization is done by each method separately: For term-based retrieval the internal tokenizer of PyTerrier is used. Each transformer-based model uses its respective tokenizer.

Retrieval Methods. As a base retrieval method, we employ BM25 as it is offered as part of PyTerrier [10] with $k_1 = 1.2$, $k_3 = 8$, and $b = 0.75$. The code snippets are concatenated with the documentation string to serve as documents. We found that using the method name, the parameter names, and the returned value as

[3] https://tree-sitter.github.io/tree-sitter/.

a representation for the code snippet works best. We chose ColBERT [8] as our second retrieval method, because it combines the semantic modeling capacities of Transformer-Encoders with fast retrieval performance due to offline indexing. This is achieved by passing each document through the Transformer-Encoder and storing its embeddings. At query processing time, the query is also passed through the encoder to compute the query embeddings, which are then used to determine the relevance between query and each document embedding. The documents with the highest relevance scores are returned by the system. As a base model for ColBERT we employ CodeBERT that models English and six programming languages [4]. We fine-tune the model for code retrieval using the CodeSearchNet Data Set [6], where the documentation comment, which describes a function, serves as the query and the method body of a code snippet serves as the document. Negative examples are sampled randomly from the entire corpus. Our CodeColBERT model can be used directly to index new documents, but it can also be adapted by further training on examples from new repositories. In addition, if the user provides positive and negative feedback for retrieval results, these can also serve as training examples. For retrieval, we use the PyTerrier bindings for ColBERT.

Code Expansion. Since not all code is always documented perfectly, SELMA can automatically add documentation for undocumented code snippets. The idea is inspired by Nogueria et al.'s Doc2Query [13] where a query is predicted for a given document. The predicted query is then appended to the documents which is then indexed. Our idea is similar: We generate documentation strings that will be included when indexing code snippets. This way we bridge the language mismatch between query and documents. For documentation generation, we include two models in the system: CodeTrans [3] and PLBART [1], which are both based on the Transformer architecture [14] and are trained on several tasks dealing with source code. We use the models fine-tuned on code summarization translating from source code to English. We provide the method body to the models, which then produce a documentation string. This string is concatenated with the method body to serve as the expanded document in the index. Experiments on the CodeSearchNet Challenge verified that BM25 with Code Expansion is almost on par with CodeColBERT while executing queries ten times faster.

3 Conclusion

This work presented SELMA, a semantic code search platform designed for local use. Its unique features are: term-based and neural code search, Code Expansion facilitated by Transformer models, and an integrated user feedback for both evaluation and refinement. We provide an already fine-tuned model[4] ready for immediate application in code search tasks without the need for further training, but it can also be adapted to the code that needs to be searched. This work was supported through grant 01IS17044 Software Campus 2.0 (TU Dresden) by the

[4] https://huggingface.co/ddrg/codecolbert.

BMBF. Furthermore, the authors are grateful for the GWK support for funding this project by providing computing time through the Center for Information Services and HPC (ZIH) at TU Dresden.

References

1. Ahmad, W.U., Chakraborty, S., Ray, B., Chang, K.W.: Unified pre-training for program understanding and generation. arXiv preprint arXiv:2103.06333 (2021)
2. Devlin, J., Chang, M.W., Lee, K., Toutanova, K.: BERT: pre-training of deep bidirectional transformers for language understanding. In: Proceedings of the 2019 Conference of the North American Chapter of the Association for Computational Linguistics: Human Language Technologies, Volume 1 (Long and Short Papers), Minneapolis, Minnesota, pp. 4171–4186. Association for Computational Linguistics (2019). https://doi.org/10.18653/v1/N19-1423. https://aclanthology.org/N19-1423
3. Elnaggar, A., et al.: Codetrans: towards cracking the language of silicon's code through self-supervised deep learning and high performance computing. arXiv e-prints pp. arXiv-2104 (2021)
4. Feng, Z., et al.: Codebert: a pre-trained model for programming and natural languages. arXiv preprint arXiv:2002.08155 (2020)
5. Guo, D., et al.: Graphcodebert: pre-training code representations with data flow. In: ICLR (2021)
6. Husain, H., Wu, H.H., Gazit, T., Allamanis, M., Brockschmidt, M.: Codesearchnet challenge: evaluating the state of semantic code search. arXiv preprint arXiv:1909.09436 (2019)
7. Kanade, A., Maniatis, P., Balakrishnan, G., Shi, K.: Learning and evaluating contextual embedding of source code. In: International Conference on Machine Learning, pp. 5110–5121. PMLR (2020)
8. Khattab, O., Zaharia, M.: Colbert: efficient and effective passage search via contextualized late interaction over bert. In: Proceedings of the 43rd International ACM SIGIR Conference on Research and Development in Information Retrieval, pp. 39–48 (2020)
9. Liu, S., Wu, B., Xie, X., Meng, G., Liu, Y.: Contrabert: enhancing code pre-trained models via contrastive learning. arXiv preprint arXiv:2301.09072 (2023)
10. Macdonald, C., Tonellotto, N.: Declarative experimentation in information retrieval using pyterrier. In: Proceedings of ICTIR 2020 (2020)
11. Neelakantan, A., et al.: Text and code embeddings by contrastive pre-training. arXiv preprint arXiv:2201.10005 (2022)
12. Nguyen, N., Nadi, S.: An empirical evaluation of github copilot's code suggestions. In: Proceedings of the 19th International Conference on Mining Software Repositories, MSR 2022, pp. 1–5. Association for Computing Machinery, New York (2022). https://doi.org/10.1145/3524842.3528470
13. Nogueira, R., Yang, W., Lin, J., Cho, K.: Document expansion by query prediction. arXiv preprint arXiv:1904.08375 (2019)
14. Vaswani, A., et al.: Attention is all you need. In: Advances in Neural Information Processing Systems, vol. 30 (2017)
15. Yates, A., Nogueira, R., Lin, J.: Pretrained transformers for text ranking: Bert and beyond. In: Proceedings of the 44th International ACM SIGIR Conference on Research and Development in Information Retrieval, pp. 2666–2668 (2021)

VADIS – A Variable Detection, Interlinking and Summarization System

Yavuz Selim Kartal[1(✉)], Muhammad Ahsan Shahid[1], Sotaro Takeshita[2,3], Tornike Tsereteli[2], Andrea Zielinski[4], Benjamin Zapilko[1], and Philipp Mayr[1]

[1] GESIS – Leibniz Institute for the Social Sciences, Cologne, Germany
{yavuzselim.kartal,ahsan.shahid,benjamin.zapilko,philipp.mayr}@gesis.org
[2] University of Mannheim, Mannheim, Germany
{sotaro.takeshita,tornike.tsereteli}@uni-mannheim.de
[3] Hochschule Mannheim, Mannheim, Germany
[4] Fraunhofer ISI, Karlsruhe, Germany
andrea.zielinski@isi.fraunhofer.de

Abstract. The VADIS system addresses the demand of providing enhanced information access in the domain of the social sciences. This is achieved by allowing users to search and use survey variables in context of their underlying research data and scholarly publications which have been interlinked with each other.

Keywords: Variable Search · Text Summarization · Variable Detection

1 Introduction

In the social sciences, as in other scientific disciplines, there is a growing necessity to support researchers by providing enhanced information access through linking of publications and underlying datasets. One of the key artifacts in social science studies are concepts, namely survey variables (SV), which have been used to measure sociological phenomena. These variables are generally the access to research data [7], but are neither represented nor interlinked in the information sources relevant for social science research, i.e., scholarly publications.

The key objective of the VADIS (VAriable Detection, Interlinking and Summarization) project [8] is to allow for searching and using survey variables in context and thereby enhance information access of scholarly publications and help to increase the reproducibility of research results. We combine text mining techniques and semantic web technologies that identify and exploit links between publications, their topics, and the specific variables that are covered in the surveys. These semantic links build the basis for the VADIS system which offers users better access to the scientific literature by, e.g., search, summarization, and linking of identified survey variables.

N. Goharian et al. (Eds.): ECIR 2024, LNCS 14612, pp. 223–228, 2024.
https://doi.org/10.1007/978-3-031-56069-9_22

In detail, the system follows three objectives: (1) serving as a demonstrator for the summarization and SV identification methods, (2) enabling search and exploration functionalities addressed in the use cases [8], and (3) serving as an evaluation test bed for user studies.

2 Related Works

Many systems have been developed concerning the main objectives of VADIS independently. For summarization, there is a line of work focusing on scientific document summarization from domain-specific datasets [3] to the models that exploit the unique properties such as citation contexts [5] or discourse structures [1,4]. Semantic Scholar and IBM Science Summarizer [6] provide extreme and extended summaries, respectively, coupled with search engines in their systems. There are various search systems in the Social Sciences. The UK Data Service, The Inter-university Consortium for Political and Social Research (ICPSR), the Understanding Society [2], the Norwegian centre for research data (NSD) and GSS Data Explorer serve as variable search systems on their specific research datasets. Google Dataset enables dataset search from a wide range of areas, and GESIS Search[1] provides an advanced search system for publications, datasets, and their variables in an integrated way.

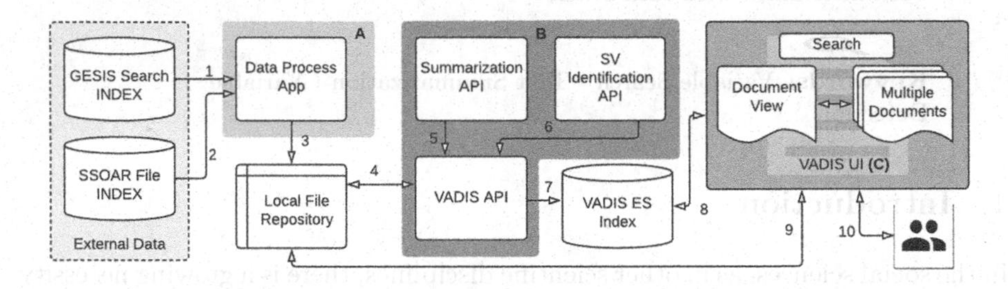

Fig. 1. VADIS Framework

3 System Overview

The current version of the VADIS system[2] contains 607 documents in English and German, which are all of the publications with digitally available research dataset linkages in GESIS Search, their extreme summaries, 7086 linkages of their sentences to variables, and their original files.

We developed a framework that integrates summarization and survey variable identification processes asynchronously and delivers their outcomes to the users.

[1] https://search.gesis.org/ .
[2] https://demo-vadis.gesis.org/.

This section explains the VADIS framework consisting of three main components: Data Preprocessing (A), VADIS Modules (B), VADIS UI (C), as shown in Fig. 1. All published modules are available via the VADIS Github project: https://github.com/vadis-project/.

3.1 Data Preprocessing (A)

We first retrieve the metadata of the publications and their PDF files from GESIS Search and SSOAR indices respectively (1, 2).[3] This includes relations between publications and research datasets with detailed variable descriptions and the abstract. All data is stored and updated on a local server (3). Grobid is used to extract plain text from the publication PDFs, while full-text data is kept for the VADIS UI search operations and survey variable identification (3).

3.2 VADIS Modules (B)

The VADIS Modules are a set of APIs and operations that yield the main outcomes of the project from the preprocessed data.

Summarization API: This API provides two features. First is an extractive summarization which selects one salient sentence as a summary in a given document. Second is an abstractive summarization that generates a short summary of a given document. The API accepts English and German documents as inputs and generates summaries in English or German. This is achieved by training models on a cross-lingual summarization dataset [10]. We also provide distilled models for a faster and memory-efficient inference that enables generating 50% more summaries per second without critical quality degradation [11].

SV Identification API: This API has two main functions. The first classifies whether a sentence primarily defines a variable by using an ensemble of a model, fine-tuned on an extension of a previously published dataset [12], and a sentence-transformers-based [9] retriever, which syntactically and semantically matches the input to similar texts from the database of variables. Given the classified sentences, the second function then uses the retriever to match a subset of variables filtered by using the publication-research dataset relation information. In combination, the functions present users with a narrow list of possible sentences defining used variables as well as a list of candidate variables.

VADIS API: This is an intermediate API that integrates other APIs, input and output data, and index. As shown in Fig. 1, it provides abstracts to Summarization API (5) and full-text and survey variables to SV Identification API (6) and it fetches their outcomes and inserts them into Elastic Search (ES) index (7). It also inserts metadata of the publications to ES for the search feature.

[3] The numbers in parentheses show the steps in Fig. 1.

3.3 VADIS User Interface (C)

The UI enables users to interact and explore the novel features available within the responsive and user-friendly demonstrator[2](10). The VADIS UI directly communicates with the ES index and local file repository containing PDF documents to retrieve data (8, 9). As shown in Fig. 2, it provides metadata, extreme summaries, and variable linkages for publications, allowing users to perform classical document search via natural language queries within the extracted full-text. Additionally, it allows users to locate extracted survey variables in the actual PDF, so that variables can be inspected in context.

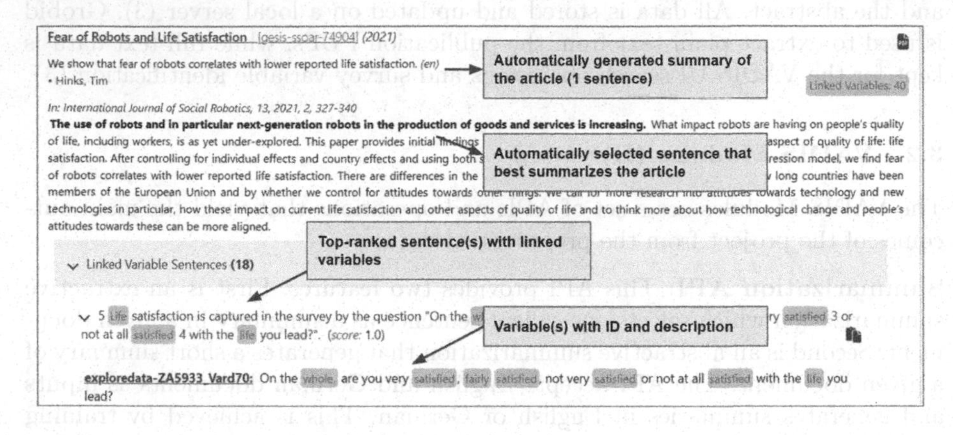

Fig. 2. VADIS User Interface

4 User Test and Conclusions

We conducted a preliminary user study with six researchers (5 social science research associates and a student; all postgraduates and experienced users in IR). The user study consisted of two search scenarios to evaluate the usability of the interface and the quality of the search results. During the tests, the users were asked to *think out loud* and in the end they should answer questions related to the search experience. Details of the search scenarios and the questions are available via the following repository: https://github.com/vadis-project/user-tests.

Finding Variables and Datasets. When searching for publications by keywords, users found useful links to survey variables (including the associated metadata), revealing a new direction to design survey questionnaires, and showing "new perspectives that I would not have found otherwise". Also, most participants found the **extreme summarization** along with metadata helpful.

Limitations. Users suggested ranking publications by the "number of detected variables", apart from classical IR measures such as relevancy and recency. In addition to the annotation of variables in the corresponding PDF of the scientific paper, students suggested a better (graphical) display of the linked variables on the result page. Moreover, it was found that the highlighting of common words in scientific documents and variables could be improved by focusing on important social science concepts.

Following a user-centred design and development process in the VADIS project, the results of the user tests give us directions for further improvements of the system. Eventually, positively evaluated features will perspectively be implemented into the GESIS search and will thus be available in a widely used search system for the social sciences.

Acknowledgement. This work was supported by the DFG project VADIS under grant numbers: ZA 939/5-1, PO 1900/5-1, EC 477/7-1, KR 4895/3-1.

References

1. Akiyama, K., Tamura, A., Ninomiya, T.: Hie-bart: document summarization with hierarchical bart. In: Proceedings of the 2021 Conference of the North American Chapter of the Association for Computational Linguistics: Student Research Workshop, pp. 159–165 (2021)
2. Buck, N., McFall, S.: Understanding society: design overview. Longitudinal Life Course Stud. **3**(1), 5–17 (2012)
3. Cachola, I., Lo, K., Cohan, A., Weld, D.S.: Tldr: Extreme summarization of scientific documents (2020). https://api.semanticscholar.org/CorpusID:216867622, ArXiv abs/ arXiv: 2004.15011
4. Cohan, A., et al.: A discourse-aware attention model for abstractive summarization of long documents. In: Proceedings of the 2018 Conference of the North American Chapter of the Association for Computational Linguistics: Human Language Technologies, Volume 2 (Short Papers). pp. 615–621. Association for Computational Linguistics, New Orleans, Louisiana (Jun 2018). https://doi.org/10.18653/v1/N18-2097, https://aclanthology.org/N18-2097
5. Cohan, A., Goharian, N.: Scientific document summarization via citation contextualization and scientific discourse. Int. J. Digit. Libr. **19**(2–3), 287–303 (2018). https://doi.org/10.1007/s00799-017-0216-8,http://link.springer.com/10.1007/s00799-017-0216-8
6. Erera, S., et al.: A summarization system for scientific documents. In: Proceedings of the 2019 Conference on Empirical Methods in Natural Language Processing and the 9th International Joint Conference on Natural Language Processing (EMNLP-IJCNLP): System Demonstrations, pp. 211–216. Association for Computational Linguistics, Hong Kong, China (Nov 2019). https://doi.org/10.18653/v1/D19-3036, https://aclanthology.org/D19-3036
7. Friedrich, T., Siegers, P.: The ofness and aboutness of survey data: improved indexing of social science questionnaires. In: Wilhelm, A.F.X., Kestler, H.A. (eds.) Analysis of Large and Complex Data. SCDAKO, pp. 629–638. Springer, Cham (2016). https://doi.org/10.1007/978-3-319-25226-1_54
8. Kartal, Y.S., et al.: Towards Automated Survey Variable Search and Summarization in Social Science Publications (2022). http://arxiv.org/abs/2209.06804

9. Reimers, N., Gurevych, I.: Sentence-bert: sentence embeddings using siamese bert-networks. In: Proceedings of the 2019 Conference on Empirical Methods in Natural Language Processing. Association for Computational Linguistics (Nov 2019). https://arxiv.org/abs/1908.10084
10. Takeshita, S., Green, T., Friedrich, N., Eckert, K., Ponzetto, S.P.: X-scitldr: cross-lingual extreme summarization of scholarly documents. In: Proceedings of the 22nd ACM/IEEE Joint Conference on Digital Libraries, pp. 1–12 (2022)
11. Takeshita, S., Green, T., Friedrich, N., Eckert, K., Ponzetto, S.P.: Cross-lingual extreme summarization of scholarly documents. Intern. J. Digital Libraries, 1–23 (2023)
12. Tsereteli, T., Kartal, Y.S., Ponzetto, S.P., Zielinski, A., Eckert, K., Mayr, P.: Overview of the SV-ident 2022 shared task on survey variable identification in social science publications. In: Proceedings of the Third Workshop on Scholarly Document Processing. pp. 229–246. Association for Computational Linguistics, Gyeongju, Republic of Korea (Oct 2022). https://aclanthology.org/2022.sdp-1.29

ARElight: Context Sampling of Large Texts for Deep Learning Relation Extraction

Nicolay Rusnachenko(✉), Huizhi Liang, Maksim Kalameyets, and Lei Shi

School of Computing, Newcastle University, Newcastle upon Tyne, UK
{nicolay.rusnachenko,huizhi.liang, maksim.kalameyets,lei.shi}@newcastle.ac.uk

Abstract. The escalating volume of textual data necessitates adept and scalable Information Extraction (IE) systems in the field of Natural Language Processing (NLP) to analyse massive text collections in a detailed manner. While most deep learning systems are designed to handle textual information as it is, the gap in the existence of the interface between a document and the annotation of its parts is still poorly covered. Concurrently, one of the major limitations of most deep-learning models is a constrained input size caused by architectural and computational specifics. To address this, we introduce ARElight[1], a system designed to efficiently manage and extract information from sequences of large documents by dividing them into segments with mentioned object pairs. Through a pipeline comprising modules for text sampling, inference, optional graph operations, and visualisation, the proposed system transforms large volumes of text in a structured manner. Practical applications of ARElight are demonstrated across diverse use cases, including literature processing and social network analysis.([1]https://github.com/nicolay-r/ARElight)

Keywords: Data Processing Pipeline · Information Retrieval · Visualisation

1 Introduction

Information Extraction (IE) in the domain of Natural Language Processing (NLP) involves a separate studies aimed on objects annotation (entities, events, etc.) in texts [3,15] and establishing connections between objects [4,8]. IE finds a significant application in text structurization, the knowledge base formation [2,6]. One of the generalized concept for structuring raw texts is to form *pipeline* of sequential text transformations, with relation extraction[1] as a module [7,26]. Another alternative to the vast number of solution follows the concept of *target-oriented systems*, aimed at applying specific machine learning architectures for the given raw input [13,18,25]. However, once texts become larger

[1] http://deepdive.stanford.edu/relation_extraction.

N. Goharian et al. (Eds.): ECIR 2024, LNCS 14612, pp. 229–235, 2024.
https://doi.org/10.1007/978-3-031-56069-9_23

or their actual amount is massive, the direct application of these systems to the entire text becomes: (i) less informative for result analysis [20], and (ii) less effective. Partitioning large texts [11] that conveyed relations between mentioned objects represents a common solution for managing long-input problems in downstream systems [16,26]. In this paper, we propose ARElight that follows bridges the gap in processing of large documents. Our system contributes by offering a scalable relation annotations, surpassing the similar slot-filling systems for processing large collections instead of single documents [11]. We demonstrate this system's ability to analyse sentiment relations in literature novel books, social networks.

2 The ARElight System Design

ARElight system represent a pipeline of further modules: (1) text sampler, (2) inference, (3) graph operations (optional), and (4) graph visualisation. Since the source of input information represent raw documents, *text sampler* module represent a core of the system. Figure 1 shows the pipeline architecture along with a detailed illustration of its core module.

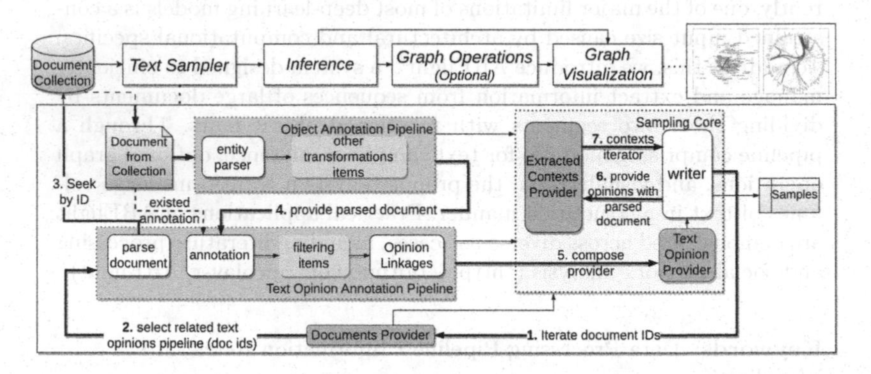

Fig. 1. ARElight-0.24.0 system design; *top*: the main application pipeline; *bottom*: architecture of the *text sampler* module with two separated pipelines for document content annotation (red blocks), data providers (yellow blocks); the process on *document collection sampling*, which is depicted in bold arrows, numbered from 1 to 7 (Color figure online)

Text Sampler. This module performs streaming extraction of context with mentioned object pairs in it from large amount of texts in the document collection. Unlike existed systems, the core module provide two separate *declarative pipelines*[2] that describe annotation for (i) objects[3] and (ii) pairs formation

[2] https://github.com/nicolay-r/AREkit/wiki/Task-Schemata.

[3] https://github.com/nicolay-r/AREkit/wiki/Pipelines:-Text-Processing.

between objects[4]. To automatically extract mentioned objects in text, our annotation pipeline provides support for models from DeepPavlov [12]. The system supports pairs forming based on: (i) document level (object value), and (ii) context level (object indices) annotations. We consider *strategy pattern* [1] for the implementation of provider components (blocks in yellow color, Fig. 1).

Inference. This module performs samples classification, followed by automatic graph serialization necessary for visualization. For samples classification, we propose supporting language models [10]. We use OpenNRE [18] as target-oriented solution for contextual relation extraction. Our system supports BERT-based text classification models [17], pre-trained with OpenNRE and distributed as PyTorch checkpoints [9]. We present classified samples as undirected graphs $G = (V, E, W_e, W_v)$ where V and E represent vertices (found objects) and edges (found pairs), and W_e, W_v denote their respective weights (frequencies in text).

Graph Operations. This module allows to serialize new graph from a pair of existing graphs. Our system supports three crucial operations:

Union ($G_1 \cup G_2$). The result graph contains all the vertices and edges that are in G_1 and G_2. The edge weight is given by $W_e = W_{e1} + W_{e2}$, and the vertex weight is its weighted degree centrality: $W_v = \sum_{e \in E_v} W_e(e)$.

Intersection ($G_1 \cap G_2$). The result graph contains only the vertices and edges common to G_1 and G_2. The edge weight is given by $W_e = \min(W_{e1}, W_{e2})$, and the vertex weight is its weighted degree centrality: $W_v = \sum_{e \in E_v} W_e(e)$.

Difference ($G_1 - G_2$). The result graph contains all the vertices from G_1 but only includes edges from E_1 that either don't appear in E_2 or have larger weights in G_1 compared to G_2. The edge weight is given by $W_e = W_{e1} - W_{e2}$ if $e \in E_1$, $e \in E_1 \cap E_2$ and $W_{e1}(e) > W_{e2}(e)$.

Visualisation. This module composes HTML page with visual user interface (UI) (Fig. 2) and launches a web server to host it. The UI consists of (a) a dataset selector, (b) visualisation options, (c) visualisation model selector, and (d) two D3JS visualisation modes – force [28] and radial [27] graph templates.

3 Experiments and Demonstration

As a demo, we propose the analysis of texts' narratives across 3 distinct use cases: (Case 1) novel "War and Peace"$_{Vol.1-3}$ by Leo Tolstoy, (Case 2) pro-Russian/Ukrainian war comments on VK social network [24], and (Case 3) X/Twitter accounts. In particular, we extract sentiment relations (pos/neg) between objects in texts [14].

Collections Preparation. We executed *text sampling+inference* scenario[5] for texts in Russian (Case 1-2), and in English (Case 3). For objects annotation[6], we consider BERT$_{mult-OntoNotes5}$ [5,12]. For samples classification we adopt

[4] https://github.com/nicolay-r/AREkit/wiki/Pipelines:-Text-Opinion-Annotation.

[5] https://github.com/nicolay-r/ARElight/wiki/Language-Specific-Application.

[6] We keep {ORG,PERSON,LOC,GPE} types and mask their values in text classification.

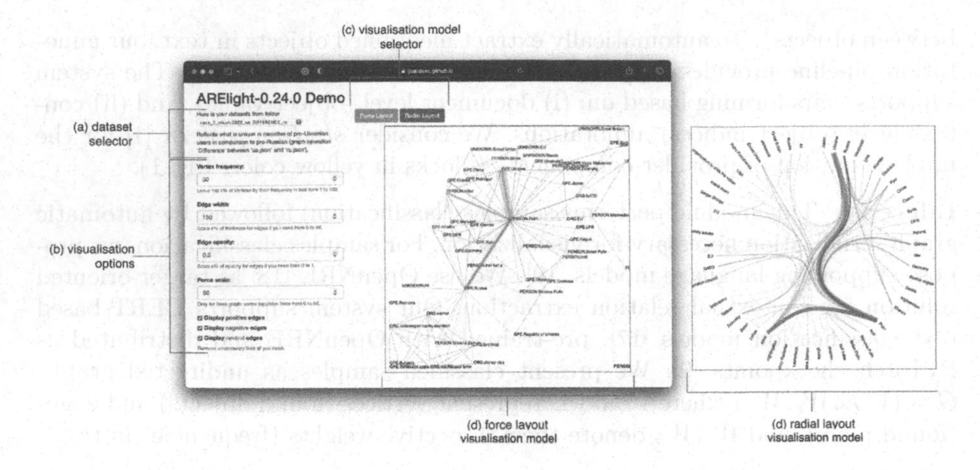

Fig. 2. The visual interface of the ARElight-0.24.0 web server

RuBERT [19] model, use fine-tuned on RuSentRel [14] and RuAttitudes [21] collections with NLI-prompt [22,23] (Case 1-2). We automatically translate these collections in English to fine-tune BERT$_{cased}$ [17] (Case 3). These models are publicly available and automatically fetched upon scenario launch.

Graph Operations. We present an example of graph analysis through various operations. For the Case 2, the aggregated pro-Russian/Ukrainian graph was obtained by employing the *Union* operation on narrative graphs of individual users, extracted from the dataset [24]. We utilized the *Intersection* operation to discern commonalities between the narratives of pro-Russian and pro-Ukrainian users (Case 2), as well as between Rishi Sunak and Boris Johnson on Platform X/Twitter (Case 3). The *Difference* operation facilitated understanding the unique aspects of one narrative compared to another. We applied it to extract differences between pro-Ukrainian and pro-Russian users' narratives and vice versa (Case 2), and to elucidate disparities between Rishi Sunak and Boris Johnson, and vice versa, on Platform X/Twitter (Case 3).

The visual interface is presented in Fig. 2. For example, one can see that pro-Ukrainian users more often mention Putin with Hitler than pro-Russians (Case 2). As for the Case 1, "War and Peace"$_{Vol.2}$ is distinctly centers on life and relations between individuals, in contrast to more war-centric themes in other volumes. In Case 3, related to the differences between B. Jonson and R. Sunak, one can observe a higher number of positive UK-France pairs and significantly more UK-Ukraine pairs. You can explore all the three cases by following the demo project link: https://guardeec.github.io/arelight_demo/template.html.

4 Conclusion and Future Work

In this paper, we introduced the ARElight system, designed to facilitate the segmentation and analysis of large documents by converting them into smaller text parts associated with mentioned object pairs, and subsequently analyzing them as graphs. The system filters and samples text segments involving such pairs throughout a sequence of large documents, constructs graphs from them, and conducts graph operations and visualisations. The aim of the proposed system was to bridge the existing gap in the programming interface between a document and its subsequent annotation. The reusability of the system components has been demonstrated through use cases and showcases its applicability to various scenarios, ranging from the analysis of books to social media accounts.

Acknowledgments. This research is partially supported by UK Research and Innovation, United Kingdom through the Strategic Priority Fund as part of the Protecting Citizens Online programme. Grant: "AGENCY: Assuring Citizen Agency in a World with Complex Online Harms", EP/W032481/2 at Newcastle University.

References

1. Gamma, E., et al.: Elements of reusable object-oriented software. Design Patterns (1995)
2. Nasukawa, I., Yi, J.: Sentiment analysis: capturing favorability using natural language processing. In: Proceedings of the 2nd International Conference on Knowledge Capture, K-CAP 2003, Sanibel Island, FL, USA, pp. 70-77. Association for Computing Machinery (2003). https://doi.org/10.1145/945645.945658.https://doi.org/10.1145/945645.945658,). isbn: 1581135831
3. Nadeau, D., Sekine, S.: "A survey of named entity recognition and classification. Lingvisticae Investigationes **30**(1), 3–26 (2007)
4. Hendrickx, I., et al.: SemEval-2010 Task 8: multi-way classification of semantic relations between pairs of nominals. In: Proceedings of the 5th International Workshop on Semantic Evaluation, Uppsala, Sweden, pp. 33-38 Association for Computational Linguistics (July 2010). https://aclanthology.org/S10-1006
5. Ralph, W., et al.: OntoNotes Release 5.0. (2012). https://doi.org/10.35111/xmhb-2b84. https://catalog.ldc.upenn.edu/docs/LDC2013T19/OntoNotes-Release-5.0.pdf
6. Walker, M., et al.: Stance classification using dialogic properties of persuasion. In: Proceedings of the 2012 Conference of the North American Chapter of the Association For Computational Linguistics: Human Language Technologies, pp. 592-596 (2012)
7. Manning, C., et al.: The Stanford CoreNLP natural language processing toolkit. In: Proceedings of 52nd Annual Meeting of the Association for Computational Linguistics: System Demonstrations, Baltimore, Maryland, pp. 55-60. Association for Computational Linguistics (June 2014). https://doi.org/10.3115/v1/P14-5010.https://aclanthology.org/P14-5010
8. Choi, E., et al.: Document-level sentiment inference with social, faction, and discourse context. In: Proceedings of the 54th Annual Meeting of the Association for Computational Linguistics (Volume 1: Long Papers). Berlin, Germany, pp. 333-343. Association for Computational Linguistics (Aug 2016). https://doi.org/10.18653/v1/P16-1032. https://aclanthology.org/P16-1032
9. Paszke, A., et al.: Automatic differentiation in PyTorch (2017)

10. Vaswani, A., et al.: Attention is all you need'. In: Guyon, I., et al. (ed.) Advances in Neural Information Processing Systems, vol. 30. Curran Associates, Inc. (2017). https://proceedings.neurips.cc/paper_files/paper/2017/file/3f5ee243547dee91fbd053c1c4a845aa-Paper.pdf
11. Adel, H., et al.: DERE: a task and domain-independent slot filling framework for declarative relation extraction. In: Proceedings of the 2018 Conference on Empirical Methods in Natural Language Processing: System Demonstrations, pp. 42-47 (2018)
12. Burtsev, M., et al.: DeepPavlov: open-source library for dialogue systems. In: Proceedings of ACL 2018, System Demonstrations, Melbourne, Australia, pp. 122-127. Association for Computational Linguistics (July 2018). https://doi.org/10.18653/v1/P18-4021. https://aclanthology.org/P18-4021
13. Xu Han, Zhiyuan Liu, and Maosong Sun. "Neural Knowledge Acquisition via Mutual Attention between Knowledge Graph and Text". In: *Proceedings of AAAI*. 2018
14. Loukachevitch, N., Rusnachenk, N.: Extracting sentiment attitudes from analytical texts. In: Proceedings of International Conference on Computational Linguistics and Intellectual Technologies Dialogue-2018, pp. 459-468 (arXiv:1808.08932) (2018)
15. Yadav, V., Bethard, S.: A survey on recent advances in named entity recognition from deep learning models'. In: Proceedings of the 27th International Conference on Computational Linguistics, Santa Fe, New Mexico, USA, pp. 2145-2158. Association for Computational Linguistics (Aug 2018). https://aclanthology.org/C18-1182
16. Zhang, S., et al.: Personalizing dialogue agents: i have a dog, do you have pets too? In: Proceedings of the 56th Annual Meeting of the Association for Computational Linguistics (Volume 1: Long Papers), Melbourne, Australia, pp. 2204-2213. Association for Computational Linguistics (July 2018). https://doi.org/10.18653/v1/P18-1205. https://aclanthology.org/P18-1205
17. Jacob Devlin et al. "BERT: Pre-training of Deep Bidirectional Transformers for Language Understanding". In: *Proceedings of the 2019 Conference of the North American Chapter of the Association for Computational Linguistics: Human Language Technologies, Volume 1 (Long and Short Papers)*. Minneapolis, Minnesota: Association for Computational Linguistics, June 2019, pp. 4171-4186. https://doi.org/10.18653/v1/N19-1423. url: https://aclanthology.org/N19-1423
18. Han, X., et al.: OpenNRE: an open and extensible toolkit for neural relation extraction. In: Padó, S., Huang, R. (ed.) Proceedings of the 2019 Conference on Empirical Methods in Natural Language Processing and the 9th International Joint Conference on Natural Language Processing (EMNLP-IJCNLP): System Demonstrations, Hong Kong, China, pp. 169-174. Association for Computational Linguistics (Nov. 2019) https://doi.org/10.18653/v1/D19-3029.https://aclanthology.org/D19-3029
19. Kuratov, Y., Arkhipov, M.: Adaptation of deep bidirectional multilingual transformers for Russian language. In: Computational Linguistics and Intellectual Technologies: Proceedings of the International Conference Dialogue 2019 (2019)
20. Rusnachenko, N., Loukachevitch, N.: Neural network approach for extracting aggregated opinions from analytical articles. In: Manolopoulos, Y., Stupnikov, S. (eds.) DAMDID/RCDL 2018. CCIS, vol. 1003, pp. 167–179. Springer, Cham (2019). https://doi.org/10.1007/978-3-030-23584-0_10
21. Rusnachenko, N., Loukachevitch, N., Tutubalina, E.: Distant supervision for sentiment attitude extraction. In: Proceedings of the International Conference on Recent Advances in Natural Language Processing (RANLP 2019) (2019)

22. Sun, C., Huang, L., Qiu, X.: Utilizing BERT for aspect-based sentiment analysis via constructing auxiliary sentence. In: Proceedings of the 2019 Conference of the North American Chapter of the Association for Computational Linguistics: Human Language Technologies, Volume 1 (Long and Short Papers), Minneapolis, Minnesota, pp. 380-385. Association for Computational Linguistics (June 2019). https://doi.org/10.18653/v1/N19-1035. https://aclanthology.org/N19-1035
23. Rusnachenko, N.: Language models application in sentiment attitude extraction task. Russian Proc. Inst. Syst. Program. RAS (Proceedings of ISP RAS) **33**(3), 199–222 (2021)
24. Kolomeets, M.: Dataset with Russian-Ukrainian war related comments from top Russian media based on their VKontakte web pages (2022). https://github.com/guardeec/datasets#mkwm2022
25. Morio, G., et al.: Hitachi at SemEval-2022 Task 10: comparing graphand Seq2Seq-based models highlights difficulty in structured sentiment analysis'. In: Proceedings of the 16th International Workshop on Semantic Evaluation (SemEval-2022), Seattle, United States, pp. 1349-1359. Association for Computational Linguistics (July 2022). https://doi.org/10.18653/v1/2022.semeval-1.188. https://aclanthology.org/2022.semeval-1.188
26. Roberts, A., et al.: Scaling up models and data with t5x and seqio. arXiv: 2203.17189 (2022)
27. Bostock, M.: D3js gallery: Hierarchical edge bundling (2023). https://observablehq.com/@d3/hierarchical-edge-bundling
28. Observable. D3js gallery: Force-directed graph (2023). https://observablehq.com/@d3/force-directed-graph/2

Translating Justice: A Cross-Lingual Information Retrieval System for Maltese Case Law Documents

Joel Azzopardi(✉)

Department of Artificial Intelligence, Faculty of ICT, University of Malta, Msida, Malta
joel.azzopardi@um.edu.mt

Abstract. In jurisdictions adhering to the Common Law system, previous court judgements inform future rulings based on the Stare Decisis principle. For enhanced accessibility and retrieval of such judgements, we introduced a cross-lingual Legal Information Retrieval system prototype focused on Malta's small claims tribunal. This system utilises Neural Machine Translation (NMT) to automatically translate Maltese judgement documents into English, enabling dual-language querying. Additionally, it employs Rhetorical Role Labelling (RRL) on sentences within the judgements, allowing for targeted searches based on specific rhetorical roles. Developed without depending on high-end resources or commercial systems, this prototype showcases the potential of AI in advancing legal research tools and making legal documents more accessible, especially for non-native speakers.

Keywords: Legal Information Retrieval · Judgements · Multi-Lingual · Neural Machine Translation · Rhetorical Role Labelling

1 Introduction

Legal case documents, pivotal in the Common Law system, are primary resources for legal professionals seeking to establish precedents and understand historical judgements based on the *Stare Decisis* principle – i.e. similar facts and situations in the past should yield similar outcomes [5,11]. These documents, forming the backbone of legal decision-making, underscore the importance of easy access and efficient retrieval for both legal experts and the general populace.

While automated tools can significantly aid in analyzing these documents, challenges arise from their extensive length, intricate structure, and lack of standardized format. Nevertheless, according to [12], case law documents tend to follow a conventional form and certain prevalent characteristics can help in overcoming these same challenges. One way of uncovering these prevalent characteristics is by automatically labelling each sentence with different Rhetorical Labels such as Facts, Previous Rulings, Arguments [13].

Case law documents from Malta are publicly available on the Maltese eCourts portal [10]. However, challenges in their analysis are further accentuated by their

N. Goharian et al. (Eds.): ECIR 2024, LNCS 14612, pp. 236–240, 2024.
https://doi.org/10.1007/978-3-031-56069-9_24

presentation in the Maltese language. Maltese has long been considered a low-resource language (in NLP) [14]. A possible solution is to utilise translation into English, more specifically, Neural Machine Translation (NMT) [3].

We present a prototype cross-lingual Information Retrieval system for legal case documents (currently limited to the small case tribunal) – legal.ir.mt. This system allows querying in either Maltese or English, and the results are presented in both Maltese and English. This functionality was possible due to the application of a publicly available NMT translation model to translate the legal case documents. Moreover, each sentence in the case law documents was automatically labelled to a legal Rhetorical role (e.g. *FACTS*, *ANALYSES*, ...). Through such labels, users can search through specific parts of interest within the case law documents (e.g. limiting the search to facts only). Our system relies on open resources, and requires only standard computational power. It exemplifies the potential of AI-driven legal research even in resource-constrained settings.

2 Similar Systems and Related Research

Research on Legal information retrieval systems, and the application of AI and NLP techniques for Legal Assistance has been garnering interest in recent years. Notable initiatives include the US's *LexisNexis* [9] and *Westlaw* [17]. *EUR-Lex* [4] offers access to European Union law and other public documents. While being multilingual, it primarily focuses on EU legislation, which makes it less suited for case law research, especially at national levels.

Noteworthy research initiatives include AILA 2020 and AILA 2021 (Shared task on Artificial Intelligence for Legal Assistance) [13]. In AILA 2021, participants were asked to perform 2 tasks: **Rhetorical Role Labelling** (RRL) – i.e. automatic labelling of sentences within case law documents into one of sevent rhetorical role categoried (Facts, Previous Rulings, Arguments, ...); and **Legal Document Summarization)** – automatic summarisation (both extractive and abstractive summarisation) of case law judgements. The best performing systems utilise a base transformer model (legal-BERT) fine-tuned on the training set provided [13].

RRL experiments reported in more recent publications report higher F1-scores. Kalamkar et al. [8] develop their own legal document corpus (larger than that used in AILA 2021), and compare the performance of a number of RRL classifiers. The best performing model is the SciBERT-HSLN model (originally developed in [2]). SciBERT-HSLN utilises SciBERT for the word embeddings, with an architecture based on the Hierarchical Sequential Labelling Network (HSLN). This achieves F1-scores of 0.79 [8]. Moreover, this code to run this model, along with the pre-trained model are made available on GitHub.[1].

A major lacuna in RRL research is that publicly-available data corpora are mainly from the Indian jurisdiction. Bhattacharya et al. [1] experimented with the generalisability of RRL models across different jurisdictions. However, results

[1] https://github.com/Legal-NLP-EkStep/rhetorical-role-baseline/tree/main accessed: October 2023.

indicate that such models suffer when transferred across different jurisdictions without fine-tuning using labelled data from the target jurisdiction.

Up to our knowledge, minimal research involving NLP and AI on Maltese legal and case law documents has been carried out – probably due to lack of resources for the Maltese language. The only publication we found describes efforts from a few years ago to digitise Maltese law in order to assist with legal research [6]. We did not find any updates on this work. Also, published research about cross-lingual retrieval that includes Maltese was not found.

3 Implementation

The legal.ir.mt prototype illustrates the potential of AI and NLP in facilitating legal research. One of its prominent features is its ability to automatically translate Maltese case law documents into English, allowing queries in both languages and presenting results bilingually. This not only renders legal texts more accessible to non-Maltese speakers but also pave to way to further NLP analysis on Maltese language documents. Moreover, users can hone their searches to target specific sections of documents, such as facts or arguments, capitalising on the Rhetorical Labelling applied to these documents.

The system has been developed to run on Linux OS. All data acquisition, data processing and server are implemented using Python. The front end interface was developed using HTML 5, CSS and Javascript. Importantly, the platform is entirely independent of commercial services. Currently, it hosts case law documents exclusively from the Maltese Small Claims Tribunal, which deals with civil monetary claims up to €5000. Given the relative simplicity of these cases (within the Maltese jurisdiction), they aptly highlight the system's capabilities.

Data extraction is done from the eCourts portal[2], which has cases dating back to 2000. Approximately 2300 documents have been acquired, available as PDFs without any underlying template. The text from the PDF documents is extracted using the *fitz* python library[3]. This library segments the contents into blocks. We noted that the Maltese case law documents generally consist of paragraphs with a single long sentence each. Therefore, we treated each block as a sentence. Unfortunately, boiler-plate text (e.g. page headers and footers) are included as seperate blocks, and these can not be filtered easily.

In order to tackle the challenges posed by the Maltese language, we employ neural machine translation to translate the documents into English. A number of online translation services (providing Maltese-English translation) exist, but they come at a cost. National (government-led) initiatives are ongoing to provide a translation platform [3], but to our knowledge, a publicly available platform is not available yet. Nevertheless, Opus [16] offers a Maltese-English parallel corpus that is accessible on HuggingFace[4]. We utilised the Marian NMT framework [7]

[2] https://ecourts.gov.mt.

[3] https://pymupdf.readthedocs.io/en/latest/module.html accessed: October 2023.

[4] https://huggingface.co/Helsinki-NLP/opus-mt-mt-en accessed: October 2023.

with this corpus. The generated translations are deemed to be of decent quality based on qualitative evaluation performed.

Upon translation, documents are processed through the RRL classifier, using the pre-trained SciBERT-HSLN classifier made available by [8]. As noted in Sect. 2, RRL classifiers struggle with cross-jurisdictional generaliation. Unfortunately, we did not have any Maltese case law documents hand-labelled with Rhetorical labels, and could not find other labelled datasets from jurisdictions that are more similar to the Maltese ones (e.g. UK).

For retrieval, we utilise the vector space model [15] with TF.IDF weights. Indices are constructed for each document as a whole, as well as for each rhetorical label found within. This allows users to search within the documents as a whole or search within specific parts (e.g. within *FACTS*). Only English texts are indexed - each sentence's English version is tokenised, case-folded, and NLTK is used to perform stopword removal and stemming (using Porter stemmer).

During querying, the query string is run through the Marian NMT translator. This translates any Maltese terms to English, but retains English terms. The query vector is then constructed using the same process described before. An inverted index with document postings list is utilised to retrieve those documents that contain at least one of the query terms. Then the relevant vector corresponding to each of these documents is compared to the query vector, and cosine similarity is calculated. If the user has opted to perform the search for any particular rhetorical label, the corresponding document vector is used.

Finally, the list of results is sorted in order of decreasing similarity, and each result is output to the user. For each result, the user is presented with those sentences (in both Maltese and English) whose English version contain one of the translated query terms. Moreover, a direct link to the relevant page on the eCourts portal is also provided.

4 Conclusion

This paper introduces the legal.ir.mt system - exemplifying how AI can enhance the accessibility and searchability of Maltese legal case documents. Through the use of neural machine translation, our system offers bilingual access to content that is otherwise solely available in Maltese. Moreover, the Rhetorical Role Label (RRL) classification optimises and refines the search process.

However, the system is not without its limitations. Some noise is resulting from the text extraction process (from PDFs), largely due to redundant boilerplate text. Additionally, the RRL classifier's precision falls short of our aspirations. Despite these challenges, we are optimistic about the future potential improvements, especially with anticipated collaboration with legal experts to create labeled datasets.

Our vision for the system's future involves extending its scope to include documents from the other courts. We also wish to incorporate other AI features such as automated judgement summaries and performing topic detection on the documents' contents.

References

1. Bhattacharya, P., Paul, S., Ghosh, K., Ghosh, S., Wyner, A.: Deeprhole: deep learning for rhetorical role labeling of sentences in legal case documents. Artif. Intell. Law **31**(1), 53–90 (2023). https://doi.org/10.1007/s10506-021-09304-5
2. Brack, A., Hoppe, A., Buschermöhle, P., Ewerth, R.: Cross-domain multi-task learning for sequential sentence classification in research papers. In: Proceedings of the 22nd ACM/IEEE Joint Conference on Digital Libraries, JCDL 2022. Association for Computing Machinery, New York (2022). https://doi.org/10.1145/3529372.3530922
3. Cortis, K., Attard, J., Spiteri, D.: Malta national language technology platform: a vision for enhancing Malta's official languages using machine translation. In: Proceedings of the First Workshop on Multimodal Machine Translation for Low Resource Languages (MMTLRL 2021), pp. 12–19. INCOMA Ltd., Online (Virtual Mode) (2021). https://aclanthology.org/2021.mmtlrl-1.3
4. EUR-Lex: Eur-lex: Access to european union law (2023). https://eur-lex.europa.eu/. Accessed Oct 2023
5. Galgani, F., Compton, P., Hoffmann, A.: Lexa. Expert Syst. Appl. **42**(17), 6391–6407 (2015). https://doi.org/10.1016/j.eswa.2015.04.022
6. Gatt, A., Pace, G.J.: Lab to life: smart search for Maltese legal professionals. In: Think Magazine, pp. 58–61 (2018)
7. Junczys-Dowmunt, M., et al.: Marian: fast neural machine translation in C++. In: Proceedings of ACL 2018, System Demonstrations, pp. 116–121. Association for Computational Linguistics, Melbourne (2018). https://doi.org/10.18653/v1/P18-4020. https://aclanthology.org/P18-4020
8. Kalamkar, P., et al.: Corpus for automatic structuring of legal documents. In: Proceedings of the Thirteenth Language Resources and Evaluation Conference, pp. 4420–4429. European Language Resources Association, Marseille (2022). https://aclanthology.org/2022.lrec-1.470
9. LexisNexis: Lexisnexis (2023). https://www.lexisnexis.com/. Accessed Oct 2023
10. Malta Court Services Agency: eCourts of Malta - Official Portal (2023). https://ecourts.gov.mt/. Accessed Oct 2023
11. Mandal, A., Chaki, R., Saha, S., Ghosh, K., Pal, A., Ghosh, S.: Measuring similarity among legal court case documents. In: Proceedings of the 10th Annual ACM India Compute Conference, Compute 2017, pp. 1–9. ACM, New York (2017). https://doi.org/10.1145/3140107.3140119
12. Moens, M.F.: Summarizing court decisions. Inf. Process. Manag. **43**(6), 1748–1764 (2007). https://doi.org/10.1016/j.ipm.2007.01.005
13. Parikh, V., et al.: Aila 2021: shared task on artificial intelligence for legal assistance. In: Proceedings of the 13th Annual Meeting of the Forum for Information Retrieval Evaluation, FIRE 2021, pp. 12–15. Association for Computing Machinery, New York (2022). https://doi.org/10.1145/3503162.3506571
14. Rosner, M., Joachimsen, J.: Il-Lingwa Maltija Fl-Era Digitali The Maltese Language in the Digital Age. META-NET White Paper Series. Springer, Heidelberg (2012). http://www.meta-net.eu/whitepapers
15. Salton, G., Wong, A., Yang, C.S.: A vector space model for automatic indexing. Commun. ACM **18**(11), 613–620 (1975)
16. Tiedemann, J., Thottingal, S.: OPUS-MT Building open translation services for the World. In: Proceedings of the 22nd Annual Conference of the European Association for Machine Translation (EAMT), Lisbon, Portugal (2020)
17. Westlaw: Westlaw (2023). https://www.westlaw.com/. Accessed Oct 2023

A Conversational Search Framework for Multimedia Archives

Anastasia Potyagalova(✉) and Gareth J. F. Jones

ADAPT Centre, School of Computing, Dublin City University, Dublin 9, Ireland
Anastasia.Potyagalova2@mail.dcu.ie, Gareth.Jones@dcu.ie

Abstract. Conversational search system seek to support users in their search activities to improve the effectiveness and efficiency of search while reducing their cognitive load. The challenges of multimedia search mean that search supports provided by conversational search have the potential to improve the user search experience. For example, by assisting users in constructing better queries and making more informed decisions in relevance feedback stages whilst searching. However, previous research on conversational search has been focused almost exclusively on text archives. This demonstration illustrates the potential for the application of conversational methods in multimedia search. We describe a framework to enable multimodal conversational search for use with multimedia archives. Our current prototype demonstrates the use of an conversational AI assistant during the multimedia information retrieval process for both image and video collections.

Keywords: Conversational search · Multimedia information retrieval · Human-Computer Interaction

1 Introduction

We present a prototype application to support research into dialogue-based multimedia search. The concept of conversational search has received considerable attention in recent years [6,8], In parallel with this, multimedia information retrieval (MIR) focusing on image and video search using conventional user-driven interaction is a long-standing area of investigation [1]. To date, very little work has examined conversational search for MIR tasks [3–5]. This is perhaps surprising since existing interactive MIR systems rely more on human engagement than traditional text retrieval systems. Searchers using MIR systems have typically been required to engage closely with the available multimodal representations of the data to construct queries, and to provide relevance feedback during the search process. The goal of the system introduced in this paper is to support research into the use of conversational search methods in MIR. In our system, a conversational agent engages with the user of an online MIR tool to support their search activities. The goals of the agent are to seek to improve user identification of potentially relevant retrieved information and to improve the user search experience in general.

N. Goharian et al. (Eds.): ECIR 2024, LNCS 14612, pp. 241–245, 2024.
https://doi.org/10.1007/978-3-031-56069-9_25

(a) The interface of search framework

(b) Image search clarification process.

Fig. 1. Search framework interface and image search process.

Engaging in a search dialogue with a user requires that a supporting conversational agent shows its understanding of the content and its potential value to the user's search process. While MIR has been under investigation for many years, little attention has been given to the active representation of content features identified by the system and its utilization in the search process. A conversational search system should exploit content features within the search process to enable the searcher to develop their queries and address their information needs. In this process, it should enable the user to learn about their topic of interest by incrementally assisting them in developing their search query, enabling them to move towards satisfying their information need. Such a dialogue can potentially reduce the cognitive load associated with achieving this with a traditional user-driven interactive search systems. Our framework combines a conversational search assistant with an interactive multimedia gallery. The user is able to engage directly with the search system while receiving suggestions from the search assistant to both help them with queries development and to guide their interaction with the retrieved content.

2 Prototype Conversational Search System

2.1 Search Interface

Figure 1a shows the search interface for our prototype conversational multimedia search system.

This interface includes the following components:

1. Agent Display: This shows the search dialogue between the agent and the user.
2. Chat Box: Enables the user to chat with the agent.
3. Image and video gallery: Represents the provided search results.
4. Clarification question: Allows the user to agree or disagree with the provided search results.
5. Multimedia clarification question: Suggests to the user an additional image for reducing irrelevant output
6. Select mode menu: Allows the user to choose the search option:

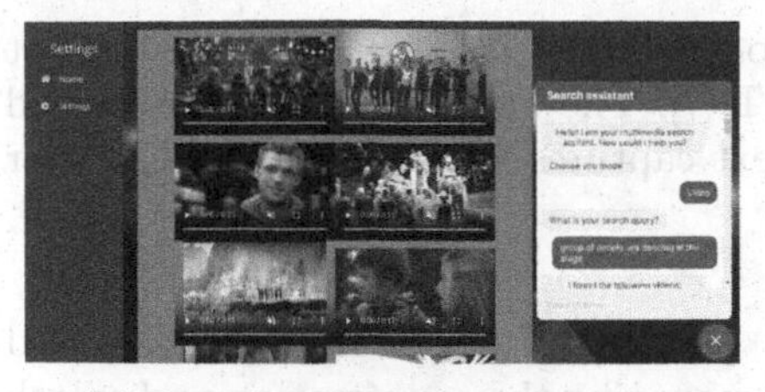

(a) Initial video search outcome.

(b) Clarification process for video search.

Fig. 2. Video search initial output and interactive clarification.

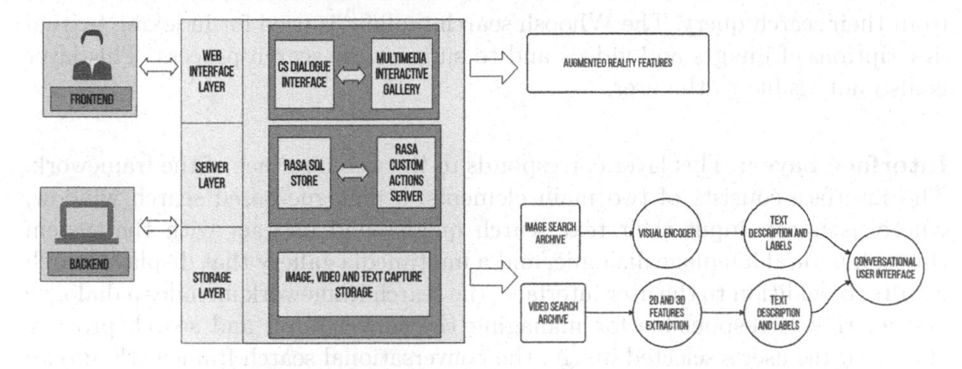

Fig. 3. Representation of framework structure.

- Image: Starts the search process for image search archive.
- Video: Starts the search process for video search archive.

7. Filter button: Allows the user to open the potential image filter in a separate window

Figure 1b shows potential agent actions, including the following: seeking clarification of search queries and suggesting images for use in revised queries. In conversational search for videos, the system asks clarifying questions based on the user's initial query or previous interactions.

The workflow of the video search scenario is presented in Fig. 2a. Agent actions include the following: seeking clarification of search queries and suggesting images for use in revised queries, as shown in Fig. 2b.

2.2 System Implementation

Figure 3 shows the architecture of our prototype conversational MIR system. The system is implemented as a web application using the Python and Flask framework. It is deployed on a virtual machine server using cloud technologies servers provided by the RASA open library[1]. The architecture can be separated into three layers:

[1] https://rasa.com/.

Storage Layer: This includes the preprocessed search datasets: image dataset Flickr30k [2] and video dataset MSR-VTT [7]. The datasets were preprocessed to extract visual features and generate text captures for the search. This layer is not visible to the user.

Server Layer: This is responsible for performing the search process, and includes a dialogue system responsible for managing the conversation and search process. The dialogue system uses a RASA API - a popular open-source framework for building conversational AI systems powered by a fine-tuned BERT model. BERT is used to identify the user's intent and extract relevant entities from their search query. The Whoosh search toolkit[2] is used for indexing textual descriptions of images and videos and to support the search process. This layer is also not visible to the user.

Interface Layer: This layer corresponds to the web interface of the framework. The interface consists of two main elements: a dialogue-based search window, where users can input their text search queries and interact with the system through natural language dialogue, and a multimedia gallery that displays search results. In addition to the user interface, the search framework includes a dialogue system that is responsible for managing the conversation and search process. Based on the user's selected image, the conversational search framework applies a filtering mechanism to narrow down the output. The framework uses visual similarity to reduce the set of images to those closely matching the selected filter image by calculating the distance between the image in dialogue and images in the multimedia gallery. The user only has access to this interface layer.

As was outlined earlier, users are able to enter text search queries into the interface, reformulate their queries as required, and interact with the search results which appear in the gallery.

2.3 Search Scenarios and Dialogue Flow

Our prototype MIR system enables users to perform a search for images and videos[3]. Potential common search scenarios for image search and video search are illustrated by the screencast video, shared at the link: https://t.ly/OSZG. The screencast includes demonstration of simple search scenarios and more complex scenarios. The framework asks clarification questions and engages in a conversational dialogue with the user to understand their intent and provide more accurate search results.

Acknowledgments. This work was conducted with the financial support of the Science Foundation Ireland Centre for Research Training in Digitally-Enhanced Reality (d-real) under Grant No. 18/CRT/6224, and partially as part of the ADAPT Centre at DCU (Grant No. 13/RC/2106_P2) (www.adaptcentre.ie). For the purpose of

[2] https://whoosh.readthedocs.io/.

[3] https://github.com/apotyagalova/conversational-search-agent.

References

1. Hanjalic, A., Lienhart, R., Ma, W.Y., Smith, J.R.: The holy grail of multimedia information retrieval: so close or yet so far away? Proc. IEEE **96**(4), 541–547 (2008). https://doi.org/10.1109/JPROC.2008.916338
2. Hodosh, M., Young, P., Hockenmaier, J.: Framing image description as a ranking task: data, models and evaluation metrics. J. Artif. Intell. Res. **47**, 853–899 (2013)
3. Kaushik, A., Jacob, B., Velavan, P.: An exploratory study on a reinforcement learning prototype for multimodal image retrieval using a conversational search interface. Knowledge **2**(1), 116–138 (2022)
4. Kim, H., Kim, D., Yoon, S., Dernoncourt, F., Bui, T., Bansal, M.: Caise: conversational agent for image search and editing. In: Proceedings of the The Thirty-Sixth AAAI Conference on Artificial Intelligence (AAAI-2022) (2022)
5. Nie, L., Jiao, F., Wang, W., Wang, Y., Tian, Q.: Conversational image search. IEEE Trans. Image Process. **30**, 7732–7743 (2021)
6. Radlinski, F., Craswell, N.: A theoretical framework for conversational search. In: Proceedings of the 2017 Conference on Conference Human Information Interaction and Retrieval, pp. 117–126 (2017)
7. Xu, J., Mei, T., Yao, T., Rui, Y.: MSR-VTT: a large video description dataset for bridging video and language. In: Proceedings of the IEEE Conference on Computer Vision and Pattern Recognition, pp. 5288–5296 (2016)
8. Zamani, H., Trippas, J.R., Dalton, J., Radlinski, F.: Conversational information seeking. arXiv preprint arXiv:2201.08808 (2022)

Building and Evaluating a WebApp for Effortless Deep Learning Model Deployment

Ruikun Wu, Jiaxuan Han, Jerome Ramos(✉), and Aldo Lipani

University College London, London, UK
{ruikun.wu.22,jiaxuan.han.22,jerome.ramos.20,aldo.lipani}@ucl.ac.uk

Abstract. In the field of deep learning, particularly Natural Language Processing (NLP), model deployment is a key process for public testing and analysis. However, developing a deployment pipeline is often difficult and time-consuming. To address this challenge, we developed SUD.DL, a web application to simplify the model deployment process for NLP researchers. Our application provides significant improvements in deployment efficiency, functionality discoverability, and deployment functionality, allowing NLP researchers to quickly deploy and test models on the web.

Keywords: Natural Language Processing · Web Applications · User Interfaces

1 Introduction

Deploying and testing natural language processing (NLP) models on the web is a crucial step in applying machine learning techniques to a wide range of real-world applications. However, the model deployment process is difficult [3] and current deployment methods often requires in-depth knowledge [2]. Acquiring deployment skills can often be time-consuming and inefficient for researchers primarily focused on researching and developing deep learning algorithms. To address this problem, we developed SUD.DL, a web application to facilitate and simplify model deployment for users with less software deployment knowledge. The primary aim of this work is to create an accessible, web-based application using open-source technologies. This application will serve as a user-friendly bridge, connecting a diverse range of users to the complex field of deep learning deployment. SUD.DL addresses the requirements of non-technical users for applying deep learning in diverse NLP tasks, including text classification, generation, translation, summarisation, question answering, and masked language modelling. Our application also allows researchers to quickly and easily deploy models to the web for testing and validation, enhancing efficiency in applied

R. Wu and J. Han—These authors contributed equally to this work.

N. Goharian et al. (Eds.): ECIR 2024, LNCS 14612, pp. 246–250, 2024.
https://doi.org/10.1007/978-3-031-56069-9_26

NLP research. By reducing the technical barriers and simplifying the model deployment process, SUD.DL aspires to facilitate a smoother integration of deep learning into various fields of application and research.

2 Methodology

SUD.DL was developed using the Python and Django framework. Specifically, the main technologies used include HTML for front-end page building and SQLite for back-end database. In addition, the Transformers package was used to load NLP models and Docker was used to enable containerised deployment of models. Note that all uploaded model files were stored in Docker's virtual environment, and other model information such as name was saved in the database. The demo video and detailed code were used to build the SUD.DL can be found and viewed at https://github.com/HanJiaxuan1/Deep-learning-APP.git.

This project also aims to optimise deployment procedures and enhance deployment efficiency. These optimisations involves many dimensions, which is included in our documentation. As stated by Sripathi et al. [5], preceding the utilisation of functionality, users often read instructional documentation to acquire information to further execute the intended task. As a result, as shown in Fig. 1, the help documentation for the deployment functionality of this web app focuses on providing short and easy-to-understand content.

Fig. 1. Help Documentation

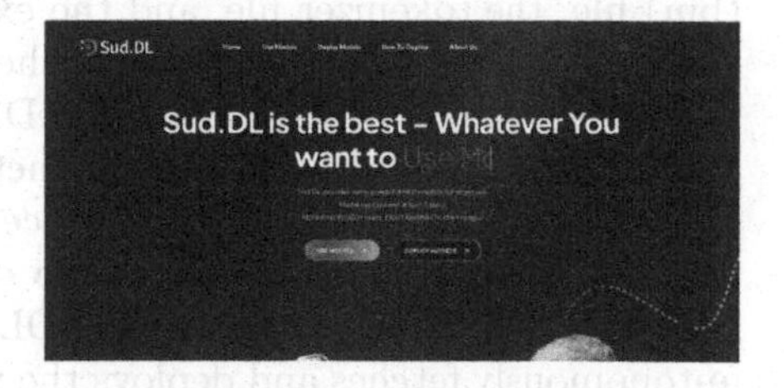

Fig. 2. SUD.DL APIs

In addition, the discoverability of functionality significantly influences its utilisation efficiency [4]. Consequently, this project provides users with clear and accessible APIs, thereby promoting the utilisation of various functionalities. As displayed in Fig. 2, all of the APIs for model deployment, model use, and deployment help documentation are easy to find.

Moreover, the reasonable layout and arrangement of elements contributes to reducing the complexity of utilising functionalities, leading to an improved user experience [1]. Thus, this project provides a user-friendly and intuitive interface for the model deployment functionality, which is achieved through progressive step-by-step guidance. Finally, streamlining the workflow of functionalities to avoid duplication and complexity is an effective approach for efficiency enhancement [4]. Therefore, the deployment functionality is designed to be finished on a single web page and consists of only four straightforward actions. To be specific, as displayed in Fig. 3, the first step is to select the NLP task for which the model is to be deployed. Subsequently, the second step involves typing the basic information of the model.

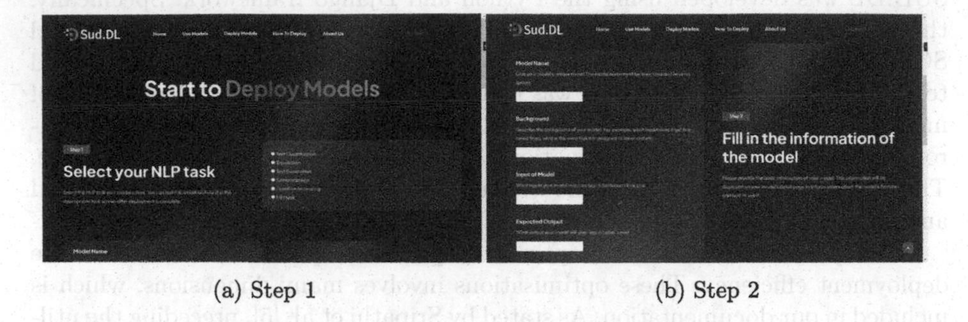

(a) Step 1 (b) Step 2

Fig. 3. SUD.DL Model Deployment Functionality

In the third step, illustrated in Fig. 4, users are required to upload key components related to their model, including but not limited to the model's binary (bin) file, the tokenizer file, and the exact versions of both PyTorch and Transformers that they employed during the model's training process. Understanding the diverse needs of its users, SUD.DL offers flexibility in the upload process. Users are presented with two distinct options: (1) they can either choose to upload all model-related files at once or (2) they can upload only the model files while also indicating the name of the pre-existing model that they fine-tuned. In the latter scenario, SUD.DL, identifies the model's requirements and autonomously fetches and deploys the relevant tokenizer files, sparing users from the complicated process of manual uploads. This feature not only streamlines the deployment process but also enhances the user experience.

In the final stage, by simply clicking on the 'DEPLOY NOW' button, the model will be deployed to the web. SUD.DL uses Docker to containerise each model to enable consistency and scalability for multiple Python environments. Once the deployment is completed, users can use their models in the model use page, as shown in Fig. 5.

Due to optimisation of the deployment process, SUD.DL distinguishes itself from the Hugging Face ecosystem with its user-friendly interface, simplified model deployment process, flexible model upload options, easy-to-understand help documentation, and the enhanced functionality discoverability. These features make SUD.DL a more intuitive and user-centric platform.

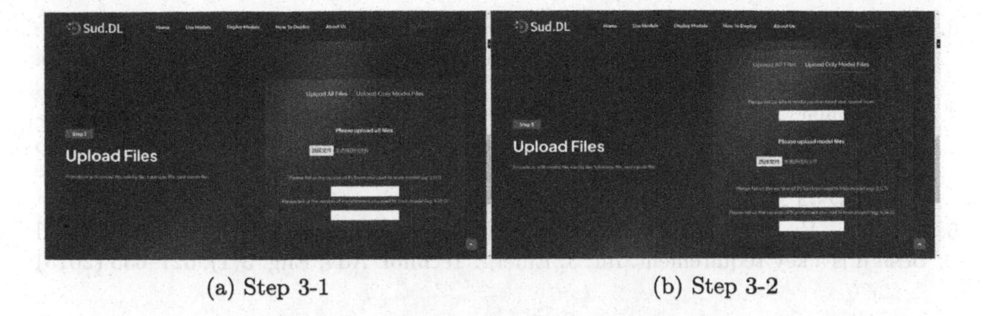

(a) Step 3-1 (b) Step 3-2

Fig. 4. SUD.DL Model Deployment Functionality

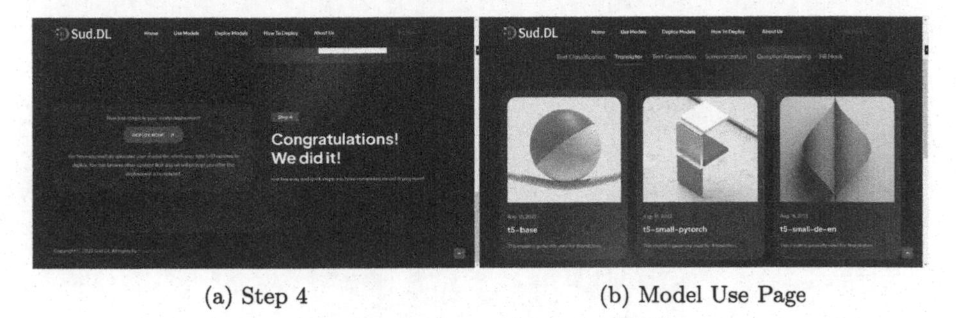

(a) Step 4 (b) Model Use Page

Fig. 5. Submit and Model Use

3 Conclusion

In summary, we developed SUD.DL, a user-friendly web application to facilitate and simplify NLP model deployment. SUD.DL assists users with limited deployment experience easily deploy models on the web. Our model deployment pipeline helps researchers shift their focus more towards model development rather than the intricacies of publishing models on the web. Despite these advancements, SUD.DL faces several limitations. Currently, support is limited to Transformer library models, suggesting a need for broader library integration. For future work, we plan to further develop our model forum and enhance the efficiency of the deployment process, particularly in file uploads and Docker image setup.

References

1. Abbasi, M.Q., Weng, J., Wang, Y., Rafique, I., Wang, X., Lew, P.: Modeling and evaluating user interface aesthetics employing iso 25010 quality standard. In: 2012 Eighth International Conference on the Quality of Information and Communications Technology, pp. 303–306 (2012). https://doi.org/10.1109/QUATIC.2012.39
2. Ma, Y., Xiang, D., Zheng, S., Tian, D., Liu, X.: Moving deep learning into web browser: how far can we go? In: The World Wide Web Conference, pp. 1234–1244 (2019). https://dl.acm.org/doi/abs/10.1145/3308558.3313639?casa_token=NkLwijBx2TsAAAAA:dksJ1pp6UnZWnTTneVJkRD8WEncG17HJZkcg5O687TqPxOhMd823yqejHVoEYVjFwG2cHBw9w_Q

3. Paleyes, A., Urma, R.G., Lawrence, N.D.: Challenges in deploying machine learning: a survey of case studies. ACM Comput. Surv. **55**(6) (2022). https://doi.org/10.1145/3533378
4. Shivakumar, S.K., Shivakumar, S.K.: Modern web platform performance principles. In: Modern Web Performance Optimization: Methods, Tools, and Patterns to Speed Up Digital Platforms, pp. 105–143 (2020). https://doi.org/10.1007/978-1-4842-6528-4_5
5. Sripathi, V., Sandru, V.: Effective usability testing-knowledge of user centered design is a key requirement. Int. J. Emerg. Technol. Adv. Eng. **3**(1), 627–635 (2013)

indxr: A Python Library for Indexing File Lines

Elias Bassani[1] and Nicola Tonellotto[2(✉)]

[1] Independent, Milan, Italy
e.bassani3@campus.unimib.it
[2] University of Pisa, Pisa, Italy
nicola.tonellotto@unipi.it

Abstract. indxr is a Python utility for indexing file lines that allows users to dynamically access specific ones, avoiding loading the entire file in the computer's main memory. indxr addresses two main issues related to working with textual data. First, users who do not have plenty of RAM at their disposal may struggle to work with large datasets. Since indxr allows accessing specific lines without loading entire files, users can work with datasets that do not fit into their computer's main memory. For example, it enables users to perform complex tasks with limited RAM without noticeable slowdowns, such as pre-processing texts and training Neural models for Information Retrieval or other tasks. Second, indxr reduces the burden of working with datasets split among multiple files by allowing users to load specific data by providing the related line numbers or the identifiers of the information they describe, thus providing convenient access to such data. This paper overviews indxr's main features. (https://github.com/AmenRa/indxr).

Keywords: Efficiency · Python · Open-Source

1 Introduction

Working with textual datasets may require the availability of a large computer's main memory to store the data to process. However, this can be prohibitive for many researchers, practitioners, and students. For example, when training a Neural model for Information Retrieval (IR), we may need to rapidly access specific documents and queries to feed them to the model. Usually, the documents are accessed multiple times in random order. Therefore, relying on a data stream may not be possible. Moreover, we may need to merge information from different sources on the fly when dealing with a dataset split among multiple files. For example, a dataset for Personalized Search [2] may comprise a file storing the document collection, one for the queries, one for the information regarding the user preferences or their interactions with the documents, and other files storing metadata. We argue that working with those datasets may be very complex from a data management perspective, discouraging early-stage IR researchers and students from tackling tasks other than ad-hoc search.[1]

[1] An example of how to use indxr in those cases is provided in the official repository.

N. Goharian et al. (Eds.): ECIR 2024, LNCS 14612, pp. 251–255, 2024.
https://doi.org/10.1007/978-3-031-56069-9_27

In this paper, we present `indxr`, a user-friendly Python utility to circumvent those issues. `indxr` allows reading specific file lines without loading the entire file into memory by providing the line numbers or the identifiers of the information they describe. In other words, `indxr` allows users to leverage disks as RAM extensions without noticeable slowdowns on downstream tasks while providing a convenient interface, as described in Sects. 2 and 3. Although the scope of our utility might seem limited, it empowers users to perform tasks they would be unable to do without upgrading their hardware. To the best of our knowledge, no available Python library is comparable to `indxr` in terms of features and scope.

2 Overview

In this section, we present the main functionalities of `indxr` and describe how it works under the hood.

Indexing and Reading. At the core of `indxr` is a Python dictionary that serves as a lookup table between line numbers or identifiers and the position of the corresponding line in the original file. `indxr` reads the given file line-by-line and stores each line position (expressed as a byte-level offset) in the lookup table. We leverage the Python's IO library `.tell()` method to get the position the cursor is pointing at. When accessing a specific line, `indxr` maps the identifier of the requested line to its position and leverages the `.seek()` function to move the cursor. Since both accessing a Python dictionary's value and the `.seek()` function are $O(1)$ operations, reading a line into memory mostly depends on the size and nature of the line. This mechanism allows us to store only the lookup table in the computer's main memory and dynamically load the line on request quite efficiently, as discussed in Sect. 3. We acknowledge this mechanism was inspired by the work of Câmara and Hauff [3].

API. `indxr` provides a common interface for all the supported file types: `txt`, `csv`, `tsv`, and `JSONl`. For the sake of brevity, we only show the implemented features considering a `JSONl` file. In a `JSONl` file each line is a `JSON` object, which roughly corresponds to a Python dictionary. We choose this type of file for showing the implemented functionalities for its increasing popularity among the IR community. Specifically, `JSONl` files are used to store document collections, where each `JSON` represent a document with an identifier and a title, a body, and/or other other metadata. `indxr` API is shown in Listing 1. In line 4, we create an Indxr object by proving a document collection's path and the property to consider as documents' identifier. As a quality-of-life feature, the user can provide a callback function to apply on-the-fly transformations when a document is accessed (e.g., extracting its title). Moreover, `indxr` automatically recognizes the file type so that the user does not need to specify one when an Indxr object is created. Lines 6 and 7 show how to access one or multiple documents by their identifiers. Users can also load documents by their line number, as shown in Lines 8 and 9. Indxr objects can iterate over the file one line at a time or even generate batches, as shown in lines 12 and 16. Finally, Indxr objects can be persisted and read from the disk, as shown in lines 19 and 20. At first glance, the

API might seem deceptively simple and narrow in its scope. However, its true strength lies in the capability to empower users to access data stored on disk with minimal delay while maintaining a low memory footprint, as discussed in Sect. 3. For advanced usage examples, we invite the reader to consult the official repository.

```
from indxr import Indxr

# Creates an Indxr object for a document collection
index = Indxr(path="path/to/collection.jsonl", key_id="id")

index.get("id_123")  # Document with id="id_123" as Python Dictionary
index.mget(["id_123", "id_321"])  # Get multiple documents
index[42]  # Document at line 43 as Python Dictionary
index[42:46]  # Documents at line 43, 44, and 45 as Python Dictionaries

# Iterate over the index sequentially
for x in index:
    # Do something...

# Iterate over the index in bacthes
for batch in index.generate_batches(batch_size=32, shuffle=True):
    # Do something...

index.write("write/to/path.json")  # Write Indxr object to file
index.read("read/from/path.json")  # Read Indxr object to file
```

Listing 1: indxr API.

3 Empirical Evaluation

In this section, we evaluate the time and space efficiency of `indxr`. The evaluation is performed on a consumer-grade workstation equipped with an AMD Ryzen™ 5950X CPU, an NVidia® RTX A6000 GPU, 64 GB of RAM, and a traditional SSD (560 MB/s read speed).

Memory Usage, Indexing Time, and Raw Access Time. We consider MSMARCO Passage v1 and MSMARCO Document v2 [1] stored as `JSONl` files to evaluate the raw access time using `indxr`. MSMARCO Passage v1 accounts for 8.8 million short documents and takes 6.5 GB of memory. MSMARCO Document v2 comprises 12 million long documents. We couldn't load it in memory as its size exceeds 100 GB on disk when uncompressed. `indxr` consumed roughly 1 and 2 GB of RAM for the two datasets, respectively, thus considerably improving memory usage. Moreover, it allowed us to overcome our hardware limitations when it comes to MSMARCO Document v2. Creating the lookup table on the CPU for MSMARCO Passage v1 and MSMARCO Document v2 took 10 s and

360 s, respectively. We monitored the average time for reading a single random document, 64 contiguous documents, and 64 non-contiguous documents. For MSMARCO Passage v1 reading a single random document took 3.82 μ s, 64 contiguous documents took 67.1 μ s, and 64 non-contiguous documents took 223 μ s. For MSMARCO Document v2 the three readings took 0.2 ms, 2.5 ms, and 11.9 ms respectively. Those reading times are very low and do not influence downstream tasks, as discussed in the next section. Moreover, they could be improved by employing an NVME drive and a faster CPU.

Neural Models Training. We considered training Neural models for IR as a real-world scenario for employing `indxr`. Specifically, we trained a CrossEncoder [6], a BiEncoder [7], and ColBERT [5] on MSMARCO Passage v1. We evaluated the training times by pre-loading the entire dataset and using `indxr` to dynamically load the texts for ⟨query, positive document, negative document⟩ triplets. We considered DistilBERT [8] as the models' backbone as it has roughly half of BERT's [4] parameters, meaning it is much more efficient and may allow us to capture training time differences between the compared approaches if any. However, we did not find any noticeable difference in the training times of the models regardless of the batch size used (we experimented with 32, 64, 128, 256, and 320). By taking a closer look at CPU and GPU usage, we found that the GPU usage was always maximized, which means we are GPU-bound despite it being quite powerful. We conclude that, for the time being, using `indxr` allows for training Neural retrievers as efficiently as pre-loading the entire MSMARCO Passage v1 dataset with a lower memory footprint.

4 Conclusion and Future Works

In this paper we presented `indxr`, a Python library for indexing file lines which allows for efficiently reading specific ones on demand, lowering RAM usage. Moreover, `indxr` reduces the burden of working with datasets split among multiple files by providing a convenient interface to access data stored on disk. Through a real-world example, we showed that `indxr` can be leveraged to use disks as RAM extensions without noticeable slowdowns on downstream tasks, thus enabling users to work with datasets that do not fit into their computers' main memory. As future works, we want to further optimize the lookup table to make it even more memory efficient and extend `indxr` to work with other file types.

Acknowledgement. This work is supported, in part, by the spoke "FutureHPC & BigData" of the ICSC - Centro Nazionale di Ricerca in High-Performance Computing, Big Data and Quantum Computing, the Spoke "Human-centered AI" of the M4C2 - Investimento 1.3, Partenariato Esteso PE00000013 - "FAIR - Future Artificial Intelligence Research", funded by European Union - NextGenerationEU, the FoReLab project (Departments of Excellence), and the NEREO PRIN project funded by the Italian Ministry of Education and Research Grant no. 2022AEFHAZ.

References

1. Bajaj, P., et al.: MS MARCO: a human generated MAchine reading COmprehension dataset (2016). http://arxiv.org/abs/1611.09268
2. Bassani, E., Kasela, P., Raganato, A., Pasi, G.: A multi-domain benchmark for personalized search evaluation. In: Proceedings of CIKM, pp. 3822–3827 (2022)
3. Câmara, A., Hauff, C.: Moving stuff around: a study on efficiency of moving documents into memory for neural IR models (2022). http://arxiv.org/abs/2205.08343
4. Devlin, J., Chang, M., Lee, K., Toutanova, K.: BERT: pre-training of deep bidirectional transformers for language understanding. In: Proceedings of NAACL-HLT, pp. 4171–4186 (2019)
5. Khattab, O., Zaharia, M.: ColBERT: efficient and effective passage search via contextualized late interaction over BERT. In: Proceedings of SIGIR, pp. 39–48 (2020)
6. Nogueira, R.F., Cho, K.: Passage re-ranking with BERT. http://arxiv.org/abs/1901.04085
7. Reimers, N., Gurevych, I.: Sentence-bert: sentence embeddings using siamese bert-networks. In: Proceedings of EMNLP, pp. 3980–3990 (2019)
8. Sanh, V., Debut, L., Chaumond, J., Wolf, T.: DistilBERT, a distilled version of BERT: smaller, faster, cheaper and lighter (2019). http://arxiv.org/abs/1910.01108

SciSpace Literature Review: Harnessing AI for Effortless Scientific Discovery

Siddhant Jain, Asheesh Kumar, Trinita Roy, Kartik Shinde, Goutham Vignesh, and Rohan Tondulkar(✉)

SciSpace, Bengaluru, India
{siddhant,asheesh,trinita,kartik,goutham,rohan}@typeset.io
http://typeset.io/

Abstract. In the rapidly evolving landscape of academia, the scientific research community barely copes with the challenges posed by a surging volume of scientific literature. Nevertheless, discovering research remains an important step in the research workflow which is also proven to be a challenging one to automate. We present Scispace Literature Review, a sophisticated, multi-faceted tool that serves as a comprehensive solution to streamline the literature review process. By leveraging the state-of-the-art methods in vector-based search, reranking, and large language models, the tool delivers features like customizable search results, data exintegration with an AI assistant, multi-language support, top papers insights, and customizable results columns to cater a researcher's requirements, and accelerate literature exploration. Resources for simplified sharing and documentation further enhance the scope and depth and breadth of research. We demonstrate the extensive use and popularity of the tool among researchers with various metrics, highlighting its value as a resource to elevate scientific literature review. This tool can be tried using this link: https://typeset.io/search.

Keywords: Information Retrieval · Vector Search · Retrieval Augmented Generation

1 Introduction and Related Works

In Academia, researchers face the growing challenge of managing the increasing volume of scientific literature. Conducting a literature survey can be demanding due to the large amount of information, the need to evaluate source quality, stay updated, handle diverse perspectives, and perform time-consuming searches. It involves critically analyzing numerous sources, synthesizing them into a clear narrative, and ensuring proper citation. In-depth search practices are crucial for meticulously exploring vast knowledge repositories to extract valuable insights, helping researchers identify relevant information and trends in their field. Several attempts have been made to streamline the integration and analysis process in literature search tools. A study by [1] explores text mining methods for keyword

N. Goharian et al. (Eds.): ECIR 2024, LNCS 14612, pp. 256–260, 2024.
https://doi.org/10.1007/978-3-031-56069-9_28

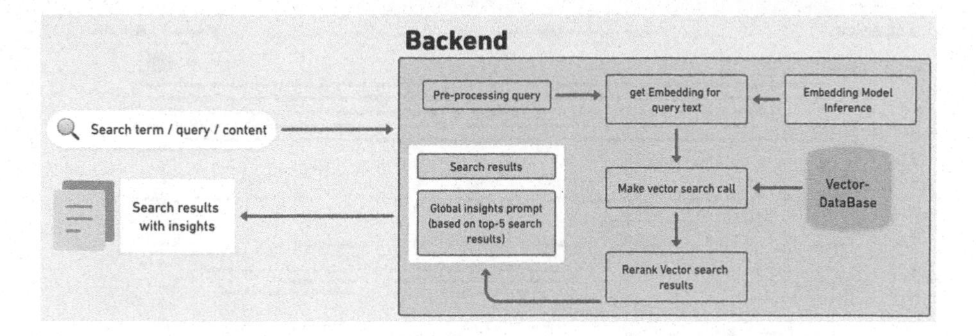

Fig. 1. Software architecture for Literature Review Pipeline.

extraction, emphasizing their importance in Computer Science. The paper also discusses the use of feature selection metrics for evaluating potential keywords. [2] presents a new method for "vector similarity searching" on dense semantic representations, which integrates with traditional fulltext engines like Elasticsearch, resulting in an efficient vector database. The effectiveness of this approach is shown through semantic searching on the English Wikipedia. The paper [3,4] introduces a technique using retrieval-augmented generation (RAG) models to improve language generation, achieving high performance in open-domain QA tasks.

SciSpace Literature Review is a tool that helps researchers efficiently navigate over 200 million academic articles across various fields. Its user-friendly interface allows easy keyword input, displays key insights, and offers customizable search options. It also integrates with an AI assistant, Copilot, for enhanced interactivity. The platform supports 75+ languages, facilitating collaboration and future reference with simplified sharing and record-keeping tools. Advanced search options enable users to refine results based on criteria like publication details, type, quality, etc. making relevant information easily accessible.

2 Methodology

In this section, we describe the architecture and methodology of the overall Literature Review Tool as depicted in Fig. 1. Users input queries are first preprocessed and transformed which is then used for further steps. The system conducts multiple vector-based searches on this query, utilizing open-source text representation models like Specter [5] and Colbert [6], and interfaces with low-latency vector databases to retrieve a few hundred results for the query. Following this integration, the Backend employs a reranking mechanism to refine the results by considering various parameters, ultimately providing the top 10 most relevant outcomes. This comprehensive approach ensures the delivery of precise and refined search results to meet the research objectives.

Post receiving the top-10 most relevant results, all the data shown on the page needs to be generated. For this, the backend initiates a call to generate a

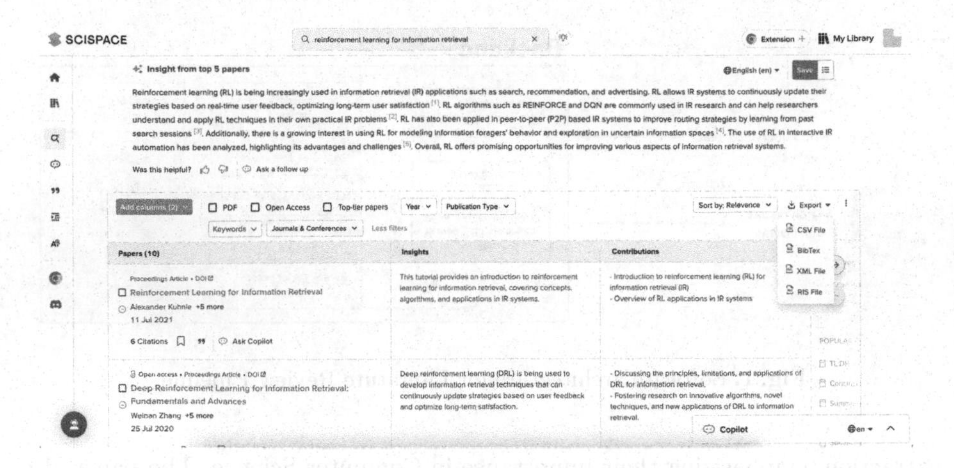

Fig. 2. SciSpace tool displaying search results, column data, and diverse features.

prompt that offers insights based on the collective knowledge extracted from the top-5 papers, effectively addressing the user's query. Furthermore, the backend proceeds to make additional calls to obtain prompts for individual paper-level insights for all ten selected papers. In the end, calls are made to various large language models to generate factual answers with citations. Similarly, outputs for various user selected columns are generated by extracting relevant context from the paper.

3 Tool Demo

The SciSpace literature review tool is a very feature rich tool. Below we provide demonstration of how the tool can be used to its full potential with the various features provided as shown in Fig. 2.

Search Suggestions: When users type their research questions in search box, we provide them few suggestions on what can be asked and nudging them towards framing better questions.

Insight from Top 5 Papers: The tool provides an answer to the query from top 5 relevant papers. This is a cohesive answer covering points across 5 papers along with citations for each sentence.

Search Results Table: For every search query, the tool provides top 10 research articles that are most relevant to the query. Along with this, various columns are provided in results table. The first column provides an insight from each paper for the given query. Rest of the columns can be used to extract any specific information from the paper.

Filtering and Sorting: Users can filter papers based on their - access, pdf availability, publication details like year, type and name. It has option to only consider top-tier journals and conferences. Keyword based filters can also be used. Users can also sort the results by relevance, citation count, alphabetical order and publication date.

Show More Papers: Users can select few papers and click on "Show more like selected" button to get more papers that are closer to the selected search result. At the bottom there is also an option to get the next 5 search results.

Copilot: The search page also has an AI assistant named Copilot that can be used to get more detailed answers about the papers in search results.

Regional Language Support: The tool supports 75+ regional languages to make research discovery accessible to every irrespective of their language.

Save and Download Results: User can save the modified search table and refer it later. Users also have the ability to export the search results and table in various formats like CSV, Bibtex, XML and RIS for future use.

4 Tool Analytics

User Feedback: 5000 users used the tool on the day of launch itself and it has been increasing ever since. The tool received an overall experience rating of 3.9/5, the quality of insights was rated at 3.8/5, and search results were rated at 4.1/5. This shows that the tool is effective in finding relevant scientific literature and also providing valuable insights. More advanced filters and new export options were added based on detailed user feedback.

Search Query Analysis: 20,000 queries were executed on day 1. From 3 months of data, the average query length, after excluding outlier queries, is 7.68 tokens. A significant majority of searches, that is, 77.64%, were conducted in the English language. Approximately 4% of searches were conducted in Spanish, Italian, and Portuguese languages respectively. German-language searches accounted for 2.7% of the total searches. ∼40% of the total searches consisted of WH-questions.

5 Conclusion

The SciSpace Literature Review tool provides a comprehensive, efficient, and user-friendly platform for navigating vast volumes of academic articles. Leveraging advanced text mining methods, retrieval-augmented generation models, and vector similarity searching, the tool delivers precise, relevant, and insightful results to complex research queries. Its features, including search suggestions, insights from top papers, advanced filters, regional language support, various export options, etc. cater to diverse user needs and preferences. The tool's high user engagement underscore its place as a critical asset for researchers. In future, we plan on improving search results further and adding new features like more filters, domain specific data, custom columns etc. based on user feedback.

References

1. Siddiqi, S., Sharan, A.: Keyword and keyphrase extraction techniques: a literature review. Int. J. Comput. Appl. **109**(2), 1–6 (2015)
2. Jan, R., Jan, P., Radim, R., Michal, R., Vit, N., Petr, S.: Semantic vector encoding and similarity search using fulltext search engines. arXiv Information Retrieval (2017)
3. Patrick, S., et al.: Retrieval-augmented generation for knowledge-intensive NLP tasks. arXiv Computation and Language (2020)
4. Cai, D., Wang, Y., Liu, L., Shi, S.: Recent advances in retrieval-augmented text generation (2022). https://doi.org/10.1145/3477495.3532682
5. Cohan, A., Feldman, S., Beltagy, I., Downey, D., Weld., D.S.: SPECTER: document-level representation learning using citation-informed transformers (2020). https://doi.org/10.18653/V1/2020.ACL-MAIN.207
6. Santhanam, K., Khattab, O., Saad-Falcon, J., Potts, C., Zaharia, M.: ColBERTv2: effective and efficient retrieval via lightweight late interaction (2022). https://doi.org/10.48550/arXiv.2112.01488

Displaying Evolving Events Via Hierarchical Information Threads for Sensitivity Review

Hitarth Narvala(✉), Graham McDonald, and Iadh Ounis

University of Glasgow, Glasgow, UK
h.narvala.1@research.gla.ac.uk,
{graham.mcdonald,iadh.ounis}@glasgow.ac.uk

Abstract. Many government documents contain sensitive (e.g. personal or confidential) information that must be protected before the documents can be released to the public. However, reviewing documents to identify sensitive information is a complex task, which often requires analysing multiple related documents that mention a particular context of sensitivity. For example, coherent information about evolving events, such as legal proceedings, is often dispersed across documents produced at different times. In this paper, we present a novel system for sensitivity review, which automatically identifies hierarchical information threads to capture diverse aspects of an event. In particular, our system aims to assist sensitivity reviewers in making accurate sensitivity judgements efficiently by presenting hierarchical information threads that provide coherent and chronological information about an event's evolution. Through a user study, we demonstrate our system's effectiveness in improving the sensitivity reviewers' reviewing speed and accuracy compared to the traditional document-by-document review process.

Keywords: Sensitivity Review · Information Threading

1 Introduction

In many countries, government documents are legislated to be released to the public to comply with Freedom of Information (FOI) laws [12]. However, government documents often contain sensitive information (e.g. personal or confidential information). Therefore, the documents must undergo an exhaustive sensitivity review to identify and protect sensitive information before the documents can be released to the public. However, identifying sensitive information is a complex task that often requires analysing hidden connections between documents, such as coherent information about an event or a discussion. For example, a chronological evolution of a legal proceeding, which mentions sensitive information about the alleged crimes or the victims' personal details, could be spread across different documents. Sensitivity reviewers need to quickly find and understand the

N. Goharian et al. (Eds.): ECIR 2024, LNCS 14612, pp. 261–266, 2024.
https://doi.org/10.1007/978-3-031-56069-9_29

complete context of an event or discussion when they make sensitivity judgements. A coherent and chronological presentation of the information thread, which captures the event's evolution, can help the reviewers achieve this goal.

For general document review tasks (e.g. e-Discovery), various systems have been proposed (e.g. [1,13]) to perform complex exploratory search tasks and efficiently assess the relevance of a document for review. However, these systems do not principally focus on sensitivity review, where all documents that should be released to the public are relevant for sensitivity review. Recently, a few sensitivity review systems (e.g. [4,5]) have been proposed for searching and navigating a collection of sensitive documents, or to assist the reviewers with automatic sensitivity classification prediction. However, these systems are not focused on presenting coherent information to the reviewers, which can help them accurately and efficiently identify sensitivities from multiple related documents. More recently, Narvala et al. [8] proposed a system for sensitivity review that can identify and prioritise coherent groups of documents, i.e., high-level semantic categories [7] and finer-grained information threads [10] to improve the efficiency of the sensitivity review process. In particular, the system by Narvala et al. [8] leveraged *sequential* information threads [10] that present a coherent and chronological *sequence* of information about an event. However, sequential information threads may not effectively capture the diverse aspects pertaining to an event's evolution [9]. In particular, an event's evolution typically forms a *hierarchical* structure [9], where each branch of the hierarchy contains a chronologically evolving sequence of documents that describe a specific aspect relating to the event.

In this paper, we present a novel system for assisting sensitivity reviewers by identifying hierarchical information threads that capture diverse aspects of an event from multiple documents. By presenting hierarchical information threads, our system enables the reviewers to quickly provide accurate sensitivity judgements through the collective inspection of coherent information about an event's evolution across multiple documents. Moreover, we present a user study to show the effectiveness of our system in improving the reviewers' review speed and accuracy compared to the traditional document-by-document review process. You can try out our system at: http://demos.terrier.org/sensreview/thread-demo.

2 Hierarchical Threading for Sensitivity Review

We now provide details about effectively leveraging hierarchical information threads for accurate and efficient sensitivity reviews. In a traditional document-by-document sensitivity review, the reviewers are presented with different documents in a *sequence*, e.g., based on the documents' creation timestamp or using the semantic relatedness between the documents [7]. In contrast, our system enables a *collective* review of coherent information from multiple documents by identifying and presenting hierarchical information threads [9] to the reviewers. In particular, our system splits the task of sensitivity reviewing documents into two stages: (1) **Thread Review**, where the reviewers can select a thread to review multiple passages from different documents that mention a particular event, and (2) **Document Review**, where the reviewers can review the entire document by taking references from the reviewed passages that were presented during the Thread Review stage.

Figure 1 shows the Thread Review interface, which presents a hierarchical information thread and allows the reviewers to provide sensitivity judgements (i.e., sensitive or non-sensitive) for each of the document passages in the thread. The reviewers can also highlight (i.e., annotate) the specific portion of text that contains any sensitive information in a passage. After submitting the review of all of the passages in the thread, the reviewers are presented with the documents that comprise any passages that were associated with the reviewed thread. Figure 2 shows the Document Review interface, which illustrates the sensitivity judgements for the passages that were deemed as sensitive (or non-sensitive) during the Thread Review stage (c.f. Fig. 1). In particular, the system enables the reviewer to refer to these passage reviews while reviewing the entire document to make more informed sensitivity judgements.

Architecture: The system is implemented in Python. Our system's review interfaces, shown in Figs. 1 and 2, are implemented using Django [3]. Our system deploys a layered architecture that comprises: (1) Data Layer, which stores the document collection, the identified threads and the reviewers' judgements, (2) Service Layer, which integrates the HINT [9] method to identify hierarchical information threads in the collection, and (3) Application Layer, which comprises the thread and document review interfaces to present the threads and documents to the reviewers, and captures their sensitivity judgements.

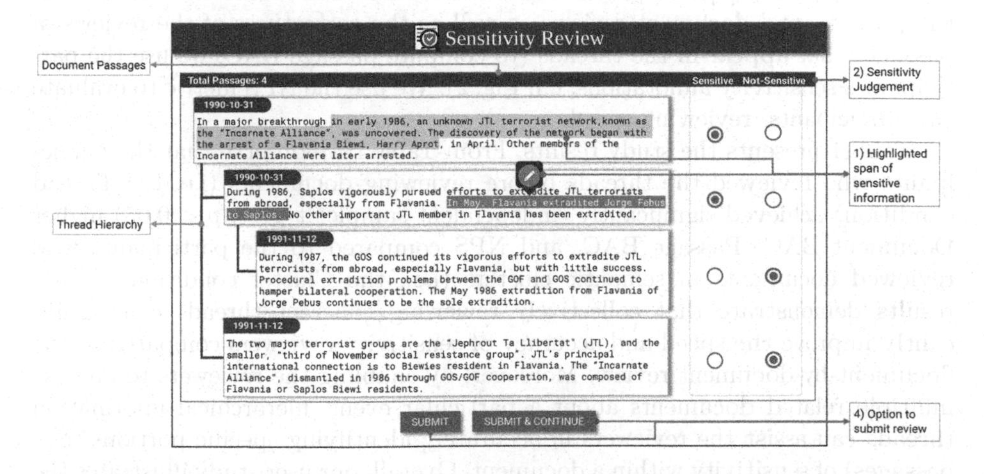

Fig. 1. Interface for collectively reviewing hierarchical information threads.

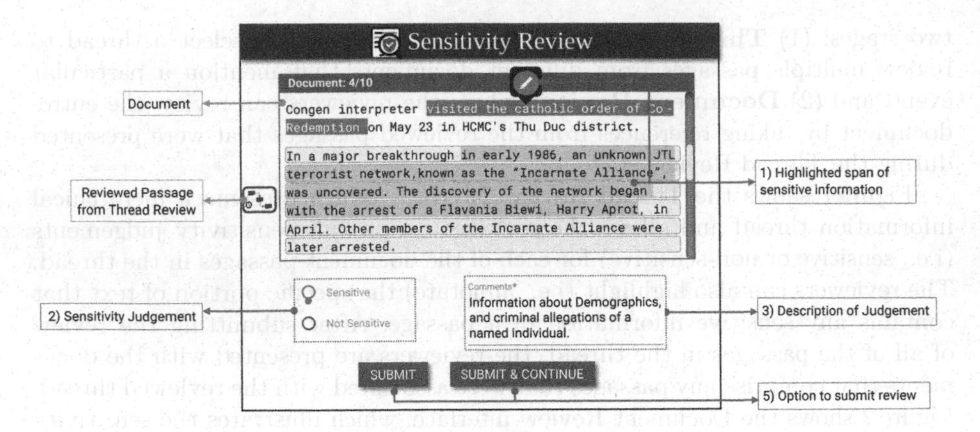

Fig. 2. Interface for reviewing a document after reviewing its corresponding threads.

3 Evaluation

We now present a user study, which evaluates the effectiveness of collectively reviewing related documents in automatically identified hierarchical information threads compared to the sequential (i.e., document-by-document) review in semantic category clusters [7]. We use the GovSensitivity collection [6] with real sensitivities to sample 25 documents (20% sensitive) to present to our study participants. We recruited 36 participants using the Prolific [11] crowdsourcing platform, and allocated the participants to review the documents using either threads or semantic clusters (i.e., 18 participants each). We measure the performance of participants using the BAC and NPS [2] metrics (same as used by Narvala et al. [7]). In particular, we use BAC to evaluate the accuracy of the participants' document reviews as well as the correctness of the reviews of passages that appear in the threads (to compute passage BAC, we use the participant's sensitivity annotations; c.f. Fig. 2). We use the NPS metric to evaluate the participants' reviewing speed.

Table 1 presents the study results. From Table 1 we observe that the participants who reviewed the threads before reviewing documents (i.e. the Thread condition) achieved significantly (independent samples t-Test; $p < 0.05$) higher Document BAC, Passage BAC, and NPS compared to the participants who reviewed documents in semantic clusters (i.e., the Cluster condition). These results demonstrate that collectively reviewing coherent threads can significantly improve the speed and accuracy of sensitivity reviewers compared to the document-by-document review. Moreover, by enabling the reviewers to analyse multiple related documents about a particular event, hierarchical information threads can assist the reviewers in accurately identifying specific portions (e.g. passages) of sensitivity within a document. Overall, our user study illustrates the promising benefits of our proposed system, where the incorporation of information threads into the sensitivity review process can improve both the effectiveness and efficiency of human sensitivity reviews.

Table 1. User study results. "⋆" denotes statistical significance (t-Test; $p < 0.05$).

Condition	Document BAC	Passage BAC	NPS (wpm)
Cluster (baseline)	0.653	0.625	209.42
Thread (our system)	**0.757**⋆	**0.709**⋆	**516.03**⋆

4 Conclusions

In this paper, we have presented a novel system to assist human sensitivity reviewers by enabling the collective review of coherent information about an event's evolution across multiple documents. Our system identifies hierarchical information threads to help the sensitivity reviewers in quickly and accurately identifying specific portions of sensitive information. Using a user study, we showed that presenting hierarchical information threads can significantly ($p < 0.05$) improve the reviewers' reviewing speed and accuracy (+146.41% NPS and +15.93% BAC) of the provided sensitivity judgements for the documents compared to traditional document-by-document review. Therefore, our system has the potential to aid government departments in complying with FOI laws by expediting the release of documents to the public while effectively protecting sensitive information.

References

1. Abualsaud, M., Ghelani, N., Zhang, H., Smucker, M.D., Cormack, G.V., Grossman, M.R.: A system for efficient high-recall retrieval. In: Proceedings of the 41st International ACM SIGIR Conference on Research and Development in Information Retrieval, pp. 1317–1320 (2018). https://doi.org/10.1145/3209978.3210176
2. Damessie, T.T., Scholer, F., Culpepper, J.S.: The influence of topic difficulty, relevance level, and document ordering on relevance judging. In: Proceedings of the 21st Australasian Document Computing Symposium, pp. 41–48 (2016). https://doi.org/10.1145/3015022.3015033
3. Django Software Foundation (2021). https://djangoproject.com. Accessed 21 Oct 2023
4. McDonald, G., Macdonald, C., Ounis, I.: How the accuracy and confidence of sensitivity classification affects digital sensitivity review. ACM Trans. Inf. Syst. **39**(1), 1–34 (2020). https://doi.org/10.1145/3417334
5. Narvala, H., McDonald, G., Ounis, I.: Receptor: a platform for exploring latent relations in sensitive documents. In: Proceedings of the 43rd International ACM SIGIR Conference on Research and Development in Information Retrieval, pp. 2161–2164 (2020). https://doi.org/10.1145/3397271.3401407
6. Narvala, H., McDonald, G., Ounis, I.: RelDiff: enriching knowledge graph relation representations for sensitivity classification. In: Findings of the 2021 Conference on Empirical Methods in Natural Language Processing, pp. 3671–3681 (2021). https://doi.org/10.18653/v1/2021.findings-emnlp.311
7. Narvala, H., McDonald, G., Ounis, I.: The role of latent semantic categories and clustering in enhancing the efficiency of human sensitivity review. In: Proceedings

of the 2022 Conference on Human Information Interaction and Retrieval, pp. 56–66 (2022). https://doi.org/10.1145/3498366.3505824
8. Narvala, H., McDonald, G., Ounis, I.: Sensitivity review of large collections by identifying and prioritising coherent documents groups. In: Proceedings of the 31st ACM International Conference on Information and Knowledge Management, pp. 4931–4935 (2022). https://doi.org/10.1145/3511808.3557182
9. Narvala, H., McDonald, G., Ounis, I.: Effective hierarchical information threading using network community detection. In: Proceedings of 45th European Conference on Information Retrieval, pp. 701–716 (2023). https://doi.org/10.1007/978-3-031-28244-7_44
10. Narvala, H., McDonald, G., Ounis, I.: Identifying chronological and coherent information threads using 5W1H questions and temporal relationships. Inf. Process. Manag. **60**(3), 103274 (2023). https://doi.org/10.1016/j.ipm.2023.103274
11. Prolific (2014). https://www.prolific.com. Accessed 21 Dec 2023
12. UNESCO: Global report on the implementation of access to information laws (2022). https://unesdoc.unesco.org/ark:/48223/pf0000383160
13. Vo, N.P.A., Guillot, F., Privault, C.: DISCO: a system leveraging semantic search in document review. In: Proceedings of the 26th International Conference on Computational Linguistics, pp. 64–68 (2016). https://aclanthology.org/C16-2014

FAR-AI: A Modular Platform for Investment Recommendation in the Financial Domain

Javier Sanz-Cruzado(✉), Edward Richards, and Richard McCreadie

University of Glasgow, Scotland, UK
javier.sanz-cruzadopuig@glasgow.ac.uk

Abstract. Financial asset recommendation (FAR) is an emerging sub-domain of the wider recommendation field that is concerned with recommending suitable financial assets to customers, with the expectation that those customers will invest capital into a subset of those assets. FAR is a particularly interesting sub-domain to explore, as unlike traditional movie or product recommendation, FAR solutions need to analyse and learn from a combination of time-series pricing data, company fundamentals, social signals and world events, relating the patterns observed to multi-faceted customer representations comprising profiling information, expectations and past investments. In this demo we will present a modular FAR platform; referred to as FAR-AI, with the goal of raising awareness and building a community around this emerging domain, as well as illustrate the challenges, design considerations and new research directions that FAR offers. The demo will comprise two components: 1) we will present the architecture of FAR-AI to attendees, to enable them to understand the how's and the why's of developing a FAR system; and 2) a live demonstration of FAR-AI as a customer-facing product, highlighting the differences in functionality between FAR solutions and traditional recommendation scenarios. The demo is supplemented by online-tutorial materials, to enable attendees new to this space to get practical experience with training FAR models. VIDEO URL.

Keywords: Recommendation · Finance · Information Retrieval · Machine Learning

1 Introducing Financial Asset Recommendation (FAR)

Financial organisations such as Banks, Fund Operators, and Fintech companies are undergoing a digital transformation, driven by the need for 24-7 online services, as well as demand for more effective automated analytic and artificial intelligence tooling to remain competitive in the market and serve the growing number of citizens requiring financial products [6]. The investment market in particular is experiencing significant disruption, as organisations transition from a model where only certified financial advisors serve customers, to mixed models

N. Goharian et al. (Eds.): ECIR 2024, LNCS 14612, pp. 267–271, 2024.
https://doi.org/10.1007/978-3-031-56069-9_30

where decision making by advisors are supported by automated services, or in some cases such services are used to directly serve customers with investment advice[1].

At its core, providing investment options to a customer is a ranking task, where the goal is to rank a set of assets (from one or more markets) based on their suitability for investment, conditioned on the current customer [2]. We refer to this ranking task as Financial Asset Recommendation (FAR). However, what makes FAR challenging is that the suitability of an investment is primarily derived from external factors, such as the short or long term market returns, the value of the currency used in the trading process, and the impact of governmental regulations or global events [7]. In addition to these external factors, FAR systems need to consider the customer circumstances, like the alignment of the recommendations with their financial risk tolerance and preferences [4]. As such, the types of solution built for FAR differ markedly from those in more traditional recommendation domains (such as movie [1] or product recommendation [5]).

More specifically, there is a stronger focus on item (asset) modelling using a diverse set of information sources. Effective solutions need to have the ability to model financial pricing data to distinguish high performing or under-valued assets. Numerical and factual data should be extracted from (textual) financial reports. Meanwhile fundamental information about asset categories is needed to enable appropriate filtering of assets based on the needs of the user [3]. Indeed, on the customer side, solutions need a means to dynamically adapt to the risk appetite of each individual customer, either by categorizing assets by risk, or developing risk-aware recommendation models. Furthermore, investments are not for forever, at some point the customer needs to realize the profits from their investment. Solutions need to consider how long the customer is expecting to invest for (the investment horizon), and rank assets according to their expected profitability for those different horizons, where the outcome of a recommendation may not be known for months or years [4].

FAR is a recommendation domain that is still its infancy, but is of growing interest both to researchers and industry. Indeed, we have been seeing a growing number of shared tasks/data challenges (e.g. FinNLP 2023 and FinArg 2024) and workshops (e.g. FinRec and FinIR) in this space over the last couple of years. The primary goal of this demo is to raise awareness of FAR as an interesting and fertile area for future researchers to work on, support community building around this emerging domain, as well as illustrate the challenges requiring further research.

2 Demonstration

In this demonstration we will present a modular platform for financial asset recommendation, referred to as FAR-AI, which illustrates how to tackle these challenges. FAR-AI is a suite of modular micro-services with an associated front-end that has been developed over the prior two years. It started research and

[1] Recommendation services are often referred to as robo-advisors in this context.

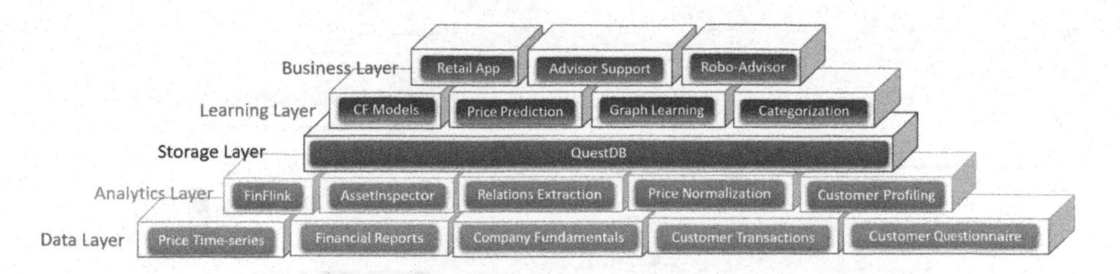

Fig. 1. FAR-AI Architecture

development within the European Commission-funded Flagship Horizon 2020 Infinitech project, and was later extended to additional markets and with new functionality through Impact Accelerator funding via UKRI. The overall architecture of FAR-AI is divided into five architectural layers, each comprised of different microservices, as shown in Fig. 1. We summarize each layer below:

- **Data Layer**: Provides a series of plug-in data ingestion services to collect information about assets, the market as a whole, customers and companies.
- **Analytics Layer**: Contains services that convert the raw input data into a form that is suitable for learning. Notable micro-services within this layer include: FinFlink, which generates customisable technical indicators from pricing time-series in real-time; AssetInspector, which collects asset descriptions and metadata for financial assets from various web sources; and relation extraction, which analyses textual financial documents and extracts factoids to be stored within a knowledge graph. This layer also is responsible for cleaning and normalising both pricing and customer data.
- **Storage Layer**: Provides persistent storage of both the raw data and outputs from the analytics layer. All data here is stored as a time-series, such that system state for past time-points can be simulated on-demand, to better facilitate back-testing.
- **Learning Layer**: This layer is responsible for batch-training the various supervised models used by the upper business layer. This includes the training of traditional collaborative filtering models (based on transaction data), as well as price-change predictors for various investment horizons and encoders for relational knowledge graphs.
- **Business Layer**: The final business layer hosts the user-facing services for different use-cases.

The demo will comprise two components. First, via an associated poster we will present the architecture of FAR-AI to attendees, to inform them about the underlying ingestion, analytics and learning layers that are needed to make effective FAR solutions. Second, we will present a live demonstration of FAR-AI as a customer-facing product, illustrating a real-world use-case for the technology, namely automatic personalised portfolio generation for financial advisors.

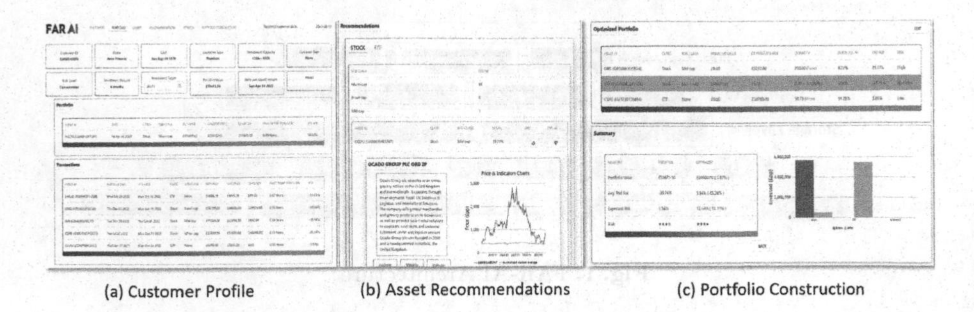

Fig. 2. FAR-AI Financial Advisor Support UI

Demo Use-Case - Financial Advisor Support: Within Banks and other financial institutions, financial advisors meet with clients to help them develop investment portfolios. The goal here is to produce a set of assets weighted by investment value for a client, where the assets should be both profitable (given a target investment horizon), and reflect the client's risk appetite while minimising risk overall. To enable this use-case, FAR-AI provides the following functionality: 1) customer profile and portfolio visualisation; 2) supervised learning of both transaction based and price change prediction-based asset recommenders; 3) financial asset search; 4) automatic portfolio construction and optimisation using the asset recommender models. The demo interface is illustrated in Fig. 2.

3 Target Audience

The primary audience for this demo is PhD students and researchers working in the areas of information retrieval and recommendation, who might be interested in the finance space as a new domain to explore. The finance domain is a growing research area and we believe that not only may there be existing ECIR attendees who might wish to adapt their research and technologies to this domain, but that there are exciting opportunities for new foundational research in this space for the next generation of students to explore. Indeed, recent relevant workshops include: FinRec at RecSys 2022, as well as FinIR and KDF at SIGIR 2020 and 2023.

4 Open Source Materials

While the full FAR-AI platform is not currently open-source we provide supplemental tutorial materials to enable attendees new to this space to get practical experience with training FAR models:

- [Tutorial iPython Notebook] Price Change Prediction
- [Tutorial iPython Notebook] Learning from Financial Transactions
- [Tutorial iPython Notebook] Financial Knowledge Graph Embedding

Acknowledgements. This work was in part carried out within the Infinitech project which is supported by the European Union's Horizon 2020 Research and Innovation programme under grant agreement no. 856632. Subsequent platform development was also financially supported via Engineering and Physical Sciences Research Council (EPSRC) Impact Accelerator, part of UK Research and Innovation (UKRI) with grant ref. number EP/X525716/1.

References

1. Bennett, J., Lanning, S.: The netflix prize. In: Proceedings of KDD Cup and Workshop 2007, San Jose, California, USA, pp. 3–6 (2007)
2. Jung, D., Dorner, V., Glaser, F., Morana, S.: Robo-advisory. Bus. Inf. Syst. Eng. **60**, 81–86 (2018)
3. McCreadie, R., et al. Next-generation personalized investment recommendations. In: Soldatos, J., Kyriazis, D. (eds.) Big Data and Artificial Intelligence in Digital Finance: Increasing Personalization and Trust in Digital Finance using Big Data and AI, pp. 171–198. Springer, Heidelberg (2022). https://doi.org/10.1007/978-3-030-94590-9_10
4. Sanz-Cruzado, J., Mccreadie, R., Droukas, N., Macdonald, C., Ounis, I.: On transaction-based metrics as a proxy for profitability of financial asset recommendations. In: Proceedings of the 3rd International Workshop on Personalization & Recommender Systems in Financial Services (FinRec 2022), Seattle, Washington, USA (2022)
5. Smith, B., Linden, G.: Two decades of recommender systems at Amazon.com. IEEE Internet Comput. **21**(3), 12–18 (2017)
6. Soldatos, J., Kyriazis, D. (eds.): Big Data and Artificial Intelligence in Digital Finance. Springer, Heidelberg (2022). https://doi.org/10.1007/978-3-030-94590-9
7. Zibriczky, D.: Recommender systems meet finance: a literature review. In: Proceedings of the 2nd International Workshop on Personalization & Recommender Systems in Financial Services (FinRec 2016), Bari, Italy, pp. 3–10 (2016)

Industry Papers

Lottery4CVR: Neuron-Connection Level Sharing for Multi-task Learning in Video Conversion Rate Prediction

Xuanji Xiao(✉), Jimmy Chen, Yuzhen Liu, Xing Yao, Pei Liu, and Chaosheng Fan

Tencent Inc., Beijing, China
growj@126.com,
{huabin.chen,yuzhen.liu,jessieyao,alexpliu,atlasfan}@tencent.com
https://scholar.google.com/citations?user=DFsMY2IAAAAJ

Abstract. As a fundamental task of industrial ranking systems, conversion rate (CVR) prediction is suffering from data sparsity problems. Most conventional CVR modeling leverages Click-through rate (CTR)&CVR multitask learning because CTR involves far more samples than CVR. However, typical coarse-grained layer-level sharing methods may introduce conflicts and lead to performance degradation, since not every neuron or neuron connection in one layer should be shared between CVR and CTR tasks. This is because users may have different fine-grained content feature preferences between deep consumption and click behaviors, represented by CVR and CTR, respectively. To address this sharing&conflict problem, we propose a neuron-connection level knowledge sharing. We start with an over-parameterized base network from which CVR and CTR extract their own subnetworks. The subnetworks have partially overlapped neuron connections which correspond to the sharing knowledge, and the task-specific neuron connections are utilized to alleviate the conflict problem. As far as we know, this is the first time that a neuron-connection level sharing is proposed in CVR modeling. Experiments on the Tencent video platform demonstrate the superiority of the method, which has been deployed serving major traffic. (The source code is available at https://github.com/xuanjixiao/onerec/tree/main/lt4rec).

Keywords: post-click conversion rate · multi-task learning · recommender system · neuron-connection level sharing · lottery ticket theory

1 Introduction

Post-click conversion rate (CVR) prediction plays a key role across industrial ranking systems, such as recommender systems (RS) [1–5], online advertising [6]. Figure 1 shows a typical scenario in Tencent Video, where the RS displays a list of similar videos when a user watches the current video. In video recommendation an RS first calls up a large number of user-related videos, then ranks and presents

N. Goharian et al. (Eds.): ECIR 2024, LNCS 14612, pp. 275–280, 2024.
https://doi.org/10.1007/978-3-031-56069-9_31

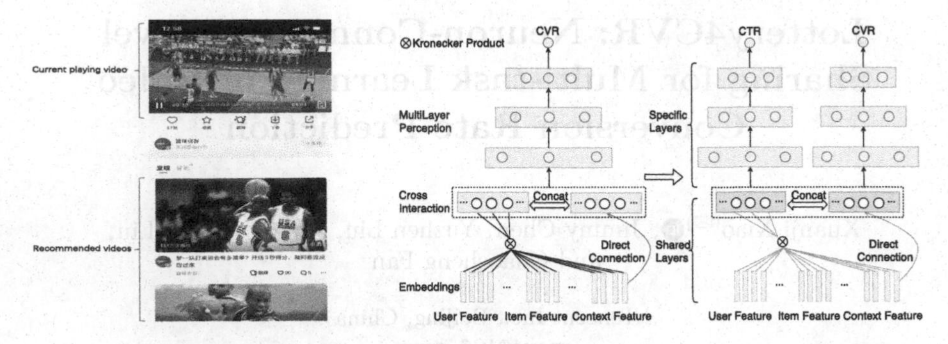

Fig. 1. A typical scenario of Tencent Video.

Fig. 2. Architectures of typical single-task CVR model, and classic layer sharing CVR model

them to users based on several metrics, such as click-through rate (CTR) or CVR. For video contents, the CVR indicator is continuous, $label \in [0, 1]$, e.g., if someone watched 2 min of a video with a length of 5 min then $CVR = 0.4$. The number of CVR samples equals the number of content clicks, i.e., the positive samples of the CTR task. CVR's data sparsity imposes a challenging model fiting process resulting in a poor generalization ability.

A feasible solution is to leverage the CTR task to share knowledge with the CVR task via multi-task learning (MTL) [1–3,7–13]. ESMM series [1,2] share the feature-embedding layer and output labels. MMOE series (MMOE, SNR, PLE [3,10,11,13]) utilize a multi-gate mixture of expert subnetworks to provide knowledge sharing at the granularity of subnetworks. As existing literature provides sharing mechanisms based on neural layers, we collectively refer to these models as layer-level CVR sharing (Fig. 2). Despite their success, these methods often introduce representation conflict [3,8]. For example, users often have random clicks or clicks for exaggerated titles or covers, which will be learned by the CTR task through specific neuron-connection weights in the CVR&CTR multi-task model. However, this learned representation shouldn't be shared with CVR because the CVR score won't be high in this situation, and vice versa.

To address the above **sharing&conflict** issue, we propose a novel neuron-connection level sharing mechanism. We start with an over-parameterized deep learning model, called the base network. CVR and CTR extract their subnetworks containing partial neuron connections from the base network. In particular, we use a network iterative magnitude pruning [14–17] to search for CVR&CTR's suitable sharing network structure. In this way, CVR and CTR can have their specific networks while sharing certain node connections, in which the overlapping part and the specific part correspond to sharing representation and conflict reduction, respectively. Our contributions can be summarized as:

- Compared with existing layer-level sharing methods, the proposed approach dramatically expands the capacity to alleviate the sharing&conflict problem through fine-grained connection sharing in CVR multi-task modeling. Besides

it being a more generalized approach to layer sharing, it also provides an automatic design of sharing parts instead of manual designs.
- To the best of our knowledge, this is the first work applying the neuron-connection level sharing to solve the CVR multi-task learning.
- We highlight the effectiveness of the approach against competitors and deploy it in Tencent video, confirming its value in real-world applications.

2 Neuron-Connection Sharing CVR

Representation Sharing by Generating Sub-networks. In MTL, Closely related tasks tend to extract similar subnets so they can use similar parts of weights, while loosely related or unrelated tasks tend to extract subnets that are different in a wide range [16,17]. Particularly, this inductive bias [18] customized to the task is embedded into the subnet structure. Ideally, tasks with similar inductive bias should be assigned similar parts of parameters. In Algorithm 1, we first use mixed samples of CVR and CTR to train a big over-parameterized mixed network as the base shared network named $sNET$. Then, each task uses its own sample executing the "mask generation" to mask the neuron connections(weights) unimportant for the task itself to construct a mask. This process aims to identify the overlapping parts of these two masks, which represent the same inductive bias held by the two tasks, i.e., knowledge that can be shared.

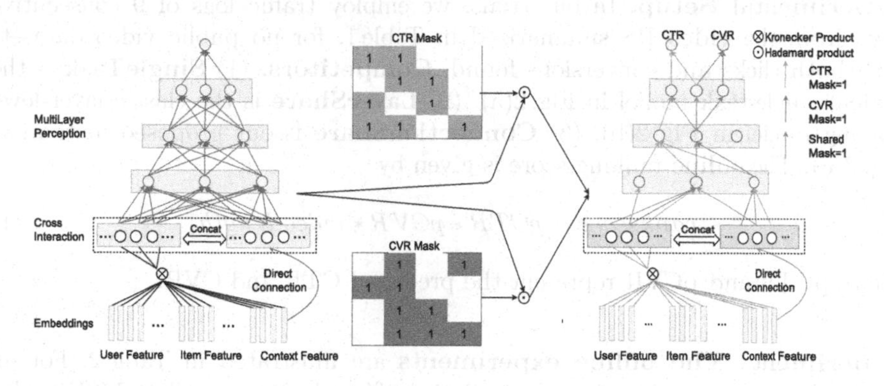

Fig. 3. Neuron-connection sharing CVR modeling.

Conflict Reduction by Automation. After we get the fully shared base network $sNET$ and each task's own matrix $mask \in \{0, 1\}$, the sub-network of the current task is obtained through $sNET \odot mask$. The exclusive part of a subnetwork represents each task's own specific inductive bias which can reduce the conflict. As shown in Fig. 3, CTR and CVR occupy the yellow part and the blue part, respectively, and share the green part in the final model. While co-occupied green part indicates the sharing representation, the yellow part and the blue part can naturally keep the unique characterization of each task, which represent their unique knowledge space.

Algorithm 1. CVR&CTR neuron-connection sharing

Input: Shared network $sNET$; masks $masks[n\_tasks]$; pruning number and rate: $n\_pruning$ and q, batch number in mixed samples of CVR and CTR $n\_batches$.
▷ **Step 1: Generate CVR/CTR subnetworks**
Train $sNET$ with mixed CVR and CTR samples to get weights θ
for tid in [CVR,CTR] **do**
  Initialize $masks[tid][0] = 1^{|\theta|}$.
  for $i = 1$ to $n\_pruning$ **do**
    Train $sNET$ for an epoch of the tid's samples after $sNET = sNET \odot masks[tid][i-1]$;
    Let x equal to the qth quantile of the absolute value of $sNET$'s weights, then update $masks[tid][i-1]$ by setting values less than x to 0 to obtain $mask[tid][i]$;
    Reset $sNET$'s weights to θ.
  end for
  Select best i of $masks[tid][i]$ based on validation set's performance as $masks[tid][best\_i]$.
end for
▷ **Step 2: Training the final sharing model**
Reinitialize $sNET$ parameters as θ.
for $batch = 1$ to $n\_batches$ **do**
  Obtain the tid of current sample batch,e.g., CTR;
  Update $sNET = sNET \odot masks[tid][best\_i]$;
  Train $sNET$ to calculate loss and do gradient updates of weights.
end for
$CVR\_model = sNET \odot masks[CVR][best\_i]$.
$CTR\_model = sNET \odot masks[CTR][best\_i]$.

3 Experiments

Experimental Setup. In our trials, we employ traffic logs of 9 consecutive days from the video RS summarized in Table 1, for no public video datasets with both clicks and conversions found. **Competitors.** (1) **SingleTask** is the typical single-task model in Fig. 2(a). (2) **LayerShare** is the classic layer-level sharing model in Fig. 2(b). (3) **ConnectionShare** is our proposed method in Fig. 3(c). The online ranking score is given by

$$rankscore = pCTR * pCVR * video\_length \tag{1}$$

where pCTR and pCVR represent the predicted CTR and CVR.

Experiment. The offline experiments are illustrated in Table 2. For an industrial recommender system, a relative 1% reduction in CVR MSE and a 0.001 absolute gain of CTR AUC are both remarkable and can acquire significant online performance improvement [1]. Specifically, The remarkable MSE reduction of 3.38% validates the effectiveness of our approach for CVR modeling, which is achieved by expanding the capacity of knowledge transfer by fine-grained sharing. **The online A/B tests** on the real-world video platform are conducted for 7 consecutive days, which is immediately after the date of the previous offline training samples. For fairness, the online ranking score (Eq.(1)) adopts the same single-task CTR model for each CVR competitor. In Table 3, ConnectionShare's online viewing time is improved by **1.5%** and **3.5%** compared to the single-task model and the layer-sharing model, respectively, which is remarkable in industrial video platforms.

Table 1. Statistics of the Experimental Dataset

Dataset	#User	#Video	#Impression	#Click	#Conversion
Tencent Video	10M	11M	297M	121M	49M

Table 2. Offline Comparison of Different Models

Model	MSE	AUC	MTL Gain	
	CVR	CTR	MSE CVR	AUC CTR
1) SingleTask	0.13688	0.78572	–	–
2) LayerShare	0.13563	0.78808	+0.00125 (–0.91%)	+0.00236 (+0.30%)
3) ConnectionShare	**0.13226**	**0.78874**	**+0.00462 (–3.38%)**	**+0.00302 (+0.38%)**

Table 3. Online A/B Test Comparison of Different Models

Rankscore		View		
pCTR_Model	pCVR_Model	Seconds	CVR	CTR
1) SingleTask	SingleTask	491.3	54.9%	24.7%
2) SingleTask	LayerShare	481.6	55.0%	**24.8%**
3) SingleTask	ConnectionShare	**498.6**	55.1%	24.7%

4 Conclusions

In this paper, we propose a neuron connection-level sharing scheme in CVR modeling. Compared to the current literature on subnetwork/layer level sharing, we achieve a finer-grained sharing to transfer knowledge discreetly and meticulously and thus further extend the capacity to address the sharing&conflict issues in MTL CVR modeling. Experiments on offline and online video recommender systems demonstrate the effectiveness of the method.

References

1. Ma, X., et al.: Entire space multi-task model: an effective approach for estimating post-click conversion rate. In: The 41st International ACM SIGIR Conference on Research & Development in Information Retrieval, pp. 1137–1140 (2018)
2. Wen, H., et al.: Entire space multi-task modeling via post-click behavior decomposition for conversion rate prediction. In: Proceedings of the 43rd International ACM SIGIR Conference on Research and Development in Information Retrieval, pp. 2377–2386 (2020)
3. Tang, H., Liu, J., Zhao, M., Gong, X.: Progressive layered extraction (ple): a novel multi-task learning (mtl) model for personalized recommendations. In: Fourteenth ACM Conference on Recommender Systems, pp. 269–278 (2020)
4. Xiao, X., He, Z.: Neighbor based enhancement for the long-tail ranking problem in video rank models. arXiv preprint arXiv:2302.08128 (2023)

5. Xiao, X., Dai, H., Dong, Q., Niu, S., Liu, Y., Liu, P.: Incorporating social-aware user preference for video recommendation. In: International Conference on Web Information Systems Engineering, pp. 544–558. Springer, Heidelberg (2023). https://doi.org/10.1007/978-981-99-7254-8_42
6. Zhu, H., et al.: Optimized cost per click in taobao display advertising. In: Proceedings of the 23rd ACM SIGKDD International Conference on Knowledge Discovery and Data Mining, pp. 2191–2200 (2017)
7. Lin, X., et al.: A pareto-efficient algorithm for multiple objective optimization in e-commerce recommendation. In: Proceedings of the 13th ACM Conference on Recommender Systems, pp. 20–28 (2019)
8. Zhang, Yu., Yang, Q.: An overview of multi-task learning. Natl. Sci. Rev. **5**(1), 30–43 (2018)
9. Zhao, Z., et al.: Recommending what video to watch next: a multitask ranking system. In: Proceedings of the 13th ACM Conference on Recommender Systems, pp. 43–51 (2019)
10. Ma, J., Zhao, Z., Yi, X., Chen, J., Hong, L., Chi, E.H.: Modeling task relationships in multi-task learning with multi-gate mixture-of-experts. In: Proceedings of the 24th ACM SIGKDD International Conference on Knowledge Discovery & Data Mining, pp. 1930–1939 (2018)
11. Ma, J., Zhao, Z., Chen, J., Li, A., Hong, L., Chi, H.: Snr: sub-network routing for flexible parameter sharing in multi-task learning. In: AAAI (2019)
12. Ouyang, K., Zheng, W., Tang, C., Xiao, X., Zheng, H.T.: Click-aware structure transfer with sample weight assignment for post-click conversion rate estimation. In: Machine Learning and Knowledge Discovery in Databases: Research Track: European Conference, ECML PKDD 2023, Turin, Italy, 18–22 September 2023, Proceedings, Part V, pp. 426–442. Springer, Heidelberg (2023). https://doi.org/10.1007/978-3-031-43424-2_26
13. Li, W., Zheng, W., Xiao, X., Wang, S.: Stan: stage-adaptive network for multi-task recommendation by learning user lifecycle-based representation. In: Proceedings of the 17th ACM Conference on Recommender Systems, RecSys 2023, pp. 602–612. Association for Computing Machinery, New York (2023)
14. Frankle, J., Dziugaite, G.K., Roy, D.M., Carbin, M.: Stabilizing the lottery ticket hypothesis. arXiv preprint arXiv:1903.01611 (2019)
15. Malach, E., Yehudai, G., Shalev-Schwartz, S., Shamir, O.: Proving the lottery ticket hypothesis: pruning is all you need. In: International Conference on Machine Learning, pp. 6682–6691. PMLR (2020)
16. Frankle, J., Carbin, M.: The lottery ticket hypothesis: finding sparse, trainable neural networks. In: International Conference on Learning Representations (2019)
17. Sun, T., Shao, Y., Li, X., Liu, P., Huang, X.: Learning sparse sharing architectures for multiple tasks. In: AAAI, pp. 1930–1939 (2020)
18. Baxter, J.: A model of inductive bias learning. J. Artif. Intell. Res. **12**, 149–198 (2000)

Semantic Content Search on IKEA.com

Mateusz Slominski, Ezgi Yıldırım(✉), and Martin Tegner

IKEA Retail (Ingka Group), Leiden, Netherlands
ezgi.yildirim@ingka.ikea.com

Abstract. In this paper, we present an approach to *content search*. The aim is to increase customer engagement with content recommendations on IKEA.com. As an alternative to Boolean search, we introduce a method based on semantic textual similarity between content pages and search queries. Our approach improves the relevance of search results by a 2.95% increase in click-through rate in an online A/B test.

Keywords: Sentence Transformers · Content Recommendations · Semantic Search

1 Introduction

Search engines typically rely on keyword search algorithms referred to as *Boolean search* [2]. In the e-commerce context, all product information, e.g. name, attributes and description, is treated as a text document. When a customer types a search query, it is divided into individual keywords. Each keyword is then searched within the document using Boolean operators and logic.

For general questions, such as *'please help me decorate my living room'*, this approach may not work well. Therefore, Boolean search is often used in conjunction with language processing, such as stemming and lemmatization, to improve search results. Boolean search is also not well-suited for handling large volumes of text data.

While product search is the most common task for many e-retailers, *content recommendations* are important for IKEA's home-furnishing products to inspire and guide customers through their home-furnishing projects. On the search page of IKEA.com, users can input free-text queries to search for both products (e.g. *sofa*) and more general content (e.g. *category page about living rooms*) as shown in Fig. 1. Due to the prior issues, the result from a Boolean search for content is not optimal. In this paper, we introduce an algorithm that searches for relevant content based on the semantic similarity between queries and content pages.

2 Content Search

Our approach has two steps, as illustrated in Fig. 2. In the first step, we map content pages to a continuous space such that each web page is summarised and represented by a real-valued vector. Queries are mapped to the same vector space. In the second step, we use a semantic similarity measure based on cosine distance to select the document(s) that are most relevant to the query.

N. Goharian et al. (Eds.): ECIR 2024, LNCS 14612, pp. 281–285, 2024.
https://doi.org/10.1007/978-3-031-56069-9_32

(a) Search page

(b) Content recommendation

Fig. 1. An example search with a list of products (Fig. 1a, top) and related content (Fig. 1a, bottom). Figure 1b shows a content page. (The search page and content can be reproduced by the search query *'stimulating baby toys'* on ikea.com/us/en.)

2.1 Data

The search space for content consists of web pages on IKEA.com with general content, that is, pages not restricted to individual products. For example, content search from a query *'please help me decorate my living room'* is likely to be directed to the product listing page with inspiration for living rooms,[1] or to a planner where users can decorate their living room in a virtual studio[2].

Each content page has a rich resource of text data, such as metadata, plain text, and image descriptions. We collect a page's text data into a document $d \in \mathcal{D}$ where $\mathcal{D}$ denotes the set of all searchable content pages/documents. Except for the selection of which text resource to include in the documents, there is no further preprocessing of the data. The number of content pages varies between market-language combinations. The UK–British English combination that we use in this paper consists of 1600 documents with an average of 452 words per document.

2.2 Semantic Search

We use pre-trained sentence-BERT models [7] to encode documents and queries into continuous representations. These are pre-trained large language models

[1] https://www.ikea.com/us/en/rooms/living-room/.

[2] https://www.ikea.com/us/en/planners/harlanda-planner/.

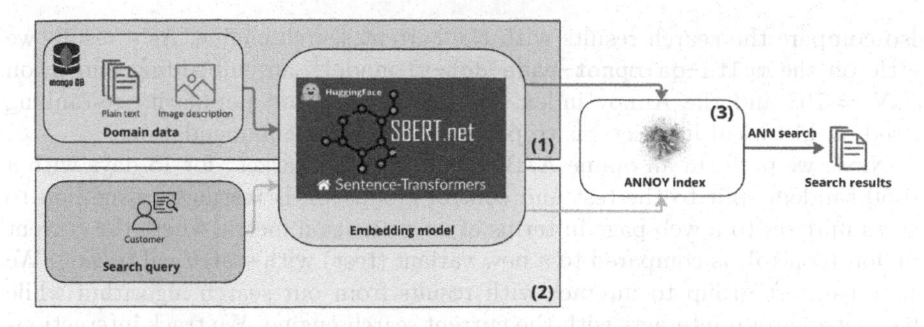

Fig. 2. Content search based on semantic similarity of search queries and content pages. (1) shows the mapping of content pages to a vector space, and (2) the mapping of search queries to the same vector space. In the retrieval step (3), we use semantic similarity measures to select the k most relevant document(s).

that capture semantic and syntactic information [3]. By transfer learning, we use their capability to represent the language and perform an information retrieval step to find the most relevant content to a query.

Sentence-BERT uses a Siamese structure where two identical networks each take an input sentence, the query q and document d respectively, and map them to N-dimensional vectors

$$q \mapsto f(q) = \mathbf{q} \in \mathbb{R}^N, \quad d \mapsto f(d) = \mathbf{d} \in \mathbb{R}^N.$$

The embeddings f are BERT networks with a pooling operation that summarises several word vectors of a sentence into a single vector. During training, Sentence-BERT networks are fine-tuned to various tasks. For semantic-similarity learning, cosine similarity predicts a score for a sentence pair, compared to the label of training data [7]. This yields (identical) embedding(s) mapping text to a vector space where semantic similarity corresponds to geometrical closeness under the cosine measure.

By using the pre-trained models, we encode all searchable web pages for a search space $D = \{\mathbf{d} = f(d), \quad \forall d \in \mathcal{D}\}$. For a given search query q, we find k documents with the most similar semantic meaning. This corresponds to finding the k vectors in D that are *closest* to the embedded query $\mathbf{q} = f(q)$ according to cosine similarity. To this end, we use approximate nearest neighbours and retrieve the nearest documents as content recommendations.

2.3 Experimental Setup and Results

We first conduct offline tests where we employ different pre-trained models and indices to retrieve documents from the search space, namely Annoy [1] and FAISS [4]. We also try different sizes of k and re-ranking methods to generate the search results.

We manually evaluate the iterations on a set of historical search queries with a qualitative comparison of the queries and the retrieved content pages. We also compare the search results with the current search engine. As a result, we settle on the `multi-qa-mpnet-base-dot-v1` model[3], an embedding dimension of $N = 768$ and the Annoy index. We use $k = 70$ and popularity re-ranking based on historical interactions to produce the final recommendations.

Next, we perform an online A/B test in the UK market[4] for 15 days with a 50-50 random split to the test and control groups. A/B testing is a method to assess updates to a web page in terms of an evaluation metric where the current version (control) is compared to a new variant (test) with statistical testing. We allow the test group to interact with results from our search algorithm while the control group interacts with the current search engine. We track interactions with content recommendations only, i.e. product clicks are not counted, and measure the total number of clicks per visitor. Table 1 shows that our semantic search (test) outperforms the existing search engine (control) by 2.95% with 99% statistical significance.

Table 1. Experimental Results

	Visitors	Change in conversions	Stat. Significance
Control	550,230	–	–
Test	549,696	+2.95%	>99%

3 Summary and Future Work

In this work, we have proposed a method that enables web visitors to search for general content on IKEA.com. The search is based on textual similarity and leverages pre-trained sentence-BERT models. The online A/B test shows promising results with a 2.95% increase in click-through rate.

There are several challenges and points for future work. IKEA operates in many markets with different languages while the performance of large language models is known to deteriorate for non-English and smaller languages in particular due to the availability of training data. Another important aspect is domain-specific words, where product names (e.g. the bookshelf *Kallax*) are unknown to the model and therefore not searchable. As for future work, we therefore propose several steps. We will collect and curate a training set with query-content interactions from user responses. With this, we can fine-tune the models to learn embeddings that are adapted to our specific search task and the IKEA context. With a ground truth, we can also make a more systematic comparison of different language models, e.g., ELMo [5] and GPT [6] to achieve better results in terms of accuracy measures. As another step, we plan to develop asymmetric models, since queries are typically much shorter in length than the documents.

Acknowledgements. Thanks to Maria Juhlin, Shahin Shahkarami and Marla Brynell for their helpful comments and suggestions.

[3] https://www.sbert.net.

[4] www.ikea.com/gb/en.

References

1. ANNOY library. https://github.com/spotify/annoy. Accessed 26 Oct 2023
2. Aliyu, M.B.: Efficiency of boolean search strings for information retrieval. Am. J. Eng. Res. **6**(11), 216–222 (2017)
3. Jiang, H., He, P., Chen, W., Liu, X., Gao, J., Zhao, T.: Smart: robust and efficient fine-tuning for pre-trained natural language models through principled regularized optimization. arXiv preprint arXiv:1911.03437 (2019)
4. Johnson, J., Douze, M., Jégou, H.: Billion-scale similarity search with GPUs. IEEE Trans. Big Data **7**(3), 535–547 (2019)
5. Peters, M.E., et al.: Deep contextualized word representations (2018)
6. Radford, A., Wu, J., Child, R., Luan, D., Amodei, D., Sutskever, I.: Language models are unsupervised multitask learners (2019). https://api.semanticscholar.org/CorpusID:160025533
7. Reimers, N., Gurevych, I.: Sentence-BERT: sentence embeddings using Siamese BERT-networks. In: Proceedings of the 2019 Conference on Empirical Methods in Natural Language Processing. Association for Computational Linguistics, November 2019. http://arxiv.org/abs/1908.10084

Let's Get It Started: Fostering the Discoverability of New Releases on Deezer

Léa Briand, Théo Bontempelli, Walid Bendada, Mathieu Morlon, François Rigaud, Benjamin Chapus, Thomas Bouabça, and Guillaume Salha-Galvan(✉)

Deezer Research, Paris, France
research@deezer.com

Abstract. This paper presents our recent initiatives to foster the discoverability of new releases on the music streaming service Deezer. After introducing our search and recommendation features dedicated to new releases, we outline our shift from editorial to personalized release suggestions using cold start embeddings and contextual bandits. Backed by online experiments, we discuss the advantages of this shift in terms of recommendation quality and exposure of new releases on the service.

Keywords: New Releases · Music Recommendation · Music Discovery

1 Introduction

Music artists release hundreds of thousands of new albums every week on the music streaming service Deezer [10]. The prompt integration of this content, along with its swift discoverability through recommender systems and search engines, holds significant importance. For Deezer, it ensures that users have immediate access to the latest music of their favorite artists while also easily coming upon new ones they might like, which is known to improve the user experience [20,25]. The proper exposure of new releases also benefits artists by amplifying their visibility, which can contribute to their success, boost their revenues, and foster the emergence of new talents [1,12]. Nonetheless, displaying the right releases to the right users remains challenging due to the limited prior information on this fresh content, especially for new artists unknown from the service [7,25].

In this paper, we present our recent efforts to better showcase new releases on Deezer, both in terms of recommendation performance and number of new releases exposed to users. Specifically, we first focus on the product context. We describe in Sect. 2 our search and recommendation features dedicated to new releases, along with their objectives and differences. We also dive into our historical semi-personalized solution for new release recommendation, based on editorial pre-selections, and its limitations in terms of catalog coverage and adaptability. Then, in Sect. 3, we detail the fully personalized systems we deployed in 2023

N. Goharian et al. (Eds.): ECIR 2024, LNCS 14612, pp. 286–291, 2024.
https://doi.org/10.1007/978-3-031-56069-9_33

to overcome these limitations, involving a cold start neural embedding model along with contextual bandits [6,8]. We provide an online evaluation of our approach on Deezer's "New releases for you" carousel of recommended new albums. Finally, we conclude in Sect. 4 by discussing our ongoing efforts to improve our models and how we present new releases on the Deezer homepage.

Fig. 1. Interface of the "New releases for you" carousel on the website version of Deezer, recommending a personalized shortlist of recently released albums to each user.

2 Listening to New Releases on Deezer

Hundreds of thousands of albums and music tracks are released on Deezer every week. Users can instantly access this content using our *search engine*, based on an Elasticsearch index plugged into a Kafka topic [11,16] refreshed in real-time. They also receive *notifications* informing them about updates from their favorite artists, which we refer to as *unmissable* releases. In addition, on the Deezer homepage, we propose "*Friday cards*", i.e., weekly playlists mixing unmissable tracks.

All these features help users retrieve releases from artists they already know. However, they hardly permit *discovering* new music from emergent or unfamiliar artists [7]. For this distinct purpose, the Deezer homepage relies on two album carousels[1] [3]. The first one, known as "*Fresh picks of the week*", presents the biggest releases according to professional editors from Deezer. The second one, called "*New releases for you*" and illustrated in Fig. 1, aims to provide personalized recommendations. Up to early 2023, it first showcased unmissable albums for each user, followed by new albums from their favorite music genres. Our data scientists were identifying each user's favorite genres, but the lists of recommended albums for each genre were manually pre-selected by Deezer editors.

This historical method for discovering new releases had limitations that motivated the work presented in this paper. To begin with, "New releases for you" carousels were only partially personalized. All fans from a genre used to receive

[1] Our carousels display swipeable ranked lists of 12 albums, followed by a "View all" button leading to a page displaying up to 100 recommended albums.

the same (non-unmissable) recommendations, regardless of their individual subgenre preferences. Moreover, editorial lists were only updated once a week, on Fridays. Carousels remained static for seven days, making the feature less engaging. New releases added during the week went unrecommended for days, and user interactions were not factored in for improved recommendations. Last but not least, only the small minority of albums handpicked by editors were recommended. Most new releases never got the chance to even be exposed to users.

3 Improving the Discoverability of New Releases

This section outlines our recent initiatives to overcome these limitations. For the sake of clarity, we focus our presentation on the discovery of *new albums*.

3.1 Cold Start Embedding Representations of New Releases

Most of our personalized recommender systems [2,3,5,6,24] leverage latent models for collaborative filtering (CF) [4,17]. By analyzing usage data on Deezer, they learn low-dimensional *embedding* vector representations of users and musical items including albums, in a vector space where proximity reflects preferences. Then, they offer recommendations based on embedding similarity metrics [5,6]. Unfortunately, such CF models often struggle to incorporate new items lacking usage data [22,25,27]. This issue, known as the *cold start* problem, explains why Deezer historically favored an editorial approach for suggesting new releases.

In 2023, we developed CF-Cold-Start. For any existing CF embedding space, CF-Cold-Start predicts future embedding vectors of items such as new albums *as early as their release date*, without requiring extensive usage data. CF-Cold-Start draws inspiration from advances in cold start research [9,18,27], including a Deezer system for *user* cold start [6]. It consists in a 3-layer neural network [13], predicting "ground truth" embedding vectors that would be computed by the CF model using a week of usage data. In the case of a new album, the input layer processes various album metadata, including information on labels and artists, as well as usage data (streams, likes...) at the release date when available. The output layer returns a *predicted embedding vector* for this album. We train CF-Cold-Start in a supervised manner using musical items for which both input data and ground truth embedding vectors are available, by minimizing the mean squared error between predicted and ground truth vectors using gradient descent [13].

We deployed CF-Cold-Start on Deezer in our "New releases for you" feature. While carousels still begin with unmissable new albums, CF-Cold-Start replaces editorial pre-selections with personalized recommendations, consisting in ranked lists of albums whose predicted embedding vectors are the most similar to each user[2] in the CF embedding space. We update embedding vector predictions every four hours through a forward pass in CF-Cold-Start, to refine

[2] Precisely, this feature uses dot product similarities and the SVD embeddings from [2].

recommendations as we collect more usage data on new albums. We use approximate nearest neighbors methods via a Golang application incorporating the Faiss library [15].

3.2 Carousel Personalization with Contextual Bandits

CF-Cold-Start offers fully personalized recommendations from the entire set of new albums, with refreshes every four hours instead of once a week. Nonetheless, it solely *exploits* the most tailored albums based on usage data. In this work, we also wanted to occasionally deviate from them, to *explore* less popular but promising albums which could be overlooked by relying on the existing usage. To this end, we developed TS-CF-Cold-Start, a CF-Cold-Start variant with a *multi-arm bandit* component for adaptive "New releases for you" album ranking. Bandit algorithms effectively handle exploration-exploitation face-offs [8,19,21]. Here, we use the *Thompson Sampling* (TS) [26] extension for contextual bandits [8]. Our *arms* are new albums, recommended by ranked batches at each bandit *round*, i.e., each carousel display on Deezer. The set of arms evolves over time, as new albums enter daily and exit the new release set after seven days.

Album embedding vectors predicted by CF-Cold-Start act as the *prior expected arm representations* [8] used for arm selection by the bandit. They are updated every four hours using TS with Gaussian distributions. Each album's mean and variance are exported in an ONNX model [23] in our service for inference and sampling. Updates are based on user click *rewards* and a *cascade-based* scheme [3] accounting for the album position in the carousel. We note that, while carousels still start with fixed unmissable releases, the stochastic nature of TS-based exploration-exploitation permits recommending different non-unmissable albums with each homepage refresh, even between two arm updates.

3.3 Online Experimental Evaluation on Deezer

We conducted large-scale A/B tests on Deezer between March and May 2023. On average, replacing editorial selections with CF-Cold-Start recommendations in "New releases for you" not only improved display-to-click rates by 6%, but also multiplied by $\times 3$ and $\times 1.5$ the weekly numbers of new albums displayed and clicked in carousels, respectively. Hence, CF-Cold-Start boosted their exposure.

Using TS-CF-Cold-Start added dynamism to carousels by sampling albums with each app refresh. However, its performance was on par with CF-Cold-Start in both clicks and exposure. This mixed result contrasts with a previous study [3] where bandits improved our *editorial playlist* carousels. We attribute our lower performance to two factors. Firstly, unmissable albums restrict bandit actions to less visible carousel slots, unlike in playlist carousels [3]. Secondly, our set of new albums constantly evolves, potentially hindering model convergence.

4 Conclusion and Future Work

In conclusion, both CF-Cold-Start and TS-CF-Cold-Start enhanced our "New releases for you" recommendations and the exposure of new albums, all of this with more dynamic updates. These initiatives foster the overall discoverability of new releases on Deezer. They also open up interesting avenues for future work. We believe our problem is conducive to the study of bandits explicitly accounting for *equity of exposure*, for improved fairness in release suggestions [14]. From a product perspective, our mixed results with TS-CF-Cold-Start also prompt us to separate unmissable albums from discoveries into two carousels, with the latter fully controlled by bandits. Lastly, we will use CS-Cold-Start to better incorporate releases in other Deezer features, including our "Flow" personalized radio [5].

Speaker and Company

Léa Briand is a Senior Data Scientist at **Deezer**, a French music streaming service created in 2007 and with over 16 million active users in 180 countries. In the Recommendation team, she develops large-scale machine learning systems to improve music recommendation on this service. Resources related to her talk will be available on: https://github.com/deezer/new-releases-ecir2024.

References

1. Aguiar, L., Waldfogel, J.: Platforms, promotion, and product discovery: evidence from Spotify playlists. Technical report, National Bureau of Economic Research (2018)
2. Bendada, W., et al.: Track mix generation on music streaming services using transformers. In: Proceedings of the 17th ACM Conference on Recommender Systems, pp. 112–115 (2023)
3. Bendada, W., Salha, G., Bontempelli, T.: Carousel personalization in music streaming apps with contextual bandits. In: Proceedings of the 14th ACM Conference on Recommender Systems, pp. 420–425 (2020)
4. Bokde, D., Girase, S., Mukhopadhyay, D.: Matrix factorization model in collaborative filtering algorithms: a survey. Procedia Comput. Sci. **49**, 136–146 (2015)
5. Bontempelli, T., Chapus, B., Rigaud, F., Morlon, M., Lorant, M., Salha-Galvan, G.: Flow moods: recommending music by moods on Deezer. In: Proceedings of the 16th ACM Conference on Recommender Systems, pp. 452–455 (2022)
6. Briand, L., Salha-Galvan, G., Bendada, W., Morlon, M., Tran, V.A.: A semi-personalized system for user cold start recommendation on music streaming apps. In: Proceedings of the 27th ACM SIGKDD Conference on Knowledge Discovery and Data Mining, pp. 2601–2609 (2021)
7. Celma, Ò., Celma, Ò.: The Long Tail in Recommender Systems. Music Recommendation and Discovery: The Long Tail, Long Fail, and Long Play in the Digital Music Space, pp. 87–107 (2010)
8. Chapelle, O., Li, L.: An empirical evaluation of Thompson sampling. Adv. Neural. Inf. Process. Syst. **24**, 2249–2257 (2011)
9. Covington, P., Adams, J., Sargin, E.: Deep neural networks for Youtube recommendations. In: Proceedings of the 10th ACM Conference on Recommender Systems, pp. 191–198 (2016)
10. Deezer (2023). https://www.deezer.com

11. Elasticsearch (2023). https://www.elastic.co/elasticsearch
12. Ferraro, A., Serra, X., Bauer, C.: What is fair? Exploring the artists' perspective on the fairness of music streaming platforms. In: Ardito, C., et al. (eds.) INTERACT 2021. LNCS, vol. 12933, pp. 562–584. Springer, Cham (2021). https://doi.org/10.1007/978-3-030-85616-8_33
13. Goodfellow, I., Bengio, Y., Courville, A.: Deep Learning. MIT Press, New York (2016)
14. Jeunen, O., Goethals, B.: Top-K contextual bandits with equity of exposure. In: Proceedings of the 15th ACM Conference on Recommender Systems, pp. 310–320 (2021)
15. Johnson, J., Douze, M., Jégou, H.: Billion-scale similarity search with GPUs. IEEE Trans. Big Data **7**(3), 535–547 (2019)
16. Kafka (2023). https://kafka.apache.org/
17. Koren, Y., Bell, R.: Advances in collaborative filtering. In: Ricci, F., Rokach, L., Shapira, B. (eds.) Recommender Systems Handbook, pp. 77–118. Springer, Cham (2015). https://doi.org/10.1007/978-1-4899-7637-6_3
18. Lee, H., Im, J., Jang, S., Cho, H., Chung, S.: MeLU: meta-learned user preference estimator for cold-start recommendation. In: Proceedings of the 25th ACM SIGKDD International Conference on Knowledge Discovery and Data Mining, pp. 1073–1082 (2019)
19. Li, L., Chu, W., Langford, J., Schapire, R.E.: A contextual-bandit approach to personalized news article recommendation. In: Proceedings of the 19th International Conference on World Wide Web, pp. 661–670 (2010)
20. Li, Y., Liu, K., Satapathy, R., Wang, S., Cambria, E.: Recent developments in recommender systems: a survey. arXiv Preprint arXiv:2306.12680 (2023)
21. McInerney, J., et al.: Explore, exploit, and explain: personalizing explainable recommendations with bandits. In: Proceedings of the 12th ACM Conference on Recommender Systems, pp. 31–39 (2018)
22. Mu, R.: A survey of recommender systems based on deep learning. IEEE Access **6**, 69009–69022 (2018)
23. Open Neural Network Exchange (ONNX) (2023). https://github.com/onnx/onnx
24. Salha-Galvan, G., Hennequin, R., Chapus, B., Tran, V.A., Vazirgiannis, M.: Cold start similar artists ranking with gravity-inspired graph autoencoders. In: Proceedings of the 15th ACM Conference on Recommender Systems, pp. 443–452 (2021)
25. Schedl, M., Zamani, H., Chen, C.W., Deldjoo, Y., Elahi, M.: Current challenges and visions in music recommender systems research. Int. J. Multimed. Inf. Retr. **7**, 95–116 (2018)
26. Thompson, W.R.: On the likelihood that one unknown probability exceeds another in view of the evidence of two samples. Biometrika **25**(3/4), 285–294 (1933)
27. Wang, J., Huang, P., Zhao, H., Zhang, Z., Zhao, B., Lee, D.L.: Billion-scale commodity embedding for E-commerce recommendation in Alibaba. In: Proceedings of the 24th ACM SIGKDD International Conference on Knowledge Discovery and Data Mining, pp. 839–848 (2018)

Variance Reduction in Ratio Metrics for Efficient Online Experiments

Shubham Baweja, Neeti Pokharna, Aleksei Ustimenko, and Olivier Jeunen(✉)

ShareChat, Edinburgh, UK
olivierjeunen@gmail.com

Abstract. Online controlled experiments, such as A/B-tests, are commonly used by modern tech companies to enable continuous system improvements. Despite their paramount importance, A/B-tests are expensive: by their very definition, a percentage of traffic is assigned an inferior system variant. To ensure statistical significance on top-level metrics, online experiments typically run for several weeks. Even then, a considerable amount of experiments will lead to *inconclusive* results (i.e. false negatives, or type-II error). The main culprit for this inefficiency is the variance of the online metrics. Variance reduction techniques have been proposed in the literature, but their direct applicability to commonly used ratio metrics (e.g. click-through rate or user retention) is limited.

In this work, we successfully apply variance reduction techniques to ratio metrics on a large-scale short-video platform: ShareChat. Our empirical results show that we can either improve A/B-test confidence in 77% of cases, or can retain the same level of confidence with 30% fewer data points. Importantly, we show that the common approach of including as many covariates as possible in regression is counter-productive, highlighting that control variates based on Gradient-Boosted Decision Tree predictors are most effective. We discuss the practicalities of implementing these methods at scale and showcase the cost reduction they beget.

1 Introduction and Motivation

Online controlled experiments (colloquially known as A/B-tests) are a crucial tool for online businesses to enable data-driven decision making [8,9]. Nevertheless, they are inherently costly and statistical nuances pose challenges when implementing them at scale and interpreting their results correctly [5,7].

The motivation behind this research is to enhance the confidence and efficiency of A/B-testing, particularly when dealing with commonly used *ratio* metrics such as click-through rates and user retention. These metrics play a crucial role in evaluating user engagement and platform performance but inhibit the use of classical statistical tools. Our goal is to increase the *sensitivity* of such ratio metrics. This can either lead to (i) more conclusive results at constant sample sizes (i.e. reduced type-II error), or (ii) a reduction of the required sample size to obtain constant p-values (i.e. saving time and experiment cost).

N. Goharian et al. (Eds.): ECIR 2024, LNCS 14612, pp. 292–297, 2024.
https://doi.org/10.1007/978-3-031-56069-9_34

Variance reduction techniques have been proposed in the literature [1,3,4,11], but their direct applicability to ratio metrics is not always straightforward. Furthermore, a myopic focus on *variance* can make us blind to *bias* [10], implying that *variance* reduction does not unequivocally lead to *sensitivity* improvements.

We discuss the practical application of a synthesis of existing variance reduction techniques to improve sensitivity (or, analogously, decrease experimentation cost) on the ShareChat platform. Our empirical observations, based on a log of historical real-world A/B-experiments, underline our theoretical insights. We find that the highest variance reduction does not map to the highest sensitivity increase. Our best-performing method can increase statistical confidence in 77% of A/B-experiments, or achieve on-par confidence with 30% fewer samples, directly impacting the cost per experiment for the business.

2 Problem Statement, Methodology and Contribution

2.1 Statistical Hypothesis Testing

Consider an online controlled experiment with deployed variants A and B, and a metric $\mathcal{M}$ to evaluate. That is, $\mathcal{M}$ is a random variable whose expectation we wish to estimate under A and B. Typically, a z-test is done to ascertain whether the difference in observed metric values is statistically significant. Denote by $\mu(\mathcal{M})$ the mean of the observed values, and by $\sigma(\mathcal{M})$ its standard deviation. A subscript μ_V indicates the mean over observed values for units assigned to V.

The z-statistic is then given by Eq. 1, which can be transformed to a two-tailed p-value for the null hypothesis $A \simeq B$ following Eq. 2, where $\Phi(\cdot)$ represents the cumulative distribution function (CDF) for a standard Gaussian.

$$z_{A \succ B}(\mathcal{M}) = \frac{\mu_A(\mathcal{M}) - \mu_B(\mathcal{M})}{\sqrt{\frac{\sigma_A(\mathcal{M})^2}{N_A} + \frac{\sigma_B(\mathcal{M})^2}{N_B}}} \quad (1) \qquad p_{A \succ B}(\mathcal{M}) = 2\Phi\left(-\left|z_{A \succ B}(\mathcal{M})\right|\right) \quad (2)$$

For a confidence level of α (typically ≈ 0.05), we can reject the null hypothesis when $p_{A \succ B}(\mathcal{M}) < \alpha$, and claim a statistically significant impact on metric $\mathcal{M}$. Many standard metrics (e.g. daily active users, event counters, et cetera) can be expressed as Bernoulli-distributed random variables. In these cases, $\mu(\mathcal{M})$ and $\sigma(\mathcal{M})$ can be straightforwardly computed using standard formulas. Aside from these metrics, modern platforms on the web typically also care about *ratio* metrics (e.g. click-through rate, retained users per active users, et cetera). In these cases, we need to apply the Delta method to estimate the variance in the denominator of Eq. 1. For a metric $\mathcal{M} \equiv \frac{\mathcal{M}_N}{\mathcal{M}_D}$, this yields:

$$\sigma(\mathcal{M})^2 \approx \frac{\mu(\mathcal{M}_N)^2}{\mu(\mathcal{M}_D)^2}\left(\frac{\sigma(\mathcal{M}_N)^2}{\mu(\mathcal{M}_N)^2} + \frac{\sigma(\mathcal{M}_D)^2}{\mu(\mathcal{M}_D)^2} - 2\frac{\text{cov}(\mathcal{M}_N, \mathcal{M}_D)}{\mu(\mathcal{M}_N)\mu(\mathcal{M}_D)}\right) \quad (3)$$

Alternatively, one can adopt a linearisation approach to obtain a new metric $L(\mathcal{M}) = \mathcal{M}_N - c\mathcal{M}_D$ that preserves directionality and statistical power [1]. Here, $c = \frac{\mu_C(\mathcal{M}_N)}{\mu_C(\mathcal{M}_D)}$, where C corresponds to the *control* variant.

2.2 Variance Reduction in Online Controlled Experiments

Lower p-values (i.e. higher z-scores) indicate higher confidence in rejecting the null hypothesis. From Eq. 1, a clear way to increase the z-score is to increase the sample size N. As this corresponds to running the experiment for a longer period or on a larger portion of traffic, this is costly. An alternative route is to instead decrease the variance of the metric $\sigma(\mathcal{M})^2$. This can be done by leveraging *control variates*. Suppose we have access to a random variable $\mathcal{M}_{\text{CV}}$ that is correlated with the metric value $\mathcal{M}$, but *independent* of the treatment. Then, we can obtain a variance-reduced metric simply by subtracting it:

$$\mathcal{M}_{\text{VR}} = \mathcal{M} - \mathcal{M}_{\text{CV}}. \tag{4}$$

The CUPED approach proposes to assign pre-experiment values of the same metric $\mathcal{M}_{\text{pre}}$ as the control variate in a linear regression model [3]:

$$\mathcal{M}_{\text{CV}} = \theta(\mathcal{M}_{\text{pre}} - \overline{\mathcal{M}_{\text{pre}}}),\ \text{with}\ \theta = \frac{\text{cov}(\mathcal{M}, \mathcal{M}_{\text{pre}})}{\sigma(\mathcal{M}_{\text{pre}})^2}\ \text{estimated on pooled data.} \tag{5}$$

As a natural extension, multiple covariates can be used in a regression model to obtain a final estimate for the outcome metric. The variance reduction this approach yields is directly proportional to the correlation between the original metric $\mathcal{M}$ and the control variate $\mathcal{M}_{\text{CV}}$. Minasayan et al. provide a characterisation of the approximation bias on the Average Treatment Effect (ATE) that is incurred by this approach [10], which is multiplicative of the form $1 - \frac{k}{N}$ with k the number of regression covariates. This implies that for fixed N, adding more covariates always leads to a *decrease* in the ATE estimate. This, in turn, leads to a *decrease* in the *numerator* of Eq. 1, which might (partially) offset variance reduction: lower variance does not unequivocally imply higher sensitivity.

Sidestepping the problems of approximation with linear models, Poyarkov et al. propose to use Gradient-Boosted Decision Trees (GBDTs) to estimate $\mathcal{M}_{\text{CV}}$ directly, reporting significant improvements over linear models and CUPED [11].

2.3 Variance Reduction for Ratio Metrics

We train a multiple linear regression model to estimate the ATE on a ratio metric $\mathcal{M}$ from past online experiments, based on multiple covariates. Following the classical CUPED method [3] extended for ratio metrics, the covariates include: the numerator from the pre-experiment period $\mathcal{M}_{N,\text{pre}}$; the denominator from the pre-experiment period $\mathcal{M}_{D,\text{pre}}$; and the linearised metric for the pre-experiment period $L(\mathcal{M}_{\text{pre}})$ (following [1]). Alternatively, we consider GBDT predictions for the numerator, denominator, and linearisation $\widehat{\mathcal{M}_N}, \widehat{\mathcal{M}_D}, \widehat{L(\mathcal{M})}$ (following [11]). The output of the regression model is then the control variate that we plug into Eq. 4 to obtain a variance-reduced metric, on which we perform a z-test following Eqs. 1 and 2.

3 Experimental Validation of Improved Sensitivity

We consider past online controlled experiments that ran on the ShareChat application, with known (conclusive) outcomes. That is, for deployed system variants A and B, we know a preference ordering to obtain $A \succ B$. All experiments ran for either 7 or 14 days, and all variants were assigned between 1.6 and 3.5 million users. In this preliminary study, we consider 13 known A/B pairs. The ratio metric we consider is *one-day retention*: the number of users who were retained from day D_0 to D_1, over the number of active users on day D_0, aggregated over the experiment period. For this metric and the variance reduction approaches introduced earlier, we measure several attributes: (i) the reduction in variance compared to the original ratio metric $\mathcal{M}$, (ii) the fraction of online experiments that have a lower p-value under the variance-reduced metric, (iii) the median relative z-score, which gives an indication of the sample size reduction that would be gained at constant sensitivity, and (iv) the type-I error measured over 220 two-week A/A-pairs on 3.5 million users per variant. Table 1 presents these results. Even though pre-experiment metric values contribute significantly to variance reduction, they also significantly impact the ATE estimate, which leads to disappointing performance when considering the percentage of experiments for which we can reject the null hypothesis with higher confidence. Indeed, when considering solely GBDT-based predictions $\widehat{\mathcal{M}}$, we obtain lower p-values for 77% of experiments. For the classical CUPED approach using pre-experiment values, this number deteriorates to 15%. Combining all covariates does not provide solace, stagnating at 31%. This is a direct result of the bias we discuss in Sect. 2.2. Furthermore, a median relative z-score of 1.19 implies that we need $1.19^2 \approx 1.42$ times fewer data points to achieve the same level of confidence as the original ratio metric ($1 - \frac{1}{1.42} \approx 30\%$; see e.g. [2,6]). This directly impacts the speed at which experiments conclude, and the number of experiments that can run concurrently, greatly improving experimental velocity. For $\alpha = 0.05$, we expect the type-I errors to converge at 5% as well. We observe that all are very close, ensuring that our encouraging results are not the result of overfitting.

Table 1. Empirical insights into variance-reduced metrics on a dataset of past A/B experiments. We consider using pre-experiment metrics $\mathcal{M}_{\text{pre}}$, GBDT-based predictions $\widehat{\mathcal{M}}$, or both. Lower variance does not unequivocally imply lower p-values, and GBDT-based predictions lead to significantly improved sensitivity with good type-I errors.

Covariates	Var. Red.	$\mathbb{P}(p\text{-value} \downarrow)$	med. rel. z	Type-I Error
$\mathcal{M}_{\text{pre}}$	–72.47%	15.38%	0.78	4.3%
$\widehat{\mathcal{M}}$	–45.66%	**76.92%**	**1.19**	4.8%
$\{\mathcal{M}_{\text{pre}}\} \cup \{\widehat{\mathcal{M}}\}$	**–72.62%**	30.77%	0.81	5.2%

4 Conclusions and Outlook

In this work, we have explored variance reduction techniques specifically tailored for ratio metrics in the context of A/B-testing. One notable takeaway from our study is a re-evaluation of the conventional CUPED method [3]. This approach, which often incorporates multiple covariates (pre-experiment metrics) can inadvertently lead to overfitting. Instead, we propose using GBDT predictors which are *unbiased* estimators and account for complex relationships among pre-experiment metrics [11]. Leveraging predicted metrics as covariates in the CUPED framework yields lower variance reduction but better performance in terms of sensitivity without inflating type-I errors, providing empirical support for the inclusion of unbiased estimators to allow for more efficient A/B-testing.

About the Presenter

Aleksei Ustimenko is a staff applied scientist at ShareChat, focusing on A/B experimentation, statistics and recommender systems. He additionally enjoys work as a theoretical mathematician, focusing on stochastic calculus.

About the Company

ShareChat (Mohalla Tech Pvt Ltd) is India's largest homegrown social media company, with 325^{+} million Monthly Active Users (MAUs) across all its platforms, and social media brands such as ShareChat App and Moj under its portfolio. Today, ShareChat App is India's leading social media platform with over 180 million MAUs spread across the country, and Moj is India's #1 short-video app with the highest MAU base of nearly 160 million.

References

1. Budylin, R., Drutsa, A., Katsev, I., Tsoy, V.: Consistent transformation of ratio metrics for efficient online controlled experiments. In: Proceedings of the Eleventh ACM International Conference on Web Search and Data Mining, WSDM 2018, pp. 55–63. ACM (2018). https://doi.org/10.1145/3159652.3159699
2. Chapelle, O., Joachims, T., Radlinski, F., Yue, Y.: Large-scale validation and analysis of interleaved search evaluation. ACM Trans. Inf. Syst. **30**(1), 1–41 (2012). https://doi.org/10.1145/2094072.2094078
3. Deng, A., Xu, Y., Kohavi, R., Walker, T.: Improving the sensitivity of online controlled experiments by utilizing pre-experiment data. In: Proceedings of the Sixth ACM International Conference on Web Search and Data Mining, WSDM 2013, pp. 123–132. ACM (2013). https://doi.org/10.1145/2433396.2433413
4. Guo, Y., Coey, D., Konutgan, M., Li, W., Schoener, C., Goldman, M.: Machine learning for variance reduction in online experiments. In: Ranzato, M., Beygelzimer, A., Dauphin, Y., Liang, P., Vaughan, J.W. (eds.) Advances in Neural Information Processing Systems, vol. 34, pp. 8637–8648. Curran Associates, Inc. (2021)

5. Jeunen, O.: A common misassumption in online experiments with machine learning models. SIGIR Forum **57**(1), 1–9 (2023). https://doi.org/10.1145/3636341.3636358
6. Kharitonov, E., Drutsa, A., Serdyukov, P.: Learning sensitive combinations of A/B test metrics. In: Proceedings of the Tenth ACM International Conference on Web Search and Data Mining, WSDM 2017, pp. 651–659. ACM (2017). https://doi.org/10.1145/3018661.3018708
7. Kohavi, R., Deng, A., Vermeer, L.: A/B testing intuition busters: common misunderstandings in online controlled experiments. In: Proceedings of the 28th ACM SIGKDD Conference on Knowledge Discovery and Data Mining, KDD 2022, pp. 3168–3177. ACM (2022). https://doi.org/10.1145/3534678.3539160
8. Kohavi, R., Tang, D., Xu, Y.: Trustworthy Online Controlled Experiments: A Practical Guide to A/B Testing. Cambridge University Press, Cambridge (2020)
9. Larsen, N., Stallrich, J., Sengupta, S., Deng, A., Kohavi, R., Stevens, N.T.: Statistical challenges in online controlled experiments: a review of A/B testing methodology. Am. Stat. 1–15 (2023). https://doi.org/10.1080/00031305.2023.2257237
10. Minasyan, V., Mirzoyan, D., Sakhnov, A.: Estimation of average treatment effect on residuals: bias derivation. Available at SSRN 3953160 (2021)
11. Poyarkov, A., Drutsa, A., Khalyavin, A., Gusev, G., Serdyukov, P.: Boosted decision tree regression adjustment for variance reduction in online controlled experiments. In: Proceedings of the 22nd ACM SIGKDD International Conference on Knowledge Discovery and Data Mining, KDD 2016, pp. 235–244. ACM (2016). https://doi.org/10.1145/2939672.2939688

Augmenting KG Hierarchies Using Neural Transformers

Sanat Sharma(✉), Mayank Poddar, Jayant Kumar, Kosta Blank, and Tracy King

Adobe Inc., San Jose, USA

sanatsha@gmail.com

Abstract. This work leverages neural transformers to generate hierarchies in an existing knowledge graph. For small (<10,000 node) domain-specific KGs, we find that a combination of few-shot prompting with one-shot generation works well, while larger KG may require cyclical generation. Hierarchy coverage increased by 98% for intents and 95% for colors.

Keywords: knowledge graphs · hierarchy generation · few-shot prompting

1 Introduction

Knowledge graphs (KG) are widely used in industry to understand user behavior and provide contextual recommendations (Fig. 1) and search results. At Adobe, we utilize a KG to understand users' creative intent and recommend Adobe assets based on the intent. While the original KG had over 4000 intent nodes, the original taxonomy was mostly flat, lacking substantial hierarchies that could amplify the semantic significance between nodes and drive additional intent-based recommendation use cases.

In this work, we present a novel approach to automatically generate intricate graph hierarchies in KGs by leveraging neural transformers. We enhance the structure of our graph by generating hierarchies for both intent (what an Adobe user wants to accomplish, e.g. create a child's birthday card or a banner for their cafe's website) and color node types, resulting in a significant increase in hierarchy coverage: 98% for intent and 95% for color. Hierarchies have key benefits to our users. **Organizational Structure**: Hierarchical relationships makes it easier to navigate and comprehend the KG. Hierarchies help maintain order and provide a clear understanding of how different concepts are related to each other. **Semantic Relationships**: Rich intent hierarchies allow us to capture the semantic relationships between concepts. They also help us unlock key features, such as powering browse and SEO relationships. **Scalability and Flexibility**: Top level categories allow for easier addition of new intents without disrupting the overall structure as the KG grows.

N. Goharian et al. (Eds.): ECIR 2024, LNCS 14612, pp. 298–303, 2024.
https://doi.org/10.1007/978-3-031-56069-9_35

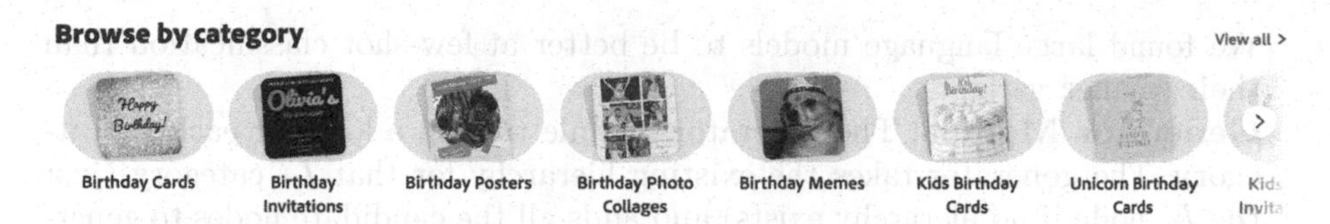

Fig. 1. Adobe Express SEO page for *birthday* with related pages powered by the intent-based KG.

2 Related Knowledge Graph Work

Knowledge graphs are widely used in industry in a variety of roles, from providing social media recommendations [11,12] to providing entity linking and semantic information between concepts [2,9]. With recent improvements in attention-based networks, specifically large transformers [5,10], there has been academic focus towards grounding language models with KGs [6], thereby providing semantic reasoning and generation based on the KG information. Recent works also investigate automated generation and completion of KGs using large transformer models [1,4]. They utilize language models like ChatGPT to add new nodes to subsets of the graph. While most works focus on adding additional nodes to KGs, our work focuses on augmenting the semantic relationships of existing nodes in the graph using large transformer models. We generate rich hierarchies and associations inside the graph, something that is novel to the field.

3 Approach

First we create top level (L_1) categories for a specific class of nodes (e.g. intent or color). L_1 categories can be selected by domain experts or by a language model. They need to be broad and expansive, as our aim is to transition from a general intent to a more specific one, progressing through multiple levels. We created the L_1 intent nodes by examining Adobe Express frequent queries and their intents, Adobe Stock content categories, and the Google open-source product type taxonomy [3]. This resulted in 26 L_1 intent categories (e.g. Business and Industry, Travel, Shopping, Beauty and Wellness, Health). These overlap with standard taxonomies but comprise a subset relevant to Adobe Stock and Express users.

After establishing the L_1 categories, a classifier module assigns all KG nodes to one or more L_1 categories. Then a generator module enhances the existing hierarchy (if any) with the newly added nodes. Finally, a scalable pipeline auto-ingests the new hierarchies into the KG and queries the KG at inference time. To generate our hierarchies, we utilize two modules (Fig. 2):

1. **Classifier Module**: The classifier module takes all nodes to be added to hierarchy and classifies them into one or more of the L_1 candidate classes.

We found large language models to be better at few-shot classification than their smaller variants.

2. **Generator Module**: The generator module runs in a loop for each L_1 category. The generator takes the existing hierarchy for that L_1 category (just the L_1 node if no hierarchy exists) and adds all the candidate nodes to generate the updated category hierarchy. We found one-shot hierarchy generation using large language models to be the best approach (Sect. 3.2).

3.1 Few-Shot Prompting

In order to use the classification module, we do few-shot learning in which we provide the language model (GPT4 with a 32K context [5]) with a few classification examples and a strict prompt. A sample prompt we used is *"You are a taxonomist, classify the given node to one or more of the provided categories. If you think the category should be its own thing, return Other. Please return a dictionary every time."* With the prompt, we provide a few sample nodes, the categories and an output prediction. Based on a few rounds of samples, we then provide the true candidates for classification to the model. Similar to other approaches in the industry [7,8], we see a significant boost of 12% in accuracy by doing few-shot learning compared to zero-shot classification.

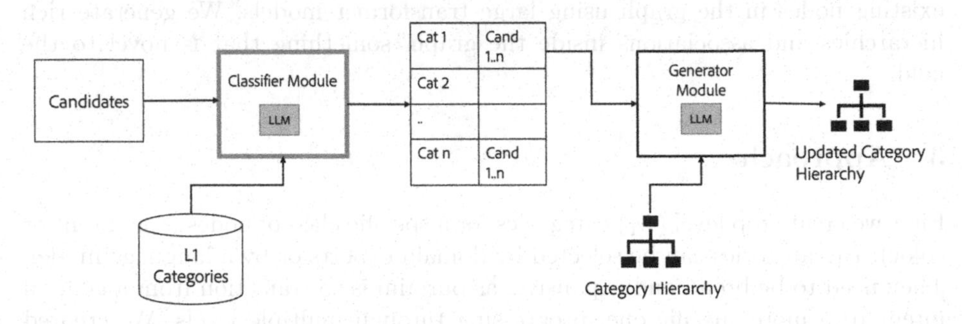

Fig. 2. Hierarchy Generation Approach

3.2 Generation Module

Once the candidates are categorized, we experimented with two techniques in the generation module to create the updated hierarchy. **Cyclical Generation**: Generate each level of the category in a loop. This means that L_2 level nodes are added first, then L_3 and so on. This is needed when the existing hierarchy is large and cannot fit in the model context. **One-Shot Generation**: All candidate nodes are added to the hierarchy in a single pass, without any cyclical generation. We found that this approach produced better results with smaller (<10,000 node) taxonomies.

Cyclical Generation. In the cyclical generation approach, we follow a cyclical pattern to generate each level of the hierarchy for an L_1 category and its children.

1. From the candidate set of nodes at level L_i, classify nodes that belong to that level. Utilize the existing nodes at that level (i.e. any nodes already present in the hierarchy) to help the language model via a few-shot approach.
2. The nodes not categorized to be part of level L_i will become part of lower levels $(L_{i+1..n})$.
3. Pass the remaining nodes as well as the existing hierarchy to the generator module to create the new updated hierarchy for level L_i. The generator module will attempt to categorize and place each of the remaining nodes under one of the L_i nodes.
4. For each of the L_i nodes and their hierarchy, repeat the process in a recursive manner to fine-tune the hierarchies. The process stops when either a specified depth (L_i) is reached or all nodes have been added to the hierarchy.

While the cyclical approach is useful, especially for larger graphs, we saw several drawbacks with it when generating our intent hierarchies. **Error Propagation**: LLMs can hallucinate or generate incorrect structured content. Having multiple steps in the generation process can lead to error propagation through the chain. This is the biggest issue with a recursive approach. **The Other Conundrum**: LLMs are bad at placing nodes into the "Other" category. This means that most nodes were assigned into a level's category (L_i) rather than being placed into the Other category (to be a part of the L3 and lower levels). **Order Importance**: Whether we pass nodes in a batch or one at a time for classification, the order of nodes plays a huge difference in the categorizing. For example, if "birthday party" was categorized as an L_i first and then the node "birthday" is shown to the LLM, it often incorrectly categorizes "birthday" as a child of "birthday party" due to their similarities. Additional checks and another overall pass is required to fix the categorizations. One-shot generation (see below) alleviates these issues (Fig. 3).

One-Shot Generation. In the one-shot generation process, we provide all the nodes to be added as well as the full existing hierarchy to the LLM and allow it

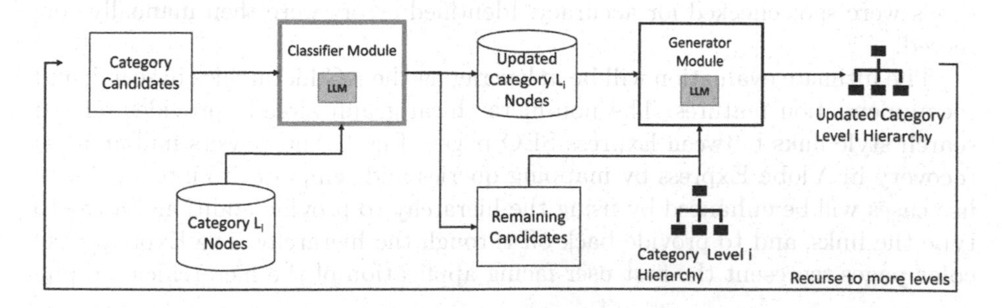

Fig. 3. Cyclical Generation requires classification of nodes and addition at each level

to generate the L_2, L_3 and lower categories with just a few examples provided. This approach worked well when there was an existing, partial hierarchy in place to guide the language model's intuition. If a large number of candidates need to be added to the hierarchies, batched generations followed by an overall pass where the language model has the chance to correct any errors is utilized. For our domain-specific use cases, we found full, one-shot generation to be more viable since our taxonomy is relatively small (<5000 intent nodes and <500 color nodes).

Updating the Graph. For new intents, the above approach is used to integrate them into the KG. When creating intents for a new domain (e.g. Adobe app tools), a subgraph is generated for the new domain and then merged into the existing KG. One algorithm improvement, suggested by an anonymous reviewer, is to examine each subgraph in the generated graph and query the LLM as a third step if it looks good. This uses the LLM for evaluation and updating, instead of generating.

4 Hierarchical KG Evaluation and Conclusions

Statistics on the hierarchical KG generated using one-shot generation and few-shot prompting are summarized below ('Lower' indicates nodes in L_5 or below categories).

KG	Nodes	In Hierarchy Before	In Hierarchy Now	% Change	L_1	L_2	L_3	L_4	Lower
Intents	4639	891	4630	419%	25	383	1826	1937	920
Colors	328	12	328	2100%	12	224	92	1	0

We evaluated the KG hierarchies through a human-in-the-loop approach. We provided a graphical interface to identify nodes that are incorrectly positioned and to offer suggestions for enhancement. We engaged 16 Adobe-internal domain experts to review the hierarchies within each L_1 category for both intent and color nodes. The hierarchies were found to be relevant >95% of the time. Lower levels were spot-checked for accuracy. Identified errors were then manually corrected.

The ultimate evaluation will be in leveraging the KG hierarchies in search and recommendation features. The non-hierarchical graph already provides related search style links between Express SEO pages (Fig. 1) and powers null and low recovery in Adobe Express by mapping queries and templates to intents. These use cases will be enhanced by using the hierarchy to provide additional links, to type the links, and to provide back-off through the hierarchy. The Express SEO color pages represent the first user-facing application of the hierarchical graph.

Adobe Company Portrait

Adobe Inc. enables customers to change the world through digital experiences and creativity. The Adobe search and discovery team supports search and recommendations across customer text, image, video, and other document types as well as over Adobe Stock assets and Adobe help and tutorials.

Main Author Bio

Sanat Sharma is a senior machine learning engineer at Adobe Inc. He earned his Master's degree from University of Texas, Austin in 2020, with a focus on NLP. Sanat's work focuses on search improvements and contextual recommendations, and his work has been published at conferences such as SIGIR and CVPR.

References

1. Carta, S., Giuliani, A., Piano, L., Podda, A.S., Pompianu, L., Tiddia, S.G.: Iterative zero-shot LLM prompting for knowledge graph construction (2023). http://arxiv.org/abs/2307.01128
2. Dong, X.L., et al.: Knowledge vault: a web-scale approach to probabilistic knowledge fusion. In: The 20th ACM SIGKDD International Conference on Knowledge Discovery and Data Mining, KDD 2014, New York, NY, USA, 24–27 August 2014, pp. 601–610 (2014). http://www.cs.cmu.edu/~nlao/publication/2014.kdd.pdf
3. Google: Google product type taxonomy. http://www.google.com/basepages/producttype/taxonomy.en-US.txt, 21 September 2021 version
4. Meyer, L.P., et al.: LLM-assisted knowledge graph engineering: experiments with ChatGPT (2023). http://arxiv.org/abs/2307.06917
5. OpenAI: GPT-4 technical report (2023). http://arxiv.org/abs/2303.08774
6. Pan, S., Luo, L., Wang, Y., Chen, C., Wang, J., Wu, X.: Unifying large language models and knowledge graphs: a roadmap (2023). http://arxiv.org/abs/2306.08302
7. Parnami, A., Lee, M.: Learning from few examples: a summary of approaches to few-shot learning (2022)
8. Song, Y., Wang, T., Mondal, S.K., Sahoo, J.P.: A comprehensive survey of few-shot learning: evolution, applications, challenges, and opportunities (2022). http://arxiv.org/abs/2205.06743
9. Stokman, F., Vries, P.: Structuring knowledge in a graph. In: van der Veer, G.C., Mulder, G. (eds.) Human-Computer Interaction, pp. 186–206. Springer, Cham (1988). https://doi.org/10.1007/978-3-642-73402-1_12
10. Touvron, H., et al.: Llama 2: open foundation and fine-tuned chat models (2023). http://arxiv.org/abs/2307.09288
11. Ugander, J., Karrer, B., Backstrom, L., Marlow, C.: The anatomy of the Facebook social graph. CoRR abs/1111.4503 (2011). http://arxiv.org/abs/1111.4503
12. Ying, R., He, R., Chen, K., Eksombatchai, P., Hamilton, W.L., Leskovec, J.: Graph convolutional neural networks for web-scale recommender systems. CoRR abs/1806.01973 (2018). http://arxiv.org/abs/1806.01973

Incorporating Query Recommendation for Improving In-Car Conversational Search

Md. Rashad Al Hasan Rony[1](✉), Soumya Ranjan Sahoo[2], Abbas Goher Khan[2], Ken E. Friedl[1], Viju Sudhi[2], and Christian Süß[1]

[1] BMW Group, Parkring 19-23, 85748 Garching in Munich, Germany
rah.rony@gmail.com

[2] Fraunhofer IAIS, Zwickauer Straße 46, 01069 Dresden, Germany

Abstract. Retrieval-augmented generation has become an effective mechanism for conversational systems in domain-specific settings. Retrieval of a wrong document due to the lack of context from the user utterance may lead to wrong answer generation. Such an issue may reduce the user engagement and thereby the system reliability. In this paper, we propose a context-guided follow-up question recommendation to internally improve the document retrieval in an iterative approach for developing an in-car conversational system. Specifically, a user utterance is first reformulated, given the context of the conversation to facilitate improved understanding to the retriever. In the cases, where the documents retrieved by the retriever are not relevant enough for answering the user utterance, we employ a large language model (LLM) to generate question recommendation which is then utilized to perform a refined retrieval. An empirical evaluation confirms the effectiveness of our proposed approaches in in-car conversations, achieving *48%* and *22%* improvement in the retrieval and system generated responses, respectively, against baseline approaches.

Keywords: Retrieval-augmented Generation · Query Recommendation

1 Introduction

With the advancement of large language models, conversational assistants have significantly improved handling diverse types of user utterances [8]. Recent approaches to retrieval-augmented generation (RAG) have been widely adopted across various industries for developing domain-specific conversational assistants [5]. In RAG-based systems, retrieval failures can lead to undesired outcomes, including hallucinated responses, incorrect answers, and disrupted user engagement. In an in-car setting, the aforementioned drawbacks may lead to unsafe handling of the vehicle by providing confusing or erroneous information. Several factors can lead to a failure in relevant document retrieval. Among them, a non-optimal retriever, ambiguous or unclear user queries, and unavailability of relevant documents in the document collection stand out as prominent reasons.

N. Goharian et al. (Eds.): ECIR 2024, LNCS 14612, pp. 304–312, 2024.
https://doi.org/10.1007/978-3-031-56069-9_36

This work focuses on the retrieval failure cases, that occurs due to ambiguous or unclear user utterances [7].

Addressing the issues, based-on the conversational-context, we propose an integrated system combining LLM-based query reformulation and clarification question recommendations. Research questions that we address are as follows:

1. **RQ1**: Can we develop a conversational search system that seamlessly integrates query reformulation and proactively recommends clarifying questions for resolving query ambiguity?
2. **RQ2**: Is the proposed prompt-oriented LLM-based approach sufficient to outperform the baseline systems?

We provide a brief description of the proposed approaches below.

2 Methodology

In a car setting, given a user utterance, the goal of the conversational system is to provide the user with information related to the car features and the car. We ignore the speech-related parts of the system to keep the discussion within the scope. Typically, a user interacts with the in-car conversational assistant in the form of a natural language query, Q (user utterance). The system then reformulates the query for better understanding and retrieves relevant documents D^* where the potential answer may exists, which the system leverages next for generating the final response. Figure 1 illustrates a high-level overview of the system architecture. We describe the approach more in detail below.

LLM-Based Query Reformulator: To combat hallucination, we employ extractive question answering model (also known as Reader) to generate answer from domain-specific documents that are already available to the system. This way, we ensure that the system does not generate anything unexpected. However, the extractive reader only takes a user utterance and document as input for generating the answer, hence, has no understanding of the conversation-context, H. To assist the reader, it is thus essential to improve the retrieval as much as possible. To tackle this issue, we incorporate prompt-oriented LLM-based Query reformulation technique that considers the user utterance and the context of the conversation and re-writes the user utterance into a question Q^* that resolves the co-references (e.g., "How to turn `it` off?" into "How to turn `seat heating` off?") for better retrieval. Experimental results confirm the effectiveness of the approach (discussed in Sect. 3).

Document Retrieval and Threshold-Based Arbitrator: Given the re-formulated user query Q^*, the retriever fetches the most semantically similar documents D^* from our *Document Storage.* Here, we leverage Sentence-Transformer [10] to obtain the contextual representation of the documents. The retrieved documents may not always suffice to answer the user query. Therefore, we compare the semantic similarity scores for the pairs (Q, D^*) to empirically find a suitable threshold. If the similarity scores are higher than the threshold, the user query Q^* together with the retrieved documents D^* are fed to the *Answer Generator* otherwise, the pair is fed to the LLM-based *Clarification Question Generator.*

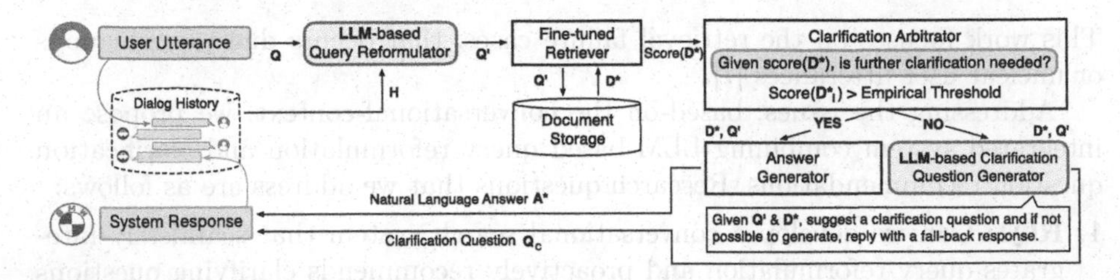

Fig. 1. A schematic overview of retrieval-oriented chain-of-thought prompting for retrieval-augmented in-car conversational system.

LLM-Based Clarification Question Generator: Conversational search often demands multiple iterations to fetch the correct information that answers the user query. When the system is uncertain about an answer, these iterations could involve, asking the user follow-up or clarification questions to understand user intents better. In this work, we propose an LLM-based zero-shot *Clarification Question Generator*, that takes Q^* and D^* as input, and generates a clarification question Q_c leveraging an LLM. While this prompts the user to affirm the Q_c in the next turn, it also makes the system response in the next turn more proactive.

Answer Generator: The answer A^* to the user query mostly lie in a short span in a relevant document. As the extractive question answering model, we fine-tuned an Albert-based [4] model with our in-house annotated question answering data to facilitate better understanding of car-related terminologies. Furthermore, an LLM-based Verbalizer which is a part of the *Answer Generator* component, generates coherent responses A^* given the query Q^*, and extracted answer span. Further implementation details can be found in Sect. A.2.

3 Experiments and Results

Experimental Setting: Our setup involves iterative development of three systems. **System 1**: the base system, consists of a retriever, reader, and LLM-based verbalizer following the standard RAG setup, **System 2**: built on top of System 1 by introducing an LLM-based query reformulator, **System 3**: extends System 2 by incorporating a *Clarification Question Generator* component, working as a recommender to the user and to improve the retrieval in the next turn. In this work, we propose System 3 and currently used in production. In our experiment, all the systems use a consistent pre-trained retriever and reader.

Dataset: We experiment with 20 multi-turn dialogs, sampled from a larger human-curated in-house dataset, comprising a total of 62 user-system interactions, and deliberately including dialogues with ambiguous user queries. Further implementation details can be found in Sect. A.1.

Results: Table 1 reports the performance of the baseline and proposed systems. We evaluate the performance of the three systems using automatic metrics: for document retrieval F1-score and ROUGE [6], and Sentence-Transformer [10] based similarity-score for natural language response generation. The results in Table 1 demonstrates noticeable incremental performance improvements. Adding the query reformulator boosted the performance over the base systems (answering RQ1), and the *Clarification Question Generator* component in System 3 further enhanced the performance over System 2 (answering RQ2). However, only *six* instances triggered clarification, with three being casual user remarks like *"That's pretty cool!"*, making the improvement less substantial. Although on a small set of pairs, the system-generated Q_c closely match human-annotated questions, with a similarity score of *0.66*, indicating high-quality contextual recommendations.

Table 1. Performance comparison of three systems in their respective settings. Scores in **bold** indicates the best performance and (Δ) the percentage of performance gain.

System	Retrieval	Response	
	F1	Rouge-L	Semantic Similarity
System 1	0.198 (+0%)	0.188 (+0%)	0.449 (+0%)
System 2	0.310 (+44%)	0.235 (+22%)	0.555 (+21%)
System 3 (proposed)	**0.324 (+4%)**	**0.237 (+1%)**	**0.561 (+1%)**

4 Discussions

Our experiments addressed the research questions in Sect. 1, improving query enrichment and resolving ambiguities by integrating query reformulation and clarification recommendation. Incorporating LLMs into the standard RAG system confirmed significant improvements over the base system, aligning with successful conversational search systems. An interesting finding is that System 3, with the Q_c component, effectively detects and handles topic drifts (as shown in Fig. 6) due to an additional retrieval step.

One limitation of the current system is that multiple invocation of LLMs introduces latency, which involves up to three LLM calls for processing a single input: query rewriting, recommendation, and answer verbalization. We anticipate that the latency could be reduced by using non-LLM-based query rewriting with a negligible performance loss. It is noteworthy that we exclusively use OpenAI GPT-3.5 Turbo as the LLM in our system, after a careful comparison against other existing LLMs.

Our analysis reveals that the current in-car dataset lacks complexity needed to fully demonstrate the advantages of system three over system two. To address this, we plan to expand the dataset with a focus on more complex scenarios highlighting misunderstanding and ambiguity. In the future, we would like to compare our zero-shot reformulation and clarification components with strong supervised baselines as often studied in the literature [1–3,9,11]. System 3's interactive nature distinguishes it from the other systems by actively considering user feedback and striving to provide helpful responses instead of default replies when unsure.

5 Conclusion

In this work, we study the incremental development of a RAG-based conversational system. With the progressive addition of LLM-based query reformulator and query clarification components, we propose a robust zero-shot domain-specific conversational search system that is capable of resolving ambiguous queries by asking context-specific Q_c aimed at understanding the user's intent by interaction. A careful set of experiments on in-house in-car dataset demonstrates the superiority of the proposed system. Further analyses highlights the effectiveness of the contributing components, thereby confirms the understanding of the system capabilities.

Company Description

BMW Group: With its four brands BMW, MINI, Rolls-Royce and BMW Motorrad, the BMW Group is the world's leading premium manufacturer of automobiles and motorcycles. The company set the course for the future at an early stage and consistently makes sustainability and efficient resource management central to its strategic direction, from the supply chain through production to the end of the use phase of all products. The BMW Group production network comprises over 30 production sites worldwide; the company has a global sales network in more than 140 countries.

Fraunhofer IAIS: Fraunhofer IAIS is the AI institute within the Fraunhofer Society, Europe's largest applied research organization. At the forefront of AI, Machine Learning, and Big Data, they drive digital transformation in Germany and Europe, with core competencies in cognitive perception, big data infrastructures, machine learning, system modeling, and visual analytics. They offer comprehensive technology and consulting services, optimizing products, services, processes, and structures, while pioneering new digital business models. Their person- and value-centric approach integrates diverse disciplines for holistic human-machine interactions.

A Appendix

A.1 Data

To evaluate the performance of our systems, we use a test collection of 20 multi-turn dialogs with the number of turns ranging from 4 to 12 (demonstrated in

Fig. 2). The annotators sampled over 45 different documents from car-related set of documents and further, crafted dialogues composed of user queries that could be answered, grounded on these paragraphs. The annotations also include (i) the relevant documents - for evaluating the Retriever performance and (ii) the reference verbalized answers - for evaluating the overall system response. In our experiments, we didn't selectively choose dialogues or construct a specialized dataset with ambiguous queries. We suggest this as a future step, as our hypothesis showed promise even with a limited number of ambiguous queries.

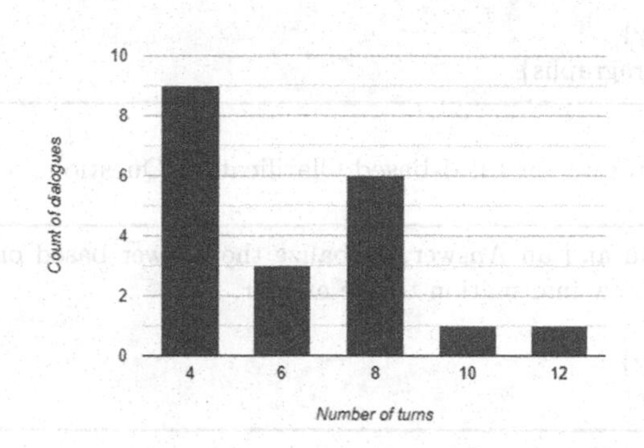

Fig. 2. Distribution of the number of turns in test dialogues.

A.2 Prompts

We used OpenAI `text-davinci-003` model as the default LLM for the *Query Reformulator*, the Q_c Generator, and the Verbalizer modules. We observed that optimizing prompts is crucial for the component's performance and may vary depending on the LLM used. Notably, as depicted in Figs. 3, 4 and 5, we explore the use of prompts for safeguarding against potential harmful inputs, subsequently diffusing and detaching from the ongoing dialog turn. This demonstrates that LLMs with effective prompt design helps in regulating diverse aspects of the overall system.

> Given a question and dialogue history, please provide an extractive summary-based question that captures the key information from the answers relevant to the question while paraphrasing the question. If there is no history, keep the query unchanged.
>
> Question: {query}
> Dialogue history: {history}

Fig. 3. Prompt for LLM-based Query Reformulator.

Given a question and the following paragraphs:
1. For unsafe or harmful questions, politely decline to answer as they are out of context. Stop any further generation.
2. If the question is safe and relevant, suggest a clarification question that demonstrates comprehension of the concept and incorporates information from the provided paragraphs. Start questions either with "Do you mean", "Are you looking for", "Would you like to".
3. If unsure about suggesting a specific clarification question, politely request more information to provide an accurate response. Stop any further generation.

Question: {query}
Paragraphs: {paragraphs}

Fig. 4. Prompt for LLM-based Clarification Question Generator.

Given a Question and an Answer, verbalize the answer based on the question. Do not add any new information to the answer.

Question: {query}
Answer: {answer}

Fig. 5. Prompt for LLM-based Verbalizer.

A.3 Empirical Threshold

The proposed system determines if the LLM-based Clarification Question Generator should be triggered based on the semantic similarity scores from the Retriever. We performed a manual inspection of different retriever scores from our in-house set and empirically determined the threshold for this decision. For our fine-tuned retriever, this valued around 0.5005. We also emphasise that this value is subjective to the chosen retriever and the chosen test sample.

A.4 Weaknesses of System 1 and System 2

As illustrated in Fig. 6, all the three systems handle single-turn dialogues reliably well. However, as the dialogue grows with multiple turns, System 1 fails to resolve co-references in the user utterance ("*them*" in the utterance "*How does my car recognize them?*") and responds with an incorrect answer. The *Query Re-formulator* module employed in System 2 and System 3 enriches the query based on the dialogue history and thereby, resolves all possible co-references in the utterance. This results in a reliable response from these systems. It is noteworthy, how System 2 fails when the user slightly switches the topic or asks for quantitative information (in this example, "*Is the parking space ... large enough?*"). System 3, instead of responding with the less confident retrieved documents, asks the user for further clarification. In the next turn, depending on the user response, the system is capable of providing a more reliable answer.

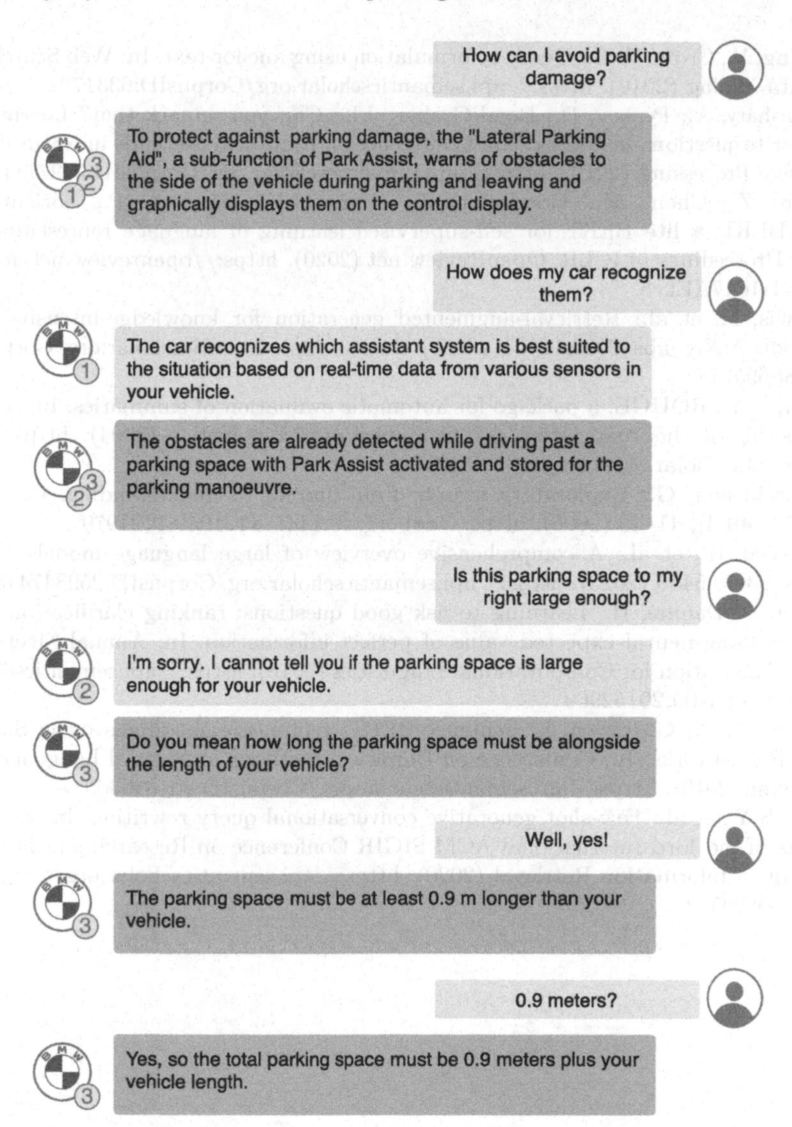

Fig. 6. Illustration of the weaknesses of System 1 and System 2 compared with the proposed System 3. Systems are indicated with numbers inside the bubbles. Green bubbles indicate correct and red bubbles indicate incorrect system responses. (Color figure online)

References

1. Bi, K., Ai, Q., Croft, W.B.: Asking clarifying questions based on negative feedback in conversational search. In: Proceedings of the 2021 ACM SIGIR International Conference on Theory of Information Retrieval (2021). https://api.semanticscholar.org/CorpusID:235829474

2. Dang, V., Croft, W.B.: Query reformulation using anchor text. In: Web Search and Data Mining (2010). https://api.semanticscholar.org/CorpusID:6331792
3. Elgohary, A., Peskov, D., Boyd-Graber, J.L.: Can you unpack that? Learning to rewrite questions-in-context. In: Conference on Empirical Methods in Natural Language Processing (2019). https://api.semanticscholar.org/CorpusID:202771124
4. Lan, Z., Chen, M., Goodman, S., Gimpel, K., Sharma, P., Soricut, R.: ALBERT: a lite BERT for self-supervised learning of language representations. In: Proceedings of ICLR. OpenReview.net (2020). https://openreview.net/forum?id=H1eA7AEtvS
5. Lewis, P., et al.: Retrieval-augmented generation for knowledge-intensive NLP tasks. ArXiv abs/2005.11401 (2020). https://api.semanticscholar.org/CorpusID:218869575
6. Lin, C.Y.: ROUGE: a package for automatic evaluation of summaries. In: Annual Meeting of the Association for Computational Linguistics (2004). https://api.semanticscholar.org/CorpusID:964287
7. Marchionini, G.: Exploratory search: from finding to understanding. Commun. ACM **49**(4), 41–46 (2006). https://doi.org/10.1145/1121949.1121979
8. Naveed, H., et al.: A comprehensive overview of large language models. ArXiv abs/2307.06435 (2023). https://api.semanticscholar.org/CorpusID:259847443
9. Rao, S., Daumé, H.: Learning to ask good questions: ranking clarification questions using neural expected value of perfect information. In: Annual Meeting of the Association for Computational Linguistics (2018). https://api.semanticscholar.org/CorpusID:29152969
10. Reimers, N., Gurevych, I.: Sentence-BERT: sentence embeddings using Siamese BERT-networks. In: Conference on Empirical Methods in Natural Language Processing (2019). https://api.semanticscholar.org/CorpusID:201646309
11. Yu, S.Y., et al.: Few-shot generative conversational query rewriting. In: Proceedings of the 43rd International ACM SIGIR Conference on Research and Development in Information Retrieval (2020). https://api.semanticscholar.org/CorpusID:219559295

Doctoral Consortium Papers

Semantic Search in Archive Collections Through Interpretable and Adaptable Relation Extraction About Person and Places

Nicolas Gutehrlé(✉)

Université de Franche-Comté, CRIT, 25000 Besançon, France
nicolas.gutehrle@univ-fcomte.fr

1 Introduction

In recent years, libraries and archives have undertaken numerous campaigns to digitise their collections. While these campaigns have increased ease of access to archival documents for a wider audience, ensuring discoverability and promoting their content remain significant challenges. Digitised documents are often unstructured, making them difficult to navigate. Accessing archive materials through search engines restricts users to keyword-based queries, leading to being overwhelmed by irrelevant documents. To enhance the exploration and exploitation of the “Big Data of the Past” [15], it is imperative to structure textual content.

2 Related Work

The automatic processing of archival documents has aroused considerable interest in recent years. The task of Named Entity Recognition (NER) applied to historical documents has received a special interest, as shown by the multiple datasets published [8,13,14,18,20,21]. These large annotated datasets allow for the training of machine-learning and deep-learning models, as in [5,7]. However, annotated historical datasets are lacking for other tasks such as Relation Extraction (RE), Event Extraction (EE) or Temporal Expression Extraction (TEE), which limits the training of machine and deep-learning based models. Moreover, the performances of NLP systems trained on contemporary datasets is highly degraded when applied to historical documents, because of domain shift, noisy input, language dynamic and semantic shift [6]. Because of the lack of available resources, rule-based systems such as [1,2,4,10,16,19] represent a good start to structure historical documents.

3 Proposed Research

We propose a method for Joint Extraction of Named Entities and Relation (JENER) based on morpho-syntactic extraction patterns. Our objective is to

N. Goharian et al. (Eds.): ECIR 2024, LNCS 14612, pp. 315–318, 2024.
https://doi.org/10.1007/978-3-031-56069-9_37

structure the textual content of historical documents by annotating the mentions of persons, places and the relations between them. These annotations would allow for example to trace the personal history of an individual, an organisation, etc. through the analysis of their occurrences in the collection. The tools and resources we develop aim to enable the exploitation of documentary funds in semantically rich search interfaces, such as automatically generated maps [12], timelines [11] or network of entities.

We have processed a dataset of periodical documents published in the 20th century from various sources in France. We collected these documents from Gallica[1], the digital archive of *Bibliothèque Nationale de France* (National Library of France, BnF). We have applied the following pre-processing steps to our dataset: *OCR post-correction, hyphen removal, Logical Layout Analysis, Article Segmentation* and *Sentence Segmentation*. The distribution of collections, issues, pages, lines and words in this dataset are shown in Table 1.

Table 1. Number of collections, issues, pages, lines and words in the dataset

Collections	Issues	Pages	Lines	Words
164	8 770	965 158	37 966 027	294 664 515

Our method relies on a Syntactic Index and a Semantic Index. The Syntactic Index stores extraction patterns, each associated with a set of possible relations, as well as the types of both entities which participate in that relation. The Syntactic Index is learned from Wikipedia and Wikidata in a distant supervision manner [17]. Given a Wikidata page, we extract the Shortest Dependency Path [3] graph between a pair of entities from the content of the related Wikipedia page. This graph is then annotated with the relation between this pair of entities. The Semantic Index stores an association score between each word and relation as a TF-IDF weight. It is inspired by Explicit Semantic Analysis (ESA) [9] and is learned from the extraction patterns stored in the Syntactic Index.

When processing a sentence, we apply the extraction patterns on the sentence's dependency graph to extract candidate dependency graphs. To annotate these candidates, we calculate the harmonic mean of the vectors found in the Semantic Index of each words in the candidate graph. The candidate graph is annotated with the relation with the highest score, unless it is smaller than a given threshold. The type of each entities is determined by the predicted relation.

We evaluate our model in terms of Precision, Recall and F1 measures. Our method is extendable to any relation described in the Wikidata ontology in any language available on Wikipedia. Moreover, it only requires to pre-process documents with part-of-speech tagging and dependency parsing. Thus, our method is adaptable to multiple languages and types of documents. Moreover, it is interpretable since it relies on explicit morpho-syntactic extraction patterns.

[1] https://gallica.bnf.fr.

References

1. Borin, L., Kokkinakis, D., Olsson, L.J.: Naming the past: named entity and animacy recognition in 19th century Swedish literature. In: Proceedings of the Workshop on Language Technology for Cultural Heritage Data (LaTeCH 2007), pp. 1–8 (2007)
2. Broux, Y., Depauw, M.: Developing onomastic gazetteers and prosopographies for the ancient world through named entity recognition and graph visualization: some examples from trismegistos people. In: Aiello, L.M., McFarland, D. (eds.) SocInfo 2014. LNCS, vol. 8852, pp. 304–313. Springer, Cham (2015). https://doi.org/10.1007/978-3-319-15168-7_38
3. Bunescu, R., Mooney, R.: A shortest path dependency kernel for relation extraction. In: Proceedings of Human Language Technology Conference and Conference on Empirical Methods in Natural Language Processing, pp. 724–731. Association for Computational Linguistics, Vancouver (2005). https://aclanthology.org/H05-1091
4. Crane, G., Jones, A.: The challenge of virginia banks: an evaluation of named entity analysis in a 19th-century newspaper collection. In: Proceedings of the 6th ACM/IEEE-CS Joint Conference on Digital Libraries, pp. 31–40 (2006)
5. Ehrmann, M.: Les Entitées Nommées, de la linguistique au TAL: Statut théorique et méthodes de désambiguïsation. Ph.D. thesis, Paris Diderot University (2008)
6. Ehrmann, M., Colavizza, G., Rochat, Y., Kaplan, F.: Diachronic evaluation of ner systems on old newspapers. In: Proceedings of the 13th Conference on Natural Language Processing (KONVENS 2016), pp. 97–107. Bochumer Linguistische Arbeitsberichte (2016)
7. Ehrmann, M., Hamdi, A., Pontes, E.L., Romanello, M., Doucet, A.: Named entity recognition and classification in historical documents: a survey. ACM Comput. Surv. **56**, 1–47 (2021)
8. Ehrmann, M., Romanello, M., Flückiger, A., Clematide, S.: Extended overview of clef hipe 2020: named entity processing on historical newspapers. In: CLEF 2020 Working Notes. Conference and Labs of the Evaluation Forum, vol. 2696. CEUR-WS (2020)
9. Gabrilovich, E., Markovitch, S.: Computing semantic relatedness using wikipedia-based explicit semantic analysis. In: International Joint Conference on Artificial Intelligence (2007). https://api.semanticscholar.org/CorpusID:5291693
10. Grover, C., Givon, S., Tobin, R., Ball, J.: Named entity recognition for digitised historical texts. In: LREC. Citeseer (2008)
11. Gutehrlé, N., Doucet, A., Jatowt, A.: Archive timeline summarization (atls): conceptual framework for timeline generation over historical document collections. In: Proceedings of the 6th Joint SIGHUM Workshop on Computational Linguistics for Cultural Heritage, Social Sciences, Humanities and Literature, pp. 13–23 (2022)
12. Gutehrlé, N., Harlamov, O., Karimi, F., Wei, H., Jean-Caurant, A., Pivovarova, L.: Spacewars: a web interface for exploring the spatio-temporal dimensions of wwi newspaper reporting. In: CEUR Workshop Proceedings (2021)
13. Hamdi, A., et al.: A multilingual dataset for named entity recognition, entity linking and stance detection in historical newspapers. In: Proceedings of the 44th International ACM SIGIR Conference on Research and Development in Information Retrieval, pp. 2328–2334 (2021)
14. Hubková, H.: Named-entity recognition in czech historical texts: using a cnn-bilstm neural network model (2019)

15. Kaplan, F., di Lenardo, I.: Big data of the past. Front. Digital Humanit. **4**, 12 (2017). https://doi.org/10.3389/fdigh.2017.00012. https://www.frontiersin.org/article/10.3389/fdigh.2017.00012
16. Milanova, I., Silc, J., Serucnik, M., Eftimov, T., Gjoreski, H.: Locale: a rule-based location named-entity recognition method for latin text. In: HistoInformatics@TPDL, pp. 13–20 (2019)
17. Mintz, M., Bills, S., Snow, R., Jurafsky, D.: Distant supervision for relation extraction without labeled data. In: Proceedings of the Joint Conference of the 47th Annual Meeting of the ACL and the 4th International Joint Conference on Natural Language Processing of the AFNLP, pp. 1003–1011. Association for Computational Linguistics, Suntec (2009). https://aclanthology.org/P09-1113
18. Neudecker, C.: An open corpus for named entity recognition in historic newspapers. In: Proceedings of the Tenth International Conference on Language Resources and Evaluation (LREC 2016), pp. 4348–4352 (2016)
19. Platas, M.L.D., Muñoz, S.R., González-Blanco, E., Fabo, P.R., Mellado, E.Á.: Medieval Spanish (12th-15th centuries) named entity recognition and attribute annotation system based on contextual information. J. Assoc. Inf. Sci. Technol. **72**, 224–238 (2020). https://api.semanticscholar.org/CorpusID:225430990
20. Ritze, D., Zirn, C., Greenstreet, C., Eckert, K., Ponzetto, S.P.: Named entities in court: the marinelives corpus. In: Language Resources and Technologies for Processing and Linking Historical Documents and Archives-Deploying Linked Open Data in Cultural Heritage-LRT4HDA Workshop Programme, pp. 26 (2014)
21. Ruokolainen, T., Kettunen, K.: À la recherche du nom perdu-searching for named entities with stanford ner in a finnish historical newspaper and journal collection. In: 13th IAPR International Workshop on Document Analysis Systems, pp. 1–2 (2018)

Document Level Event Extraction from Narratives

Luís Filipe Cunha[1,2(✉)]

[1] FCUP - University of Porto, Porto, Portugal
luis.f.cunha@inesctec.pt
[2] LIAAD - INESC TEC, Porto, Portugal

1 Motivation

One of the fundamental tasks in Information Extraction (IE) is Event Extraction (EE), an extensively studied and challenging task [13,15], which aims to identify and classify events from the text. This involves identifying the event's central word (trigger) and its participants (arguments) [1]. These elements capture the event semantics and structure, which have applications in various fields, including biomedical texts [42], cybersecurity [24], economics [12], literature [32], and history [33]. Structured knowledge derived from EE can also benefit other downstream tasks such as Question Answering [20,30], Natural Language Understanding [21], Knowledge Base Graphs [3,37], summarization [8,10,41] and recommendation systems [9,18].

Despite the existence of several English EE systems [2,22,25,26], they face limited portability to other languages [4] and most of them are designed for closed domains, posing difficulties in generalising. Furthermore, most current EE systems restrict their scope to the sentence level, assuming that all arguments are contained within the same sentence as their corresponding trigger. However, real-world scenarios often involve event arguments spanning multiple sentences, highlighting the need for document-level EE.

2 Related Work

Several approaches have been proposed to tackle EE [13,15]. For instance, Lu et al. (2021) [22] introduced Text2Event, a Transformer-based model, employing a sequence-to-structure approach using a T5 model [29]. Lyu et al. (2021) used Zero-Shot Learning, employing Semantic Role Labeling (SRL) and QA with BERT-based models [23]. Nguyen and Grishman (2018) [27] employed Graph Neural Networks, using the words as nodes and dependency tree links as edges. Yan et al. (2019) [39] proposed a Multi-Order Graph Attention Network using Graph Attention Networks (GATs), which employ self-attention to compute the nodes relative importance.

Regarding document-level EE, Du and Cardie (2020) [5] studied context length's impact on EE model performance using multi-granularity contextualized

N. Goharian et al. (Eds.): ECIR 2024, LNCS 14612, pp. 319–324, 2024.
https://doi.org/10.1007/978-3-031-56069-9_38

word encodings. Huang and Peng (2021) [11] proposed the Document-level Event Extraction with DVN (DEED) model, using Deep Value Networks to capture cross-event dependencies. Li et al. (2021) [16] performed argument extraction using generative models. Xu et al. (2021) [38] presented a Heterogeneous Graph-based Interaction Model with a Tracker (GIT) for global interaction features among sentences, addressing arguments dispersed across sentences.

3 Research Questions

RQ1: Can we shorten the gap between Portuguese and English current State-of-the-Art (SOTA) event extraction systems? We intend to employ automatic translation to generate Portuguese versions of existing event-driven corpora. In this regard, we have recently secured funding through the PTicola project funded by the Portuguese Foundation for Science and Technology. This has provided us with Google Cloud Platform credits, enabling access to essential resources for the translation task. Additionally, we intend to explore alternative methods, such as data augmentation [7,35] to generate new labelled data.

RQ2: Can we exploit the events' syntactic features to generalise event extraction models to other languages and domains? We plan to leverage syntactic features associated with the events' structure, which might exhibit similarities across different languages and domains. Previous research [28] suggests that such similarities exist as events with different types often share a similar structure. Furthermore, it was also observed [34] that cross-lingual syntactic features for EE such as POS and dependency tree links can be learned on the source language corpus and transferred to other target languages.

RQ3: Can we perform EE on the document-level by improving models' comprehensive understanding of inter-sentence contextual information? Document-level EE can be enhanced through various techniques. For example, coreference resolution can be used to identify all the mentions within the text that refer to the same argument [5]. The syntactic dependency tree links can be used to bring the trigger and its arguments into closer proximity in the document, as the distance between words in the dependency tree is shorter [31]. Finally, encoding tokens on the document level enhances word representations with inter-sentence contextual information when compared to sentence-level encodings. This can improve the model's overall understanding of the whole narrative [11,36,38].

RQ4: Can we leverage other neural architectures to improve the effectiveness of current SOTA event extraction systems? We plan to leverage the power of Graph Neural Networks, integrating both semantic and syntactic features to create a more comprehensive representation of the text. In fact, in recent years, the adoption of graphs to represent events has been widely used [17,40] and has proven to outperform several previous architectures. Another methodology that has been showing promising results in IE tasks consists of using Question Answering and Question Generation models [6,14,19].

Acknowledgements. This work is financed by National Funds through the FCT - Fundação para a Ciência e a Tecnologia, I.P. (Portuguese Foundation for Science and Technology) within the project StorySense, with reference 2022.09312.PTDC (DOI 10.54499/2022.09312.PTDC).

References

1. English annotation guidelines for events. Linguistic Data Consortium (2005). https://www.ldc.upenn.edu/sites/www.ldc.upenn.edu/files/english-events-guidelines-v5.4.3.pdf
2. Balali, A., Asadpour, M., Campos, R., Jatowt, A.: Joint event extraction along shortest dependency paths using graph convolutional networks. Knowl.-Based Syst. **210**, 106492 (2020). https://doi.org/10.1016/j.knosys.2020.106492. https://www.sciencedirect.com/science/article/pii/S0950705120306213
3. Bosselut, A., Bras, R.L., Choi, Y.: Dynamic neuro-symbolic knowledge graph construction for zero-shot commonsense question answering (2020)
4. Cunha, L.F., Campos, R., Jorge, A.: Event extraction for Portuguese: a qa-driven approach using ace-2005. In: Springer's LNAI - Lecture Notes in Artificial Intelligence. Springer, Heidelberg (2023). https://doi.org/10.1007/978-3-031-49008-8_32
5. Du, X., Cardie, C.: Document-level event role filler extraction using multi-granularity contextualized encoding. In: Proceedings of the 58th Annual Meeting of the Association for Computational Linguistics, pp. 8010–8020. Association for Computational Linguistics, Online (2020). https://doi.org/10.18653/v1/2020.acl-main.714. https://aclanthology.org/2020.acl-main.714
6. Du, X., Cardie, C.: Event extraction by answering (almost) natural questions. In: Proceedings of the 2020 Conference on Empirical Methods in Natural Language Processing (EMNLP), pp. 671–683. Association for Computational Linguistics, Online (2020). https://doi.org/10.18653/v1/2020.emnlp-main.49. https://aclanthology.org/2020.emnlp-main.49
7. Ferguson, J., Lockard, C., Weld, D., Hajishirzi, H.: Semi-supervised event extraction with paraphrase clusters. In: Proceedings of the 2018 Conference of the North American Chapter of the Association for Computational Linguistics: Human Language Technologies, vol. 2 (Short Papers), pp. 359–364. Association for Computational Linguistics, New Orleans (2018). https://doi.org/10.18653/v1/N18-2058. https://aclanthology.org/N18-2058
8. Filatova, E., Hatzivassiloglou, V.: Event-based extractive summarization. In: Text Summarization Branches Out, pp. 104–111. Association for Computational Linguistics, Barcelona (2004). https://aclanthology.org/W04-1017
9. Gao, L., Wu, J., Qiao, Z., Zhou, C., Yang, H., Hu, Y.: Collaborative social group influence for event recommendation. In: Proceedings of the 25th ACM International on Conference on Information and Knowledge Management, CIKM 2016, pp. 1941–1944. Association for Computing Machinery, New York (2016). https://doi.org/10.1145/2983323.2983879
10. Hsi, A.: Event Extraction for Document-Level Structured Summarization (2022). https://doi.org/10.1184/R1/21610728.v1. https://kilthub.cmu.edu/articles/thesis/Event_Extraction_for_Document-Level_Structured_Summarization/21610728

11. Huang, K.H., Peng, N.: Document-level event extraction with efficient end-to-end learning of cross-event dependencies. In: Proceedings of the Third Workshop on Narrative Understanding, pp. 36–47. Association for Computational Linguistics, Virtual (2021). https://doi.org/10.18653/v1/2021.nuse-1.4. https://aclanthology.org/2021.nuse-1.4
12. Jacobs, G., Hoste, V.: Extracting fine-grained economic events from business news. In: Proceedings of the 1st Joint Workshop on Financial Narrative Processing and MultiLing Financial Summarisation, pp. 235–245. COLING, Barcelona (Online) (2020). https://aclanthology.org/2020.fnp-1.36
13. Lai, V.D.: Event extraction: a survey (2022)
14. Li, F., et al.: Event extraction as multi-turn question answering. In: Findings of the Association for Computational Linguistics: EMNLP 2020, pp. 829–838. Association for Computational Linguistics, Online (2020). https://doi.org/10.18653/v1/2020.findings-emnlp.73. https://aclanthology.org/2020.findings-emnlp.73
15. Li, Q., et al.: A survey on deep learning event extraction: approaches and applications. IEEE Trans. Neural Netw. Learn. Syst. 1–21 (2022). https://doi.org/10.1109/TNNLS.2022.3213168
16. Li, S., Ji, H., Han, J.: Document-level event argument extraction by conditional generation. CoRR arxiv:2104.05919 (2021)
17. Lin, Y., Ji, H., Huang, F., Wu, L.: A joint neural model for information extraction with global features. In: Proceedings of the 58th Annual Meeting of the Association for Computational Linguistics, pp. 7999–8009. Association for Computational Linguistics, Online (2020). https://doi.org/10.18653/v1/2020.acl-main.713. https://aclanthology.org/2020.acl-main.713
18. Liu, C.Y., Zhou, C., Wu, J., Xie, H., Hu, Y., Guo, L.: Cpmf: a collective pairwise matrix factorization model for upcoming event recommendation. In: 2017 International Joint Conference on Neural Networks (IJCNN), pp. 1532–1539 (2017). https://doi.org/10.1109/IJCNN.2017.7966033
19. Lu, D., Ran, S., Tetreault, J., Jaimes, A.: Event extraction as question generation and answering. In: Proceedings of the 61st Annual Meeting of the Association for Computational Linguistics, vol. 2: Short Papers, pp. 1666–1688. Association for Computational Linguistics, Toronto (2023). https://doi.org/10.18653/v1/2023.acl-short.143. https://aclanthology.org/2023.acl-short.143
20. Lu, J., Tan, X., Pergola, G., Gui, L., He, Y.: Event-centric question answering via contrastive learning and invertible event transformation. In: Findings of the Association for Computational Linguistics: EMNLP 2022, pp. 2377–2389. Association for Computational Linguistics, Abu Dhabi (2022). https://aclanthology.org/2022.findings-emnlp.176
21. Lu, Y., et al.: Text2event: controllable sequence-to-structure generation for end-to-end event extraction (2021)
22. Lu, Y., et al.: Text2Event: controllable sequence-to-structure generation for end-to-end event extraction. In: Proceedings of the 59th Annual Meeting of the Association for Computational Linguistics and the 11th International Joint Conference on Natural Language Processing, vol. 1: Long Papers, pp. 2795–2806 (2021)
23. Lyu, Q., Zhang, H., Sulem, E., Roth, D.: Zero-shot event extraction via transfer learning: challenges and insights. In: Proceedings of the 59th Annual Meeting of the Association for Computational Linguistics and the 11th International Joint Conference on Natural Language Processing, vol. 2: Short Papers (2021)

24. Man Duc Trong, H., Trong Le, D., Pouran Ben Veyseh, A., Nguyen, T., Nguyen, T.H.: Introducing a new dataset for event detection in cybersecurity texts. In: Proceedings of the 2020 Conference on Empirical Methods in Natural Language Processing (EMNLP), pp. 5381–5390. Association for Computational Linguistics, Online (2020). https://doi.org/10.18653/v1/2020.emnlp-main.433. https://aclanthology.org/2020.emnlp-main.433
25. Nguyen, M.V., Lai, V.D., Nguyen, T.H.: Cross-task instance representation interactions and label dependencies for joint information extraction with graph convolutional networks. In: Proceedings of the 2021 Conference of the North American Chapter of the Association for Computational Linguistics: Human Language Technologies, pp. 27–38 (2021). https://doi.org/10.18653/v1/2021.naacl-main.3
26. Nguyen, M.V., Min, B., Dernoncourt, F., Nguyen, T.: Joint extraction of entities, relations, and events via modeling inter-instance and inter-label dependencies. In: Proceedings of the 2022 Conference of the North American Chapter of the Association for Computational Linguistics: Human Language Technologies, United States, pp. 4363–4374 (2022). https://doi.org/10.18653/v1/2022.naacl-main.324
27. Nguyen, T., Grishman, R.: Graph convolutional networks with argument-aware pooling for event detection. In: Proceedings of the AAAI Conference on Artificial Intelligence, vol. 32, no. 1 (2018). https://doi.org/10.1609/aaai.v32i1.12039. https://ojs.aaai.org/index.php/AAAI/article/view/12039
28. Pustejovsky, J.: The syntax of event structure. Cognition **41**(1), 47–81 (1991). https://doi.org/10.1016/0010-0277(91)90032-Y
29. Raffel, C., et al.: Exploring the limits of transfer learning with a unified text-to-text transformer. J. Mach. Learn. Res. **21**(1) (2020)
30. Schiffman, B., McKeown, K., Grishman, R., Allan, J.: Question answering using integrated information retrieval and information extraction. In: Human Language Technologies 2007: The Conference of the North American Chapter of the Association for Computational Linguistics; Proceedings of the Main Conference, pp. 532–539. Association for Computational Linguistics, Rochester (2007). https://aclanthology.org/N07-1067
31. Sha, L., Qian, F., Chang, B., Sui, Z.: Jointly extracting event triggers and arguments by dependency-bridge rnn and tensor-based argument interaction. In: Proceedings of the AAAI Conference on Artificial Intelligence, vol. 32, no. 1 (2018). https://doi.org/10.1609/aaai.v32i1.12034. https://ojs.aaai.org/index.php/AAAI/article/view/12034
32. Sims, M., Park, J.H., Bamman, D.: Literary event detection. In: Proceedings of the 57th Annual Meeting of the Association for Computational Linguistics, pp. 3623–3634. Association for Computational Linguistics, Florence (2019). https://doi.org/10.18653/v1/P19-1353. https://aclanthology.org/P19-1353
33. Sprugnoli, R., Tonelli, S.: Novel event detection and classification for historical texts. Comput. Linguist. **45**(2), 229–265 (2019). https://doi.org/10.1162/coli_a_00347. https://aclanthology.org/J19-2002
34. Subburathinam, A., et al.: Cross-lingual structure transfer for relation and event extraction. In: Proceedings of the 2019 Conference on Empirical Methods in Natural Language Processing and the 9th International Joint Conference on Natural Language Processing (EMNLP-IJCNLP), pp. 313–325. Association for Computational Linguistics, Hong Kong (2019). https://doi.org/10.18653/v1/D19-1030. https://aclanthology.org/D19-1030

35. Tong, M., et al.: Improving event detection via open-domain trigger knowledge. In: Proceedings of the 58th Annual Meeting of the Association for Computational Linguistics, pp. 5887–5897. Association for Computational Linguistics, Online (2020). https://doi.org/10.18653/v1/2020.acl-main.522. https://aclanthology.org/2020.acl-main.522
36. Wan, Q., et al.: Joint document-level event extraction via token-token bidirectional event completed graph. In: Proceedings of the 61st Annual Meeting of the Association for Computational Linguistics, vol. 1: Long Papers, pp. 10481–10492. Association for Computational Linguistics, Toronto (2023). https://doi.org/10.18653/v1/2023.acl-long.584. https://aclanthology.org/2023.acl-long.584
37. Wu, X., Wu, J., Fu, X., Li, J., Zhou, P., Jiang, X.: Automatic knowledge graph construction: a report on the 2019 icdm/icbk contest. In: Wang, J., Shim, K., Wu, X. (eds.) Proceedings - 19th IEEE International Conference on Data Mining, ICDM 2019, pp. 1540–1545. ICDM, Institute of Electrical and Electronics Engineers (IEEE), United States (2019). https://doi.org/10.1109/ICDM.2019.00204
38. Xu, R., Liu, T., Li, L., Chang, B.: Document-level event extraction via heterogeneous graph-based interaction model with a tracker. In: Proceedings of the 59th Annual Meeting of the Association for Computational Linguistics and the 11th International Joint Conference on Natural Language Processing, vol. 1: Long Papers, pp. 3533–3546. Association for Computational Linguistics, Online (2021). https://doi.org/10.18653/v1/2021.acl-long.274. https://aclanthology.org/2021.acl-long.274
39. Yan, H., Jin, X., Meng, X., Guo, J., Cheng, X.: Event detection with multi-order graph convolution and aggregated attention. In: Proceedings of the 2019 Conference on Empirical Methods in Natural Language Processing and the 9th International Joint Conference on Natural Language Processing (EMNLP-IJCNLP), pp. 5766–5770. Association for Computational Linguistics, Hong Kong (2019). https://doi.org/10.18653/v1/D19-1582. https://aclanthology.org/D19-1582
40. You, H., Samuel, D., Touileb, S., Øvrelid, L.: EventGraph: event extraction as semantic graph parsing. In: Proceedings of the 5th Workshop on Challenges and Applications of Automated Extraction of Socio-political Events from Text (CASE), pp. 7–15. Association for Computational Linguistics, Abu Dhabi (Hybrid) (2022). https://aclanthology.org/2022.case-1.2
41. Zhang, Z., Elfardy, H., Dreyer, M., Small, K., Ji, H., Bansal, M.: Enhancing multi-document summarization with cross-document graph-based information extraction. In: Proceedings of the 17th Conference of the European Chapter of the Association for Computational Linguistics, pp. 1696–1707. Association for Computational Linguistics, Dubrovnik (2023). https://aclanthology.org/2023.eacl-main.124
42. Zhu, L., Zheng, H.: Biomedical event extraction with a novel combination strategy based on hybrid deep neural networks. BMC Bioinf. **21**, 1–12 (2020)

Effective and Efficient Transformer Models for Sequential Recommendation

Aleksandr V. Petrov(✉)

University of Glasgow, Glasgow, UK
a.petrov.1@research.gla.ac.uk

Motivation and Background. The focus of our work is *sequential recommender systems*. Sequential recommender systems use ordered sequences of user-item interactions to predict future interactions of the user. The state-of-the-art sequential recommendation models, such as SASRec [6], BERT4Rec [15] are based on the Transformer [16] architecture, which was originally designed for the natural language processing domain. To adapt the Transformer architecture for the sequential recommendation problem, researchers use item IDs in the sequences instead of tokens in the sentences. However, this mapping has a number of limitations that make the use of Transformer-based models problematic in industrial applications.

First, Transformer-based models require a lot of computational resources to train. For example, our experiments with the BERT4Rec model show that training the original implementation of the model on a relatively small MovieLens-1M dataset may require up to 20 h on a modern GPU in order to reproduce results reported in the original paper [15] (see details in our reproducibility paper [10]). In the Natural Language Processing domain, this large training time is less problematic because, usually, Transformer models are trained once on a large corpus of texts and then fine-tuned for every specific task. However, the application of fine-tuning is limited in recommender systems: most recommendation datasets contain a unique set of items (in contrast to the sets of tokens being the same in different Language tasks), so the models trained on different sets of items are incompatible with each other. Therefore, we have to train a new instance of a model for every new dataset from scratch.

Second, the number of items may be several orders of magnitude larger than the vocabulary size of a language model. For example, YouTube has more than 800 million videos available. In contrast, most of the Transformer-based language models, such as BERT [3], have a vocabulary of not more than 100,000 tokens. Finally, while existing generation of Transformer-based models achieves state-of-the-art results for *ranking accuracy*, typically measured by such metrics as NDCG or Recall@K, many researchers recently argued that a good recommender system should also optimise for beyond-accuracy goals [4], such as diversity or novelty. Optimising for beyond-accuracy goals remains unsolved, and most existing solutions rely on greedy reranking techniques, such as Maximal Marginal Reranking [1] or Serendipity Oriented Greedy [7].

Despite these challenges, the progress in language processing models, information retrieval, and machine learning suggests that there can be a path to

N. Goharian et al. (Eds.): ECIR 2024, LNCS 14612, pp. 325–327, 2024.
https://doi.org/10.1007/978-3-031-56069-9_39

resolving the scalability and beyond-accuracy effectiveness issues. For example, Clark et al. [2] showed that choosing an appropriate training objective may significantly reduce required computations for training a Transformer model. Jean et al. [5] showed that language models with a large vocabulary can be trained effectively and efficiently when using a *Negative Sampling* technique together with an adjusted loss function. Zhan et al. [17] demonstrated that the memory footprint of Transformer-based dense retrieval models can be reduced by using *Joing Product Quantisation* (JPQ) technique. Ouyang et al. [8] proposed to use *autoregressive generation* optimised with Reinforcement Learning-based techniques for tuning Transformers for intricate goals. These advancements lead us to the Thesis Statement, which we pose in the next section.

Thesis Statement and Research Questions. The Thesis states that Transformer-based models can be used for large-scale sequential recommender systems efficiently and effectively for both training and inference, even when effectiveness includes beyond-accuracy objectives. In particular, we can make the training of a Transformer-based model more efficient with recency-based sampling. Moreover, we can scale Transformer models to millions of items using negative sampling coupled with an improved loss function to avoid unnecessary computations and, therefore, improve training efficiency without degrading inference effectiveness. Additionally, we can reduce the memory footprint of the model using quantisation techniques for embedding tensor compression, which can also help scale the model to millions of items. Finally, we hypothesise that we can make Transformer-based models effective, even when effectiveness includes beyond-accuracy goals, such as increasing diversity or decreasing popularity bias, by using a generative approach for sequential recommendation and using reinforcement learning to optimise these goals.

To structure our research toward the posed Thesis Statement, we address the following research questions: **RQ1**: Can we improve training time and effectiveness of Transformer-based sequential recommendations by modifying the training task and the loss function? **RQ2**: Can we effectively and efficiently train Transformer-based recommenders with large-catalogue datasets by using negative sampling? **RQ3**: Can the memory footprint of Transformer recommenders with large item catalogues be reduced using JPQ? **RQ4**: Can generative recommendation models match the effectiveness of traditional score-and-rank models for sequential recommendation? **RQ5**: Can generative recommendation models provide a better tradeoff between accuracy and beyond-accuracy goals compared to traditional reranking approaches?

We have already answered some of the research questions in a series of our publications, while others are still in progress. Indeed, we analysed state-of-the-art in paper [10], answered our first research question on training objectives and proposed an efficient and effective training objective in a conference paper [9], which has been extended as a journal paper [13] (**RQ1**). We analysed negative sampling and proposed an effective method for training negatively sampled models in the conference paper [12] (**RQ2**). We showed that item embedding tensor can be compressed without losing model effectiveness [14] (**RQ3**). We demonstrated the effectiveness of generative models in a workshop paper [11] (**RQ4**). The work on other research questions is still in progress or under review. This concludes the brief overview of the research.

References

1. Carbonell, J., Goldstein, J.: The use of MMR, diversity-based reranking for reordering documents and producing summaries. In: SIGIR (1998)
2. Clark, K., Luong, M.T., Le, Q.V., Manning, C.D.: ELECTRA: pre-training text encoders as discriminators rather than generators. In: ICLR (2020)
3. Devlin, J., Chang, M.W., Lee, K., Toutanova, K.: BERT: pre-training of deep bidirectional transformers for language understanding. In: NAACL-HLT (2019)
4. Ge, M., Delgado-Battenfeld, C., Jannach, D.: Beyond accuracy: evaluating recommender systems by coverage and serendipity. In: RecSys (2010)
5. Jean, S., Cho, K., Memisevic, R., Bengio, Y.: On using very large target vocabulary for neural machine translation. In: ACL-IJCNLP (2015)
6. Kang, W.C., McAuley, J.: Self-attentive sequential recommendation. In: ICDM (2018)
7. Kotkov, D., Veijalainen, J., Wang, S.: How does serendipity affect diversity in recommender systems? A serendipity-oriented greedy algorithm. Comput. **102**(2), 393–411 (2018). https://doi.org/10.1007/s00607-018-0687-5
8. Ouyang, L., et al.: Training language models to follow instructions with human feedback (2022)
9. Petrov, A.V., Macdonald, C.: Effective and efficient training for sequential recommendation using recency sampling. In: RecSys (2022)
10. Petrov, A.V., Macdonald, C.: A systematic review and replicability study of BERT4Rec for sequential recommendation. In: RecSys (2022)
11. Petrov, A.V., Macdonald, C.: Generative sequential recommendation with GPTRec. In: Gen-IR@SIGIR (2023)
12. Petrov, A.V., Macdonald, C.: gSASRec: reducing overconfidence in sequential recommendation trained with negative sampling. In: RecSys (2023)
13. Petrov, A.V., Macdonald, C.: RSS: effective and efficient training for sequential recommendation using recency sampling. ACM Trans. Recommender Syst. (2023)
14. Petrov, A.V., Macdonald, C.: RecJPQ: training Large-Catalogue Sequential Recommenders. In: WSDM (2024)
15. Sun, F., et al.: BERT4Rec: sequential recommendation with bidirectional encoder representations from transformer. In: CIKM (2019)
16. Vaswani, A., et al.: Attention is All you Need. In: NeurIPS (2017)
17. Zhan, J., Mao, J., Liu, Y., Guo, J., Zhang, M., Ma, S.: Jointly optimizing query encoder and product quantization to improve retrieval performance. In: CIKM (2021)

Reproduction and Simulation of Interactive Retrieval Experiments

Jana Isabelle Friese(✉)

University of Duisburg-Essen, Duisburg, Germany
jana.friese@uni-due.de

1 Motivation

The reproducibility crisis, spanning across various scientific fields, substantially affects information retrieval research [1]. In interactive IR, the more active role of the user and additional interactions pose further challenges to reproducibility [8]. User experiments provide user-centric evaluation, but given the unpredictable nature of users, abstracting user behavior is crucial for reproducible results [9].

User simulation offers a controlled, cost-effective alternative to real-world testing, eliminating user-induced unpredictability. Still, the efficacy of simulations depends on their realism [2]. Current simulations consider a limited set of interactions, disregarding functions with a real-world impact on user experiences.

Markov models provide a means of capturing these additional functionalities and extending simulations accordingly. Through Markov models, user interactions and their transitions can be modeled based on observed behavior, facilitating realistic predictions [14]. Realistic interaction models enhance simulations, improve reproducibility and can be employed to assess reproduction quality.

2 Related Work

Reproducibility. PRIMAD is a framework for reproducibility, outlining the components of a standard IR experiment [6]. This model has been expanded to PRIMAD-U for a more user-oriented evaluation by introducing a distinct user component, recognizing the user's vital role in interactive IR [5].

Petras and colleagues identified 17 reporting elements for IIR studies and judged current metadata practices as an inadequate reporting standard [12].

To the best of our knowledge, there have been no attempts to reproduce interactive information retrieval experiments.

Reproducibility Measures. Identifying a successful reproduction requires a clear definition of success. Maistro and colleagues examined common reproducibility measures and highlighted challenges in quantifying reproducibility. They observed that a single score is insufficient for identifying successful reproductions, and the choice of measures should depend on the specific goal [10].

N. Goharian et al. (Eds.): ECIR 2024, LNCS 14612, pp. 328–330, 2024.
https://doi.org/10.1007/978-3-031-56069-9_40

Interaction Modeling. As a quantitative model for interactive IR, IPRP extends the Probability Ranking Principle, considering user interactions and dynamic information needs, assessing the costs and benefits of decisions in search situations [7]. However, statements about user behavior remain theoretical.

For practical application, Markov models can be used. Vu Tran developed Markov models for various task types and user groups by analyzing retrieval sessions. These models were used for evaluation and simulation [13].

Simulation. Balog and colleagues provided a comprehensive overview of simulation in IR, viewing user simulation as a tool to enhance evaluations and possibly the only way to achieve reproducibility for interactive experiments. [3].

To determine if simulations replicate real interactions, Breuer and colleagues validated query simulations [4].

A practical toolkit for user simulation in interactive retrieval is SimIIR [11], further enhanced in SimIIR 2.0 with features like Markov model-based interactions and dynamic, user type-specific simulations [15].

3 Research Goals and Methodology

We aim to address the following research questions:

RQ1.1: What are the key elements in creating reliable user models for accurate reproductions in IIR?
RQ1.2: How can corresponding user models be leveraged for valid simulations of user behavior?
RQ2: What are the benefits and limitations of capturing previously disregarded functionalities by Markov models, and how do they impact the reproducibility of IIR experiments?
RQ3: How do user simulations compare to real users, and how can such comparisons be conducted using accordingly extended quantitative user models?
RQ4: How can models that are tailored for different user types improve the realism of simulations and their ability to reproduce real user behavior?

To answer RQ1.1 and RQ1.2, we divide logged session data into randomly stratified groups, one representing the original and the other representing a reproduction, and determine if one group reproduces the other. We create Markov models for both groups, comparing model parameters and retrieval results to identify crucial elements for reproduction. Using data from different experiments enables us to assess the reproducibility of the initial results and identify cross-experiment model elements crucial for reliably describing user behavior.

Subsequently, to answer RQ2 and RQ3, we use these models as a basis for user simulation, including often disregarded functionalities. By reproducing the original experiment with simulated user interactions, we can demonstrate the suitability of the models for simulating human-like interactions. This allows us to assess how simulations compare to user experiments and how such comparisons should be conducted.

The variability in user behavior hinders a one-size-fits-all approach. To address this issue, models can be customized for different user types or even individuals. This enhances realism and facilitates comparisons based on altered user behavior. Building on Tran's work [13], we seek to answer RQ4 by investigating the impact of incorporating different user types on the reproducibility and simulation of IIR experiments.

References

1. Baker, M.: 1,500 Scientists lift the lid on reproducibility. Nature **533**,(7604) (2016)
2. Balog, K., Maxwell, D., Thomas, P., Zhang, S.: Report on the 1st simulation for information retrieval workshop (Sim4IR 2021) at SIGIR2021. In: ACM SIGIR Forum vol. 55, no. 2 (2022)
3. Balog, K., Zhai, C.: User Simulation for Evaluating Information Access Systems. arXiv preprint arXiv:2306.08550 (2023)
4. Breuer, T., Fuhr, N., Schaer, P.: Validating simulations of user query variants. In: Hagen, M., Verberne, S., Macdonald, C., Seifert, C., Balog, K., Nørvåg, K., Setty, V. (eds.) ECIR 2022. LNCS, vol. 13185, pp. 80–94. Springer, Cham (2022). https://doi.org/10.1007/978-3-030-99736-6_6
5. Breuer, T.: Reproducible Information Retrieval Research: From Principled System-Oriented Evaluations Towards User-Oriented Experimentation, University of Duisburg-Essen (2023)
6. Ferro, N., Fuhr, N., Jarvelin, K., Kando, N., Lippold, M., Zobel, J.: Increasing reproducibility in IR: findings from the dagstuhl seminar on" reproducibility of data-oriented experiments in e-science. In: ACM SIGIR Forum. vol. 50, no. 1, pp. 68–82. ACM, New York, NY, USA (2016)
7. Fuhr, N.: A probability ranking principle for interactive information retrieval. Inf. Retrieval **11**(3), 251–265 (2008)
8. Kelly, D.: Methods for evaluating interactive information retrieval systems with users. Found. Trends in Inf. Retrieval **3**(1–2), 1–224 (2009)
9. Liu, J., Shah, C.: Interactive IR user study design, evaluation, and reporting. Morgan & Claypool Publishers (2019)
10. Maistro, M., Breuer, T., Schaer, P., Ferro, N.: An in-depth investigation on the behavior of measures to quantify reproducibility. Inf. Process.& Manage **60**(3), 103332 (2023)
11. Maxwell, D., Azzopardi, L.: Simulating interactive information retrieval: SimIIR: A framework for the simulation of interaction. In: Proceedings of the 39th International ACM SIGIR conference on Research and Development in Information Retrieval, SIGIR 2016, pp. 1141–1144. ACM, Pisa, Italy (2016)
12. Petras, V., Bogers, T., Gade, M.: Elements of IIR studies: a review of the 2006-2018 IIiX and CHIIR conferences. In: CEUR Workshop Proceedings, vol. 2337, pp. 37–41 (2019)
13. Tran, V.: Modellierung von Benutzerverhalten im interaktiven Retrieval mit Markov-Ketten, University of Duisburg-Essen (2020)
14. Yang, G.H., Dong, X., Luo, J., Zhang, S.: Session search modeling by partially observable Markov decision process. Inform. Retrieval J. **21**(1), 56–80 (2017). https://doi.org/10.1007/s10791-017-9316-8
15. Zerhoudi, S., et al.: The SimIIR 2.0 framework: user types, markov model-based interaction simulation, and advanced query generation. In: Proceedings of the 31st ACM International Conference on Information and Knowledge Management, pp. 4661–4666 (2022)

Cascading Ranking Pipelines for Sensitivity-Aware Search

Jack McKechnie(✉)

University of Glasgow, Glasgow, Scotland
j.mckechnie.1@research.gla.ac.uk

Abstract. Search engines are designed to make information accessible. However, some information should not be accessible, such as documents concerning citizenship applications or personal information. This *sensitive information* is often found interspersed with other potentially useful non-sensitive information. As such, collections containing sensitive information cannot be made searchable due to the risk of revealing sensitive information. The development of search engines capable of safely searching collections containing sensitive information to provide relevant and *non-sensitive* information would allow previously hidden collections to be made available. This work aims to develop sensitivity-aware search engines via two-stage cascading retrieval pipelines.

1 Motivation

It is in the interest of organisations to make the documents that they create publicly available to build trust and public image [2,6]. However, the public release of sensitive information is highly undesirable, for example, due to privacy or security concerns. Sensitive information is often unstructured, free-text information that is interspersed with useful non-sensitive information. This, coupled with the contextual nature of sensitivities makes automatic sensitivity detection a challenging task, traditionally taking expert-level human judgement. Digital humanities scholars and the UK government have identified that the inability to release collections that potentially contain sensitive information hinders access and innovation in multiple sectors, impeding economic growth and research output [3–5].

Sensitivity-aware search (SAS), the integration of sensitive information identification into a search engine's retrieval model, would allow document collections that potentially contain sensitive information to be directly indexed and searched without exposing any sensitive information. In SAS, the retrieval model ranks documents by balancing relevance and sensitivity such that the user's information need is satisfied without revealing sensitive information.

2 Background and Related Work

SAS aims to allow users to search collections containing both sensitive and non-sensitive documents without sensitive information being revealed [7]. Thus, SAS

N. Goharian et al. (Eds.): ECIR 2024, LNCS 14612, pp. 331–333, 2024.
https://doi.org/10.1007/978-3-031-56069-9_41

seeks to appropriately balance relevance and sensitivity such that the *information provider* (IP) is confident enough that sensitive information will not be revealed and the *information consumer* (IC) has their information need satisfied.

Sayed et al. [9] proposed a SAS model by optimising a learning-to-rank (LTR) model for a SAS metric. However, the model proposed by Sayed et al. [9] uses LTR algorithms and does not leverage the advances that neural IR has brought to traditional search tasks, which we aim to do via cascading ranking pipelines.

3 Proposed Research

The proposed research involves the development of a multi-stage pipeline for SAS. A cascading pipeline prepares the ranking for the subsequent stage, culminating in a high-quality ranking satisfying the IP and IC. An initial ranker is augmented with sensitivity-aware pseudo-relevance feedback (SA-PRF), increasing the retrieval of relevant non-sensitive documents. This ranking is then fed to a sensitivity-aware reranker. At each stage risk and utility can be calculated. The research to be carried out aims to answer the following three research questions.

RQ1: Can Recall of Sensitive Documents be Decreased via SA-PRF Feedback Approaches? In the ad-hoc search task, PRF is used to augment queries to help overcome vocabulary mismatch. By developing solutions to rewrite queries using terms/embeddings that are indicative of relevant non-sensitive documents, we hypothesise that recall of relevant non-sensitive documents can be increased.

RQ2: Can an Initial Ranking be Reranked to Maximise the Number of Relevant Documents Shown at a Ranking Depth k, Without Revealing Sensitive Documents? Given an initial ranking, expensive rerankers are frequently used to improve the quality of the top k ranking [1]. We propose to develop rerankers that consider both relevance and sensitivity to return the most relevant non-sensitive documents in the top k as possible. We have built a neural reranker that models both relevance and sensitivity simultaneously via a multi-task learning approach that outperforms the LTR approach from Sayed et al. [9].

RQ3: Does a Cascading Retrieval Pipeline Architecture Provide a Trade-off Satisfactory to Both the Information Consumer and Provider? Cascading retrieval pipelines are used to apply increasingly expensive models to a decreasing number of documents. Using inexpensive first-stage rankers, augmented with the proposed SA-PRF approaches, and the (more expensive) reranking approaches developed, we propose to build a sensitivity-aware cascading retrieval pipeline. Each stage prepares the ranking for the next stage to culminate in a final ranking that is within the risk tolerance of the IP and satisfies the information need of the IC. Risk and utility can be measured at each stage to gather an overall view of how the ranking satisfies each stakeholder.

4 Research Methodology

The development of SAS engines requires test collections that are labelled for both relevance and different sensitivities. Two such collections exist; SARA [10], which we have developed, and Avocado [8]. RQs 1 and 2 are being investigated by carrying out experiments to test how the strategies perform compared to their relevance-only equivalents and to filtering classified sensitive documents from their rankings. Finally, RQ3 is being investigated by combining the approaches proposed into one pipeline. The ability of this pipeline to satisfy both the IP and the IC will be verified with SAS metrics and a user study.

References

1. Asadi, N., Lin, J.: Effectiveness/efficiency tradeoffs for candidate generation in multi-stage retrieval architectures. In: Proceedings of SIGIR, pp. 997–1000 (2013)
2. Cossins, A.: Revisiting open government: recent developments in shifting the boundaries of government secrecy under public interest immunity and freedom of information law. Federal Law Rev. **23**(2), 226–276 (1995)
3. Department for Science, Innovation and Technology, Department for Business and Trade, Office for Artificial Intelligence, Department for Digital, Culture, Media & Sport, and Department for Business, Energy & Industrial Strategy. Artificial intelligence sector deal. https://www.gov.uk/government/publications/artificial-intelligence-sector-deal (2019)
4. Jaillant, L., Caputo, A.: Unlocking digital archives: cross-disciplinary perspectives on ai and born-digital data. AI Soc. **37**(3), 823–835 (2022)
5. Jaillant, L., Rees, A.: Applying AI to digital archives: trust, collaboration and shared professional ethics. DSH **38**(2), 571–585 (2023)
6. Kang, J., Hustvedt, G.: Building trust between consumers and corporations: the role of consumer perceptions of transparency and social responsibility. J. Bus. Ethics **125**, 253–265 (2014)
7. Oard, D.W., Shilton, K., Lin, J.: Evaluating search among secrets. In: Proceedings of EVIA@NTCIR (2016)
8. Sayed, M.F., et al.: A test collection for relevance and sensitivity. In: Proceedings of SIGIR, pp. 1605–1608 (2020)
9. Sayed, M.F., Oard, D.W.: Jointly modeling relevance and sensitivity for search among sensitive content. In: Proceedings of SIGIR, pp. 615–624 (2019)
10. McKechnie, J., McDonald, G.: Sara: a collection of sensitivity-aware relevance assessments (2024). arxiv.org/abs/2401.05144

Analyzing Mathematical Content for Plagiarism and Recommendations

Ankit Satpute[1,2](✉)

[1] Georg August University of Göttingen, Göttingen, Germany
[2] FIZ Karlsruhe Leibniz Institute for Information Infrastructure, Berlin, Germany
Ankit.Satpute@fiz-karlsruhe.de

1 Motivation

Defined as "the use of ideas, concepts, words, or structures without appropriately acknowledging the source to benefit in a setting where originality is expected" [6], plagiarism poses a severe concern in the rapidly increasing number of scientific publications. The Vroniplag has documented plagiarism in 212 dissertations [19], and zbMATH Open pointed plagiarised research papers in mathematics [17]. The easily recognizable copy-paste type plagiarism [12] will likely diminish due to the accessibility of AI-powered models like ChatGPT. Plagiarism among researchers is more concealed, which is challenging for existing Plagiarism Detection Systems (PDS) [1]. Scientific research often builds upon the foundations laid by existing literature, discovering similar works as an integral part of research. Recommender systems (RS) assist users in coping with many scientific documents by showing similar ones the user might be interested in. RS has become a crucial filtering and discovery tool that many users of digital libraries rely on.

Even though both PDS and RS have different final objectives, they share the goal of finding similar content. However, existing PDS and RS focus primarily on textual content and do not utilize non-textual elements, specifically mathematical content, to their full potential [7,16]. The significance of mathematical content is much higher and valued in scientific documents from STEM (Science, Technology, Engineering, and Mathematics). Despite this, efforts to utilize mathematical content for document similarity remain in infancy. This thesis addresses this gap by analyzing and utilizing mathematical content for content similarity.

2 Background and Related Work

There are two main reasons for the lack of methods utilizing mathematical content to find similar scientific documents. *Reason 1* is the unavailability of large-scale, annotated datasets of similar mathematical content in a machine-processable format like LaTeX or MathML. PAN Datasets [18] are a frequently used resource to develop and evaluate PDS [2,5,15,20]. Gienapp et al. [8] presented the Webis-STEREO-21 dataset containing reused text passages (without math) from scientific publications. For plagiarised mathematical content,

N. Goharian et al. (Eds.): ECIR 2024, LNCS 14612, pp. 334–337, 2024.
https://doi.org/10.1007/978-3-031-56069-9_42

resources as comprehensive as those for text reuse are missing. There are Math Information Retrieval (MIR) datasets such as NTCIR [21] and ARQMath [11], but they have very limited similar math content pairs. No existing RS datasets consider mathematical contents.

Reason 2 is the sole analysis of the presentational similarity of mathematical content, such as matching math symbol occurrences [13,14]. Considering presentational mathematical similarity at all is a valuable starting point, but there is a need to develop advanced methods analyzing the semantics of mathematical formulae [17]. Moreover, current approaches analyze text and mathematical formulae separately. However, identifying semantic similarity of mathematical content requires a combined analysis because mathematical formulae are mostly context-dependent [9]. Only two works analyzed mathematical content in scientific documents to identify plagiarism [13,14]. Both studied primary mathematical symbol occurrences and used a small evaluation dataset of 10 document pairs. No PDS thus far considers semantic textual and non-textual content similarity [7,10]. Search engines, such as Searchonmath [4], Approach0 [24], etc., are tailored towards mathematical formulae. Even though they allow textual and mathematical terms for searching, the content is not considered semantically. Language models are considered for finding math similarities [3,22,23]. Most of these language models are not trained on mathematical content, thus questioning the extensibility of these approaches to consider math semantic similarity.

3 Proposed Research

The objective of this doctoral thesis is to:

Conceive, devise, and evaluate robust approaches for math content similarity capable of identifying obfuscated plagiarism and generating relevant recommendations for scientific documents.

To achieve this objective, we will perform the following research tasks:

- T1: Investigate the strengths and limitations of state-of-the-art mathematical content similarity detection approaches.
- T2: Formulate features of mathematical contents, develop and evaluate detection approaches for locating semantically and syntactically similar math.
- T3: Devise a similarity assessment that combines text, math, and citations to detect similar scientific documents.
- T4: Implement a PDS and an RS with the best-performing developed approach to demonstrate its applicability in a real-world document collection.
- T5: Evaluate the proposed approach by assessing the implemented PDS and RS's effectiveness, computational efficiency, and usability.

Acknowledgement. This work is funded by the Deutsche Forschungsgemeinschaft (DFG, German Research Foundation) - 437179652 and the Deutscher Akademischer Austauschdienst (DAAD, German Academic Exchange Service - 57515245).

References

1. Alzahrani, S.M., Salim, N., Abraham, A.: Understanding plagiarism linguistic patterns, textual features, and detection methods. IEEE Trans. Syst. Man Cybern. Part C Appl. Rev. **42**(2), 133–149 (2012). https://doi.org/10.1109/TSMCC.2011.2134847
2. Arabi, H., Akbari, M.: Improving plagiarism detection in text document using hybrid weighted similarity. Expert Syst. Appl. **207**, 118034 (2022)
3. Dadure, P., Pakray, P., Bandyopadhyay, S.: BERT-based embedding model for formula retrieval. In: CLEF (Working Notes), pp. 36–46 (2021)
4. Diaz, Y., Nishizawa, G., Mansouri, B., Davila, K., Zanibbi, R.: The mathdeck formula editor: interactive formula entry combining latex, structure editing, and search. In: Extended Abstracts of the 2021 CHI Conference on Human Factors in Computing Systems, pp. 1–5 (2021)
5. El-Rashidy, M.A., Mohamed, R.G., El-Fishawy, N.A., Shouman, M.A.: Reliable plagiarism detection system based on deep learning approaches. Neural Comput. Appl. **34**(21), 18837–18858 (2022)
6. Fishman, T.: We know it when we see it is not good enough: toward a standard definition of plagiarism that transcends theft, fraud, and copyright (2009)
7. Foltýnek, T., Meuschke, N., Gipp, B.: Academic plagiarism detection: a systematic literature review. ACM Comput. Surv. **52**(6), 1–42 (2020). https://doi.org/10.1145/3345317
8. Gienapp, L., Kircheis, W., Sievers, B., Stein, B., Potthast, M.: A large dataset of scientific text reuse in open-access publications. Sci. Data **10**(1), 58 (2023). https://doi.org/10.1038/s41597-022-01908-z
9. Greiner-Petter, A., Schubotz, M., Breitinger, C., Scharpf, P., Aizawa, A., Gipp, B.: Do the math: making mathematics in Wikipedia computable. IEEE Trans. Pattern Anal. Mach. Intell. **45**(4), 4384–4395 (2022). https://doi.org/10.1109/TPAMI.2022.3195261
10. Lovepreet, Gupta, V., Kumar, R.: Survey on plagiarism detection systems and their comparison. In: Behera, H.S., Nayak, J., Naik, B., Pelusi, D. (eds.) Computational Intelligence in Data Mining: Proceedings of the International Conference on ICCIDM 2018, pp. 27–39. Springer, Singapore (2020). https://doi.org/10.1007/978-981-13-8676-3_3
11. Mansouri, B., Agarwal, A., Oard, D.W., Zanibbi, R.: Advancing math-aware search: the ARQMath-3 lab at CLEF 2022. In: Hagen, M., et al. (eds.) ECIR 2022. LNCS, vol. 13186, pp. 408–415. Springer, Cham (2022). https://doi.org/10.1007/978-3-030-99739-7_51
12. McCabe, D.L.: Cheating among college and university students: a North American perspective. Int. J. Educ. Integrity **1**(1) (2005). https://doi.org/10.21913/IJEI.v1i1.14
13. Meuschke, N., Schubotz, M., Hamborg, F., Skopal, T., Gipp, B.: Analyzing mathematical content to detect academic plagiarism. In: Proceedings of the International Conference on Information and Knowledge Management (CIKM) (2017). https://doi.org/10.1145/3132847.3133144
14. Meuschke, N., Stange, V., Schubotz, M., Kramer, M., Gipp, B.: Improving academic plagiarism detection for stem documents by analyzing mathematical content and citations. In: Proceedings of the ACM/IEEE Joint Conference on Digital Libraries (JCDL) (Jun 2019). https://doi.org/10.1109/JCDL.2019.00026

15. Potthast, M., Stein, B., Eiselt, A., Barrón-Cedeño, A., Rosso, P.: PAN plagiarism corpus 2011 (PAN-PC-11) (Jun 2011). https://doi.org/10.5281/zenodo.3250095, https://doi.org/10.5281/zenodo.3250095
16. Scharpf, P., Mackerracher, I., Schubotz, M., Beel, J., Breitinger, C., Gipp, B.: AnnoMathTex - a formula identifier annotation recommender system for stem documents. In: Proceedings of the 13th ACM Conference on Recommender Systems (RecSys 2019). ACM, Copenhagen, Denmark (Sept 2019). https://doi.org/10.1145/3298689.3347042
17. Schubotz, M., Teschke, O., Stange, V., Meuschke, N., Gipp, B.: Forms of plagiarism in digital mathematical libraries. In: Intelligent Computer Mathematics - 12th International Conference, CICM 2019, Prague, Czech Republic, July 8–12, 2019, Proceedings (2019). https://doi.org/10.1007/978-3-030-23250-4_18
18. Stein, B., Koppel, M., Stamatatos, E.: Plagiarism analysis, authorship identification, and near-duplicate detection PAN'07. ACM SIGIR Forum **41**(2), 68–71 (2007). https://doi.org/10.1145/1328964.1328976
19. Weber-Wulff, D.: Talking to a wall: the response of German universities to documentations of plagiarism in doctoral theses. In: Bjelobaba, S., Foltýnek, T., Glendinning, I., Krásničan, V., Dlabolová, D.H. (eds.) Academic Integrity: Broadening Practices, Technologies, and the Role of Students: Proceedings from the European Conference on Academic Integrity and Plagiarism 2021, pp. 363–371. Springer, Cham (2022). https://doi.org/10.1007/978-3-031-16976-2_20
20. Yu, W., Pang, L., Xu, J., Su, B., Dong, Z., Wen, J.R.: Optimal partial transport based sentence selection for long-form document matching. In: Proceedings of the 29th International Conference on Computational Linguistics, pp. 2363–2373 (2022)
21. Zanibbi, R., Aizawa, A., Kohlhase, M., Ounis, I., Topic, G., Davila, K.: NTCIR-12 MathIR task overview. In: NTCIR (2016)
22. Zhong, W., Xie, Y., Lin, J.: Applying structural and dense semantic matching for the ARQMath lab 2022, CLEF. Proc. Working Notes CLEF **2022**, 5–8 (2022)
23. Zhong, W., Yang, J.H., Lin, J.: Evaluating token-level and passage-level dense retrieval models for math information retrieval. arXiv preprint arXiv:2203.11163 (2022)
24. Zhong, W., Zhang, X., Xin, J., Zanibbi, R., Lin, J.: Approach zero and Anserini at the CLEF-2021 ARQMath track: Applying substructure search and BM25 on operator tree path tokens. In: Proceedings CLEF 2021 (CEUR Working Notes) (2021)

Shuffling a Few Stalls in a Crowded Bazaar: Potential Impact of Document-Side Fairness on Unprivileged Info-Seekers

Seán Healy(✉)

ADAPT Centre, Dublin City University, Dublin, Ireland
sean.healy@adaptcentre.ie

Abstract. Information systems rely on algorithmic ranking to ascertain expected relevance. Concerns about this strategy have resulted in the emergence of a field of inquiry referred to as *fair ranking.* Within this field, the aim varies between one-sided and two-sided fairness across automatically generated rankings. But research has focused primarily on fairness among document providers as opposed to fairness among searchers. Concerns have already been raised about the present framing of fairness. In the following line of research, a novel framing concern is introduced, whereby researchers may fail to consider the broader context of search engine usage among protected groups of searchers.

Keywords: Fairness · Exposure · Privilege

An algorithmic ranking system—without the fairness interventions of the past decade, e.g. [1,7]—brings immense value to those seeking information, and scarce, narrowly distributed value to those providing the information, most of whom are buried beyond the end of the ranking. Given that only a small handful of items can appear on a limited display area, it's *winner takes all*, leading to knock-on fairness issues for document providers. There has been a strong push towards provider-side fairness over the past several years, most recently producing NTCIR's Fair Web tasks, and many different strands of fairness research, outlined in surveys, papers and book chapters.

Orthogonal to this push, concerns have been raised regarding the *framing* of fair ranking problems [3,6]. My research adds another voice in that chorus of framing concerns. Here, the concern relates to real-world identity-based privilege, and the related consequences, when utility-lowering interventions, despite what simple metrics may indicate, impact users unequally in a broader context. I introduce readers to this notion of *unequal impact* through the image of a bazaar.

Fair search efforts to date have often aimed to address fairness for both the searchers and providers of documents, but only within the scope of a centrally controlled search system. What if that centralised system is just one modestly sized stall in an entire *bazaar* of information seeking services and toolboxes?

N. Goharian et al. (Eds.): ECIR 2024, LNCS 14612, pp. 338–340, 2024.
https://doi.org/10.1007/978-3-031-56069-9_43

Furthermore, what if the remaining stalls forego fairness interventions? In this scenario, those with the greatest social capital within the 'bazaar' would be least affected by the lowered utility, whereas those with lowest social capital would have no choice but to accept the lower utility at the standalone, fair stall.

My work considers the state of affairs before highly accurate, universally available search solutions entered the information seeking toolbox of the masses. Before this time, finding good quality information was still possible, but only for the greatly privileged. Without a line of inquiry into searcher impact in utility-lowering measures, some services may be modified in such a way that pushes things back into this earlier state of affairs: wherein the marginalised have less access to the highest quality information, and the powerful, with stronger positioning in information networks, experience a much lighter impact once fairness-for-providers negatively impacts utility.

Proposals for provider-side fairness interventions are plentiful. Many of these are outlined in a recent survey [9], which classifies fairness efforts into various distributive justice philosophies, ranging from *Formal or meritocratic equal opportunity* (Formal EO) through Luck-Egalitarian EO, described by Roemer as *"more egalitarianism than utilitarianism, less egalitarianism than Rawls"* [5, p. 460], to Rawlsian EO [4].

Under some EO frameworks, utility reductions are inevitable in these efforts; some drop in utility is permitted in order to enable the fairness sought for document providers. In previous work which explicitly considered utility to various protected groups, i.e. multi-stakeholder work, equal levels of utility (or utility drop) are generally considered an observation of fairness at work. This is the case, for example, in personalised music recommendation, where fairness to a diverse set of listeners is one of the goals [2,8]. However, after all fairness interventions are applied, work to date has failed to consider the contextualised impact of search quality drop given different backgrounds and privileges.

A system-wide drop in utility for a childcare search engine, for e.g., may be of far greater detriment to single expat mothers as compared to a native married couple. Failing to consider such cases would constitute a *framing trap* [3,6]. My research investigates the possibility that marginalised users may rely much more on gratis search tools, by reason of lower social capital in a dominant culture. If this is the case, fairness interventions focussing excessively on document providers, to the detriment of marginalised searchers, might be considered a new form of *poverty premium.*

Research Questions

RQ1 *"Under what minimal set of assumptions does fairness to document providers lead to negative outcomes for a protected group of information seekers?"*

RQ2 *"What evidence is there to reject or support the validity of these assumptions from* **RQ1***?"*

References

1. Biega, A.J., Gummadi, K.P., Weikum, G.: Equity of attention: amortizing individual fairness in rankings. In: The 41st International ACM SIGIR Conference on Research & Development in Information Retrieval, pp. 405–414 (2018)
2. Dinnissen, K., Bauer, C.: Fairness in music recommender systems: a stakeholder-centered mini review. Front. Big Data **5**, 319608 (2022)
3. de Jonge, T., Hiemstra, D.: UNFair: Search engine manipulation, undetectable by amortized inequity. In: Proceedings of the 2023 ACM Conference on Fairness, Accountability, and Transparency, pp. 830–839. FAccT '23, Association for Computing Machinery, New York, NY, USA (2023). https://doi.org/10.1145/3593013.3594046,
4. Rawls, J.: A theory of justice: Revised edition. Harvard University Press (1999)
5. Roemer, J.E.: Equality of opportunity: A progress report. Social Choice and Welfare, pp. 455–471 (2002)
6. Selbst, A.D., Boyd, D., Friedler, S.A., Venkatasubramanian, S., Vertesi, J.: Fairness and abstraction in sociotechnical systems. In: Proceedings of the Conference on Fairness, Accountability, and Transparency, pp. 59–68. FAT* '19, Association for Computing Machinery, New York, NY, USA (2019)
7. Singh, A., Joachims, T.: Fairness of exposure in rankings. In: Proceedings of the 24th ACM SIGKDD International Conference on Knowledge Discovery and Data Mining, pp. 2219–2228. KDD '18, Association for Computing Machinery, New York, NY, USA (2018)
8. Wu, H., Ma, C., Mitra, B., Diaz, F., Liu, X.: A multi-objective optimization framework for multi-stakeholder fairness-aware recommendation. ACM Trans. Inf. Syst. **41**(2), 1–29 (dec 2022). https://doi.org/10.1145/3564285
9. Zehlike, M., Yang, K., Stoyanovich, J.: Fairness in ranking. ACM Comput. Surv. **55**(6) (2022). https://doi.org/10.1145/3533379

Knowledge Transfer from Resource-Rich to Resource-Scarce Environments

Negin Ghasemi(✉)

Radboud University, Nijmegen, The Netherlands
negin.ghasemitaheri@ru.nl

1 Motivation

Resource-scarce environments have limited data, creating barriers and suboptimal experiences for users, while resource-rich environments are well-stocked with comprehensive information. Disparities in resource availability across digital platforms present significant challenges for users seeking valuable information. My research addresses the critical need to enhance information retrieval in resource-scarce environments by exploring innovative scenarios and approaches for knowledge transfer. Two key scenarios investigated include unified mobile app search, addressing dominance issues among certain apps, and resource-scarce e-commerce marketplaces, where question answering needs improvement. The research also discusses challenges in creating new dataset labels with partial judgments and proposes addressing them through automatic labeling.

2 Related Work

Target Apps Selection. Prior research on unified mobile search focuses on creating a single channel for user access to various mobile applications, aiming to identify the most relevant apps for user queries [2]. The initial step involves selecting target apps, similar to the resource selection task in federated search [10]. However, mobile app environments pose challenges due to heterogeneity, uncooperativeness, and data sparsity, leading to biased target app selection [1,2]. Recent research [2,3] has emphasized modeling target app selection as a query classification problem, with various models developed to address data sparsity concerns. In our research, we frame the target apps selection as a ranking problem, utilizing a BERT-based ranker to overcome data sparsity and represent uncooperative resources.

Product-Related Question Answering. Previous works on product-related question answering (PQA) in e-commerce focus on answering consumer queries using resources like catalogs, reviews, and Q&A sections. PQA has received extensive research attention [8], categorized into abstraction, extraction, and retrieval-based methods, with recent emphasis on the latter [7]. Retrieval-based

N. Goharian et al. (Eds.): ECIR 2024, LNCS 14612, pp. 341–344, 2024.
https://doi.org/10.1007/978-3-031-56069-9_44

methods use past questions to assess review relevance [8], evolving to incorporate pre-trained language models like BERT for improved answerability [11,12]. While there is conducted work closely aligned with our research [9], they concentrate on products within a single marketplace. Our work centers on parallel marketplaces in different countries.

3 Proposed Research and Methodology

How Can We Leverage the Success of Applying LLMs to Different Ranking Tasks in Target Apps Selection to Decrease the Bias Towards Resource-Rich Apps? As voice-based search interfaces gain popularity, the traditional format of using search browser apps has evolved into a unified mobile search framework. The critical challenge is identifying the most relevant target apps for queries. Previous research suggests generating resource descriptions based on apps' previous queries [5]. However, varying data availability leads to sparse representations and biased selection toward popular apps. I started this research by applying BERT as a re-ranker to address these limitations, with initial findings published in [5]. The research demonstrates that pre-trained transformers, like BERT, can mitigate bias and outperform other models in predicting less popular apps, thus improving the quality and diversity of app selection.

How Can We Leverage the Unique Features of Cross-Market Items to Enhance Product-Related Question Answering in Resource-Scarce Markets? Online shops like Amazon, eBay, and Etsy are expanding globally, encountering limited user interaction data in new resource-scarce markets, resulting in products with few reviews and unanswered questions. Addressing this challenge, the paper [4] introduces Cross-Market Question Answering, which involves finding relevant answers to questions posted in resource-scarce main marketplaces utilizing data from resource-rich auxiliary marketplaces. I started working on this area by creating a large-scale Q&A dataset, XMarket-QA, based on Amazon's US and UK marketplaces, revealing a significant temporal gap between the first answered question in the UK compared to the US. Moreover, the majority of questions in the US are posted before the first question is answered in the UK, highlighting the potential of leveraging data from the resource-rich marketplace when data is scarce in another marketplace. Two approaches are explored to enhance product-related question answering: identifying similar questions for identical items in the auxiliary marketplace and jointly ranking questions for similar items in the main and auxiliary marketplaces. The proposed model, CMJim, outperforms baselines, emphasizing the effectiveness of leveraging resource-rich marketplace data even with limited auxiliary data.

How Can We Leverage the LLMs to Deal with Lack of Judgment in Newly Created Datasets for Resource-Scarce Environments? Creating new dataset labels poses various challenges that can significantly impact the dataset's quality, reusability, and scalability. Annotating datasets is a resource-intensive task, requiring human expertise and time. Existing datasets often have partial relevance labels. However, Previous work demonstrates that with more available judgment comes a better judgment of a new model, and there is a lower risk of the dataset becoming outdated [6]. I aim to incorporate automatic labeling models [8, 11, 12] along with newer LLMs to automate part of the labeling process in dataset creation. The benefits of automatic labeling include cost-efficiency for large-scale tasks, crucial in academia with limited budgets, time efficiency, and scalability, particularly when leveraging open-source LLMs. The approach aims to optimize resource usage while achieving research objectives effectively.

References

1. Aliannejadi, M., Zamani, H., Crestani, F., Croft, W.B.: In situ and context-aware target apps selection for unified mobile search. In: Proceedings of the 27th ACM International Conference on Information and Knowledge Management, CIKM 2018, pp. 1383–1392. Association for Computing Machinery, New York (2018)
2. Aliannejadi, M., Zamani, H., Crestani, F., Croft, W.B.: Target apps selection: towards a unified search framework for mobile devices. In: The 41st International ACM SIGIR Conference on Research & Development in Information Retrieval, SIGIR 2018, pp. 215–224. Association for Computing Machinery, New York (2018)
3. Aliannejadi, M., Zamani, H., Crestani, F., Croft, W.B.: Context-aware target apps selection and recommendation for enhancing personal mobile assistants. ACM Trans. Inf. Syst. **39**(3) (2021)
4. Ghasemi, N., Aliannejadi, M., Bonab, H., Kanoulas, E., de Vries, A.P., Allan, J., Hiemstra, D.: Cross-market product-related question answering. In: Proceedings of the 46th International ACM SIGIR Conference on Research and Development in Information Retrieval, SIGIR 2023, pp. 1293–1302. Association for Computing Machinery (2023)
5. Ghasemi, N., Aliannejadi, M., Hiemstra, D.: Bert for target apps selection: Analyzing the diversity and performance of bert in unified mobile search. arXiv preprint arXiv:2109.06306 (2021)
6. Ghasemi, N., Hiemstra, D.: BERT meets cranfield: Uncovering the properties of full ranking on fully labeled data. In: Proceedings of the 16th Conference of the European Chapter of the Association for Computational Linguistics: Student Research Workshop, pp. 58–64. Association for Computational Linguistics, Online (Apr 2021)
7. Kratzwald, B., Eigenmann, A., Feuerriegel, S.: Rankqa: Neural question answering with answer re-ranking. In: Proceedings of the 57th Annual Meeting of the Association for Computational Linguistics, pp. 6076–6085 (2019)
8. McAuley, J., Yang, A.: Addressing complex and subjective product-related queries with customer reviews. In: Proceedings of the 25th International Conference on World Wide Web, pp. 625–635 (2016)
9. Rozen, O., Carmel, D., Mejer, A., Mirkis, V., Ziser, Y.: Answering product-questions by utilizing questions from other contextually similar products. In: Proceedings of the 2021 Conference of the North American Chapter of the Association for Computational Linguistics: Human Language Technologies, pp. 242–253. Association for Computational Linguistics, Online (2021)

10. Shokouhi, M., Si, L.: Federated search. Found. Trends Inf. Retr. **5**(1), 1–102 (2011)
11. Zhang, S., Lau, J.H., Zhang, X., Chan, J., Paris, C.: Discovering relevant reviews for answering product-related queries. In: 2019 IEEE International Conference on Data Mining (ICDM), pp. 1468–1473. IEEE (2019)
12. Zhang, S., Zhang, X., Lau, J.H., Chan, J., Paris, C.: Less is more: rejecting unreliable reviews for product question answering. In: Hutter, F., Kersting, K., Lijffijt, J., Valera, I. (eds.) ECML PKDD 2020. LNCS (LNAI), vol. 12459, pp. 567–583. Springer, Cham (2021). https://doi.org/10.1007/978-3-030-67664-3_34

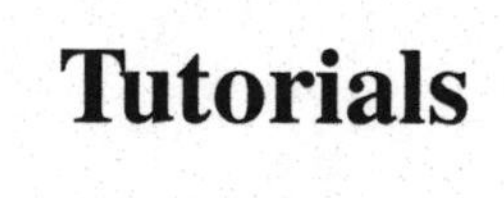
Tutorials

PhD Candidacy: A Tutorial on Overcoming Challenges and Achieving Success

Johanne R. Trippas[1](✉) and David Maxwell[2]

[1] RMIT University, Melbourne, Australia
j.trippas@rmit.edu.au
[2] Booking.com, Delft, The Netherlands
maxwelld90@acm.org

Abstract. Undertaking a PhD is a demanding yet rewarding experience. PhD candidates develop a deep understanding of their research topic and acquire a wide range of skills, including *(i)* formulating research questions; *(ii)* conducting research ethically and rigorously; *(iii)* communicating research findings effectively to both academic and non-academic audiences alike; *(iv)* forging a profile as an independent researcher; and *(v)* developing a teaching portfolio. PhD candidates inevitably experience challenges during their candidature. These challenges can be overcome by applying various techniques to adapt and learn from these experiences. This tutorial introduces strategies to help them advance in the PhD process. It is presented by two early career researchers in information retrieval, who have the unique perspective of being close enough to their time as PhD candidates to remember the highs and lows of PhD life yet far enough removed from the process to reflect on their experiences and provide insights. The tutorial will empower attendees to share, review, and refine productivity methods for their PhD journey. It provides a non-judgemental platform for open discussions led by the presenters.

Keywords: PhD research · Productivity · Success · Experiences

1 Introduction and Motivation

Earning a PhD is an achievement. The privilege of appending the letters *PhD* to one's name symbolises expertise in a specific field and acts as a testament to their determination and resilience. PhD candidates embark on a journey, aspiring to join the select few with a doctoral degree. In 2018, just 1.1% of people within the 25–64 age group were reported to possess a doctoral degree across all *OECD* nations [8, Figure B7.1]. This fraction of the population gains the opportunity to refine their skills within the structured confines of a university, benefiting from invaluable support and mentorship from their supervisor(s) [2,9].

D. Maxwell—The work undertaken by this author is not related to Booking.com's activities.

N. Goharian et al. (Eds.): ECIR 2024, LNCS 14612, pp. 347–351, 2024.
https://doi.org/10.1007/978-3-031-56069-9_45

Doctoral candidates are provided with valuable support for acquiring new skills, building meaningful connections, and honing their analytical thinking abilities, all within a relatively condensed time frame [4]. To realise these objectives, substantial financial, intellectual, and interpersonal resources are channelled into nurturing the candidate's growth. These resources must be maximised, ensuring the highest likelihood of the candidate's success. This investment is pivotal for the individual and represents a vital component of the broader ecosystem of *"giving back to society"* by cultivating efficient researchers.

This tutorial is designed as a platform to explore PhD candidates' diverse and unique journeys, highlighting the presenters' transitions across academic institutions and between academia and industry. Targeting current and aspiring PhD students, we aim to foster a dynamic peer-to-peer dialogue, drawing insights from academics and industry professionals. Instead of replacing formal university guidance, our goal is to *supplement it* by facilitating discussions on varied success strategies in academia. A pivotal aspect of our tutorial will be examining the impact of *Large Language Models (LLMs)* on PhD research, particularly addressing worries regarding the possible devaluation of one's work and suggesting adaptive, resilient approaches. In doing so, we strive to create a collaborative environment for openly addressing contemporary challenges and seizing opportunities in PhD trajectories.

The tutorial's goal is to catalyse current and future PhD candidates, empowering them to establish connections with their peers, fostering engagement with senior researchers, celebrating accomplishments, handling setbacks with resilience, and encouraging active involvement within the academic community. While this tutorial may not encompass all aspects of the PhD experience, it is designed to be a foundational resource for building a well-connected and strengthened ECIR community, particularly among junior scholars.

2 Syllabus

2.1 Key Presenter Themes

Everyday Practical Tips (90 Min.). This first theme introduces and discusses practical tips; we begin with time management and writing tools — including good working practices. We introduce how to plan, keep track, and optimise available time. We share resources, including tips and tricks on writing or editing research papers and theses [3]. We also encourage attendees to take ownership of their projects, including steering and framing their work. Afterwards, we draw attention to mental health, sharing several techniques useful to develop our relationship with stress or dealing with negative emotions.[1]

An ECIR PhD can present unique research issues that candidates must understand. While such assistance is already available (e.g., work by Kelly [5]), we highlight some lessons we learned during our respective PhDs, including insights into managing student/supervisor relationships.

[1] This is ***not*** a replacement for professional help.

Presenting and Communicating Your Research (90 Min.). Research communication is the process of interpreting and presenting research findings to a broader audience and is a crucial aspect of being a scientist. In this theme, we provide tips on explaining research and how to present research in presentations. We provide frameworks to summarise and story-tell research findings and examples of noteworthy presentations.

For this theme, we draw on previous materials [7] as a case study to discuss lessons learned for presenting and communicating research. We provide tools, tips and resources to improve and increase a PhD student's sense of capability to finish their thesis. Furthermore, we share experiences of the transition from being a PhD student to a post-PhD.

2.2 Panel

As part of our tutorial, we host a panel discussion featuring senior researchers from the ECIR community. This session allows attendees to gain insights from seasoned experts in information retrieval, offering valuable perspectives on research, academia, and career development. The panel discussion promises to empower attendees with actionable advice and strategic guidance, equipping them to confidently navigate the intricacies of an information retrieval PhD and academia. The tutorial presenters will lead this panel and will start with a round of questions before a Q&A with the audience.

2.3 Breakout Groups, 'Speed-Dating', and Other Researchers

In the breakout groups, our primary aim is to foster meaningful discussions and collaboration among participants. We intend to encourage attendees to explore topics relevant to their work and research interests tied to the conference themes.

2.4 Learning Outcomes

This tutorial is designed to enhance a PhD candidate's sense of purpose, capability, community engagement, and resourcefulness, contributing to developing a vibrant academic culture [6]. Furthermore, it underscores that the PhD journey is not a solitary pursuit of research and personal growth but rather a *shared endeavour*. Students will receive encouragement in their research pursuits, with a focus on highlighting available support systems. By participating in this tutorial, attendees will gain knowledge, establish a peer support network, and acquire essential skills and competencies in time management, strategies to address challenges, and communication skills.

3 Presenters

Johanne Trippas.[2] is a Vice-Chancellor's Research Fellow at RMIT University, specialising in intelligent systems, focusing on digital assistants and

[2] Johanne Trippas is the primary contact person.

conversational information seeking. Her research aims to enhance information accessibility through conversational systems, interactive information retrieval, and human-computer interaction. Additionally, Johanne is currently part of the NIST TREC program committee and is an ACM CHIIR steering committee member. She serves as vice-chair of the SIGIR Artifact Evaluation Committee, tutorial chair for ECIR'24, general chair of the ACM CUI'24, and ACM SIGIR-AP'23 proceedings chair. She has organised workshops (CHIIR'20-22, ECIR'24) [1,10], a TREC Track (CAsT'22), and tutorials (CHIIR'21, SIGIR'22, and WebConf'23) [11].

David Maxwell. David is currently a Data Scientist at *Booking.com*, based in Amsterdam, The Netherlands, where he tackles a range of data science challenges. Despite his move from academia, he remains in touch with the Information Retrieval community. Prior to this, he was a postdoctoral researcher at *Delft University of Technology (TU Delft)* in the Netherlands, collaborating with Dr Claudia Hauff. He earned his PhD in *Interactive Information Retrieval* from the *University of Glasgow*, Scotland, in June 2019, affiliated with the School of Computing Science. His research interests include understanding user behaviours when undertaking complex search tasks, and modelling such behaviours. Over the years, David has delivered well-received talks on his experiences as a PhD student—from discussing his research, to sharing the positive and negative experiences he faced. He has released an online guide[3] highlighting the challenges (and potential solutions) that students may face when writing up [7].

References

1. Buchanan, G., McKay, D., Clarke, C.L., Azzopardi, L., Trippas, J.: Made to measure: A workshop on human-centred metrics for information seeking. In: Proceedings of 5th ACM CHIIR, pp. 484–487 (2020)
2. DesJardins, M.: How to succeed in graduate school: a guide for students and advisors: part i of ii. XRDS **1**(2), 3–9 (1994)
3. Evans, D., Gruba, P., Zobel, J.: How to write a better thesis. Melbourne Univ, Publishing (2011)
4. Fischer, B.A., Zigmond, M.J.: Survival skills for graduate school and beyond. N. Dir. High. Educ. **1998**(101), 29–40 (1998)
5. Kelly, D.: Methods for evaluating interactive information retrieval systems with users. Found. Trends Inf. Retr. **3**(1–2), 1–224 (2009)
6. Lizzio, A.: Designing an orientation and transition strategy for commencing students: A conceptual summary of research and practice. Griffith University, First year experience project (2006)
7. Maxwell, D.: Writing up: the most intense thing you'll ever do? (2019). https://www.dmax.scot/things/phd/writing-up/
8. OECD: Education at a Glance 2019 (2019)
9. Phillips, E., Pugh, D.: How to get a PhD: a handbook for students and their supervisors. McGraw-Hill Education (2010)

[3] http://www.dmax.scot/things/phd/writing-up/.

10. Trippas, J.R., Thomas, P., Spina, D., Joho, H.: Third cair workshop. In: Proceedings of 5th ACM CHIIR, pp. 492–494 (2020)
11. Trippas, J.R., Maxwell, D.: The phd journey: reaching out and lending a hand. In: Proceedings of the 2021 Conference on Human Information Interaction and Retrieval, CHIIR 2021, pp. 345–346. (2021)

Explainable Recommender Systems with Knowledge Graphs and Language Models

Giacomo Balloccu, Ludovico Boratto, Gianni Fenu, Francesca Maridina Malloci, and Mirko Marras(✉)

University of Cagliari, Cagliari, Italy
{gballoccu,ludovico.boratto,mirko.marras}@acm.org,
{fenu,francescam.malloci}@unica.it

Abstract. In this tutorial, we delve into recent advances in explainable recommendation using Knowledge Graphs (KGs). The session begins by introducing the fundamental principles behind the increasing adoption of KGs in modern recommender systems. Then, the tutorial explores recent techniques that leverage KGs as an input for language models tailored to explainable recommendation, describing also data types, methods, and evaluation protocols and metrics. Conceptual elements are complemented with hands-on sessions, providing practical implementations using open-source tools and public datasets. Concluding with a comprehensive case study in the education domain as a recap, the tutorial analyses emerging issues and outlines prospective trajectories in this field. The tutorial website is available at https://explainablerecsys.github.io/ecir2024/.

1 Introduction

Motivation. Modern regulations, exemplified by GDPR, emphasise the *right to explanation* to enhance transparency in decision-making processes based on artificial intelligence [12]. Indeed, explanations play a crucial role for individuals and businesses, influencing user trust, engagement, retention, enjoyment, and decision speed in modern intelligent systems, including recommender systems (RSs) [1,9,15,21]. With RSs often lacking transparency, current research focuses on making recommendation processes more transparent [24]. To this end, explainable RSs often combine external knowledge, especially from Knowledge Graphs (KGs), with user-item interaction data [3,17]. A notable class of approaches that makes this combination possible, usually referred to as path-reasoning, aims to extracts explainable paths connecting recommended items and users in the KG and then leverage such paths to create textual explanations [5,22,23,26]. Moreover, recent advances leverage causal language modeling for path reasoning, requiring to sample paths from the KG as training instances for auto-regressive models. Key representatives include PLM [11] and PEARLM [2], with the latter ensuring integrity by preventing language model (LM) hallucination.

N. Goharian et al. (Eds.): ECIR 2024, LNCS 14612, pp. 352–357, 2024.
https://doi.org/10.1007/978-3-031-56069-9_46

Scope and Relevance. The increasing significance of the abovementioned emerging techniques in the field motivates this tutorial. Our tutorial serves as a comprehensive instructional session, offering an introduction to the topic of end-user explainability and, for this reason, aiming to provide attendees with a hands-on exploration of KGs and LMs for explainable recommendation. With RSs playing a prominent role and the focus on explainability, this tutorial aligns well with the interests and goals of the European information retrieval community.

Learning Objectives. Upon completion of this tutorial, participants can acquire a comprehensive understanding of explainable RSs and their collaborative potential with LMs empowered with KGs. Practical proficiency is developed by handling augmented interaction datasets, articulating fundamental approaches in explainable recommendation, and constructing models grounded in KGs, specifically emphasising path reasoning methods based on language models. Additionally, participants can master strategies for generating and assessing textual explanations derived from model-generated reasoning paths.

2 Tutorial Format and Content Outline

Format. Our tutorial combines lecture slides and hands-on sessions facilitated by Jupyter notebooks, available through Google Colab, to describe the field of explainable recommendation using KGs and LMs. To ensure active engagement, short quizzes and practical exercises are in both the slides and the source codes. The tutorial website provides the necessary links to the slides, Jupyter notebooks, reference lists, pre-processed datasets, and any utilized source code.

Outline. The tutorial's outline includes the following sections.

- **Introduction to explainable recommendation (slides, 20 min.):** Historical overview and foundational concepts behind explainable recommendations, emphasising their impact on utility, coverage, trust, and broader societal implications [4,6,8,10,13,14,18–21,25].
- **Hands on interaction data with KGs (notebook, 25 min.):** Public datasets from movie (ML1M [13]), music (LASTFM [19]) and e-commerce (Amazon [16]) domains, demonstrating the loading and preprocessing steps of the corresponding KGs used for explainability in recommendation models.
- **Explainable RSs based on path-reasoning methods (slides, 20 min.):** Overview of the recommendation pipeline and methodological families in KG-based recommendations, followed by a detailed exploration of path reasoning methods like PGPR [22] and CAFE [23].
- **Path reasoning: PGPR & CAFE (notebook, 25 min.):** Practical aspects of implementing PGPR and CAFE, therefore utilising reinforcement learning and neural-symbolic reasoning to optimise recommendations and generate textual explanations for end-users.

- **Explainable RSs based on language models (slides, 20 min.):** Language modeling paradigms and their applications in recommendation, with a deep dive into the integrating KGs and LMs with PLM [11] and PEARLM [2].
- **Language modeling: PLM & PEARLM (notebook, 25 min.):** Practical aspects of creating training data and performing training models based on PLM and PEARLM, leveraging pretrained models to generate top-k recommendations and address hallucination issues.
- **Deep dive on explainable RSs in education (notebook, 25 min.):** Theoretical and practical knowledge applied to the educational sector using the COCO dataset [7,8], providing a recap of the recommender system workflow and challenges in complex KG creation, model training, and evaluation.
- **Challenges and open issues (slides, 20 min.):** Current challenges and open issues in the field of explainable recommendation using KGs and LMs.

3 Targeted Audience and Prerequisites

This tutorial, designed for individuals at a beginner to intermediate level, is suitable for early-stage researchers, industry technologists, and practitioners. It includes essential background material for those unfamiliar with RSs, explanations, and knowledge graphs. No prior expertise in RSs with KGs is assumed. While a basic understanding of Python programming and common libraries like Pandas and NumPy is preferable, the tutorial is designed to be accessible to those with varying levels of programming knowledge. Notably, the tutorial's focus on explainability makes it an interdisciplinary subject, extending beyond algorithms and appealing to a diverse audience interested in this perspective.

4 Presenters' Biography

Giacomo Balloccu (webpage: https://giacoballoccu.github.io) is a PhD candidate at the Department of Mathematics and Computer Science of the University of Cagliari (Italy). His research interests focus on the social aspects of recommender systems, with emphasis on user experience and business impact. Recent research works aim to combine language models and knowledge graphs for more authentic, knowledge-based explanations. His research activity, particularly within the domain of knowledge-based explainable recommendation, has been featured in top-tier conferences, such as SIGIR and ECIR, and journals, such as Elsevier's Knowledge-Based Systems. He also gave a tutorial on explainable recommendation at RecSys. He also served as a student volunteer at RecSys in 2021 and 2023. In 2023, he completed an internship at Amazon as an Applied Scientist, focusing on contextual targeting by means of language models.

Ludovico Boratto (webpage: https://www.ludovicoboratto.com) is an Assistant Professor at the Department of Mathematics and Computer Science of the University of Cagliari (Italy). His research interests focus on recommender systems and their impact on stakeholders. He has co-authored over 60 papers published in top-tier conference proceedings and journals. His research brought him to give talks and tutorials at top-tier research centres and conferences, including CIKM, UMAP, RecSys, ICDE, ECIR, WSDM, ICDM, DSAA, and ECAI. He is the editor of the book "Group Recommender Systems: An Introduction" by Springer. He is an editorial board member of the "Information Processing & Management" journal (Elsevier) and "Journal of Intelligent Information Systems" (Springer), and guest editor of several journals' special issues. He is part of the program committees of the main Web conferences, where he received four outstanding contribution awards. In 2012, he got his Ph.D. at the University of Cagliari. From May 2016 to April 2021, he joined Eurecat as a Senior Research Scientist in the Data Science and Big Data Analytics research group. In 2010 and 2014, he visited Yahoo! Research in Barcelona.

Gianni Fenu (webpage: https://web.unica.it/unica/it/ateneo_s07_ss01.page?contentId=SHD30371) is Full Professor at the Department of Mathematics and Computer Science of the University of Cagliari (Italy). His research interests focus on responsible recommender systems, digital education, and personalization. He has authored more than 120 papers in conferences and journals. He has been the scientific responsible a wide range of projects, such as ILEARNTV MIUR-UE (2014-2017, 10 ME, 6 partners) and the European Research M-Commerce and Development.

Francesca Maridina Malloci (webpage: https://francescamalloci.com) is Assistant Professor at the Department of Mathematics and Computer Science of the University of Cagliari (Italy). Her research interests are focused on predictive analytics, with several papers published in international journal and conference proceedings. Current research activities delved into decision-making algorithms for multi-stakeholders contexts, with attention to recommender systems and education. She received the BSc (2016) and MSc (2018) degrees in Computer Science, both cum laude, from the University of Cagliari. In 2019, she has been visiting scientist at the EURECAT Technology Centre (Spain), collaborating with the Data Science and Big Data Analytics Unit.

Mirko Marras (webpage: https://www.mirkomarras.com) is an Assistant Professor at the Department of Mathematics and Computer Science of the University of Cagliari (Italy). Before that, he was a postdoctoral researcher at EPFL (Switzerland) and a visiting scholar at Eurecat (Spain) and New York University (USA). His research ranges across various domains impacted by user modelling and personalization. He has co-authored more than 80 papers in top-tier conferences and journals and given tutorials at ECML-PKDD, RecSys, ICDE, ECIR, WSDM, ICDM, and UMAP. He has also co-chaired workshops on related themes at ECIR, WSDM, ICCV, EDM, and ECML-PKDD. He is part of the program committees of top-tier conferences, where he received three outstanding reviewer awards. He is an associate editor for Springer's Journal of Ambient Intelligence and Humanized Computing and Neural Processing Letters.

Acknowledgement. We acknowledge financial support under the National Recovery and Resilience Plan (NRRP), Miss. 4 Comp. 2 Inv. 1.5 - Call for tender No.3277 published on Dec 30, 2021 by the Italian Ministry of University and Research (MUR) funded by the European Union - NextGenerationEU. Prj. Code ECS0000038 eINS

Ecosystem of Innovation for Next Generation Sardinia, CUP F53C22000430001, Grant Assignment Decree N. 1056, Jun 23, 2022 by the MUR.

References

1. Atzori, A., Fenu, G., Marras, M.: Explaining bias in deep face recognition via image characteristics. In: Proceedings of the IEEE International Joint Conference on Biometrics, IJCB, pp. 1–10. IEEE (2022)
2. Balloccu, G., Boratto, L., Cancedda, C., Fenu, G., Marras, M.: Faithful path language modelling for explainable recommendation over knowledge graph. CoRR abs/ arXiv: 2310.16452 (2023)
3. Balloccu, G., Boratto, L., Cancedda, C., Fenu, G., Marras, M.: Knowledge is power, understanding is impact: utility and beyond goals, explanation quality, and fairness in path reasoning recommendation. In: Proc. of the 45th European Conference on Information Retrieval, ECIR. LNCS, vol. 13982, pp. 3–19. Springer (2023). https://doi.org/10.1007/978-3-031-28241-6_1
4. Balloccu, G., Boratto, L., Fenu, G., Marras, M.: Post processing recommender systems with knowledge graphs for recency, popularity, and diversity of explanations. In: Proceedings of the 45th International ACM SIGIR Conference on Research and Development in Information Retrieval, SIGIR, pp. 646–656. ACM (2022)
5. Balloccu, G., Boratto, L., Fenu, G., Marras, M.: Reinforcement recommendation reasoning through knowledge graphs for explanation path quality. Knowl.-Based Syst. **260**, 110098 (2023)
6. Choi, Y., et al.: EdNet: a large-scale hierarchical dataset in education. In: Bittencourt, I.I., Cukurova, M., Muldner, K., Luckin, R., Millán, E. (eds.) AIED 2020. LNCS (LNAI), vol. 12164, pp. 69–73. Springer, Cham (2020). https://doi.org/10.1007/978-3-030-52240-7_13
7. Dessì, D., Fenu, G., Marras, M., Reforgiato Recupero, D.: Leveraging cognitive computing for multi-class classification of e-learning videos. In: Blomqvist, E., Hose, K., Paulheim, H., Ławrynowicz, A., Ciravegna, F., Hartig, O. (eds.) ESWC 2017. LNCS, vol. 10577, pp. 21–25. Springer, Cham (2017). https://doi.org/10.1007/978-3-319-70407-4_5
8. Dessì, D., Fenu, G., Marras, M., Reforgiato Recupero, D.: COCO: semantic-enriched collection of online courses at scale with experimental use cases. In: Rocha, Á., Adeli, H., Reis, L.P., Costanzo, S. (eds.) WorldCIST'18 2018. AISC, vol. 746, pp. 1386–1396. Springer, Cham (2018). https://doi.org/10.1007/978-3-319-77712-2_133
9. Fenu, G., Galici, R., Marras, M.: Experts' view on challenges and needs for fairness in artificial intelligence for education. In: Proc. of the 23rd International Conference on Artificial Intelligence in Education, AIED. LNCS, vol. 13355, pp. 243–255. Springer (2022). https://doi.org/10.1007/978-3-031-11644-5_20
10. Ge, M., Delgado-Battenfeld, C., Jannach, D.: Beyond accuracy: evaluating recommender systems by coverage and serendipity. In: Proceedings of the 2010 ACM Conference on Recommender Systems, RecSys, pp. 257–260. ACM (2010)
11. Geng, S., Fu, Z., Tan, J., Ge, Y., de Melo, G., Zhang, Y.: Path language modeling over knowledge graphs for explainable recommendation. In: Proceedings of the ACM Web Conference 2022, TheWebConf, pp. 946–955. ACM (2022)
12. Goodman, B., Flaxman, S.R.: European union regulations on algorithmic decision-making and a "right to explanation". AI Mag. **38**(3), 50–57 (2017)

13. Harper, F.M., Konstan, J.A.: The movielens datasets: history and context. ACM Trans. Interact. Intell. Syst. **5**(4), 19:1–19:19 (2016)
14. Helberger, N., Karppinen, K., D'acunto, L.: Exposure diversity as a design principle for recommender systems. Inform. Commun. Soc. **21**(2), 191–207 (2018)
15. Miller, T.: Explanation in artificial intelligence: insights from the social sciences. Artif. Intell. **267**, 1–38 (2019)
16. Ni, J., Li, J., McAuley, J.J.: Justifying recommendations using distantly-labeled reviews and fine-grained aspects. In: Proceedings of the 2019 Conference on Empirical Methods in Natural Language Processing and the 9th International Joint Conference on Natural Language Processing, EMNLP-IJCNLP, pp. 188–197. ACL (2019)
17. Oramas, S., Ostuni, V.C., Noia, T.D., Serra, X., Sciascio, E.D.: Sound and music recommendation with knowledge graphs. ACM Trans. Intell. Syst. Technol. **8**(2) (2016)
18. Ricci, F., Rokach, L., Shapira, B.: Recommender systems: introduction and challenges. In: Ricci, F., Rokach, L., Shapira, B. (eds.) Recommender Systems Handbook, pp. 1–34. Springer, Boston, MA (2015). https://doi.org/10.1007/978-1-4899-7637-6_1
19. Schedl, M.: The lfm-1b dataset for music retrieval and recommendation. In: Proceedings of the 2016 ACM on International Conference on Multimedia Retrieval, ICMR, pp. 103–110. ACM (2016)
20. Shin, D.: The effects of explainability and causability on perception, trust, and acceptance: implications for explainable ai. Int. J. Hum Comput Stud. **146**, 102551 (2021)
21. Tintarev, N., Masthoff, J.: A survey of explanations in recommender systems. In: Proceedings of the 23rd International Conference on Data Engineering Workshops, ICDE Workshops, pp. 801–810. IEEE (2007)
22. Xian, Y., Fu, Z., Muthukrishnan, S., de Melo, G., Zhang, Y.: Reinforcement knowledge graph reasoning for explainable recommendation. In: Proceedings of the 42nd International ACM SIGIR Conference on Research and Development in Information Retrieval, pp. 285–294. ACM (2019)
23. Xian, Y., et al.: Cafe: coarse-to-fine neural symbolic reasoning for explainable recommendation. In: Proceedings of the 29th ACM International Conference on Information & Knowledge Management, CIKM, pp. 1645–1654 (2020)
24. Zhang, Y., Chen, X.: Explainable recommendation: a survey and new perspectives. Found. Trends Inf. Retr. **14**(1), 1–101 (2020)
25. Zhang, Y., Lai, G., Zhang, M., Zhang, Y., Liu, Y., Ma, S.: Explicit factor models for explainable recommendation based on phrase-level sentiment analysis. In: Proceedings of the 37th International ACM SIGIR Conference on Research and Development in Information Retrieval, SIGIR, pp. 83–92. ACM (2014)
26. Zhao, Y., et al.: Time-aware path reasoning on knowledge graph for recommendation. ACM Trans. Inf. Syst. **41**(2) (2022)

Quantum Computing for Information Retrieval and Recommender Systems

Maurizio Ferrari Dacrema[1], Andrea Pasin[2](✉), Paolo Cremonesi[1], and Nicola Ferro[2]

[1] Politecnico di Milano, Milan, Italy
{maurizio.ferrari,paolo.cremonesi}@polimi.it
[2] University of Padua, Padua, Italy
andrea.pasin.1@phd.unipd.it, nicola.ferro@unipd.it

Abstract. *Quantum Computing (QC)* is a research field that has been in the limelight in recent years. In fact, many researchers and practitioners believe that it can provide benefits in terms of efficiency and effectiveness when employed to solve certain computationally intensive tasks. In *Information Retrieval (IR)* and *Recommender Systems (RS)* we are required to process very large and heterogeneous datasets by means of complex operations, it is natural therefore to wonder whether QC could also be applied to boost their performance. The goal of this tutorial is to show how QC works to an audience that is not familiar with the technology, as well as how to apply the QC paradigm of *Quantum Annealing (QA)* to solve practical problems that are currently faced by IR and RS systems. During the tutorial, participants will be provided with the fundamentals required to understand QC and to apply it in practice by using a *real* D-Wave quantum annealer through APIs.

Keywords: Quantum Computing · Quantum Annealing · Information Retrieval · Recommender Systems

1 Motivation

Quantum Computing (QC) is a rapidly growing field, involving an increasing number of researchers and practitioners from different backgrounds to develop new methods using quantum computers to perform faster computations. With QC it has been possible to tackle practical problems achieving good results in terms of efficiency and effectiveness [2,4,6,11,12]. Since *Information Retrieval (IR)* and *Recommender Systems (RS)* face big challenges due to the scalability of complex algorithms on top of very large datasets, QC seems very promising but its application to IR and RS has been explored only to a limited extent.

For these reasons, we propose a tutorial covering the fundamental concepts of QC, with a focus on the practical application of the *Quantum Annealing (QA)* paradigm through interactive coding sessions teaching participants how to use the cutting-edge quantum annealers to solve realistic problems. The QA paradigm offers a good trade-off between computational power and a low

N. Goharian et al. (Eds.): ECIR 2024, LNCS 14612, pp. 358–362, 2024.
https://doi.org/10.1007/978-3-031-56069-9_47

access barrier since it requires to formulate the task one wishes to solve as an optimization problem using the *Quadratic Unconstrained Binary Optimization (QUBO)* [3] formulation, thus hiding the complexity of the underlying QC system and making it easily accessible even for people without a background in quantum physics. QA has already been applied to IR and RS tasks [1,2,8,9] showing that, although not always superior to classical methods, quantum annealers have matured enough to reliably tackle realistic problems.

The tutorial will also introduce QuantumCLEF[1] [10], a lab organized at CLEF 2024 that offers an infrastructure to develop and evaluate QA algorithms.

Overall, we aim at lowering the often perceived barrier-of-entry to QC and at providing participants the basic knowledge to develop and code QC algorithms.

2 Format

The tutorial will cover theoretical and practical aspects underneath QC (and especially QA) by allowing participants to code and use real quantum annealers to solve optimization problems usually faced by many computer systems, including IR and RS systems. The duration of the tutorial will be half-day (3 h) plus breaks and it will be subdivided into 4 parts.

2.1 Materials

The tutorial will include slides and Jupyter Notebooks. The materials will be available on GitHub prior to the tutorial and openly available afterwards. A guide on how to create a free D-Wave account, set up the environment, and connect to quantum annealers will be provided before the tutorial. Participants will also have access to the infrastructure made available by QuantumCLEF.

2.2 Outline

The tutorial will be divided into 4 parts, starting from the theoretical aspects of QC, QA and the QUBO problem formulation. Then there will be a practical part where participants can code and solve some problems using quantum annealers.

Part 1: QC Foundations (30 min). The first part consists in a gentle introduction to QC, showing its potential benefits but also limitations. We will also delve into the QA paradigm. It comprises:

- overview and basic understanding of QC and its potential benefits;
- description of the paradigms of Quantum Circuit model and Adiabatic model.
- the relations between the classical optimizations meta-heuristics *Simulated Annealing* [13] and *Quantum-inspired Annealing* [5] to the Adiabatic model;
- how to represent the energy configuration of a quantum system (i.e., Hamiltonian) using the Ising model;

[1] https://qclef.dei.unipd.it/.

- the evolution of a quantum system to a state of minimal energy;
- the similarity between the Hamiltonian of a system and the QUBO representation for optimizing (NP-hard) problems.

Part 2: QUBO Formulation (30 min). This part will show how to represent classical binary optimization problems in QUBO formulation [7]. In particular, it will explain how to write NP-complete binary decision problems (e.g., number partitioning) and NP-hard binary optimization problems (e.g., quadratic assignment) in QUBO formulation, describing constraints and loss functions.

Part 3: QC for IR and RS and their Evaluation (30 min). This part will introduce IR and RS problems which can be solved by using QA, namely Feature Selection, Clustering and Model Boosting. Moreover, we will discuss how to evaluate such QA algorithms from both the efficiency and effectiveness point of view. Finally, it will introduce the QuantumCLEF lab and explain how to use its development and evaluation infrastructure.

Part 4: Hands-On (90 Min). This part discusses how to use quantum annealers, which are available as a cloud service. It involves:

- the architecture and topology of a quantum annealer;
- how to use the QUBO formulation of a problem to program the quantum annealer via Minor Embedding;
- how the density of the QUBO problem impacts the number of variables required on the quantum annealer;
- how to program a quantum annealer and read the result (**hands-on**);
- Feature Selection and Clustering (**hands-on**);
- execution and evaluation of one of the above algorithms on the QuantumCLEF infrastructure (**hands-on**).

3 Audience

This tutorial is intended for people coming from IR and RS but also from other fields, such as Machine Learning, Big Data, Operations Research, and Optimization. In fact, QC and QA can be applied to solve problems in different domains and, even if the practical part is focused in using QA for IR and RS systems, the considered problems are very general and common to several research areas.

Targeted audience: Due to the topic's novelty, the target audience is of researchers and industry practitioners mainly belonging to IR and RS.

Prerequisite knowledge: This tutorial will be self-contained and has minimal prerequisite knowledge, mainly consisting on being familiar with the concept of decision and optimization problems. For those interested in the hands-on part, basic Python programming skills are required to interact with quantum annealers through the tools provided by D-Wave[2].

[2] https://docs.ocean.dwavesys.com/en/stable/.

4 Tutorial History

To the best of our knowledge, we are not aware of other similar tutorials in the IR and RS fields.

5 Presenters

Maurizio Ferrari Dacrema is Assistant Professor at Politecnico di Milano, Italy. His main research focus is the application of Quantum Computing to machine learning tasks as well as the use of machine learning to improve the effectiveness of current generation quantum computers. He also has significant experience on reproducibility and evaluation of recommender systems. He won the ACM Best Paper Award at ACM RecSys 2019, has been teaching assistant at Politecnico di Milano for the MSc course on Recommender Systems since 2017, for the MSc course on Quantum Computing since 2023 and was Lecturer of the PhD course Applied Quantum Machine Learning in 2021. He is Demo and Late-Breaking Results Co-Chair of RecSys 2024 and was co-organizer of the LERI workshop at RecSys 2023.

Andrea Pasin is a PhD student at University of Padua, Italy, currently studying and investigating the possible applications of Quantum Annealing for Information Access systems to improve their performance.

Paolo Cremonesi is Full Professor of Recommender Systems and Quantum Computing at Politecnico di Milano, Italy. He has extensive experience on Quantum Computing, and he is the co-leader of the Quantum Computing research activities within the Italian National Research Centre for High Performance Computing, Big Data and Quantum Computing. He has also extensive experience in the reproducibility and evaluation of recommender systems. He has been member of the Steering Committee of ACM RecSys, among the most important conferences on Recommender Systems, since 2017. He was General Chair of ACM RecSys 2017 and Program Chair of ACM TvX 2013. Paolo co-authored and presented several tutorials: on "Recommender Systems" at TvX 2012, on "Evaluation of Recommender Systems" at UMAP 2013, on "Cross-domain Recommender Systems" at RecSys 2014, on "Sequence-aware Recommender Systems" at WWW 2019, RecSys 2019 and UMAP 2019. Paolo was also co-organizer of the LERI workshop at RecSys 2023.

Nicola Ferro is Full Professor in Computer Science at the Department of Information Engineering at the University of Padua, Italy. His main research interests are information retrieval, data management and representation, and their evaluation. He chairs the Steering Committee of CLEF, the European evaluation initiative on multimodal and multilingual information access systems, and the Steering Committee of ESSIR, the European Summer School on Information Retrieval. He is Senior PC Member in top-tier conferences, like ECIR, ACM SIGIR, ACM CIKM, WSDM. He was General Chair of ESSIR 2016 and Associate Editor for ACM TOIS. He was awarded the SIGIR Academy in 2023.

References

1. Ferrari Dacrema, M., Felicioni, N., Cremonesi, P.: Optimizing the selection of recommendation carousels with quantum computing. In: RecSys 2021: Fifteenth ACM Conference on Recommender Systems, Amsterdam, The Netherlands, 27 September 2021–1 October 2021, pp. 691–696. ACM (2021). https://doi.org/10.1145/3460231.3478853
2. Ferrari Dacrema, M., Moroni, F., Nembrini, R., Ferro, N., Faggioli, G., Cremonesi, P.: Towards feature selection for ranking and classification exploiting quantum annealers. In: SIGIR 2022: The 45th International ACM SIGIR Conference on Research and Development in Information Retrieval, Madrid, Spain, 11–15 July 2022, pp. 2814–2824. ACM (2022). https://doi.org/10.1145/3477495.3531755
3. Glover, F.W., Kochenberger, G.A., Du, Y.: Quantum bridge analytics I: a tutorial on formulating and using QUBO models. 4OR **17**(4), 335–371 (2019). https://doi.org/10.1007/s10288-019-00424-y
4. Jaschek, T., Bucyk, M., Oberoi, J.S.: A quantum annealing-based approach to extreme clustering. In: Arai, K., Kapoor, S., Bhatia, R. (eds.) FICC 2020. AISC, vol. 1130, pp. 169–189. Springer, Cham (2020). https://doi.org/10.1007/978-3-030-39442-4_15
5. Kadowaki, T., Nishimori, H.: Quantum annealing in the transverse ising model. Phys. Rev. E **58**, 5355–5363 (1998). https://doi.org/10.1103/PhysRevE.58.5355
6. Kerenidis, I., Prakash, A.: Quantum recommendation systems. In: Papadimitriou, C.H. (ed.) 8th Innovations in Theoretical Computer Science Conference, ITCS 2017, Berkeley, CA, USA, 9–11 January 2017, LIPIcs, vol. 67, pp. 49:1–49:21. Schloss Dagstuhl - Leibniz-Zentrum für Informatik (2017). https://doi.org/10.4230/LIPICS.ITCS.2017.49
7. Lucas, A.: Ising formulations of many NP problems. Front. Phys. **2**, 5 (2014)
8. Nembrini, R., Carugno, C., Ferrari Dacrema, M., Cremonesi, P.: Towards recommender systems with community detection and quantum computing. In: RecSys 2022: Sixteenth ACM Conference on Recommender Systems, Seattle, WA, USA, 18–23 September 2022, pp. 579–585. ACM (2022). https://doi.org/10.1145/3523227.3551478
9. Nembrini, R., Ferrari Dacrema, M., Cremonesi, P.: Feature selection for recommender systems with quantum computing. Entropy **23**(8), 970 (2021). https://doi.org/10.3390/E23080970
10. Pasin, A., Ferrari Dacrema, M., Cremonesi, P., Ferro, N.: qCLEF: a proposal to evaluate quantum annealing for information retrieval and recommender systems. In: Arampatzis, A., et al. (eds.) CLEF 2023. LNCS, vol. 14163, pp. 97–108. Springer, Cham (2023). https://doi.org/10.1007/978-3-031-42448-9_9
11. Pilato, G., Vella, F.: A survey on quantum computing for recommendation systems. Inf. **14**(1), 20 (2023). https://doi.org/10.3390/info14010020
12. Ramezani, S.B., Sommers, A., Manchukonda, H.K., Rahimi, S., Amirlatifi, A.: Machine learning algorithms in quantum computing: a survey. In: 2020 International Joint Conference on Neural Networks (IJCNN), pp. 1–8 (2020). https://doi.org/10.1109/IJCNN48605.2020.9207714
13. Van Laarhoven, P.J., Aarts, E.H., van Laarhoven, P.J., Aarts, E.H.: Simulated Annealing. Springer, Dordrecht (1987). https://doi.org/10.1007/978-94-015-7744-1_2

Recent Advances in Generative Information Retrieval

Yubao Tang[1,2(✉)], Ruqing Zhang[1,2], Zhaochun Ren[3], Jiafeng Guo[1,2], and Maarten de Rijke[4]

[1] CAS Key Lab of Network Data Science and Technology, ICT, CAS, Beijing, China
{tangyubao21b,zhangruqing,guojiafeng}@ict.ac.cn
[2] University of Chinese Academy of Sciences, Beijing, China
[3] Leiden University, Leiden, The Netherlands
z.ren@liacs.leidenuniv.nl
[4] University of Amsterdam, Amsterdam, The Netherlands
m.derijke@uva.nl

Abstract. Generative retrieval (GR) has become a highly active area of information retrieval that has witnessed significant growth recently. Compared to the traditional "index-retrieve-then-rank" pipeline, the GR paradigm aims to consolidate all information within a corpus into a single model. Typically, a sequence-to-sequence model is trained to directly map a query to its relevant document identifiers (i.e., docids). This tutorial offers an introduction to the core concepts of the novel GR paradigm and a comprehensive overview of recent advances in its foundations and applications. We start by providing preliminary information covering foundational aspects and problem formulations of GR. Then, our focus shifts towards recent progress in docid design, training approaches, inference strategies, and applications of GR. We end by outlining remaining challenges and issuing a call for future GR research. This tutorial is intended to be beneficial to both researchers and industry practitioners interested in developing novel GR solutions or applying them in real-world scenarios.

Keywords: Generative retrieval · Comprehensive overview · Recent progress

1 General Information

Information retrieval (IR) is a core task in a wide range of real-world applications, such as web search [21,24] and question answering [9,10]. It aims to retrieve information from a large repository that is relevant to an information need. Most existing IR methods follow a common pipeline paradigm of "index-retrieve-then-rank," which includes (i) building an index for each document in the corpus [14]; (ii) retrieving an initial set of candidate documents for a query [17]; and (iii) determining the relevance degree of each candidate [14]. Despite its wide usage, this paradigm has limitations: (i) during training, heterogeneous

N. Goharian et al. (Eds.): ECIR 2024, LNCS 14612, pp. 363–368, 2024.
https://doi.org/10.1007/978-3-031-56069-9_48

modules with different optimization objectives may lead to sub-optimal performance, and capturing fine-grained relationships between queries and documents is challenging; and (ii) during inference, a large document index is needed to search over the corpus, which may come with substantial memory and computational requirements.

Recently, a fundamentally different paradigm, known as *generative retrieval* (GR) [16], has garnered attention to replace the long-standing pipeline paradigm. The key idea of the GR paradigm is to parameterize the indexing, retrieval, and ranking components of traditional IR systems into a single consolidated model. Specifically, a sequence-to-sequence (Seq2Seq) model is trained to directly map queries to their relevant document identifiers (docids). Such a single-step generative model dramatically simplifies the search process, could be optimized in an end-to-end manner, and could better leverage the capabilities of large language models (LLMs). Based on [19], there are two families of GR, namely *closed-book* GR and *open-book* GR. Closed-book GR refers to the scenario where the language model that is used for directly generating relevant information resources for an information need, and the model is the only source of knowledge leveraged during generation. Open-book GR, on the other hand, allows the language model to draw on external memory prior to, during, and after generation. Here, our main focus is on closed-book GR.

Many publications have emerged in reputable conferences, e.g., SIGIR [3,6], CIKM [4,5,29], KDD [26], NeurIPS [2,25,27,28], ICLR [8], and ACL [7,11,13,23], in Gen-IR@SIGIR2023 [15,20,22,33], in journals [31], and on arXiv [12,18,30,32]. At SIGIR'23, Marc Najork, serving as the keynote speaker, provided a comprehensive summary of existing GR systems and discussed many open challenges in this emerging field [19]. The first workshop on generative information retrieval at SIGIR'23 (Gen-IR@SIGIR2023) [1] welcomed many submissions and attendees, underscoring the IR community's current keen interest in GR.

The time is right to offer a tutorial on the topic of GR. Therefore, we have organized and presented a tutorial dedicated to GR at the 1st International ACM SIGIR Conference on Information Retrieval in the Asia Pacific (SIGIR-AP 2023) on November 26, 2023, in Beijing, China. At ECIR'24 we offer a new edition of the tutorial that has been revised based on the feedback received and incorporates coverage of new relevant work. We hope this tutorial will generate the interest of more researchers and help them gain a better understanding of this novel field.

2 Tutorial Information

2.1 Format and Length

This is a 3-h and lecture-style tutorial.

2.2 Tutorial Outline

1. **Introduction** (15 min)
   - An overview of the tutorial
   - Why generative retrieval?
2. **Preliminaries** (15 min)
   - Retrieval task formulation: generative models vs. discriminative models
   - Basic concepts in generative retrieval
3. **Generative retrieval: Docid design** (30 min)
   - Pre-defined static docids
     - Single docids: number-based and word-based docids
     - Multiple docids
   - Learnable docids: jointly with retrieval tasks
4. **Generative retrieval: Training approaches** (40 min)
   - Static corpora: supervised learning with labeled data, and pre-training with unlabeled data
   - Dynamic corpora: continual learning
5. **Generative retrieval: Inference strategies** (25 min)
   - For a single docid: constrained beam search, constrained greedy search and FM-index
   - For multiple docids: heuristic scoring functions
6. **Generative retrieval: Applications** (35 min)
   - Offline application: e.g., entity retrieval, fact checking, recommender systems, multi-hop retrieval and code generation
   - Industry applications
7. **Conclusions and future directions** (20 min)

3 Target Audience and Prerequisites

The tutorial will be accessible to anyone who has a basic knowledge of IR and NLP. The topic will be of interest to both IR and NLP researchers in academia and practitioners in the industry.

4 Presenters

Yubao Tang is a Ph.D. student at the Institute of Computing Technology, Chinese Academy of Sciences. She obtained her M.Sc. degree from the Institute of Information Engineering, Chinese Academy of Sciences, and her B.Eng. from Sichuan University. Her research focuses on information retrieval, and she is the first author of a full paper on generative retrieval at KDD'23 [26].

Ruqing Zhang is an Associate Researcher at the Institute of Computing Technology, Chinese Academy of Sciences. Her recent research focuses on information retrieval, with a particular emphasis on generative information retrieval, the robustness of neural ranking models, and trustworthy retrieval through the lens of causality. She has authored several papers in the field of generative retrieval [3–6,15,26]. Additionally, Ruqing co-organized the first workshop on generative information retrieval at SIGIR'23 (Gen-IR@SIGIR23) to foster discussions and innovations in GR.

Zhaochun Ren is an Associate Professor at Leiden University. His research interests focus on research problems at the interface of information retrieval and natural language processing, with an emphasis on generative retrieval, recommender systems, and conversational information seeking. He aims to develop intelligent systems that can address complex user requests and solve core challenges in both information retrieval and natural language processing towards that goal. He has been working on various topics related to generative retrieval research. In addition to his academic experience, he worked on e-commerce search and recommendation at JD.com for 2+ years. He has been invited to give tutorials at SIGIR'18 and NLPCC'22.

Jiafeng Guo is a Researcher at the Institute of Computing Technology, Chinese Academy of Sciences (CAS) and a Professor at the University of Chinese Academy of Sciences. He is the director of the CAS key lab of network data science and technology. He has worked on a number of topics related to web search and data mining, with a current focus on neural models for information retrieval and natural language understanding. He has received multiple best paper (runner-up) awards at leading conferences (CIKM'11, SIGIR'12, CIKM'17, WSDM'22). He has been (co)chair for many conferences, e.g., reproducibility track co-chair of SIGIR'23, workshop co-chair of SIGIR'21 and short paper co-chair of SIGIR'20. He serves as an associate editor for ACM Transactions on Information Systems and Information Retrieval Journal. Jiafeng has previously taught tutorials at ACML, CCIR and CIPS ATT.

Maarten de Rijke is a Distinguished University Professor of Artificial Intelligence and Information Retrieval at the University of Amsterdam. His research is focused on designing and evaluating trustworthy technology to connect people to information, particularly search engines, recommender systems, and conversational assistants. He is the scientific director of the Innovation Center for Artificial Intelligence and a former editor-in-chief of ACM Transactions on Information Systems and of Foundations and Trends in Information Retrieval, and a current co-editor-in-chief of Springer's Information Retrieval book series, (associate) editor for various journals and book series. He has been general (co)chair or program (co)chair for CIKM, ECIR, ICTIR, SIGIR, WSDM, WWW, and has previously taught tutorials at these same venues and AAAI.

5 Tutorial Materials

We plan to share the following materials on this website:[1] (i) Slides: All slides are made publicly available. (ii) Annotated bibliography: An annotated compilation of references that lists all works discussed in the tutorial and provides a good basis for further study. (iii) Code: An annotated list of pointers to open source code bases and datasets for the work discussed in the tutorial. (iv) Videos: A video recording of the presentation will be made available.

References

1. Bénédict, G., Zhang, R., Metzler, D.: Gen-IR@SIGIR 2023: the first workshop on generative information retrieval. In: SIGIR, pp. 3460–3463 (2023)
2. Bevilacqua, M., Ottaviano, G., Lewis, P., Yih, W.T., Riedel, S., Petroni, F.: Autoregressive search engines: generating substrings as document identifiers. In: NeurIPS, pp. 31668–31683 (2022)
3. Chen, J., Zhang, R., Guo, J., Fan, Y., Cheng, X.: GERE: generative evidence retrieval for fact verification. In: SIGIR, pp. 2184–2189 (2022)
4. Chen, J., Zhang, R., Guo, J., Liu, Y., Fan, Y., Cheng, X.: CorpusBrain: pre-train a generative retrieval model for knowledge-intensive language tasks. In: CIKM, pp. 191–200 (2022)
5. Chen, J., et al.: Continual learning for generative retrieval over dynamic corpora. In: CIKM, pp. 306–315 (2023)
6. Chen, J., et al.: A unified generative retriever for knowledge-intensive language tasks via prompt learning. In: SIGIR, pp. 1448–1457 (2023)
7. Chen, X., Liu, Y., He, B., Sun, L., Sun, Y.: Understanding differential search index for text retrieval. In: Findings of ACL, pp. 10701–10717 (2023)
8. De Cao, N., Izacard, G., Riedel, S., Petroni, F.: Autoregressive entity retrieval. In: ICLR (2021)
9. Guu, K., Lee, K., Tung, Z., Pasupat, P., Chang, M.: Retrieval augmented language model pre-training. In: ICML, pp. 3929–3938 (2020)
10. Kwiatkowski, T., et al.: Natural questions: a benchmark for question answering research. Trans. Assoc. Comput. Linguist. **7**, 452–466 (2019)
11. Lee, H., et al.: Nonparametric decoding for generative retrieval. In: Findings of the ACL 2023, pp. 12642–12661 (2023)
12. Li, Y., Yang, N., Wang, L., Wei, F., Li, W.: Learning to rank in generative retrieval. arXiv preprint arXiv:2306.15222 (2023)
13. Li, Y., Yang, N., Wang, L., Wei, F., Li, W.: Multiview identifiers enhanced generative retrieval. In: ACL, pp. 6636–6648 (2023)
14. Liu, S., Xiao, F., Ou, W., Si, L.: Cascade ranking for operational e-commerce search. In: KDD, pp. 1557–1565 (2017)

[1] https://ecir2024-generative-ir.github.io.

15. Liu, Y.A., Zhang, R., Guo, J., Chen, W., Cheng, X.: On the robustness of generative retrieval models: an out-of-distribution perspective. In: Gen-IR@SIGIR (2023)
16. Metzler, D., Tay, Y., Bahri, D., Najork, M.: Rethinking search: making domain experts out of dilettantes. SIGIR Forum **55**(1), 1–27 (2021)
17. Mitra, B., Nalisnick, E., Craswell, N., Caruana, R.: A dual embedding space model for document ranking. arXiv preprint arXiv:1602.01137 (2016)
18. Nadeem, U., Ziems, N., Wu, S.: CodeDSI: differentiable code search. arXiv preprint arXiv:2210.00328 (2022)
19. Najork, M.: Generative information retrieval. In: SIGIR, p. 1 (2023)
20. Nguyen, T., Yates, A.: Generative retrieval as dense retrieval. In: Gen-IR@SIGIR (2023)
21. Nguyen, T., et al.: MS MARCO: a human generated machine reading comprehension dataset. In: Workshop on Cognitive Computation: Integrating Neural and Symbolic Approaches (2016)
22. Pradeep, R., et al.: How does generative retrieval scale to millions of passages? In: Gen-IR@SIGIR (2023)
23. Ren, R., Zhao, W.X., Liu, J., Wu, H., Wen, J.R., Wang, H.: TOME: a two-stage approach for model-based retrieval. In: ACL, pp. 6102–6114 (2023)
24. Rose, D.E., Levinson, D.: Understanding user goals in web search. In: WWW, pp. 13–19 (2004)
25. Sun, W., et al.: Learning to tokenize for generative retrieval. In: NeurIPS (2023)
26. Tang, Y., et al.: Semantic-enhanced differentiable search index inspired by learning strategies. In: KDD, pp. 4904–4913 (2023)
27. Tay, Y., et al.: Transformer memory as a differentiable search index. In: NeurIPS, vol. 35, pp. 21831–21843 (2022)
28. Wang, Y., et al.: A neural corpus indexer for document retrieval. In: NeurIPS, vol. 35, pp. 25600–25614 (2022)
29. Wang, Z., Zhou, Y., Tu, Y., Dou, Z.: NOVO: learnable and interpretable document identifiers for model-based IR. In: CIKM, pp. 2656–2665 (2023)
30. Zhang, P., Liu, Z., Zhou, Y., Dou, Z., Cao, Z.: Term-sets can be strong document identifiers for auto-regressive search engines. arXiv preprint arXiv:2305.13859 (2023)
31. Zhou, Y.J., Yao, J., Dou, Z.C., Wu, L., Wen, J.R.: DynamicRetriever: a pre-trained model-based IR system without an explicit index. Mach. Intell. Res. **20**(2), 276–288 (2023)
32. Zhou, Y., Yao, J., Dou, Z., Wu, L., Zhang, P., Wen, J.R.: Ultron: an ultimate retriever on corpus with a model-based indexer. arXiv preprint arXiv:2208.09257 (2022)
33. Zhuang, S., et al.: Bridging the gap between indexing and retrieval for differentiable search index with query generation. In: Gen-IR@SIGIR (2023)

Transformers for Sequential Recommendation

Aleksandr V. Petrov(✉) and Craig Macdonald

University of Glasgow, Glasgow, UK
a.petrov.1@research.gla.ac.uk, craig.macdonald@glasgow.ac.uk

Abstract. Sequential recommendation is a recommendation problem that aims to predict the next item in the sequence of user-item interactions. Sequential recommendation is similar to language modelling in terms of learning sequence structure; therefore, variants of the Transformer architecture, which has recently become mainstream in language modelling, also achieved state-of-the-art performance in sequential recommendation. However, despite similarities, training Transformers for recommendation models may be tricky: most recommendation datasets have their unique item sets, and therefore, the pre-training/finetuning approach, which is very successful for training language models, has limited applications for recommendations. Moreover, a typical recommender system has to work with millions of items, much larger than the vocabulary size of language models. In this tutorial, we cover adaptations of Transformers for sequential recommendation and techniques that help to mitigate the training challenges. The half-day (3 h + a break) tutorial consists of two sessions. The first session provides a background of the Transformer architecture and its adaptations to Recommendation scenarios. It covers classic Transformer-based models, such as SASRec and BERT4Rec, their architectures, training tasks and loss functions. In this session, we also discuss the specifics of training these models with large datasets and discuss negative sampling and the mitigation problem of the overconfidence problem caused by negative sampling. We also discuss the problem of the large item embedding tensor and the approaches to mitigate this problem, allowing training of the models even with very large item catalogues. In the second part of the tutorial, we focus specifically on modern generative transformer-based models for sequential recommendation. We discuss specifics of generative models for sequential recommending, such as item ID representation and recommendation list generation strategies. We also cover modern adaptations of large language models (LLMs) to recommender systems and discuss concrete examples, such as the P5 model. We conclude the session with our vision for the future development of the recommender systems field in the era of Large Language Models.

1 Aims and Objectives

The overall aim of this tutorial is to give attendees a theoretical overview and practical recommendations for building sequential recommender systems using

N. Goharian et al. (Eds.): ECIR 2024, LNCS 14612, pp. 369–374, 2024.
https://doi.org/10.1007/978-3-031-56069-9_49

recent advancements in deep learning, language models and their adaptations to recommendation problems. By starting with the introduction of the sequential recommendation problem and overview of the transformer models (such as BERT [4] and GPT [19]) models, the content will be accessible for attendees with introductory knowledge of recommender systems or language models. The tutorial also aims to provide information on the practical aspects of scaling sequential recommender systems to large catalogues with millions of items, which are common in industrial applications such as e-commerce websites and music streaming services; therefore, it will help industry practitioners that attend ECIR how to adapt state-of-the-art models to their real-world problems. The tutorial also covers the most recent advancements in generative recommender models and intends to provide the attendees with the vision and ideas for further research.

Overall, after attending the tutorial, the attendees will achieve the following intended learning objectives:

ILO 1A. Describe the problem of Sequential Recommendation and draw the parallels with the problem of language modelling.

ILO 1B. Understand the Transformer architecture, Transformer Encoder-based models (BERT) and Transformer Decoder-based models (GPT).

ILO 1C. Describe classic transformer-based sequential recommender models, SASRec and BERT4Rec, and understand their similarities and differences.

ILO 1D. Understand the techniques of training the models for large catalogues: training objectives, negative sampling, model compression.

ILO 2A. Describe sequential recommendation as a generative task and draw parallels with language generation and generative transformers.

ILO 2B. Describe the item tokenisation problem and understand tokenisation techniques, such as SVD tokenisation.

ILO 2C. Describe the whole recommendation list generation problem and optimization for beyond-accuracy goals and describe listwise generative models, such as GPTRec and P5.

ILO 2D. Understand the role of Large Language Models (LLMs) in the future development of recommender systems.

2 Instructors

Aleksandr V. Petrov (m) (https://asash.github.io) is a third-year PhD candidate at the University of Glasgow's Information Retrieval Group, specializing in the usage of Transformer models for recommender systems with large catalogues. His scholarly contributions include multiple papers in premier recommender system venues, such as RecSys, WSDM, and ACM TORS. His work earned him a Best Paper award at RecSys 2023 and a Best Paper nomination at RecSys 2022. Before his PhD, Aleksandr accumulated over ten years of industry experience, with senior roles at large tech companies such as Amazon and Yandex, as well as a co-founder of the recommender systems startup E-Contenta.

Prof. Craig Macdonald (m) (http://www.dcs.gla.ac.uk/ craigm/) is a Professor within the Information Retrieval Group at the University of Glasgow. He has co-authored over 230 publications in information retrieval, including on efficient and effective query processing, as well as on the practical deployments of learning-to-rank approaches. He has received best paper awards at ECIR (2014), SIGIR (2015) and RecSys (2023). Craig has been joint coordinator of the TREC Blog, Microblog and Web tracks and is the lead maintainer of the PyTerrier platform. He has presented tutorials at ECIR 2008, ECIR 2017, SIGIR 2018, ECIR 2021 and CIKM 2022. He has lectured on Database Systems, Recommender Systems and Information Retrieval at the University of Glasgow to 1st year- and BSc/Master-level cohorts.

3 Motivation of the Tutorial

Sequential Recommendation is one of the most active research tasks in the field of recommender systems. For example, in the recent ACM RecSys 2023 conference, the premier venue for presenting recommender systems research, sequential recommendation was one of the most popular topics, with two sessions dedicated to sequential recommender systems. Sequential recommendation research papers are among the most cited papers in many IR-related conferences - for instance, SIGIR [8], WSDM [3,25], CIKM [6,23,29], ICDM [9], RecSys [27] all have papers with more than 100 citations (according to Google Scholar on 25/10/2023) addressing sequential recommendation. The Best Paper Awards in both RecSys 2022 [2] and RecSys 2023 [15] were given for the works on Transformers for Sequential Recommendation.

However, despite overall popularity, there is little research on the topic from the European community. For example, despite Recommender Systems being clearly stated as a topic of interest at the ECIR conference, there are no Sequential Recommendation papers among the most cited at ECIR, according to Google Scholar. The RecSys summer school 2023, which took place in Denmark and was organised mostly by European researchers, did not cover topics related to sequential recommendation and transformer models.

Therefore, we argue that sequential recommendation is an important and active area of research which is underrepresented in the European IR community. This ECIR tutorial intends to help close this gap and introduce Europeans, and indeed anyone else interested, to this active area of research.

Additionally, despite achieving great success in academia, Transformer-based sequential recommender systems are not common in industry because of efficiency problems when training these models for real-world datasets. This tutorial intends to close this gap and provide industry practitioners with practical techniques of scaling these models to industry-scale datasets.

Finally, generative large language models, such as ChatGPT, have recently achieved success in solving many tasks [1], and we believe that they will play a key role in the near-term development of the recommender systems. Transformer-based sequential recommender models are closely related to LLMs, as they utilise the same Transformer architecture and training tasks. We believe that sequential recommender systems will benefit a lot from the developments in large language models, and this tutorial intends to provide insights and ideas into future developments and motivate the attendees for future research on the topic. Moreover, the use of language models for generative retrieval [17,24,26] is of growing interest, and there are interesting parallels between sequential recommendation and generative retrieval.

4 Brief Description of Topics and Structure

The topic of the tutorial lies in the intersection of language models and recommender systems, and during the tutorial, we will cover the key papers from both domains and draw the parallels between them. The first session contains the introductory material to the problem of sequential recommendation and Transformer models, as well as describes the challenges of scaling them to real-world datasets. The second session covers the most recent research on generative models and Large Language models for recommendation. Below we provide the structure of the sessions and key references.
Session 1: 90 min - Classic Transformers For Sequential Recommendation: In this session, we provide the introductory material to the topic of sequential recommendation, including drawing the parallels with language models and describing the most popular models in both language models and sequential recommendation; in this session, we also highlight the differences between the language modelling and recommendation task and discuss challenges specific to recommendation problems, such as large catalogue.
Introduction: 20 min [ILO 1A] [5,7,10,11,21].
Transformers for Language Modelling: 20 min [ILO 1B] [4,18]
Popular Transformers for Sequential Recommendation: 20 min [ILO 1C] [9,14, 15,23].
Scaling Transformers to Large Catalogues: 30 min [ILO 1D] [13,15,16,22].
Session 2: 90 min - Generative Transformers and LLMs for Sequential Recommendation In the second session, we cover the most recent advancements in the domain of sequential recommendation with transformers and discuss how generative models and Large Language models can be used to solve some of the challenges of sequential recommender systems and provide our vision of the future development of sequential recommender systems.
Sequential recommendation as a Generative task: 30 min [ILO 2A] [12,18].
Items tokenisation: 20 min [ILO 2B] [12,20].
List-level generation for beyond-accuracy goals: 20 min [ILO 2C] [12].
Large Language Models and the future of recommendations: 20 min [ILO 2D] [1,28].

5 Target Audience

This tutorial is targeted at ECIR attendees who have at least introductory knowledge of the Recommender Systems or Information Retrieval and are familiar with the basics of machine learning (e.g., loss function, gradient descent). In particular, this tutorial will be of interest to PhD students who work on modern recommender systems, industry practitioners who work with large-scale recommender systems, and lecturers who are interested in including state-of-the-art material and examples for their university course.

References

1. Brown, T., et al.: Language models are few-shot learners. In: Proceedings of NeurIPS, vol. 33, pp. 1877–1901 (2020)
2. Chen, H., et al.: Denoising self-attentive sequential recommendation. In: Proceedings of the 16th ACM Conference on Recommender Systems, pp. 92–101 (2022)
3. Chen, X., et al.: Sequential recommendation with user memory networks. In: Proceedings of WSDM, pp. 108–116 (2018)
4. Devlin, J., Chang, M.W., Lee, K., Toutanova, K.: BERT: pre-training of deep bidirectional transformers for language understanding. In: Proceedings of NAACL-HLT, pp. 4171–4186 (2019)
5. Gunawardana, A., Shani, G., Yogev, S.: Evaluating recommender systems. In: Ricci, F., Rokach, L., Shapira, B. (eds.) Recommender Systems Handbook, pp. 547–601. Springer, New York (2022). https://doi.org/10.1007/978-1-0716-2197-4_15
6. Hidasi, B., Karatzoglou, A.: Recurrent neural networks with top-k gains for session-based recommendations. In: Proceedings of the CIKM, pp. 843–852 (2018)
7. Hidasi, B., Karatzoglou, A., Baltrunas, L., Tikk, D.: Session-based recommendations with recurrent neural networks. In: Proceedings of the ICLR (2016)
8. Huang, J., Zhao, W.X., Dou, H., Wen, J.R., Chang, E.Y.: Improving sequential recommendation with knowledge-enhanced memory networks. In: Proceedings of the SIGIR, pp. 505–514 (2018)
9. Kang, W.C., McAuley, J.: Self-attentive sequential recommendation. In: Proceedings of the ICDM, pp. 197–206 (2018)
10. Krichene, W., Rendle, S.: On sampled metrics for item recommendation. Commun. ACM **65**(7), 75–83 (2022)
11. Meng, Z., McCreadie, R., Macdonald, C., Ounis, I.: Exploring data splitting strategies for the evaluation of recommendation models. In: Proceedings of the RecSys, pp. 681–686 (2020)
12. Petrov, A.V., Macdonald, C.: Generative sequential recommendation with GPTRec. In: Proceedings of the Gen-IR@SIGIR (2023)
13. Petrov, A.V., Macdonald, C.: Effective and efficient training for sequential recommendation using recency sampling. In: Proceedings of the RecSys, pp. 81–91 (2022)
14. Petrov, A.V., Macdonald, C.: A Systematic Review and Replicability Study of BERT4Rec for Sequential Recommendation. In: Proc. RecSys. pp. 436–447 (2022)
15. Petrov, A.V., Macdonald, C.: gSASRec: reducing overconfidence in sequential recommendation trained with negative sampling. In: Proceedings of the RecSys, pp. 116–128 (2023)
16. Petrov, A.V., Macdonald, C.: RecJPQ: training large-catalogue sequential recommenders. In: Proceedings of the WSDM (2024)
17. Pradeep, R., et al.: How does generative retrieval scale to millions of passages? In: Proceedings of the Gen-IR@SIGIR (2023)
18. Radford, A., Narasimhan, K., Salimans, T., Sutskever, I.: Improving language understanding by generative pre-training (2018)

19. Radford, A., Wu, J., Child, R., Luan, D., Amodei, D., Sutskever, I.: Language models are unsupervised multitask learners (2019)
20. Rajput, S., et al.: Recommender systems with generative retrieval. In: Proceedings of the NeurIPS (2023)
21. Rendle, S., Freudenthaler, C., Schmidt-Thieme, L.: Factorizing personalized Markov chains for next-basket recommendation. In: Proceedings of the WWW, p. 811 (2010)
22. Shi, H.J.M., Mudigere, D., Naumov, M., Yang, J.: Compositional embeddings using complementary partitions for memory-efficient recommendation systems. In: Proceedings of the KDD, pp. 165–175 (2020)
23. Sun, F., et al.: BERT4Rec: sequential recommendation with bidirectional encoder representations from transformer. In: Proceedings of the CIKM, pp. 1441–1450 (2019)
24. Sun, W., et al.: Learning to tokenize for generative retrieval. In: Proceedings of the NeurIPS (2023)
25. Tang, J., Wang, K.: Personalized top-N sequential recommendation via convolutional sequence embedding. In: Proceedings of the WSDM, pp. 565–573 (2018)
26. Tay, Y., et al.: Transformer memory as a differentiable search index (2022)
27. Wu, L., Li, S., Hsieh, C.J., Sharpnack, J.: SSE-PT: sequential recommendation via personalized transformer. In: Proceedings of the RecSys, pp. 328–337 (2020)
28. Yuan, Z., et al.: Where to go next for recommender systems? ID- vs. modality-based recommender models revisited (2023)
29. Zhou, K., et al.: S3-Rec: self-supervised learning for sequential recommendation with mutual information maximization. In: Proceedings of the CIKM, pp. 1893–1902 (2020)

Affective Computing for Social Good Applications: Current Advances, Gaps and Opportunities in Conversational Setting

Priyanshu Priya[1(✉)], Maujama Firdaus[2], Gopendra Vikram Singh[1], and Asif Ekbal[1(✉)]

[1] Indian Institute of Technology Patna, Dayalpur Daulatpur, India
{priyanshu_2021cs26,gopendra_1921cs15,asif}@iitp.ac.in,
priyanshu528priya@gmail.com, asif.ekabl@gmail.com
[2] University of Alberta, Edmonton, Canada

Abstract. Affective computing involves examining and advancing systems and devices capable of identifying, comprehending, processing, and emulating human emotions, sentiment, politeness and personality characteristics. This is an ever-expanding multidisciplinary domain that investigates how technology can contribute to the comprehension of human affect, how affect can influence interactions between humans and machines, how systems can be engineered to harness affect for enhanced capabilities, and how integrating affective strategies can revolutionize interactions between humans and machines. Recognizing the fact that affective computing encompasses disciplines such as computer science, psychology, and cognitive science, this tutorial aims to delve into the historical underpinnings and overarching objectives of affective computing, explore various approaches for affect detection and generation, its practical applications across diverse areas, including but not limited to social good (like persuasion, therapy and support, etc.), address ethical concerns, and outline potential future directions.

Keywords: Affect · Emotion · Sentiment · Empathy · Persuasion · Politeness · Mental Health

1 Motivation

Affect is a vital component for ensuring humanly essence to human-machine interactions because of its ability to simulate empathy and other empathy-related aspects. Consequently, the affective computing research works have witnessed a great surge of interest in recent times owing to their promising potential and alluring commercial and social importance. Affect-aware systems that are capable of comprehending users' affective states help humans achieve their desired goals whilst acting as a companion to them.

N. Goharian et al. (Eds.): ECIR 2024, LNCS 14612, pp. 375–380, 2024.
https://doi.org/10.1007/978-3-031-56069-9_50

The current advances in affective computing span numerous human aspects catering to a large pool of research. These aspects include emotion, sentiment, emotion cause, politeness, multimodality and many more. It illustrates the wide applicability of affective systems but also makes it more challenging for beginners to get started. The beginners should have knowledge of these aspects in addition to being conversant with the latest developments in neural natural language processing (NLP) techniques. A thorough tutorial could offer them valuable insights. With this objective, we offer this tutorial, which functions as a comprehensive manual for research and advancement in affective computing and proposes potential future pathways within this field. An in-depth comprehension of affective computing highlights the importance of understanding and responding to user emotions, which can lead to more relevant and effective information retrieval.

2 Tutorial Details

Tutorial Format. This tutorial is designed to provide comprehensive understanding on affective computing for social good applications. It will delve deep into specific topics such as relevance of empathy, politeness, etc. in conversational agents for social good providing attendees with a thorough knowledge base. We intend to present well-organized tutorial following a structured outline that would include presentations, demonstrations and practical examples. The tutorial aims to encourage active participation from attendees, which may involve asking question, engaging in discussions and future collaborations, if possible.

Length. Half day (3 h plus break)

Detailed Outline. The tutorial will be outlined as follows:

1. **Fundamentals of Affective Computing (15 min)** We will introduce the fundamental concepts of affective computing, including the different models of affect and various typology of affective states [13,17].
2. **Need for Affect in Human-Machine Interaction (15 min)** We will precisely talk about the works suggesting that the incorporation of affect in machines could make the machine emotionally intelligent and enhance the user experience, eventually bridging the human-machine gap [9].
3. **Affect in Text, Voice, and Facial Expressions (30 min)** Affect, an intrinsic aspect of humans, is evident through visual, vocal, and verbal signals. We will discuss leveraging information from different modalities like text, audio, and video for enhanced affective computing.
4. **Affect for Social Good Application Areas (90 min)** This part covers the relevance of affect-aware systems for social good application areas:
   Therapy and Support: Mental health issues are significant public health concerns and have been the subject of extensive research, including computational investigations. The authors of [7,14,19] introduced a deep learning approach for categorizing adverse emotional states associated with mental

health issues. Recently, there have been attempts to build socially and/or emotionally-intelligent systems [11,18].

Persuasion: Persuasion helps individuals convince someone to agree with their point of view, motivate them to take action, or resolve a conflict. Recently, there have been a few works that explored persuasion for charity donation task [12,16,20], mental health counseling and legal aid [15].

Suicide Prevention: Affective computing plays a vital role in suicide prevention, offering various advantages. It enables continuous real-time monitoring of individuals' emotional well-being, prompting alerts for timely intervention [3–6]. AI-powered systems equipped with emotion recognition capabilities promptly assist those in need. Predictive analytics discern individuals at higher risk based on historical emotional patterns, while sentiment-based triage prioritizes those experiencing immediate distress.

E-commerce: Affective computing has a transformative impact on e-commerce, enhancing user experiences through personalized product recommendations and real-time satisfaction assessment. By harnessing affective information, it can build trust, drive sales, and improve customer satisfaction [1,2,8,10].

5. **Ethical and Social Implications of Affective Computing (15 min)** Affective systems are ubiquitous in everyday life. Research on how we use empathy in our lives spans several applications like customer care, therapy, etc. While we think about these applications, it is also crucial to keep several ethical considerations in mind, including: **(i)** Privacy and Confidentiality, **(ii)** Fairness and Bias, **(iii)** Transparency and Explainability, **(iv)** Empathy and Emotional Intelligence, and **(v)** User Empowerment and Control to foster responsible AI development.
6. **Conclusion and Future Directions (15 min)** Affective computing holds promise for more empathetic, persuasive, and socially responsible AI systems, but ethical considerations and responsible deployment remain crucial as we harness their full potential for society's benefit. The future of affective computing involves refining emotional detection accuracy, especially in suicide prevention. It also entails integrating NLP and computer vision for comprehensive multimodal empathy detection. Besides, there is potential for real-time applications in persuasion analysis and politeness assessment, particularly in social media monitoring and content optimization.

3 Target Audience

We believe that the potential target audience for this tutorial could be students at all levels (Doctorals, Masters, Bachelors), and anyone who is associated with e-commerce, customer care, and related application areas, and researchers. At the end of the tutorial, the audience will be able to answer what affective computing is and its scope, grasp the significance of emotional artificial intelligence (AI) and how it contributes to improved human-computer interactions, different deep learning-based approaches to identify and/or generate affective information, various challenges associated with affective computing, including privacy concerns and ethical considerations, and its diverse range of applications in healthcare, marketing, customer service, and more.

The comprehensive reading materials list and our relevant expertise are available at https://github.com/priyanshu528priya/ECIR2024-Tutorial.

4 Presenters

1. **Priyanshu Priya** is a Research Scholar with the Department of Computer Science and Engineering, Indian Institute of Technology (IIT) Patna, India. Her research interests include natural language processing with a focus on politeness and empathy in conversations. Priyanshu has published several journal and conference papers in the reputed foras, such as Expert Systems With Applications, LREC, AAAI, IJCAI, ACL, IJCNLP-AACL, EMNLP. Priyanshu has presented a tutorial– Empathetic Conversational Artificial Intelligence Systems: Recent Advances and New Frontiers, in IJCAI 2023.
2. **Mauajama Firdaus** is a PostDoc researcher in the Department of Computer Science at the University of Alberta, Canada. She completed her Ph.D. from IIT Patna in 2021. Her main area of research is Natural Language Processing and Dialogue Generation. She has published papers in various peer-reviewed conferences and journals of international repute and has been a reviewer for top-tier conferences like AAAI, IJCAI, EMNLP, etc. Mauajama has also presented a tutorial, Empathetic Conversational Artificial Intelligence Systems: Recent Advances and New Frontiers, in IJCAI 2023.
3. **Gopendra Vikram Singh** is a Research Scholar with the Department of Computer Science and Engineering, IIT Patna, Patna, India. His research is focused on recognizing sentiment, emotion, humor, and sarcasm within conversations, predicting multilabel emojis, detecting cognitive distortion in clinical conversations, and identifying gender-biased content within social media posts.
4. **Asif Ekbal**, Indian Institute of Technology Patna, India.
   Webpage: http://www.iitp.ac.in/~asif/
   Asif Ekbal is currently an Associate Professor in the Department of Computer Science and Engineering, IIT Patna. He has been pursuing research in NLP, Information Extraction, Text Mining and ML applications for the last 18 years. He has authored around 280 papers in top-tier conferences (AAAI, IJCAI, ACL, EMNLP, etc.) and journals (PLOS One, ACM Transactions, IEEE Transaction on Affective Computing etc.) Asif has been involved in several sponsored research projects funded by different private agencies (Elsevier, Accenture, Samsung, Flipkart); and Govt. agencies (MeiTY, Govt. of India, MHRD, Govt. of India, and SERB, Govt. of India etc.). He has presented tutorials in IJCAI 2023 (Conversational AI), IJCAI 2019 (Sentiment Analysis) and in Flipkart Data Science Conference (Machine Translation, and has delivered around 55 invited talks in conferences, workshops, and in Industries, such as Accenture, Microsoft, Huawei, Wipro etc. He is an awardee of the "Best Innovative Project Award from the Indian National Academy

of Engineering", Govt. of India, "JSPS Invitation Fellowship" from Govt of Japan and "Visvesvaraya Young Faculty Research Fellowship Award" of the Govt. of India. Google Scholar Citation which is the benchmark for Computer Science and Engineering shows 9084 career citations to Dr. Asif's papers. His h-index: 47 and i-10 index: 208.

References

1. Firdaus, M., Ekbal, A., Bhattacharyya, P.: Incorporating politeness across languages in customer care responses: towards building a multi-lingual empathetic dialogue agent. In: Proceedings of the Twelfth Language Resources and Evaluation Conference, pp. 4172–4182 (2020)
2. Firdaus, M., Ekbal, A., Bhattacharyya, P.: PoliSe: reinforcing politeness using user sentiment for customer care response generation. In: Proceedings of the 29th International Conference on Computational Linguistics, pp. 6165–6175 (2022)
3. Ghosh, S., Ekbal, A., Bhattacharyya, P.: Cease, a corpus of emotion annotated suicide notes in English. In: Proceedings of The 12th Language Resources and Evaluation Conference, pp. 1618–1626 (2020)
4. Ghosh, S., Ekbal, A., Bhattacharyya, P.: Am i no good? Towards detecting perceived burdensomeness and thwarted belongingness from suicide notes. arXiv preprint arXiv:2206.06141 (2022)
5. Ghosh, S., Ekbal, A., Bhattacharyya, P.: A multitask framework to detect depression, sentiment and multi-label emotion from suicide notes. Cogn. Comput. **14**, 1–20 (2022)
6. Ghosh, S., Maurya, D.K., Ekbal, A., Bhattacharyya, P.: Em-persona: emotion-assisted deep neural framework for personality subtyping from suicide notes. In: Proceedings of the 29th International Conference on Computational Linguistics, pp. 1098–1105 (2022)
7. Ghosh, S., Singh, G.V., Ekbal, A., Bhattacharyya, P.: COMMA-DEER: commonsense aware multimodal multitask approach for detection of emotion and emotional reasoning in conversations. In: Proceedings of the 29th International Conference on Computational Linguistics, pp. 6978–6990 (2022)
8. Golchha, H., Firdaus, M., Ekbal, A., Bhattacharyya, P.: Courteously yours: inducing courteous behavior in customer care responses using reinforced pointer generator network. In: Proceedings of the 2019 Conference of the North American Chapter of the Association for Computational Linguistics: Human Language Technologies, Volume 1 (Long and Short Papers), pp. 851–860 (2019)
9. Kühnlenz, B., Sosnowski, S., Buß, M., Wollherr, D., Kühnlenz, K., Buss, M.: Increasing helpfulness towards a robot by emotional adaption to the user. Int. J. Soc. Robot. **5**, 457–476 (2013)
10. Mishra, K., Firdaus, M., Ekbal, A.: Please be polite: towards building a politeness adaptive dialogue system for goal-oriented conversations. Neurocomputing **494**, 242–254 (2022)
11. Mishra, K., Priya, P., Ekbal, A.: Help me heal: a reinforced polite and empathetic mental health and legal counseling dialogue system for crime victims. In: Proceedings of the AAAI Conference on Artificial Intelligence, vol. 37, pp. 14408–14416 (2023)
12. Mishra, K., Samad, A.M., Totala, P., Ekbal, A.: PEPDS: a polite and empathetic persuasive dialogue system for charity donation. In: Proceedings of the 29th International Conference on Computational Linguistics, pp. 424–440 (2022)

13. Picard, R.W.: Affective Computing. MIT Press, Cambridge (2000)
14. Priya, P., Firdaus, M., Ekbal, A.: A multi-task learning framework for politeness and emotion detection in dialogues for mental health counselling and legal aid. Expert Syst. Appl. **224**, 120025 (2023)
15. Priya, P., Mishra, K., Totala, P., Ekbal, A.: Partner: a persuasive mental health and legal counselling dialogue system for women and children crime victims. In: Elkind, E. (ed.) Proceedings of the Thirty-Second International Joint Conference on Artificial Intelligence, IJCAI 2023, pp. 6183–6191. International Joint Conferences on Artificial Intelligence Organization, August 2023. https://doi.org/10.24963/ijcai.2023/686, https://doi.org/10.24963/ijcai.2023/686, aI for Good
16. Samad, A.M., Mishra, K., Firdaus, M., Ekbal, A.: Empathetic persuasion: reinforcing empathy and persuasiveness in dialogue systems. In: Findings of the Association for Computational Linguistics: NAACL 2022, pp. 844–856 (2022)
17. Scherer, K.R.: Emotion as a multicomponent process: a model and some cross-cultural data. J. Pers. Soc. Psychol. **5**, 37–63 (1984)
18. Sharma, A., Miner, A.S., Atkins, D.C., Althoff, T.: A computational approach to understanding empathy expressed in text-based mental health support. arXiv preprint arXiv:2009.08441 (2020)
19. Singh, G.V., Ghosh, S., Ekbal, A., Bhattacharyya, P.: Decode: detection of cognitive distortion and emotion cause extraction in clinical conversations. In: Kamps, J., et al. (eds.) ECIR 2023. LNCS, vol. 13981, pp. 156–171. Springer, Cham (2023). https://doi.org/10.1007/978-3-031-28238-6_11
20. Wang, X., et al.: Persuasion for good: towards a personalized persuasive dialogue system for social good. arXiv preprint arXiv:1906.06725 (2019)

Query Performance Prediction: From Fundamentals to Advanced Techniques

Negar Arabzadeh[1(✉)], Chuan Meng[2], Mohammad Aliannejadi[2], and Ebrahim Bagheri[3]

[1] University of Waterloo, Waterloo, Canada
narabzad@uwaterloo.ca

[2] University of Amsterdam, Amsterdam, The Netherlands
{c.meng,m.aliannejadi}@uva.nl

[3] Toronto Metropolitan University, Toronto, Canada
Bagheri@torontomu.ca

Abstract. Query performance prediction (QPP) is a core task in information retrieval (IR) that aims at predicting the retrieval quality for a given query without relevance judgments. QPP has been investigated for decades and has witnessed a surge in research activity in recent years; QPP has been shown to benefit various aspects, e.g., improving retrieval effectiveness by selecting the most effective ranking function per query [5,7]. Despite its importance, there is no recent tutorial to provide a comprehensive overview of QPP techniques in the era of pre-trained/large language models or in the scenario of emerging conversational search (CS); In this tutorial, we have three main objectives. First, we aim to disseminate the latest advancements in QPP to the IR community. Second, we go beyond investigating QPP in ad-hoc search and cover QPP for CS. Third, the tutorial offers a unique opportunity to bridge the gap between theory and practice; we aim to equip participants with the essential skills and insights needed to navigate the evolving landscape of QPP, ultimately benefiting both researchers and practitioners in the field of IR and encouraging them to work around the future avenues on QPP.

1 Motivation and Objectives

Query performance prediction (QPP) is a core task in information retrieval (IR). The task of QPP aims to predict the effectiveness of retrieved results for a given query in the absence of relevance judgments [14,18,24,33]. The importance of the QPP task cannot be overstated because of numerous applications of QPP. QPP serves as a guiding compass at query time. QPP can choose the most effective ranking function per query and select the best variant from multiple query reformulations [51], delivering more relevant and tailored results to users [11,15]. QPP can be instrumental in optimizing the retrieval efficiency by choosing the appropriate number of documents to process in a multi-stage retrieval setting [19]. Moreover, in the scenario of emerging conversational search (CS) [25,37,39], QPP can predict the retrieval quality for each user query in a conversation and

N. Goharian et al. (Eds.): ECIR 2024, LNCS 14612, pp. 381–388, 2024.
https://doi.org/10.1007/978-3-031-56069-9_51

assist an intelligent system in determining when to take the initiative (e.g., ask a clarifying question) [1] to get more information from a user [6,43]. Apart from the application at query time, a recent study [29] has shown the application of QPP in creating a collection; this study uses QPP to predict adaptive pool depth for each query so as to reduce relevance judgment costs.

Despite the importance of QPP to the IR community, to the best of our knowledge, there has been only one tutorial [42] about QPP at ICTIR 2020. We believe that it is the right time to revisit the idea of having a QPP tutorial for the following three reasons: (i) with the rapid development of advanced neural-based techniques in recent years, QPP has witnessed a surge in proposing neural-based QPP methods [4,16,23,35,36,47,51] or predicting the performance of neural-based retrievers [3,22,26,27,47]; (ii) conversational search (CS) [30,38,52] has been recognized as an emerging research area in IR, and research [25,37,39] into QPP for CS have been conducted very recently.

This tutorial has three main aims to provide a platform to disseminate the latest advancements in QPP to the IR community. Moreover, we go beyond investigating QPP in ad-hoc search and delve into studying QPP for CS, which has remained relatively under-explored. Last but not least, this tutorial offers a unique opportunity to bridge the gap between theory and practice. We will provide practical insights and hands-on experience to empower researchers, practitioners, and enthusiasts to effectively apply QPP techniques in their work. In summary, our main objectives in this tutorial are as follows:

- **Highlighting the importance of QPP:** we aim to underscore the pivotal role that QPP plays in enhancing IR, by illuminating various applications of QPP.
- **Providing a comprehensive overview of latest QPP methods:** our tutorial endeavors to provide participants with a thorough and up-to-date overview of state-of-the-art QPP techniques. This comprehensive survey spans both lexical-based methods and the latest neural-based QPP approaches, ensuring that attendees gain insights into the diverse landscape of QPP methodologies.
- **Exploring the impact of cutting-edge neural-based techniques advancements on QPP:** we shed light on the latest developments within the QPP task, specifically, how predicting the performance for neural-based retrievers differs from conventional sparse retrievers. We also cover the work done exploring the potential of leveraging potential of leveraging large language models (LLMs) to enhance QPP methods. Our goal is to foster a forward-looking perspective, inspiring participants to contribute to the ongoing evolution of QPP research in the IR community.
- **Exploring QPP for CS:** we first summarize the findings and insights from very recent studies into QPP for CS, and then show the drawbacks of current QPP methods in the scenario of CS, and discuss open avenues for future research.
- **Facilitating hands-on experience with supporting materials:** to enhance the learning experience, we are committed to equipping participants with supporting materials that enable practical application.

2 Format and Schedule

This tutorial will span a half-day, totaling 3 h plus breaks, and will be presented in person. We propose the following program:

Introduction to QPP [25 min]. This opening section provides (i) a foundational understanding of the task of QPP in the context of ad-hoc search and CS, (ii) applications of QPP (why QPP is vital) in these domains, (iii) different categorizations of QPP methods, including pre-retrieval and post-retrieval QPP methods, and (iv) various evaluation methods [28] relevant to QPP task.

Pre-retrieval QPP Methods [30 min]. This section focuses on strategies for conducting QPP before the retrieval process. We include (i) statistical-based pre-retrieval QPP methods [32–34,53], and (ii) recently introduced neural-based ones [8,9,44,45,50] We highlight the drawbacks and advantages of both categories.

Post-retrieval QPP Methods [55 min]. Since post-retrieval QPP methods have attracted more attention compared to pre-retrieval QPP methods, we will cover both lexical-based QPP methods and neural-based methods, separately. We will explore the principles and techniques that underlie lexical-based QPP, gaining insights into how lexical features and linguistic analysis can be harnessed to assess and enhance QPP after retrieval, covering state-of-the-art retrieval score-based QPP methods [10,20,40,46,48,54]. In addition, We dive into the cutting-edge domain of neural-based post-retrieval QPP methods. Participants will acquire a profound understanding of how neural networks and advanced machine learning techniques are employed to forecast query performance following the retrieval process. Our exploration covers QPP methods based on basic neural networks [2,21,51], along with methods based on pre-trained language models [4,16,23,31,36].

Impact of Retriever Types on QPP Effectiveness [25 min]. Many studies in QPP have primarily focused on predicting the performance of traditional high-dimensional sparse retrievers like BM25. However, recent research has expanded the scope by investigating how predicting the performance of other retriever types, such as dense or learned sparse retrievers, can differ from sparse retrievers [3,22,26,27,47]. In this section, we first demonstrate and discuss how different retriever types impact the effectiveness of QPP methods, paving the way for a deeper understanding of this evolving landscape; and we pose research questions that hold the potential to shape the future of QPP in this domain, encouraging exploration in this field.

QPP and LLMs [15 min]. LLMs have already been successfully used in many IR tasks, such as re-ranking [41] and query expansion [49]. However, QPP using LLMs has been little studied. In this section, we will explore and discuss the potential ways of harnessing LLMs to enhance QPP effectiveness.

QPP for CS [15 min]. This section delves into QPP for CS. We aim to (i) summarize the findings and insights from very recent studies [25,37,39] into

QPP for CS, (ii) show the drawbacks of current QPP methods in the scenario of CS, and (iii) discuss open avenues for future research.

Limitations and Future Work [10 min]. In the final segment, we will (i) engage in a thoughtful discussion about the major theoretical and conceptual limitations that currently exist in QPP research and practice, and (ii) present exciting avenues for future work, inspiring participants to contribute to the ongoing evolution of QPP in IR. Last but not least, throughout the whole tutorial, we hold interactive *hands-on experience sessions* in which, participants will receive access to interactive Google Colab Notebooks, built on top of our recent comprehensive QPP repository,[1] allowing them to apply their knowledge in practice. Attendees will have the opportunity to implement QPP methods in real-world scenarios, reinforcing their understanding and refining their practical skills, providing a more tangible experience of prediction outcomes and enabling participants to compare them from computational perspectives.

3 Intended Audience

Our tutorial is thoughtfully designed to accommodate a diverse audience, catering to individuals with varying levels of familiarity with IR and related subjects. Our intended audience for this tutorial consists of two main groups: (i) professionals who already possess a solid understanding of fundamental IR techniques; our tutorial serves as a valuable extension of their knowledge by providing deeper insights and practical knowledge; and (ii) people who may be relatively new to these advanced topics; we offer them a friendly entry point by starting with the basics of problem formulation; while we assume that our participants have a foundational understanding of topics typically covered in an undergraduate IR course, our tutorial will provide the necessary details and explanations to ensure that all participants can comfortably access and comprehend the content.

4 Presenters

Negar Arabzadeh is a Ph.D. student at the University of Waterloo. Her research is aligned with ad-hoc search and CS in IR. Negar's Master's thesis was focused on neural-based pre-retrieval QPP. She has published relevant papers in SIGIR, CIKM, ECIR, and IP&M. Negar has previously conducted tutorials in SIGIR 2022, ECIR 2023, and WSDM 2023 [12,13,17].

Chuan Meng is a Ph.D. student at IRLab, University of Amsterdam, supervised by Maarten de Rijke and Mohammad Aliannejadi. His main research topic is CS and QPP. Chuan has published papers in prestigious proceedings such as SIGIR, EMNLP, CIKM, and AAAI. Moreover, he serves as a committee member for conferences including ACL, WWW, EMNLP, WSDM, COLING, SIGKDD, AAAI, ECIR, and a reviewer for journals including TOIS and IP&M.

[1] https://github.com/ChuanMeng/QPP4CS.

Mohammad Aliannejadi is an Assistant Professor at IRLab, University of Amsterdam. His research interests include conversational information access, recommender systems, and QPP. Mohammad has co-organized various evaluation campaigns such as TREC CAsT, TREC iKAT, ConvAI3, and IGLU. Moreover, Mohammad has held multiple tutorials and lectures on CS, such as CHIIR, SIKS, and ASIRF.

Ebrahim Bagheri is a Professor and the Director for the Laboratory for Systems, Software, and Semantics (LS^3) at Toronto Metropolitan University. He holds a Canada Research Chair (Tier II) in social information retrieval and an NSERC industrial research chair in social media analytics. He currently leads the NSERC program on responsible AI (http://responsible-ai.ca). He is an associate editor for ACM transactions on intelligent systems and technology (TIST) and Wiley's computational intelligence.

References

1. Aliannejadi, M., Kiseleva, J., Chuklin, A., Dalton, J., Burtsev, M.: Building and evaluating open-domain dialogue corpora with clarifying questions. In: EMNLP (2021)
2. Arabzadeh, N., Bigdeli, A., Zihayat, M., Bagheri, E.: Query performance prediction through retrieval coherency. In: Hiemstra, D., Moens, M.-F., Mothe, J., Perego, R., Potthast, M., Sebastiani, F. (eds.) ECIR 2021, Part II. LNCS, vol. 12657, pp. 193–200. Springer, Cham (2021). https://doi.org/10.1007/978-3-030-72240-1_15
3. Arabzadeh, N., Hamidi Rad, R., Khodabakhsh, M., Bagheri, E.: Noisy perturbations for estimating query difficulty in dense retrievers. In: CIKM (2023)
4. Arabzadeh, N., Khodabakhsh, M., Bagheri, E.: BERT-QPP: contextualized pre-trained transformers for query performance prediction. In: CIKM (2021)
5. Arabzadeh, N., Mitra, B., Bagheri, E.: MS MARCO chameleons: challenging the MS MARCO leaderboard with extremely obstinate queries. In: Proceedings of the 30th ACM International Conference on Information & Knowledge Management, pp. 4426–4435 (2021)
6. Arabzadeh, N., Seifikar, M., Clarke, C.L.: Unsupervised question clarity prediction through retrieved item coherency. In: CIKM, pp. 3811–3816 (2022)
7. Arabzadeh, N., Yan, X., Clarke, C.L.: Predicting efficiency/effectiveness trade-offs for dense vs. sparse retrieval strategy selection. In: Proceedings of the 30th ACM International Conference on Information & Knowledge Management, pp. 2862–2866 (2021)
8. Arabzadeh, N., Zarrinkalam, F., Jovanovic, J., Al-Obeidat, F., Bagheri, E.: Neural embedding-based specificity metrics for pre-retrieval query performance prediction. IP&M **57**(4), 102248 (2020)
9. Arabzadeh, N., Zarrinkalam, F., Jovanovic, J., Bagheri, E.: Neural embedding-based metrics for pre-retrieval query performance prediction. In: Jose, J.M., et al. (eds.) ECIR 2020, Part II. LNCS, vol. 12036, pp. 78–85. Springer, Cham (2020). https://doi.org/10.1007/978-3-030-45442-5_10
10. Arabzadeh, N., Zarrinkalam, F., Jovanovic, J., Bagheri, E.: Geometric estimation of specificity within embedding spaces. In: Proceedings of the 28th ACM International Conference on Information and Knowledge Management, pp. 2109–2112 (2019)

11. Bellogín, A., Castells, P.: Predicting neighbor goodness in collaborative filtering. In: Andreasen, T., Yager, R.R., Bulskov, H., Christiansen, H., Larsen, H.L. (eds.) FQAS 2009. LNCS (LNAI), vol. 5822, pp. 605–616. Springer, Heidelberg (2009). https://doi.org/10.1007/978-3-642-04957-6_52
12. Bigdeli, A., Arabzadeh, N., SeyedSalehi, S., Zihayat, M., Bagheri, E.: Gender fairness in information retrieval systems. In: SIGIR (2022)
13. Bigdeli, A., Arabzadeh, N., Seyedsalehi, S., Zihayat, M., Bagheri, E.: Understanding and mitigating gender bias in information retrieval systems. In: Kamps, J., et al. (eds.) ECIR 2023. LNCS, vol. 13982, pp. 315–323. Springer, Cham (2023). https://doi.org/10.1007/978-3-031-28241-6_32
14. Carmel, D., Yom-Tov, E.: Estimating the query difficulty for information retrieval. Synthesis Lectures on Information Concepts, Retrieval, and Services (2010)
15. Carmel, D., Yom-Tov, E., Roitman, H.: Enhancing digital libraries using missing content analysis. In: Proceedings of the 8th ACM/IEEE-CS Joint Conference on Digital Libraries, pp. 1–10 (2008)
16. Chen, X., He, B., Sun, L.: Groupwise query performance prediction with BERT. In: Hagen, M., et al. (eds.) ECIR 2022. LNCS, vol. 13186, pp. 64–74. Springer, Cham (2022). https://doi.org/10.1007/978-3-030-99739-7_8
17. Clarke, C.L., Diaz, F., Arabzadeh, N.: Preference-based offline evaluation. In: WSDM, pp. 1248–1251 (2023)
18. Cronen-Townsend, S., Zhou, Y., Croft, W.B.: Predicting query performance. In: SIGIR, pp. 299–306 (2002)
19. Culpepper, J.S., Clarke, C.L., Lin, J.: Dynamic cutoff prediction in multi-stage retrieval systems. In: Proceedings of the 21st Australasian Document Computing Symposium, pp. 17–24 (2016)
20. Cummins, R., Jose, J., O'Riordan, C.: Improved query performance prediction using standard deviation. In: SIGIR (2011)
21. Datta, S., Ganguly, D., Greene, D., Mitra, M.: Deep-QPP: a pairwise interaction-based deep learning model for supervised query performance prediction. In: WSDM, pp. 201–209 (2022)
22. Datta, S., Ganguly, D., Mitra, M., Greene, D.: A relative information gain-based query performance prediction framework with generated query variants. TOIS **41**(2), 1–31 (2022)
23. Datta, S., MacAvaney, S., Ganguly, D., Greene, D.: A 'pointwise-query, listwise-document' based query performance prediction approach. In: SIGIR, pp. 2148–2153 (2022)
24. Faggioli, G., Ferro, N., Mothe, J., Raiber, F.: QPP++ 2023: query-performance prediction and its evaluation in new tasks. In: Kamps, J., et al. (eds.) ECIR 2023. LNCS, vol. 13982, pp. 388–391. Springer, Cham (2023). https://doi.org/10.1007/978-3-031-28241-6_42
25. Faggioli, G., Ferro, N., Muntean, C.I., Perego, R., Tonellotto, N.: A geometric framework for query performance prediction in conversational search. In: SIGIR, pp. 1355–1365 (2023)
26. Faggioli, G., et al.: Towards query performance prediction for neural information retrieval: challenges and opportunities. In: ICTIR, pp. 51–63 (2023)
27. Faggioli, G., Formal, T., Marchesin, S., Clinchant, S., Ferro, N., Piwowarski, B.: Query performance prediction for neural IR: Are we there yet? In: Kamps, J., et al. (eds.) ECIR 2023. LNCS, vol. 13980, pp. 232–248. Springer, Cham (2023). https://doi.org/10.1007/978-3-031-28244-7_15

28. Faggioli, G., Zendel, O., Culpepper, J.S., Ferro, N., Scholer, F.: sMARE: a new paradigm to evaluate and understand query performance prediction methods. Inf. Retr. J. **25**(2), 94–122 (2022)
29. Ganguly, D., Yilmaz, E.: Query-specific variable depth pooling via query performance prediction. In: SIGIR, pp. 2303–2307 (2023)
30. Gao, J., Xiong, C., Bennett, P., Craswell, N.: Neural approaches to conversational information retrieval. arXiv preprint arXiv:2201.05176 (2022)
31. Hashemi, H., Zamani, H., Croft, W.B.: Performance prediction for non-factoid question answering. In: SIGIR, pp. 55–58 (2019)
32. Hauff, C.: Predicting the effectiveness of queries and retrieval systems. In: SIGIR Forum, vol. 44, p. 88 (2010)
33. Hauff, C., Hiemstra, D., de Jong, F.: A survey of pre-retrieval query performance predictors. In: CIKM, pp. 1419–1420 (2008)
34. He, B., Ounis, I.: Inferring query performance using pre-retrieval predictors. In: International Symposium on String Processing and Information Retrieval (2004)
35. Khodabakhsh, M., Bagheri, E.: Semantics-enabled query performance prediction for ad hoc table retrieval. Inf. Process. Manage. **58**(1), 102399 (2021)
36. Khodabakhsh, M., Bagheri, E.: Learning to rank and predict: multi-task learning for ad hoc retrieval and query performance prediction. Inf. Sci. **639**, 119015 (2023)
37. Meng, C., Aliannejadi, M., de Rijke, M.: Performance prediction for conversational search using perplexities of query rewrites. In: QPP++2023, pp. 25–28 (2023)
38. Meng, C., Aliannejadi, M., de Rijke, M.: System initiative prediction for multi-turn conversational information seeking. In: CIKM, pp. 1807–1817 (2023)
39. Meng, C., Arabzadeh, N., Aliannejadi, M., de Rijke, M.: Query performance prediction: From ad-hoc to conversational search. In: SIGIR, pp. 2583–2593 (2023)
40. Pérez-Iglesias, J., Araujo, L.: Standard deviation as a query hardness estimator. In: Chavez, E., Lonardi, S. (eds.) SPIRE 2010. LNCS, vol. 6393, pp. 207–212. Springer, Heidelberg (2010). https://doi.org/10.1007/978-3-642-16321-0_21
41. Pradeep, R., Sharifymoghaddam, S., Lin, J.: RankVicuna: zero-shot listwise document reranking with open-source large language models. arXiv preprint arXiv:2309.15088 (2023)
42. Roitman, H.: ICTIR tutorial: modern query performance prediction: theory and practice. In: ICTIR, pp. 195–196 (2020)
43. Roitman, H., Erera, S., Feigenblat, G.: A study of query performance prediction for answer quality determination. In: SIGIR, pp. 43–46 (2019)
44. Roy, D., Ganguly, D., Mitra, M., Jones, G.J.: Estimating gaussian mixture models in the local neighbourhood of embedded word vectors for query performance prediction. IP&M **56**(3), 1026–1045 (2019)
45. Salamat, S., Arabzadeh, N., Seyedsalehi, S., Bigdeli, A., Zihayat, M., Bagheri, E.: Neural disentanglement of query difficulty and semantics. In: CIKM, pp. 4264–4268 (2023)
46. Shtok, A., Kurland, O., Carmel, D., Raiber, F., Markovits, G.: Predicting query performance by query-drift estimation. TOIS **30**, 1–35 (2012)
47. Singh, A., Ganguly, D., Datta, S., McDonald, C.: Unsupervised query performance prediction for neural models with pairwise rank preferences. In: SIGIR, pp. 2486–2490 (2023)
48. Tao, Y., Wu, S.: Query performance prediction by considering score magnitude and variance together. In: CIKM (2014)
49. Wang, L., Yang, N., Wei, F.: Query2doc: query expansion with large language models. arXiv preprint arXiv:2303.07678 (2023)

50. Zamani, H., Bendersky, M.: Multivariate representation learning for information retrieval. arXiv preprint arXiv:2304.14522 (2023)
51. Zamani, H., Croft, W.B., Culpepper, J.S.: Neural query performance prediction using weak supervision from multiple signals. In: SIGIR (2018)
52. Zamani, H., Trippas, J.R., Dalton, J., Radlinski, F.: Conversational information seeking. arXiv preprint arXiv:2201.08808 (2022)
53. Zhao, Y., Scholer, F., Tsegay, Y.: Effective pre-retrieval query performance prediction using similarity and variability evidence. In: Macdonald, C., Ounis, I., Plachouras, V., Ruthven, I., White, R.W. (eds.) ECIR 2008. LNCS, vol. 4956, pp. 52–64. Springer, Heidelberg (2008). https://doi.org/10.1007/978-3-540-78646-7_8
54. Zhou, Y., Croft, W.B.: Query performance prediction in web search environments. In: SIGIR, pp. 543–550 (2007)

Workshops

The 7th International Workshop on Narrative Extraction from Texts: Text2Story 2024

Ricardo Campos[1,2,3](✉), Alípio Jorge[1,4], Adam Jatowt[5], Sumit Bhatia[6], and Marina Litvak[7]

[1] LIAAD – INESCTEC, Porto, Portugal
[2] University of Beira Interior, Covilhã, Portugal
ricardo.campos@ubi.pt
[3] Ci2 - Smart Cities Research Center - Polytechnic Institute of Tomar, Tomar, Portugal
[4] FCUP, University of Porto, Porto, Portugal
amjorge@fc.up.pt
[5] University of Innsbruck, Innsbruck, Austria
adam.jatowt@uibk.ac.at
[6] MDSR Lab, Adobe Systems, India
Sumit.Bhatia@adobe.com
[7] Shamoon College of Engineering, Beer Sheva, Israel
marinal@sce.ac.il

Abstract. The Text2Story Workshop series, dedicated to Narrative Extraction from Texts, has been running successfully since 2018. Over the past six years, significant progress, largely propelled by Transformers and Large Language Models, has advanced our understanding of natural language text. Nevertheless, the representation, analysis, generation, and comprehensive identification of the different elements that compose a narrative structure remains a challenging objective. In its seventh edition, the workshop strives to consolidate a common platform and a multidisciplinary community for discussing and addressing various issues related to narrative extraction tasks. In particular, we aim to bring to the forefront the challenges involved in understanding narrative structures and integrating their representation into established frameworks, as well as in modern architectures (e.g., transformers) and AI-powered language models (e.g., chatGPT) which are now common and form the backbone of almost every IR and NLP application. Text2Story encompasses sessions covering full research papers, work-in-progress, demos, resources, position and dissemination papers, along with keynote talks. Moreover, there is dedicated space for informal discussions on methods, challenges, and the future of research in this dynamic field.

1 Motivation

The continuous expansion of an array of social networks, from Facebook and Instagram to the trendsetting dynamism of TikTok and "X", coupled with the escalating presence of traditional news outlets on the Web, has fundamentally transformed the dynamics of information generation, dissemination and consumption. In this interconnected digital

N. Goharian et al. (Eds.): ECIR 2024, LNCS 14612, pp. 391–397, 2024.
https://doi.org/10.1007/978-3-031-56069-9_52

ecosystem, the traditional paradigm of relying on a limited set of sources for information has given way to a range of voices, spanning journalists, subject matter experts, and the influential realm of social media personas. News about significant events, like the Russia-Ukraine war or the Israeli-Palestinian conflict, is no longer confined to structured news websites; rather, it unfolds across diverse channels, from Facebook posts to social media influencers and comment sections, where discussions persist and evolve over days, weeks, or even months. Moreover, as the digital landscape evolves, people increasingly engage with news from the past through the corridors of web archives [14]. These archives serve as temporal gateways, allowing individuals to traverse historical moments, revisit past narratives, and explore the evolution of public discourse over time. The complexity of this continuous and evolving flow poses a significant challenge for readers seeking to navigate and process information across various dimensions of a given topic in a short amount of time.

Automated narrative extraction from text emerges as a compelling solution to this challenge that goes beyond merely identifying interconnected raw documents. Such a process, involves extracting essential narrative elements, compiling coherent storylines and presenting them in accessible forms such as timelines. While the field has witnessed substantial progress in information extraction and natural language processing, particularly in the realm of automatic text interpretation [30], the nuanced identification and analysis of narrative elements within documents or document sets remain a frontier marked by unresolved challenges [27, 28]. Although large language models have showcased their adeptness in capturing diverse linguistic nuances and generating high-quality text [12], they encounter limitations when dealing with more extensive texts, such as chapters in fictional literary works, where handling longer-range dependencies and contextual information remains a hurdle [32]. The transformative potential of narrative extraction spans across diverse industries, with finance [3, 11, 13, 35], business [16], news outlets [22], and healthcare [24] at the forefront. The overarching aim is to empower users with tools that facilitate swift comprehension of information within economic and financial reports, patient records, and large volumes of data, while providing more engaging and alternative ways to explore shared narratives through interactive visualizations [15].

Beyond the realms of information comprehension, the application of infographics and timelines [26], knowledge graphs [2], text summarization [21] and keyword extraction systems [5] emerge as a streamlined approach to represent automatically identified narrative chains, guiding human readers in navigating complex stories with distinct turning points and character networks. In the pursuit of computational creativity, the automatic generation of text [34] showcases impressive results. However, the narrative intent of generated output and a profound understanding of the methods employed by humans pose ongoing challenges in the realm of explainable AI. Within this expansive landscape, various open problems persist, ranging from issues such as hallucination in generated text [23] to the pervasive influence of bias [17]. Moreover, the transparency and explainability of generation techniques [1], the reliability of extracted facts [29, 33], and the quest for efficient and objective evaluation methods for generated narratives [4, 10] remain at the forefront of research endeavors. The discussing of the challenges associated with text annotation and the complexities inherent in identifying and marking narrative elements [31], upon which the robustness and accuracy of automated systems are built, is also at the forefront. All the above referred topics underscore the multifaceted nature of the field and the ongoing exploration of avenues to enhance the robustness and applicability of narrative generation technologies. As we explore the transformative potential of narrative extraction from text in this workshop, we challenge participants to shape the future narrative of information comprehension, by navigating not only the currents of present-day information, but also the historical texts that form our collective understanding.

2 Scientific Objectives

The Text2Story workshop, now in its seventh edition [https://text2story24.inesctec.pt], aims to provide a common forum to consolidate the multi-disciplinary efforts and foster discussions to identify the wide-ranging issues related to the narrative extraction task. In the first six editions [6–9, 18, 19], we had an approximate total number of 300 participants, 81 papers accepted for presentation, 13 keynote speakers and a Special Issue on IPM Journal [20] devoted to this matter, which demonstrates the growing activity of this research area. In this year's edition, we aim to bring together scientists conducting relevant research in the field of identifying and producing narratives/stories from textual sources, encompassing journalistic texts, scientific articles, and even social networks. Researchers will showcase their most recent advancements, emphasizing the practical applications of their findings across diverse domains, such as information extraction, information retrieval, natural language processing, text mining, artificial intelligence, machine learning, and natural language generation. Overall, the workshop has the following main objectives: (1) raise awareness within the Information Retrieval (IR) community regarding the challenges posed by narrative extraction and comprehension; (2) bridge the gap and foster connections between academic research, practitioners, and industrial applications; (3) discuss new methods, recent advances, and emerging challenges; (4) share experiences from research projects, case studies, and scientific outcomes structured around fundamental research questions related to narrative understanding; (5) identify dimensions that might be influenced by the automation of the narrative process; (6) highlight tested hypotheses that did not result in the expected outcomes.

Our topics revolves around fundamental inquiries pertinent to narrative understanding, encompassing the following questions: How can reliable and accurate narratives be efficiently extracted/generated from extensive, multi-genre, and multi-lingual datasets? What methodologies can be employed to annotate data and assess novel approaches effectively? How can one guarantee the explainability, interpretability and coherence of narratives? To what extent can a novel approach be adapted to new tasks, genres, and languages with minimal effort? What ethical considerations and safeguards need to be implemented to ensure responsible narrative extraction? How can collaboration between human annotators and automated systems be optimized to achieve more accurate and nuanced narrative understanding? What role do cultural and contextual nuances play in narrative extraction, and how can these elements be effectively incorporated into automated processes? Are there innovative approaches to handle challenges posed by ambiguous or contradictory information within narratives?

We welcomed contributions from interested researchers on all aspects related to narrative information extraction aspects, narrative representation, and analysis and generation, including the extraction and formal representation of events, their temporal aspects and intrinsic relationships, comprehension of generated narratives, and more. In addition to this, we seek contributions related to alternative means of presenting the information, narrative applications, verification of extracted facts and on the formal aspects of evaluation, including the proposal of new datasets, annotation schema, and evaluation metrics. Special attention will be given to multilingual approaches and resources. Finally, we challenge the interested researchers to consider submitting a paper that makes use of the tls-covid19 dataset [25].

3 Organizing Team

Ricardo Campos is an assistant professor at the University of Beira Interior. He is an integrated researcher of LIAAD-INESC TEC and a collaborator of Ci2 (Polytechnic of Tomar). He is PhD in Computer Science (CS) by the University of Porto (UP), with over ten years of research experience in IR and NLP. He is an editorial board member of the International Journal of Data Science and Analytics (Springer) and of the IPM Journal (Elsevier), being also a co-chair and a PC member of several top conferences.

Alípio M. Jorge works in the areas of data mining, ML, recommender systems and NLP. He is PhD in Computer Science (CS) by the University of Porto (UP), Full Professor at the same University and head of the CS department since 2017. He is the coordinator of the research lab LIAAD-INESC TEC, and a principal researcher of several research projects. He was the coordinator for the Portuguese Strategy on Artificial Intelligence "AI Portugal 2030".

Adam Jatowt is Full Professor at the University of Innsbruck. He has received his Ph.D. in Information Science and Technology from the U. Tokyo, Japan in 2005. His research interests lie in an area of IR, knowledge extraction from text and in digital history. Adam has been serving as a PC co-chair and general chair of several top conferences. He was also a co-organizer of over 20 international workshops.

Sumit Bhatia is a Senior Machine Learning Scientist at Media and Data Science Research Lab, Adobe Systems, India. He received his Ph.D. from the Pennsylvania State University in 2013. With primary research interests in the fields of Knowledge Management, IR and Text Analytics, Sumit is a co-inventor of more than a dozen patents. He has served on program committees of several top conferences and journals.

Marina Litvak is a Senior Lecturer at the department of Software Engineering, Shamoon College of Engineering (SCE), Israel. Marina received her PhD degree from Information Sciences dept. at Ben Gurion University at the Negev, Israel in 2010. Marina's research focuses mainly on Multilingual Text Analysis, Social Networks, Knowledge Extraction from Text, and Summarization. She has published over 70 academic papers, including journal and top-level conference publications.

Acknowledgements. Ricardo Campos is financed by National Funds through the Portuguese funding agency, FCT - Fundação para a Ciência e a Tecnologia, within the project StorySense (10.54499/2022.14409312.PTDC) and Alípio Jorge within project LA/P/0063/2020.

References

1. Alonso, J.M., et al.: Interactive natural language technology for explainable artificial intelligence. In: Heintz, F., Milano, M., O'Sullivan, B. (eds.) TAILOR 2020. LNCS (LNAI), vol. 12641, pp. 63–70. Springer, Cham (2021). https://doi.org/10.1007/978-3-030-73959-1_5
2. Amorim, E., et al.: Brat2Viz: a tool and pipeline for visualizing narratives from annotated texts. In: Proceedings of the 4th International Workshop on Narrative Extraction from Texts (Text2Story@ECIR 2021). Online. April 1, pp. 49–56 (2021)
3. Athanasakou, V., et al.: Proceedings of the 1st Joint Workshop on Financial Narrative Processing and MultiLing Financial Summarisation (FNP-FNS'20) co-located to Coling'20, Barcelona, Spain (Online). Dec 12, pp. 1–245 (2020)
4. Ayed, A.B., Biskri, I., Meunier, J.-G.: An efficient explainable artificial intelligence model of automatically generated summaries evaluation: a use case of bridging cognitive psychology and computational linguistics. In: Sayed-Mouchaweh, M. (ed.) Explainable AI Within the Digital Transformation and Cyber Physical Systems, pp. 69–90. Springer, Cham (2021). https://doi.org/10.1007/978-3-030-76409-8_5
5. Campos, R., Mangaravite, V., Pasquali, A., Jorge, A.M., Nunes, C., Jatowt, A.: A text feature based automatic keyword extraction method for single documents. In: Pasi, G., Piwowarski, B., Azzopardi, L., Hanbury, A. (eds.) ECIR 2018. LNCS, vol. 10772, pp. 684–691. Springer, Cham (2018). https://doi.org/10.1007/978-3-319-76941-7_63
6. Campos, R., Jorge, A., Jatowt, A., Sumit, B., Litvak, M.: Sixth international workshop on narrative extraction from texts (Text2Story'23). In: Caputo, A., et al. (eds.) Advances in Information Retrieval. ECIR 2023. LNCS, vol. 13982, pp. 377–383. Springer, Cham (2023)
7. Campos, R., Jorge, A., Jatowt, A., Sumit, B., Litvak, M.: Fifth international workshop on narrative extraction from texts (Text2Story'22). In: Hagen, M. (eds.) Advances in Information Retrieval. ECIR 2022. LNCS, vol. 13186, pp. 552–556. Springer, Cham (2022)
8. Campos, R., Jorge, A., Jatowt, A., Bhatia, S., Finlayson, M.: Fourth international workshop on narrative extraction from texts (Text2Story'21). In: Hiemstra, D., Moens, M.-F., Mothe, J., Perego, R., Potthast, M., Sebastiani, F. (eds.) Advances in Information Retrieval. ECIR 2021. LNCS, vol. 12657, pp. 701–704. Springer, Cham (2021)
9. Campos, R., Jorge, A., Jatowt, A., Sumit, B.: Third international workshop on narrative extraction from texts (Text2Story'20). In: Jose, J., et al. (eds.) ECIR 2020. LNCS, vol. 12036, pp. 648–653. Springer, Cham (2020)
10. Celikyilmaz, A., Clark, E., Gao, J.: Evaluation of text generation: a survey. arXiv preprint arXiv:2006.14799 (2020)
11. El-Haj, M., Litvak, M., Pittaras, N., Giannakopoulos, G.: The financial narrative summarisation shared task (FNS'20). In: Proceedings of the 1st Joint Workshop on Financial Narrative Processing and MultiLing Financial Summarisation, pp. 1–12 (2020)
12. Elkins, K., Chun, J.: Can GPT-3 pass a Writer's turing test? J. Cult. Anal. **5**(2), 17212 (2020)
13. El-Haj, M., Rayson, P., Zmandar, N.: Proceedings of the 4th Financial Narrative Processing Workshop. ACL (2022)
14. Gomes, D., Demidova, E., Winters, J., Risse, T.: The Past Web: Exploring Web Archives (2021)
15. Gonçalves, F., Campos, R., Jorge, A.: Text2Storyline: generating enriched storylines from text. In: Caputo, A., et al. (eds.) Advances in Information Retrieval. ECIR 2023. LNCS, vol. 13982, pp. 248–254. Springer, Cham (2023). https://doi.org/10.1007/978-3-031-28241-6_22
16. Grobelny, J., Smierzchalska, J., Krzysztof, K.: Narrative gamification as a method of increasing sales performance: a field experimental study. Int. J. Acad. Res. Bus. Soc. Sci. **8**(3), 430–447 (2018)

17. Guo, W., Caliskan, A.: Detecting emergent intersectional biases: contextualized word embeddings contain a distribution of human-like biases. In: Proceedings of the 2021 AAAI/ACM Conference on AI, Ethics, and Society, pp. 122–133 (2021)
18. Jorge, A., Campos, R., Jatowt, A., Bhatia, S.: Second international workshop on narrative extraction from texts (Text2Story'19). In: Azzopardi, L., Stein, B., Fuhr, N., Mayr, P., Hau, C., Hiemstra, D. (eds.) ECIR 2019. LNCS, vol. 11438, pp. 389–393 (2019)
19. Jorge, A., Campos, R., Jatowt, A., Nunes, S.: First international workshop on narrative extraction from texts (Text2Story'18). In: Pasi, G., et al. (eds.) Advances in Information Retrieval. ECIR 2018. LNCS, vol. 10772, pp. 833–834 (2018)
20. Jorge, A., Campos, R., Jatowt, A., Nunes, S.: Special issue on narrative extraction from texts (Text2Story): preface. IPM J. **56**(5), 1771–1774
21. Liu, S., et al.: TIARA: interactive, topic-based visual text summarization and analysis. ACM Trans. Intell. Syst. Technol. **3**(2), Article 25, 28 pages (2012)
22. Martinez-Alvarez, M., et al.: First international workshop on recent trends in news information retrieval (NewsIR'16). In: Ferro, N., et al. (eds.) ECIR 2016. LNCS, vol. 9626, pp. 878–882. Springer, Cham (2016). https://doi.org/10.1007/978-3-319-30671-1_85
23. Maynez, J., Narayan, S., Bohnet, B., McDonald, R.: On faithfulness and factuality in abstractive summarization. In: Proceedings of the 58th Annual Meeting of the Association for Computational Linguistics, pp. 1906–1919 (2020)
24. Özlem, U., Amber, S., Weiyi, S.: Chronology of your health events: approaches to extracting temporal relations from medical narratives. Biomedical Inf. **46**, 1–4 (2013)
25. Pasquali, A., Campos, R., Ribeiro, A., Santana, B., Jorge, A., Jatowt, A.: TLS-Covid19: a new annotated corpus for timeline summarization. In: Hiemstra, D., Moens, M.-F., Mothe, J., Perego, R., Potthast, M., Sebastiani, F. (eds.) ECIR 2021. LNCS, vol. 12656, pp. 497–512. Springer, Cham (2021). https://doi.org/10.1007/978-3-030-72113-8_33
26. Pasquali, A., Mangaravite, V., Campos, R., Jorge, A.M., Jatowt, A.: Interactive system for automatically generating temporal narratives. In: Azzopardi, L., Stein, B., Fuhr, N., Mayr, P., Hauff, C., Hiemstra, D. (eds.) ECIR 2019. LNCS, vol. 11438, pp. 251–255. Springer, Cham (2019). https://doi.org/10.1007/978-3-030-15719-7_34
27. Piper, A.: Computational narrative understanding: a big picture analysis. In: Proceedings of the Big Picture Workshop, Singapore, pp. 28–39. Association for Computational Linguistics (2023)
28. Ranade, P., Dey, S., Joshi, A., Finin, T.: Computational understanding of narratives: a survey. IEEE Access **10**, 101575–101594 (2022). https://doi.org/10.1109/ACCESS.2022.3205314
29. Saakyan, A., Chakrabarty, T., Muresan, S.: COVID-Fact: Fact Extraction and Verification of Real-World Claims on COVID-19 Pandemic. arXiv preprint arXiv:2106.03794 (2021)
30. Santana, B., Campos, R., Amorim, E., Alípio, J., Purificação, S., Nunes, S.: A survey on narrative extraction from textual data. Artif. Intell. Rev. (2023). https://doi.org/10.1007/s10462-022-10338-7
31. Sun, W., Rumshisky, A., Uzener, O.: Annotating temporal information in clinical narratives. J. Biomed. Inform. **46**, S5–S12 (2013)
32. Sun, S., Krishna, K., Mattarella-Micke, A., Iyyer, M.: Do long-range language models actually use long-range context? In: Proceedings of the 2021 Conference on Empirical Methods in Natural Language Processing, pp. 807–822, Online and Punta Cana, Dominican Republic. ACL (2021)
33. Vo, N., Lee, K.: Learning from fact-checkers: analysis and generation of fact-checking language. In: Proceedings of the 42nd International ACM SIGIR Conference on Research and Development in Information Retrieval, Paris, France, 21–25 July, pp. 335–344 (2019)

34. Wu, Y.: Is automated journalistic writing less biased? An experimental test of auto-written and human-written news stories. Journal. Pract. **14**(7), 1–21 (2019)
35. Zmandar, N., El-Haj, M., Rayson, P., Litvak, M., Giannakopoulos, G., Pittaras, N.: The financial narrative summarisation shared task FNS 2021. In: Proceedings of the 3rd Financial Narrative Processing Workshop, pp. 120–125 (2021)

KEIR @ ECIR 2024: The First Workshop on Knowledge-Enhanced Information Retrieval

Zaiqiao Meng[1](✉), Shangsong Liang[2], Xin Xin[3], Gianluca Moro[4], Evangelos Kanoulas[5], and Emine Yilmaz[6]

[1] University of Glasgow, Glasgow, UK
Zaiqiao.Meng@glasgow.ac.uk
[2] MBZUAI, Abu Dhabi, UAE
Shangsong.Liang@mbzuai.ac.ae
[3] Shandong University, Jinan, China
XinXin@sdu.edu.cn
[4] University of Bologna, Bologna, Italy
Gianluca.Moro@unibo.it
[5] University of Amsterdam, Amsterdam, The Netherlands
E.Kanoulas@uva.nl
[6] University College London, London, UK
Emine.Yilmaz@ucl.ac.uk

Abstract. The infusion of external knowledge bases into IR models can provide enhanced ranking results and greater interpretability, offering substantial advancements in the field. The first workshop on Knowledge-Enhanced Information Retrieval (KEIR @ ECIR 2024) will serve as a platform to bring together researchers from academia and industry to explore and discuss various aspects of knowledge-enhanced information retrieval systems, such as models, techniques, data collection and evaluation. The workshop aims to not only deliberate upon the advantages and hurdles intrinsic to the development of knowledge-enhanced pretrained language models, IR models and recommendation models but also to facilitate in-depth discussions concerning the same.

Keywords: Information Retrieval · Knowledge Graph · Recommendation System · Large Language Models

1 Motivation

Pretrained language models (PLMs), such as ChatGPT and GPT4 [3], are generating significant impact across a wide spectrum of artificial intelligence domains, including natural language processing (NLP), information retrieval (IR) and recommendation systems (RecSys). Concurrently, knowledge graphs (KGs) such as Freebase [2], DBpedia [8] and Wikidata [12], provide a structured way of storing knowledge about the real world. These KGs enhance PLMs by infusing external knowledge, which in turn enhances inference and interpretability [6,10,13,14].

N. Goharian et al. (Eds.): ECIR 2024, LNCS 14612, pp. 398–402, 2024.
https://doi.org/10.1007/978-3-031-56069-9_53

However, despite the remarkable progress in PLMs and KGs, current IR systems still exhibit certain weaknesses. Indeed, traditional IR systems, including retrieval and recommendation systems, often encounter challenges in addressing semantic nuances [4,11], context relevance [5,15], and handling domain-specific intricacies [7,9], leading to imprecise results. This gap in understanding can hinder their effectiveness in delivering accurate and contextually relevant results. Hence, these limitations highlight the pressing requirement for innovative approaches capable of harnessing external knowledge to bridge these gaps.

While recent progress of NLP has boosted the research of integrating knowledge graphs into PLMs, integrating external knowledge into the retrieval and recommendation tasks has not yet fully harnessed the potential offered by these advancements. In fact, the utilisation of external knowledge, particularly domain-specific knowledge, to enhance retrieval processes has been a subject of research since the early 1990s [1,11]. Despite the long-standing interest in this area, the recent surge in advancements in PLMs has sparked new possibilities for integrating such external knowledge with NLP techniques, presenting a timely and opportune moment to revisit and revitalise these research efforts.

Following this evolution of ideas, we propose this Knowledge-Enhanced Information Retrieval workshop scheduled for ECIR 2024 (KEIR @ ECIR 2024), as a forum for the discussion of efficient and effective approaches to explore the various knowledge-enhanced information retrieval problems. This workshop addresses the growing need for more effective ways to retrieve and utilise external information in a rapidly expanding technological landscape. In particular, we wish to promote and encourage the IR community to raise and debate questions on the following main themes: knowledge-enhanced retrieval models, knowledge-enhanced recommendation models and knowledge-enhanced language models. Aligned with these core themes and driving motivations, this workshop holds a steadfast commitment to fostering collaboration among researchers engaged in the realm of knowledge integration for IR, RecSys and NLP.

2 Workshop Topics and Goals

The workshop will focus on models, techniques, data collection, and evaluation methodologies for various knowledge-enhanced information retrieval problems. These include but are not limited to:

- Knowledge-enhanced information retrieval models.
- Knowledge-enhanced approaches for query processing, including query parsing, query expansion, relevance feedback, and query reformulation.
- Knowledge-enhanced recommendation models.
- Knowledge-enhanced language models for retrieval.
- Data augmentation for knowledge-enhanced information retrieval.
- Knowledge retrieval from unstructured data and structured data.
- Applications of knowledge-enhanced retrievals, such as dialogue systems, question answering, summarisation and other domain-specific applications.

- Data collection for knowledge-enhanced information retrieval.
- Evaluation methodologies for knowledge-enhanced retrieval.
- The interpretability and analysis of knowledge-enhanced models for IR, including potential biases and ethical considerations.

The goal of this workshop will serve as a platform to not only deliberate upon the advantages and hurdles intrinsic to the development of knowledge-enhanced PLMs, IR models and RecSys models but also to facilitate in-depth discussions concerning the same.

3 Format and Planned Activities

We plan to organise a full-day in-person workshop, and will be structured in four sessions, including a blend of keynote talks, light talks, accepted workshop presentations and breakout/panel discussion, promoting vibrant interactions on emerging topics among participants.

The workshop's call for papers invites submissions with lengths ranging from a minimum of 6 pages to a maximum of 12 pages. Authors of accepted papers in this workshop will be offered the opportunity to deliver paper presentations. Additionally, authors with relevant papers accepted in the main conference may be invited to give a light talk during the workshop.

4 Expected Audience

Our workshop is anticipated to attract a diverse audience of researchers, academics, industry professionals and students from the fields of IR, NLP, RecSys and related domains. For example, those PhD students who are interested in advancing retrieval techniques by integrating external knowledge sources and using PLMs. Industry practitioners who are exploring novel approaches to enhance their search engines or recommendation systems through their domain knowledge graphs will find valuable insights and opportunities for collaboration. In addition, academics seeking to extend their understanding of knowledge-driven models and their real-world applications will benefit from engaging discussions.

5 Organisers

The organisation team consists of active IR, NLP and RecSys researchers.

Zaiqiao Meng. Zaiqiao is a Lecturer at the Information Retrieval Group of the University of Glasgow. His research interests include IR, NLP, KG and RecSys. He is particularly interested in combining LLMs with KGs. Zaiqiao was previously working as a Postdoctoral Researcher at the Language Technology Laboratory at the University of Cambridge, and at the Information Retrieval Group at the University of Glasgow, respectively. Zaiqiao obtained his Ph.D. in computer science from Sun Yat-sen University in December 2018. He served as the area chair for many NLP conferences, such as ACL 2022 and EMNLP 2023.

Shangsong Liang. Shangsong is an Assistant Professor of MBZUAI. He got his PhD degree from the University of Amsterdam, the Netherlands. Prior to joining MBZUAI, he was a research scientist at KAUST and an associate researcher at University College London. Shangsong is an editor member of the journal of Information Processing and Management. He is the PC member and reviewer in several conferences and journals. Shangsong has received various awards/honours such as the SIGIR 2017 Outstanding Reviewer Award, and Outstanding Contribution for instructing Data Mining course from the International Petroleum Engineers, the Kingdom of Saudi Arabia Section.

Xin Xin. Xin is an Assistant Professor at the School of Computer Science and Technology of Shandong University, as a member of the Information Retrieval Lab. Before that, he got his PhD degree in computing science from the University of Glasgow. Formerly, he got his master's degree from the School of Software Engineering at SJTU and his bachelor's degree from XJTU. His research interests span recommender systems, reinforcement learning, graph learning, causal inference for recommendation and NLP. His work appeared in several top-tier ML & IR conferences including SIGIR, WSDM, CIKM, IJCAI, ACL and UAI.

Gianluca Moro. Gianluca is an Associate Professor of the University of Bologna. His research focuses on text mining, natural language processing, data mining and data science, with particular interest in Artificial Intelligence methods based on machine learning and deep neural networks, among which generative language models, to transform data into knowledge, both in structured and unstructured data domains and big data. He develops research in national and international research projects, he co-authored over ninety publications, mainly in international journals and top congresses, among which AAAI, EMNLP, IJCAI, ECAI, ACL, COLING. He has organised workshops at prestigious international conferences and collaborates on scientific projects with research centres and public and private companies. He leads the research team in NLP at DISI in Cesena.

Evangelos Kanoulas. Evangelos is a professor of computer science at the University of Amsterdam, leading the Information Retrieval Lab at the Informatics Institute. His research lies in developing evaluation methods and algorithms for search, and recommendation, with a focus on learning robust models of language that can be used to understand noisy human language, retrieve textual data from large corpora, generate faithful and factual text, and converse with the user. His research has been published at SIGIR, CIKM, KDD, WWW, WSDM, EMNLP and other venues in the fields of IR and NLP. He has proposed and organized numerous search benchmarking competitions as part of the Text Retrieval Conference (TREC) and the Conference and Labs of the Evaluation Forum (CLEF). Furthermore, he is a member of the Ellis society.

Emine Yilmaz. Emine is a Professor and EPSRC Fellow at University College London, Department of Computer Science. She is also a faculty fellow at the Alan

Turing Institute and an ELLIS fellow. Her research interests lie in the fields of IR and NLP. She has been awarded two best paper awards (ACM ICTIR 2022 Best Student Paper and ACM CHIIR 2017 Best Paper) and was nominated for the best paper award several times (ACM WSDM 2011, ACM SIGIR 2010 and 2009). She has held various senior roles in journals and conferences, including co-editor-in-chief of the Information Retrieval Journal, editorial board member of the AI Journal, elected executive committee member of ACM SIGIR, PC Chair for ACM CIKM 2022 Applied Track, and leadership positions for conferences such as ECIR 2020, ACM SIGIR 2018, ACM ICTIR 2017, ACM SIGIR 2023 Industry Chair, WWW 2021 Panels Chair, ACM WSDM 2017 Practice and Experience Chair, and ECIR 2017 Doctoral Consortium Chair.

References

1. Aronson, A.R., Rindflesch, T.C., Browne, A.C.: Exploiting a large thesaurus for information retrieval. In: RIAO, vol. 94, pp. 197–216 (1994)
2. Bollacker, K., Evans, C., Paritosh, P., Sturge, T., Taylor, J.: Freebase: a collaboratively created graph database for structuring human knowledge. In: Proceedings of SIGMOD, pp. 1247–1250 (2008)
3. Bubeck, S., et al.: Sparks of artificial general intelligence: early experiments with GPT-4. arXiv preprint arXiv:2303.12712 (2023)
4. Bulathwela, S., Pérez-Ortiz, M., Yilmaz, E., Shawe-Taylor, J.: Leveraging semantic knowledge graphs in educational recommenders to address the cold-start problem. In: Semantic AI in Knowledge Graphs, pp. 1–20 (2023)
5. Chandradevan, R., Yang, E., Yarmohammadi, M., Agichtein, E.: Learning to enrich query representation with pseudo-relevance feedback for cross-lingual retrieval. In: SIGIR, pp. 1790–1795 (2022)
6. Das, R., et al.: Knowledge base question answering by case-based reasoning over subgraphs. In: Proceedings of ICML, pp. 4777–4793 (2022)
7. Frisoni, G., Mizutani, M., Moro, G., Valgimigli, L.: BioReader: a retrieval-enhanced text-to-text transformer for biomedical literature. In: Proceedings of EMNLP, pp. 5770–5793 (2022)
8. Lehmann, J., et al.: DBpedia-a large-scale, multilingual knowledge base extracted from Wikipedia. Semant. Web **6**(2), 167–195 (2015)
9. Luo, M., Mitra, A., Gokhale, T., Baral, C.: Improving biomedical information retrieval with neural retrievers. In: Proceedings of AAAI, vol. 36, pp. 11038–11046 (2022)
10. Meng, Z., Liu, F., Clark, T., Shareghi, E., Collier, N.: Mixture-of-partitions: infusing large biomedical knowledge graphs into BERT. In: Proceedings of EMNLP, pp. 4672–4681 (2021)
11. Rindflesch, T.C., Aronson, A.R.: Semantic processing in information retrieval. In: Proceedings of SCAMC, p. 611 (1993)
12. Vrandečić, D., Krötzsch, M.: Wikidata: a free collaborative knowledgebase. Commun. ACM **57**(10), 78–85 (2014)
13. Wang, H., et al.: Knowledge-adaptive contrastive learning for recommendation. In: Proceedings of WSDM, pp. 535–543 (2023)
14. Wang, X., He, X., Cao, Y., Liu, M., Chua, T.S.: KGAT: knowledge graph attention network for recommendation. In: Proceedings of SIGKDD, pp. 950–958 (2019)
15. Zhang, X., et al.: Variational reasoning over incomplete knowledge graphs for conversational recommendation. In: Proceedings of WSDM, pp. 231–239 (2023)

ROMCIR 2024: Overview of the 4th Workshop on Reducing Online Misinformation Through Credible Information Retrieval

Marinella Petrocchi[1,2] and Marco Viviani[3](✉)

[1] Institute of Informatics and Telematics, National Research Council (IIT – CNR), Pisa, Italy
marinella.petrocchi@iit.cnr.it
[2] IMT Scuola Alti Studi Lucca, Lucca, Italy
[3] Department of Informatics, Systems, and Communication, University of Milano-Bicocca (DISCo – UNIMIB), Milan, Italy
marco.viviani@unimib.it

Abstract. In the realm of the Social Web, we are continuously surrounded by information pollution, posing significant threats to both individuals and society as a whole. Instances of false news, for instance, wield the power to sway public opinion on matters of politics and finance. Deceptive reviews can either bolster or tarnish the reputation of businesses, while unverified medical advice may steer people toward harmful health practices. In light of this challenging landscape, it has become imperative to ensure that users have access to both topically relevant and truthful information that does not warp their perception of reality, and there has been a surge of interest in various strategies to combat disinformation through different contexts and multiple tasks. The purpose of the ROMCIR Workshop, for some years now, is precisely that of engaging the Information Retrieval community to explore potential solutions that extend beyond conventional misinformation detection approaches. Key objectives include integrating information truthfulness as a fundamental dimension of relevance within Information Retrieval Systems (IRSs) and ensuring that truthful search results are also explainable to IRS users. Moreover, it is essential to evaluate the role of generative models such as Language Models (LLMs) in inadvertently amplifying misinformation problems, and how they can be used to support IRSs.

Keywords: Information Retrieval · Information Disorder · Information Truthfulness · Misinformation · Explainability · Large Language Models

1 Motivation and Relevance to ECIR

Technology is so much fun but we can drown in our technology.
The fog of information can drive out knowledge.

N. Goharian et al. (Eds.): ECIR 2024, LNCS 14612, pp. 403–408, 2024.
https://doi.org/10.1007/978-3-031-56069-9_54

This quote by American historian Daniel J. Boorstin, about the computerization of libraries, is from the July 1983 New York Times. Some 40 years later, it is probably still more relevant than it was then. In fact, the process of disintermediation that Web 2.0 technologies brought in the generation and dissemination of online content within the Social Web has led to the well-known problems of information overload [1,6] and the spread of misinformation [4], which make it difficult for users to find information that is truly useful for their purposes [11,17]. These problems risk being exacerbated by the recent development of generative models such as Large Language Models (LLMs) [7], which can produce texts that are apparently indistinguishable from those written by a human being but which may be entirely devoid of truthfulness [10,19].

Hence, the ROMCIR Workshop concerns the study and the development of IR solutions (or parts thereof) to provide access to users to (topically) relevant and truthful information (as key measures of relevance, but not the only ones), to mitigate the human-generated or AI-generated information disorder phenomenon with respect to distinct domains. By "information disorder" we mean all forms of communication pollution, from misinformation made out of ignorance, automatically built on the basis of biased content, to intentional sharing of false content (generated both manually and automatically) [18]. This phenomenon can be additionally exacerbated in the presence of filter bubbles [2] and echo chambers [5,16] that often characterize the online digital ecosystem. Hence, tackling the information disorder issue is very complex, as it concerns different contents (e.g., Web pages, news, reviews, medical information, online accounts, etc.), different Web and social media platforms (e.g., microblogging platforms, social networking services, social question-answering systems, etc.), different purposes (e.g., identifying false information, accessing information based on its truthfulness, retrieving truthful information, etc.), and different open issues related in particular to AI (e.g., explainability of search results, assessment of the truthfulness of automatically generated content, generative models to support IRSs, etc.) [3], and data confidentiality [9]. In the context of developing IR systems that take into consideration the issues introduced above, the study and development of appropriate experimental evaluation paradigms is also of paramount importance [8,15].

2 Aim and Topics of Interest

The Workshop aims at considering the issues related to online information disorder in the context of Information Retrieval, also considering related Artificial Intelligence fields such as Natural Language Processing (NLP), Natural Language Understanding (NLU), Computer Vision, Machine and Deep Learning. Hence, the topics of interest of ROMCIR 2024 include, but are not limited to:

- Artificial Intelligence and information truthfulness assessment
- Bot/spam/troll detection
- Computational fact-checking/truthfulness assessment

- Crowdsourcing for information truthfulness assessment
- Disinformation/misinformation and bias detection
- Generative models and information truthfulness assessment
- Harassment/bullying/hate speech detection
- Information polarization in online communities, echo chambers
- Propaganda identification/analysis
- Retrieval and evaluation of truthful information
- Security, privacy, and information truthfulness
- Sentiment/emotional analysis and stance detection
- Societal reaction to misinformation
- Trust, reputation, and misinformation

3 Past Editions

The first three editions of the ROMCIR Workshop, all co-located with the ECIR conference, led to fervent discussion and presentation of innovative work concerning a variety of open issues related to information disorder and IR. The first edition took place in online mode on April 1, 2021. The second edition took place both in presence in Stavanger, Norway, and online, on April 10, 2022. The third edition took place in presence in Dublin, Ireland, on April 2, 2023. Papers submitted to the three editions of ROMCIR are collected in CEUR Proceedings [12–14], which are freely accessible. Further and updated information on past and current ROMCIR editions can be found on the official website.[1]

4 Workshop Format and Intended Audience

Format. The ROMCIR Workshop usually lasts for one day. It includes invited talks by experts on various aspects of the problems considered in the context of IR, followed by the presentation of accepted papers, interspersed and/or followed by discussions involving all participants and addressing specific open questions.

Audience. The ROMCIR Workshop is aimed at different scientific communities working in interconnected research fields. Definitely the Information Retrieval community, without forgetting those who work in the field of Text Mining and Natural Language Processing, and, in general, Artificial Intelligence. The problem faced is so transversal and concerns both textual and multi-modal content that the Workshop can be an opportunity to create an interdisciplinary context in which to propose different solutions (both model-driven and data-driven) to guarantee access to truthful information to users.

5 Workshop Organizers

The 4th edition of the ROMCIR Workshop is organized by Marinella Petrocchi and Marco Viviani.

[1] https://romcir.disco.unimib.it/.

Marinella Petrocchi is a Senior Researcher at the Institute of Informatics and Telematics of the National Research Council (IIT-CNR) in Pisa, Italy, under the Trust, Security, and Privacy research unit. She also collaborates with the Sysma unit at IMT School for Advanced Studies, in Lucca, Italy. Her field of research lies between Cybersecurity, Artificial Intelligence, and Data Science. Specifically, she studies novel techniques for online fake news/fake accounts detection and automated methods to rank the reputability of online news media. She is the author of several international publications on these themes and she usually gives talks and lectures on the topic. She is in the core team of the TOFFEe project (TOols for Fighting FakEs), funded by IMT. *Website*: https://www.iit.cnr.it/marinella.petrocchi/

Marco Viviani is an Associate Professor at the University of Milano-Bicocca, Department of Informatics, Systems, and Communication (DISCo). He works in the Information and Knowledge Representation, Retrieval and Reasoning (IKR3) Lab. He has been Co-organizer of several Special Tracks and Workshops at International Conferences, and General Co-chair of MDAI 2019. He is an Associate Editor of Social Network Analysis and Mining, an Area Editor (Web Intelligence and E-Services) of the International Journal of Computational Intelligence Systems, an Editorial Board Member of Online Social Networks and Media, and a Guest Editor of several Special Issues in International Journals related to the issue of information disorder online. His main research activities include Social Computing, Information Retrieval, Text Mining, Natural Language Processing, Trust and Reputation Management, and User Modeling. On these topics, he has written several international publications. *Website*: https://ikr3.disco.unimib.it/people/marco-viviani/

5.1 Program Committee Members

- John Bianchi, IMT Scuola Alti Studi Lucca, Italy
- Edoardo Di Paolo, Università degli Studi di Roma "La Sapienza", Italy
- Carlos A. Iglesias, Universidad Politécnica de Madrid, Spain
- Udo Kruschwitz, Universität Regensburg, Germany
- David Losada, Universidad de Santiago de Compostela, Spain
- Gabriella Pasi, Università degli Studi di Milano-Bicocca, Italy
- Marinella Petrocchi, Istituto di Informatica e Telematica – CNR, Italy
- Manuel Pratelli, IMT Scuola Alti Studi Lucca, Italy
- Daisy Romanini, Istituto di Informatica e Telematica – CNR, Italy

- Paolo Rosso, Universitat Politècnica de València, Spain
- Irene Sanchez-Rodriguez, IMT Scuola Alti Studi Lucca, Italy
- Fabio Saracco, Centro Ricerche Enrico Fermi, Italy
- Marco Viviani, Università degli Studi di Milano-Bicocca, Italy

Acknowledgements. Partially supported by re-DESIRE: *DissEmination of ScIentific REsults* 2.0, funded by IIT-CNR; by SERICS (PE00000014) under the NRRP MUR program funded by the EU - NGEU; by the PNRR ICSC National Research Centre for High Performance Computing, Big Data and Quantum Computing (CN00000013), under the NRRP MUR program funded by the NextGenerationEU.

References

1. Bawden, D., Holtham, C., Courtney, N.: Perspectives on information overload. In: Aslib proceedings, vol. 51, pp. 249–255. MCB UP Ltd. (1999)
2. Bozdag, E., Van Den Hoven, J.: Breaking the filter bubble: democracy and design. Ethics Inf. Technol. **17**, 249–265 (2015)
3. Cabitza, F., Ciucci, D., Pasi, G., Viviani, M.: Responsible AI in healthcare. arXiv preprint arXiv:2203.03616 (2022)
4. Chen, S., Xiao, L., Kumar, A.: Spread of misinformation on social media: what contributes to it and how to combat it. Comput. Hum. Behav. **141**, 107643 (2023)
5. Del Vicario, M., et al.: The spreading of misinformation online. Proc. Natl. Acad. Sci. **113**(3), 554–559 (2016)
6. Khaleel, I., et al.: Health information overload among health consumers: a scoping review. Patient Educ. Couns. **103**(1), 15–32 (2020)
7. Kojima, T., Gu, S.S., Reid, M., Matsuo, Y., Iwasawa, Y.: Large language models are zero-shot reasoners. Adv. Neural. Inf. Process. Syst. **35**, 22199–22213 (2022)
8. Lioma, C., Simonsen, J.G., Larsen, B.: Evaluation measures for relevance and credibility in ranked lists. In: Proceedings of the ACM SIGIR International Conference on Theory of Information Retrieval, pp. 91–98 (2017)
9. Livraga, G., Viviani, M.: Data confidentiality and information credibility in on-line ecosystems. In: Proceedings of the 11th International Conference on Management of Digital Ecosystems, pp. 191–198 (2019)
10. Monteith, S., Glenn, T., Geddes, J.R., Whybrow, P.C., Achtyes, E., Bauer, M.: Artificial intelligence and increasing misinformation. Br. J. Psychiatry **224**, 1–3 (2023)
11. Pasi, G., Viviani, M.: Information credibility in the social web: contexts, approaches, and open issues. arXiv preprint arXiv:2001.09473 (2020)
12. Petrocchi, M., Viviani, M.: Overview of ROMCIR 2022: the 2nd workshop on reducing online misinformation through credible information retrieval. In: ROMCIR 2022 CEUR Workshop Proceedings, vol. 3138, pp. i–vii (2022)
13. Petrocchi, M., Viviani, M.: Overview of ROMCIR 2023: the 3rd workshop on reducing online misinformation through credible information retrieval. In: ROMCIR 2023 CEUR Workshop Proceedings, vol. 3406, pp. i–ix (2023)
14. Saracco, F., Viviani, M.: Overview of ROMCIR 2021: workshop on reducing online misinformation through credible information retrieval. In: ROMCIR 2021 CEUR Workshop Proceedings vol. 2838, pp. i–vii (2021)
15. Suominen, H., et al.: Overview of the CLEF eHealth evaluation lab 2021. In: Candan, K.S., et al. (eds.) CLEF 2021. LNCS, vol. 12880, pp. 308–323. Springer, Cham (2021). https://doi.org/10.1007/978-3-030-85251-1_21

16. Villa, G., Pasi, G., Viviani, M.: Echo chamber detection and analysis: a topology- and content-based approach in the COVID-19 scenario. Soc. Netw. Anal. Min. **11**(1), 78 (2021)
17. Viviani, M., Pasi, G.: Credibility in social media: opinions, news, and health information-a survey. Wiley Interdisc. Rev. Data Mining Knowl. Discov. **7**(5), e1209 (2017)
18. Wardle, C., Derakhshan, H.: Information disorder: toward an interdisciplinary framework for research and policy making. Counc. Europe **27** (2017)
19. Xu, D., Fan, S., Kankanhalli, M.: Combating misinformation in the era of generative AI models. In: Proceedings of the 31st ACM International Conference on Multimedia, pp. 9291–9298 (2023)

1st Workshop on Information Retrieval for Understudied Users (IR4U2)

Maria Soledad Pera[1(✉)], Federica Cena[2], Theo Huibers[3], Monica Landoni[4], Noemi Mauro[2], and Emiliana Murgia[5]

[1] Web Information Systems - TU Delft, Delft, The Netherlands
m.s.pera@tudelft.nl
[2] University of Turin, Turin, Italy
{federica.cena,noemi.mauro}@unito.it
[3] University of Twente, Enschede, The Netherlands
t.w.c.huibers@utwente.nl
[4] Università della Svizzera italiana, Lugano, Switzerland
monica.landoni@usi.ch
[5] Università di Genova, Genoa, Italy
emiliana.murgia@unige.it

Abstract. Information Retrieval (IR) remains an active, fast-paced area of research. However, most advances in IR have predominantly benefited the so-called "classical" users, e.g., English-speaking adults. We envision IR4U2as a forum to spotlight efforts that, while sparse, consider *diverse*, and often *understudied*, user groups when designing, developing, assessing, and deploying the IR technologies that directly impact them. The key objectives for IR4U2 are: (1) raise awareness about ongoing efforts focused on IR technologies designed for and used by often understudied user groups, (2) identify challenges and open issues impacting this area of research, (3) ignite discussions to identify common frameworks for future research, and (4) enable cross-fertilization and community-building by sharing lessons learned from research catering to different audiences by researchers and (industry) practitioners across various disciplines.

Keywords: Understudied users · Information Access · Information Retrieval

1 Motivation

Information retrieval (**IR**) is the dominant form of information access [9]. As such, it is not surprising that technological advances that directly and indirectly foster information access continue to be of interest to researchers and industry practitioners alike. Over the past few decades, developments in search, recommender, and question-answering systems, along with complementary IR tasks such as clustering, filtering, and text processing, have been well studied and documented; so are the efforts that help IR keep up with the fast-paced era of Large Language Models (LLM), Artificial Intelligence (AI), new modalities of

N. Goharian et al. (Eds.): ECIR 2024, LNCS 14612, pp. 409–414, 2024.
https://doi.org/10.1007/978-3-031-56069-9_55

interactions, and platforms enabling information access. We see, however, that works at SIGIR, ECIR, RecSys, and other related venues for the most part aid or study the so-called "classical," e.g., English-speaking adults.

Almost 20 years ago, Anderson et al. [1] discussed the concept of digital democratization of society and its connection to affordable Internet and enhanced information access to users from all backgrounds [4]. From reports from Statista [13], we know that today there are more than 5 billion Internet users *worldwide*. The most popular reasons for using the Internet, among users 16 years old and up, pertain to information access: places and travel, following news and events, finding information, along with education and studying [14]–the last two are also common reasons among younger user groups [3]. Has the IR community, however, genuinely reflected on whether existing IR technologies genuinely support the right to information access of *all* these billions of users? Consider popular search engines that are a portal to a rich and up-to-date set of resources: Do they account for individuals with low literacy levels who might not comprehend the resources retrieved? What about stereotypes that inadvertently might make them into their sets of retrieved results? Are these platforms naturally accessible and ready to prioritize retrieval of resources suited, for example, to those who are visually impaired [3,11,15]? Similar issues affect recommender systems, as recommendation algorithms used on platforms like TikTok or YouTube might inadvertently expose children to disturbing materials with far-reaching consequences [2]. What about AI technologies powering varied IR solutions: Are they inclusive and ensure fair representation of understudied populations?

Although IR works that aim to foster information access among people with diverse abilities and needs exist, they are very sporadic, and target, for instance, older adults, young children and adolescents, individuals afflicted by mental health disorders, users with autism spectrum disorder (ASD), individuals with intellectual disabilities, or those with specific learning needs [5,6,10,12,16,19]. Venues like CHI, UMAP, IDC, and ASSETS have recognized the impact of technology on these understudied populations. In the case of IR, the community has started to look at fairness and bias constructs, bringing awareness to issues of gender and minority representation. Still more is required to get a comprehensive mosaic of how IR technology can and should be shaped to better serve all its users and what that means. To elevate access to information, it is key to account for the environments in which these technologies are deployed and the nuances of individuals using IR technologies. The former is something the IR community is already undertaking, evidenced by endeavors targeting advancements in domain-specific IR, including scholarly works and workshops focused on the design, evaluation, and use of IR systems (and complementary IR technology) in medical [18], legal [17], fashion [8], and financial [7] domains, etc. Still, the wide range of users with different skill sets and needs has been mostly overlooked.

We argue that to understand how and whether IR technology can truly serve a diverse population it is critical to reflect on different use cases and take a human-driven approach that puts the different needs of real users at the center. Doing so requires multidisciplinary discussion and mutual understanding among the broad and sometimes conflicting perspectives if solutions will, in turn, fit the real-world setting and respond to the complex (information) requirements of a wide range of individuals. IR4U2 is meant to serve as a forum to bring together researchers and practitioners who have consistently allocated efforts to advance knowledge in this area but have seen their contributions accepted at non-core IR venues. By co-locating IR4U2 at ECIR we intend to make understudied groups more visible among the IR community.

2 Workshop Overview

The **goals** for IR4U2 are to: (1) Build a community involving multidisciplinary Ph.D. students, academics, and industry practitioners aiming to advance understanding of core IR technology and its impact on a broad range of understudied users; (2) Take inventory of the different kinds of use cases, research work, ongoing efforts, and existing resources; (3) Identify open challenges and opportunities for future research directions; and (4) Outline a general approach to facilitate research endeavors driven by and at the service of specific user groups.

We expect contributions that will help construct a snapshot of the works in this area. For this, we will elicit two types of submissions: **peer-reviewed** contributions (research papers and position/vision papers), as well as **editorially-reviewed** contributions (to enable potential attendees to articulate their views on the topics of the workshop, share already-published works, bring awareness to ongoing European projects in this area). **Topics** of interest include: (i) User modeling to enable IR and recommendation technologies tailored to understudied populations, (ii) Data collection and benchmark development of IR catering to understudied populations, (iii) IR applications targeting understudied populations, (iv) UI/UX for search, recommender, and question-answering systems for understudied populations, (v) IR-related technology (clustering, classification, text processing, text complexity) and their impact on understudied populations, (vi) Different perspectives of evaluation, (vii) Design of the user interaction with IR systems, and (viii) Ethical issues associated with IR technologies for understudied users; e.g., their right to be represented, acknowledged, have access, and be served by existing and to-be-developed systems. This is not an exhaustive list, as we are also interested in contributions discussing challenges inherent to designing IR technology at the service of understudied user groups–from the need for multidisciplinary, multi-stakeholder collaborations to how to build datasets.

We will host a **highly participatory, full-day, in-person** workshop, which will include a panel where we will invite participants championing different understudied users and discuss the importance of adopting a user-driven approach and integrating a wide range of perspectives to advance IR in a meaningful manner; brief presentations of accepted contributions; small group discussions; and joint discussions to merge findings from work in small groups that will result in plans for future editions of the IR4U2 workshop. A complete program can be found on the workshop website: https://ir4u2workshop.wixsite.com/ir4u2.

3 Organizers

Sole Pera is an Associate Professor at the Web Information Systems group (EEMCS) at TU Delft. Sole's research focuses on IR, with a special emphasis on enhancing information access for typically underserved user groups. She serves as (S)PC for conferences including SIGIR, UMAP, CHIIR, RecSys, and ECIR. She was General Chair for RecSys '18 and Program Chair for UMAP '23. Among others, she co-organized the Workshop on Educational Recommender Systems, the ComplexRec workshop, and 6 editions of the KidRec workshop.

Federica Cena is an Associate Professor at the Computer Science Department of the University of Turin. She works on the intersection of Artificial Intelligence and Human-Computer Interaction. Her recent research has studied the implications of the Internet of Things for user modeling and personalization, with a special focus on assistive applications for cognitive disabilities and frailty.

Theo Huibers has been researching information retrieval and human media interaction for over 30 years. Since 2002, he has been a professor in Human Media Interaction & Computer Science at the University of Twente and co-founder of Wizenoze, an international eTech company founded in 2013.

Monica Landoni is a titular professor at the faculty of Informatics at Università della Svizzera Italiana. She is vice chair of the ACM IDC Conference Steering Committee and an active member of EUGAIN, the European Network For Gender Balance in Informatics. She has worked on national and European projects investigating how technology can support children when searching, writing, and reading for education and pleasure. While doing that, she has happily designed and conducted many collaborative design sessions in formal and informal settings, carefully taking into account the needs, requests, roles, and points of view of real users, often from understudied communities.

Noemi Mauro is an Assistant Professor at the Computer Science Department of the University of Torino. Her research interests concern user modeling, recommender systems, cultural heritage, information filtering, and information visualization. She won the best paper award at UMAP '20 with the paper "Personalized Recommendation of PoIs to People with Autism". She is a PC member of the top conferences in her research areas and a reviewer for several related journals. She co-edited the special issue "Intelligent Systems for People with Diverse Cognitive Abilities" in the Human-computer Interaction journal. She is an Editorial Board Member of the UMUAI journal.

Emiliana Murgia has a degree in Literature and various training courses, including Communication and Management of School and Training Institutions. She transitioned from a career in communication to teaching at the primary school in 1999/2000; actively promoting technology in education, she has been a "digital animator" at school, managing various (inter)national projects, and has collaborated with the University of Milano Bicocca since 2014 on teaching with technologies. In 2018, she joined a multidisciplinary research team focusing on online information access, including AI. Currently on secondment to participate in a National Doctorate program in Learning Sciences and Digital Technologies.

References

1. Anderson, R.H., Bikson, T.K., Law, S.A., Mitchell, B.M.: Universal access to e-mail: feasibility and societal implications (1995)
2. Archie, A.: A.U.K. agency has fined TikTok nearly $16 million for handling of children's data (2023). https://www.npr.org/2023/04/05/1168114842/tik-tok-uk-fine
3. Azpiazu, I.M., Dragovic, N., Pera, M.S., Fails, J.A.: Online searching and learning: yum and other search tools for children and teachers. Inf. Retrieval J. **20**, 524–545 (2017)
4. Baker, P., Potts, A.: 'why do white people have thin lips?'google and the perpetuation of stereotypes via auto-complete search forms. Crit. Discourse Stud. **10**(2), 187–204 (2013)
5. Danovitch, J.H.: Growing up with Google: how children's understanding and use of internet-based devices relates to cognitive development. Hum. Behav. Emerg. Technol. **1**(2), 81–90 (2019)
6. Delgado, P., Ávila, V., Fajardo, I., Salmerón, L.: Training young adults with intellectual disability to read critically on the internet. J. Appl. Res. Intellect. Disabil. **32**(3), 666–677 (2019)
7. Feng, F., Luo, C., He, X., Liu, Y., Chua, T.S.: FinIR 2020: the first workshop on information retrieval in finance. In: Proceedings of the 43rd International ACM SIGIR Conference on Research and Development in Information Retrieval, pp. 2451–2454 (2020)
8. Jaradat, S., Dokoohaki, N., Corona Pampin, H.J., Shirvany, R.: Workshop on recommender systems in fashion and retail. In: Proceedings of the 15th ACM Conference on Recommender Systems, pp. 810–812 (2021)
9. Manning, C.D.: An introduction to information retrieval. Cambridge university press (2009)
10. Mauro, N., Ardissono, L., Cena, F.: Personalized recommendation of PoIs to people with autism. In: Proceedings of the 28th ACM Conference on User Modeling, Adaptation and Personalization, pp. 163–172, July 2020
11. Meyer, G., Wassyng, A., Lawford, M., Sabri, K., Shirani, S.: Literature review of computer tools for the visually impaired: a focus on search engines. Artif. Intell. Healthcare Med., 237–259 (2022)
12. Milton, A., Pera, M.S.: Into the unknown: exploration of search engines' responses to users with depression and anxiety. ACM Trans. Web **17**(4), 1–29 (2023)
13. Internet and social media users in the world 2023 (2023). https://www.statista.com/statistics/617136/digital-population-worldwide/
14. Reasons for using the internet worldwide by age 2022 (2023). https://www.statista.com/statistics/1387376/internet-using-global-reasons-by-age/
15. Noble, S.U.: Algorithms of oppression. In: Algorithms of Oppression. New York university press (2018)
16. Tsiakas, K., Barakova, E., Khan, J.V., Markopoulos, P.: BrainHood: towards an explainable recommendation system for self-regulated cognitive training in children. In: Proceedings of the 13th ACM International Conference on PErvasive Technologies Related to Assistive Environments, pp. 1–6 (2020)
17. Verberne, S., Kanoulas, E., Wiggers, G., Piroi, F., de Vries, A.P.: ECIR 2023 workshop: legal information retrieval. In: In: Kamps, J., et al. Advances in Information Retrieval. European Conference on Information Retrieval, pp. 412–419. Springer, Cham (2023). https://doi.org/10.1007/978-3-031-28241-6_46

18. White, R.W., Yom-Tov, E., Horvitz, E., Agichtein, E., Hersh, W.: Report on the SIGIR 2013 workshop on health search and discovery. In: ACM SIGIR Forum, vol. 47, pp. 101–108. ACM New York, NY, USA (2013)
19. Xie, B., Charness, N., Fingerman, K., Kaye, J., Kim, M.T., Khurshid, A.: When going digital becomes a necessity: Ensuring older adults' needs for information, services, and social inclusion during COVID-19. In: Older Adults and COVID-19, pp. 181–191. Routledge (2021)

First International Workshop on Graph-Based Approaches in Information Retrieval (IRonGraphs 2024)

Ludovico Boratto[1], Daniele Malitesta[2], Mirko Marras[1](✉), Giacomo Medda[1], Cataldo Musto[3], and Erasmo Purificato[4]

[1] University of Cagliari, Cagliari, Italy
{ludovico.boratto,mirko.marras}@acm.org, giacomo.medda@unica.it
[2] Polytechnic University of Bari, Bari, Italy
daniele.malitesta@poliba.it
[3] University of Bari, Bari, Italy
cataldo.musto@uniba.it
[4] Otto von Guericke University Magdeburg, Magdeburg, Germany
erasmo.purificato@ovgu.de

Abstract. In the dynamic field of information retrieval, the adoption of graph-based approaches has become a notable research trend. Fueled by the growing research on Knowledge Graphs and Graph Neural Networks, these approaches rooted in graph theory have shown significant promise in enhancing the effectiveness and relevance of information retrieval results. With this motivation in mind, this workshop serves as a platform, bringing together researchers and practitioners from diverse backgrounds, to delve into and discuss the integration of modern graph-based methodologies into information retrieval methods. The workshop website is available at https://irongraphs.github.io/ecir2024/.

Keywords: Graphs · Graph Neural Networks · Knowledge Graphs · Information Retrieval · Algorithms · Search · Recommendation

1 Motivation and Relevance

In recent years, Information Retrieval (IR) has witnessed a shift in focus, driven by the increasing complexity of data structures, user demands for personalised experiences, and the growth of interconnected information sources. Traditional linear models, while foundational, often struggle to leverage rich interconnections that define the digital age. In this scenario, graph-based approaches have proven their ability to model complex relationships and semantic connections [11,13,19].

Graph-based approaches, such as those driven by Graph Neural Networks (GNNs) [14,16,24,28,29], allow us to extract knowledge from networks of connections, changing the way we approach IR [4,12,15,17,18,22,23,25–27].For

N. Goharian et al. (Eds.): ECIR 2024, LNCS 14612, pp. 415–421, 2024.
https://doi.org/10.1007/978-3-031-56069-9_56

instance, by traversing the graph through its edges, we can better understand the context and meaning embedded in the data. This semantic understanding enables personalised and context-aware IR systems. As another example, by analysing user interactions within the graph, we can better investigate beyond-accuracy aspects, such as fairness and explainability [1–3,5–10,20,21].

Given the importance of graph-based approaches and the rapidly-changing techniques driving search and recommendation, the *First International Workshop on Graph-based Approaches in Information Retrieval* (IRonGraphs 2024) represents the first ECIR's workshop that aims to collect novel contributions in this continuously growing field and provide a common ground for interested researchers and practitioners within the European IR community.

2 Workshop Vision, Objectives, and Outcomes

The workshop vision is centred around the idea that graphs are a crucial data structure for modeling various domains. The success of graph representation learning has prompted the proposal for a dedicated event focused on IR topics. The resulting objectives include raising awareness of graph-based approaches in the IR community, identifying areas within IR that can benefit from such approaches, soliciting contributions addressing graph-based approaches, familiarising the IR community with graph-based research and practices, and uncovering gaps in IR research based on real-world needs. As outcomes, the workshop includes compiling and publishing workshop proceedings in the *CCIS* series by *Springer* as well as sharing slides and recordings associated with accepted papers. Additional outcomes will be represented by a strengthened community of researchers in graph-based approaches within IR, fostering future collaboration.

3 Workshop Format and Content Outline

The workshop involves a series of 40-minute spot thematic and interactive sessions (possibly 4 in total), wherein accepted contributions are grouped into 3–4 papers based on their topic or applicative domain. Each session comprises paper presentations lasting 7 to 10 min each (30 min in total), followed by a 10-minute discussion on the papers. Workshop organisers will facilitate these discussions, extracting lessons learned and providing brainstorming points on the session's theme. Thematic sessions will be complemented by two 60-minute keynote talks, one from academia and another from industry. Insights from thematic sessions and keynote talks will inform a final discussion lasting 60 min, possibly with renowned experts, to establish a roadmap for shared initiatives.

4 Intended Audience

This workshop targets a diverse audience, encompassing individuals interested in exploring innovative contributions to graph-based approaches. This audience includes researchers specialising in information retrieval, machine learning, and deep learning, as well as practitioners from both academia and industry. The workshop's emphasis on graphs as ubiquitous data structures across various domains highlights its interdisciplinary nature, extending beyond algorithmic considerations, fostering engagement among participants with varied expertise.

5 Workshop Organisers' Biography

Ludovico Boratto (webpage: https://www.ludovicoboratto.com/) is an Assistant Professor at the Dept. of Mathematics and Computer Science of the University of Cagliari (Italy). His research interests focus on recommender systems and their impact on stakeholders. He has co-authored over 60 papers published in top-tier conference proceedings and journals. His research brought him to give talks and tutorials at top-tier research centres and conferences, including CIKM, UMAP, RecSys, ICDE, ECIR, WSDM, ICDM, DSAA, and ECAI. He is the editor of the book "Group Recommender Systems: An Introduction" by Springer. He is an editorial board member of the "Information Processing & Management" journal (Elsevier) and "Journal of Intelligent Information Systems" (Springer), and guest editor of several journals' special issues. He is part of the program committees of the main Web conferences, where he received four outstanding contribution awards. In 2012, he got his Ph.D. at the University of Cagliari. From May 2016 to April 2021, he joined Eurecat as a Senior Research Scientist in the Data Science and Big Data Analytics research group. In 2010 and 2014, he visited Yahoo! Research in Barcelona.

Daniele Malitesta (webpage: https://danielemalitesta.github.io) is a PhD candidate at the Polytechnic University of Bari (Italy). His current research focuses on recommendation algorithms leveraging side information, with a focus on graph- and multimodal-based recommender systems. He has published papers at top-tier conferences, such as SIGIR, ECIR, RecSys, and MM, and served as a reviewer at venues such as ICLR, NeurIPS, SIGIR, RecSys, ECIR, and LoG. He is a co-developer of Elliot, a framework for the rigorous evaluation and reproducibility of recommender systems. Last summer, he visited Pasquale Minervini at the University of Edinburgh during a PhD internship. Recently, he presented a tutorial entitled "Graph Neural Networks for Recommendation: Reproducibility, Graph Topology, and Node Representation" at LoG 2023.

Mirko Marras (webpage: https://www.mirkomarras.com/) is an Assistant Professor at the Dept. of Mathematics and Computer Science of the University of Cagliari (Italy). Before that, he was a postdoctoral researcher at EPFL (Switzerland) and a visiting scholar at Eurecat (Spain) and New York University (USA). His research ranges across various domains impacted by user modelling and personalization, including business, education, entertainment, and healthcare. He has co-authored more than 80 papers in top-tier conferences and journals and has given tutorials at ECML-PKDD, RecSys, ICDE, ECIR, WSDM, ICDM, and UMAP. He has also co-chaired several workshops on related themes at ECIR,

WSDM, ICCV, EDM, and ECML-PKDD. He is part of the program committees of top-tier conferences in the field, where he received three outstanding reviewer awards. He is an associate editor for Springer's Journal of Ambient Intelligence and Humanized Computing and Neural Processing Letters.

Giacomo Medda (webpage: https://jackmedda.github.io/) is a PhD candidate at the Dept. of Mathematics and Computer Science of the University of Cagliari (Italy). In 2022, he has been a visiting scholar at Eurecat (Spain), where he worked on graph-based recommender systems with Dr. Francesco Fabbri. His research has focused on analysing and developing beyond-accuracy algorithms to improve the fairness and explainability of recommender and speaker recognition systems. He has published papers in top-tier conferences, such as CIKM and ECIR, and journals, such as IPM and PRLetters. He has served as a reviewer for flagship workshops, such as BIAS and KaRS, and journals, such as Neural Processing Letters and TOIS.

Cataldo Musto (webpage: http://www.di.uniba.it/~swap/musto) is an Assistant Professor at the Dept. of Informatics, University of Bari. He completed his PhD in 2012 under the supervision of Prof. Giovanni Semeraro. His research focuses on the adoption of natural language processing techniques for fine-grained semantic content representation in recommender systems and user modelling platforms. He acts as a program committee member for RecSys and UMAP, and he organized several events on user modelling and recommender systems. In 2016 and 2017, he gave a tutorial at UMAP conference on semantics-aware representation in content-based personalized systems. Recently, he organized the workshop on Explainable User Modeling (ExUM), jointly held with UMAP 2019–2023.

Erasmo Purificato (webpage: https://erasmopurif.com/) has been a Research Assistant in the "Human-Centred AI" group at the Otto von Guericke University Magdeburg (OVGU) and in the "Human-Centred Technologies for Educational Media" department at the Leibniz Institute for Educational Media | Georg Eckert Institute (GEI), since February 2020. He is doing his PhD in Computer Science at OVGU with a project entitled "Human-Centred Fairness Analysis on Graph Neural Network-Based Models for Behavioural User Profiling". From July 2021, he has been appointed by the Guglielmo Marconi University of Rome, Italy, as an Adjunct Professor in "Software Engineering". He co-organised the HCAI4U Workshop at CHItaly, the APEx-UI Workshop at IUI, and the IEEE Autumn School ISACT at ICHMS. He has given tutorials at CIKM and UMAP. He has been part of the program committee of several conferences and workshops, such as SIGIR, RecSys, UMAP, HT and ExUM, and served as a reviewer for journals, such as TORS and TITS.

References

1. Abdelrazek, M., Purificato, E., Boratto, L., De Luca, E.W.: FairUP: a framework for fairness analysis of graph neural network-based user profiling models. In: Proceedings of the 46th International ACM SIGIR Conference on Research and Development in Information Retrieval, SIGIR, p. 3165-3169. SIGIR 2023, ACM (2023)
2. Anelli, V.W., Deldjoo, Y., Noia, T.D., Malitesta, D., Paparella, V., Pomo, C.: Auditing consumer- and producer-fairness in graph collaborative filtering. In: Proceedings of the 45th European Conference on Information Retrieval, ECIR. LNCS, vol. 13980, pp. 33–48. Springer, Cham (2023). https://doi.org/10.1007/978-3-031-28244-7_3
3. Anelli, V.W., et al.: How neighborhood exploration influences novelty and diversity in graph collaborative filtering. In: Proceedings of the 2nd Workshop on Multi-Objective Recommender Systems co-located with 16th ACM Conference on Recommender Systems, RecSys. CEUR Workshop Proceedings, vol. 3268. CEUR-WS.org (2022)
4. Anelli, V.W., Malitesta, D., Pomo, C., Bellogín, A., Sciascio, E.D., Noia, T.D.: Challenging the myth of graph collaborative filtering: a reasoned and reproducibility-driven analysis. In: Proceedings of the 17th ACM Conference on Recommender Systems, RecSys, pp. 350–361. ACM (2023)
5. Balloccu, G., Boratto, L., Cancedda, C., Fenu, G., Marras, M.: Faithful path language modelling for explainable recommendation over knowledge graph. CoRR **abs/2310.16452** arXiv:2310.16452 (2023)
6. Balloccu, G., Boratto, L., Cancedda, C., Fenu, G., Marras, M.: Knowledge is power, understanding is impact: Utility and beyond goals, explanation quality, and fairness in path reasoning recommendation. In: Proceedings of the 45th European Conference on Information Retrieval, ECIR. LNCS, vol. 13982, pp. 3–19. Springer, Cham (2023). https://doi.org/10.1007/978-3-031-28241-6_1
7. Balloccu, G., Boratto, L., Fenu, G., Marras, M.: Post processing recommender systems with knowledge graphs for recency, popularity, and diversity of explanations. In: Proceedings of the 45th International ACM SIGIR Conference on Research and Development in Information Retrieval, SIGIR, pp. 646–656. ACM (2022)
8. Balloccu, G., Boratto, L., Fenu, G., Marras, M.: Reinforcement recommendation reasoning through knowledge graphs for explanation path quality. Knowl. Based Syst. **260**, 110098 (2023)
9. Boratto, L., Fabbri, F., Fenu, G., Marras, M., Medda, G.: Counterfactual graph augmentation for consumer unfairness mitigation in recommender systems. In: Proceeding of the 32nd ACM International Conference on Information and Knowledge Management, CIKM, pp. 3753–3757. ACM (2023)
10. Boratto, L., Fenu, G., Marras, M., Medda, G.: Practical perspectives of consumer fairness in recommendation. Inf. Process. Manag. **60**(2), 103208 (2023)
11. Cambria, E., Mao, R., Han, S., Liu, Q.: Sentic parser: a graph-based approach to concept extraction for sentiment analysis. In: Proceeding of the IEEE International Conference on Data Mining Workshops, ICDM - Workshops, pp. 1–8. IEEE (2022)
12. Chen, J., Zhu, G., Hou, H., Yuan, C., Huang, Y.: AutoGSR: neural architecture search for graph-based session recommendation. In: Proceeding of the 45th International ACM SIGIR Conference on Research and Development in Information Retrieval, SIGIR, pp. 1694–1704. ACM (2022)

13. Halilaj, L., Dindorkar, I., Lüttin, J., Rothermel, S.: A knowledge graph-based approach for situation comprehension in driving scenarios. In: Verborgh, R., et al. (eds.) ESWC 2021. LNCS, vol. 12731, pp. 699–716. Springer, Cham (2021). https://doi.org/10.1007/978-3-030-77385-4_42
14. Hamilton, W.L., Ying, Z., Leskovec, J.: Inductive representation learning on large graphs. In: Proceedings of the Annual Conference on Neural Information Processing Systems, NIPS, pp. 1024–1034 (2017)
15. Kamphuis, C.: Graph databases for information retrieval. In: Jose, J.M., et al. (eds.) ECIR 2020. LNCS, vol. 12036, pp. 608–612. Springer, Cham (2020). https://doi.org/10.1007/978-3-030-45442-5_79
16. Kipf, T.N., Welling, M.: Semi-supervised classification with graph convolutional networks. In: Proceeding of the 5th International Conference on Learning Representations, ICLR. OpenReview.net (2017)
17. Liu, S., Ounis, I., Macdonald, C.: An MLP-based algorithm for efficient contrastive graph recommendations. In: Proceedings of the 45th International ACM SIGIR Conference on Research and Development in Information Retrieval, SIGIR, pp. 2431–2436. ACM (2022)
18. Malitesta, D., Pomo, C., Anelli, V.W., Mancino, A.C.M., Sciascio, E.D., Noia, T.D.: A topology-aware analysis of graph collaborative filtering. CoRR **abs/2308.10778** arXiv:2308.10778 (2023)
19. Mayank, M., Sharma, S., Sharma, R.: DEAP-FAKED: knowledge graph based approach for fake news detection. In: Proceedings of the IEEE/ACM International Conference on Advances in Social Networks Analysis and Mining, ASONAM, pp. 47–51. IEEE (2022)
20. Medda, G., Fabbri, F., Marras, M., Boratto, L., Fenu, G.: GNNUERS: fairness explanation in GNNs for recommendation via counterfactual reasoning. CoRR **abs/2304.06182** arXiv:2304.06182 (2023)
21. Purificato, E., Boratto, L., De Luca, E.W.: Do graph neural networks build fair user models? Assessing disparate impact and mistreatment in behavioural user profiling. In: Proceedings of the 31st ACM International Conference on Information & Knowledge Management, CIKM, p. 4399–4403. CIKM 2022, ACM (2022)
22. Spillo, G., Musto, C., Polignano, M., Lops, P., de Gemmis, M., Semeraro, G.: Combining graph neural networks and sentence encoders for knowledge-aware recommendations. In: Proceedings of the 31st ACM Conference on User Modeling, Adaptation and Personalization, UMAP, p. 1–12. UMAP 2023, ACM (2023)
23. Thonet, T., Renders, J.-M., Choi, M., Kim, J.: Joint personalized search and recommendation with hypergraph convolutional networks. In: Hagen, M., et al. (eds.) ECIR 2022. LNCS, vol. 13185, pp. 443–456. Springer, Cham (2022). https://doi.org/10.1007/978-3-030-99736-6_30
24. Velickovic, P., Cucurull, G., Casanova, A., Romero, A., Liò, P., Bengio, Y.: Graph attention networks. In: Proceedings of the 6th International Conference on Learning Representations, ICLR. OpenReview.net (2018)
25. Witschel, H.F., Riesen, K., Grether, L.: KvGR: a graph-based interface for explorative sequential question answering on heterogeneous information sources. In: Jose, J.M., et al. (eds.) ECIR 2020. LNCS, vol. 12035, pp. 760–773. Springer, Cham (2020). https://doi.org/10.1007/978-3-030-45439-5_50
26. Yi, Z., Ounis, I., Macdonald, C.: Graph contrastive learning with positional representation for recommendation. In: Proceedings of the 45th European Conference on Information Retrieval, ECIR. Lecture Notes in Computer Science, vol. 13981, pp. 288–303. Springer, Cham (2023). https://doi.org/10.1007/978-3-031-28238-6_19

27. Yu, H.C., Dai, Z., Callan, J.: PGT: pseudo relevance feedback using a graph-based transformer. In: Hiemstra, D., Moens, M.-F., Mothe, J., Perego, R., Potthast, M., Sebastiani, F. (eds.) ECIR 2021. LNCS, vol. 12657, pp. 440–447. Springer, Cham (2021). https://doi.org/10.1007/978-3-030-72240-1_46
28. Zhang, C., Song, D., Huang, C., Swami, A., Chawla, N.V.: Heterogeneous graph neural network. In: Proceedings of the 25th ACM SIGKDD International Conference on Knowledge Discovery & Data Mining, SIGKDD, pp. 793–803 (2019)
29. Zhang, Z., Cui, P., Zhu, W.: Deep learning on graphs: a survey. IEEE Trans. Knowl. Data Eng. **34**(1), 249–270 (2022)

The Search Futures Workshop

Leif Azzopardi[1(✉)], Charles L. A. Clarke[2], Paul B. Kantor[3], Bhaskar Mitra[4], Johanne R. Trippas[5], and Zhaochun Ren[6]

[1] University of Strathclyde, Glasgow, UK
leif.azzopardi@strath.ac.uk
[2] University of Waterloo, Waterloo, Canada
[3] Rutgers University, New Brunswick, USA
[4] Microsoft Research, Montreal, Canada
[5] RMIT University, Melbourne, Australia
[6] Leiden University, Leiden, The Netherlands

Abstract. The field and community of Information Retrieval (IR) are changing and evolving in response to the latest developments and advances in Artificial Intelligence (AI) and research culture. As the field and community re-oriented and re-consider its positioning within computing and information sciences more generally – it is timely to gather and discuss more seriously our field's vision for the future – the challenges and threats that the community and field faces – along with the bold new research questions and problems that are arising and emerging as we re-imagine search. This workshop aims to provide a forum for the IR community to voice and discuss their concerns and pitch proposals for building and strengthening the field and community.

Keywords: Search · Information retrieval · Artificial intelligence

1 Introduction and Motivation

The field of Information Retrieval (IR) is undergoing a profound transformation, spurred by the continual evolution and breakthroughs in the realm of artificial intelligence and the broader changing research landscape. This reformation period finds our field and community in a state of uncertainty, as we contemplate and reevaluate our role and significance within the broader context of computing and information sciences. This juncture in our journey serves as an opportune moment to convene and engage in a deep and purposeful dialogue concerning the future trajectory of our field. We must collectively confront the myriad challenges and potential threats that loom on the horizon, all while embracing the newfound opportunities and bold research inquiries that emerge as we embark on a re-imagined quest for the next generation "memex machine"[1].

The purpose of this workshop is to serve as a dedicated platform for the IR community to candidly express and deliberate upon the issues that weigh on our collective conscience. It is a forum where we can voice our concerns and

[1] https://doi.org/10.1145/1897816.1897840.

N. Goharian et al. (Eds.): ECIR 2024, LNCS 14612, pp. 422–425, 2024.
https://doi.org/10.1007/978-3-031-56069-9_57

brainstorm and present innovative proposals aimed at fortifying and enriching our field and the community that sustains it.

As we stand at the intersection of technological innovation and scholarly introspection, we find ourselves confronted with a multitude of pertinent questions. *How can we harness the power of AI to enhance the effectiveness of information retrieval? What safeguards do we need to put in place to protect the integrity and privacy of the data we handle? How can we ensure that the fruits of our research are accessible and beneficial to all members of society?* These are just a few examples of the pressing issues we face as we navigate this dynamic, new IR landscape.

In our pursuit of "*search futures*", this workshop aims to provide a forum for the community to discuss and contribute to our collective agenda for the research directions and field. We hope that together, we can chart a course that not only safeguards IR's continued relevance and vitality but also propels it into uncharted territories of discovery and exploration.

2 Workshop Goals and Objectives

The Search Futures Workshop aims to provide a much-needed forum for the IR community to discuss the emerging challenges to the field and community. Our goals are to:

- Provide a forum at ECIR to discuss the pressing and emerging issues our field faces, and,
- Produce a report detailing the initial outcomes of this first workshop on Search Futures.
- Continue this ambitious series for Search Future workshops at subsequent IR conferences to include further and wider perspectives.

2.1 Topics of Interest

Short position statements from participants will be solicited through direct invitation and an open call to the ECIR community. We will select a diverse and representative subset from the position statements submitted to present their position/perspective during the workshop. We would like to attract a broad range of positions about the future of search. Topics of interest may include, but are not limited to:

- IR and related fields
  - What the field of IR is tackling, should be tackling, is not tackling,... and is such research is even important?
  - What is IR any more in the context of recommender systems, NLP, ML, AI, etc.?
  - What are the core research questions we should be answering?
- IR in the age of generative AI
  - How generative AI is changing the nature and relevance of search?

    - How can wc distinguish originals from derivatives, real from fake, etc.?
    - When everything can be generated, what is a document? What are we retrieving?
- IR and community
    - How can we build and grow the IR community?
    - How can we support newer community members?
    - Is a bigger, more diverse community better?
    - What are our conferences turning into?
    - Where are all the core IR papers?
    - What should the scope and remit of IR conferences and journals be?
- IR and the business
    - What are the new economics of IR?
    - How does conversational search change current business models?
    - How do traditional media/content-based models fit into the emerging landscape?
    - How can IR further optimize workplace productivity?
- IR ethics, trust and responsibility
    - What is the duty/responsible of an IR system?
    - If I want to see more, should the system give it to me?
    - How can we trust IR systems if they make up everything?
    - How environmentally responsible are the IR systems we are making?
- IR and people, users, consumers, creators, ...
    - Are creators still needed in the age of generative AI?
    - Is IR helping overcome the digital divide?
    - Is IR addressing the disparity in information access, especially in marginalized communities?
    - Should IR systems protect users from information overload?
    - What is the future of results presentation?
    - How can we develop IR systems integrated with IoT?
    - How do information systems influence user's emotions?

3 Organisers

To run the workshop, we have six organizers, five of whom can confirm that they will attend ECIR in person. Our organization team aims to bring together the IR Oldies with the up-and-coming stars in our field from industry, academia, and around the globe.

Leif Azzopardi is an Associate Professor in the Department of Computer and Information Sciences at the University of Strathclyde, Glasgow. His research focuses on building models and metrics for interactive information retrieval with a focus on model users in the lab and in the wild. Recently, he has been working with Microsoft Search and AI on (conversational) search. He has organized numerous IR events e.g. PC Chair of FDIA (2008–2015), PC Chair of IIiX 2014, PC Chair SimInt 2010 @ ACM SIGIR, and PC Chair of ICTIR 2008, General Chair of ACM CHIIR 2019 and PC Chair of ECIR 2019.

Charles L. A. Clarke is a Professor in the School of Computer Science at the University of Waterloo, Canada. His research focuses on data intensive tasks and efficiency, including search, ranking, question answering, and other problems involving human language data at scale. In addition to his academic experience, he has worked on search engine technology for both Microsoft Bing and Facebook Search. He has previously co-organized workshops at ECIR (2014, 2011), SIGIR (2016, 2015, 2013, 2012), WSDM (2012) and CHIIR (2023, 2020).

Paul Kantor is Distinguished Professor (Emeritus) of Information Science at Rutgers, and an Honorary Associate in the Department of Industrial and Systems Engineering at the University of Wisconsin Madison. His work has primarily focused on evaluation of Information (Retrieval) Systems, with an emphasis on relating that evaluation to the specific needs of the system's user at the moment. He also developed early recommendation systems called ANLI (pre WWW) and AntWorld, which have vanished without a trace. That research has been supported by the US NSF, Department of Education, DARPA, and NATO.

Bhaskar Mitra is a Principal Researcher at Microsoft Research based in Montreal, Canada. His research focuses on AI-mediated information and knowledge access and questions of fairness and ethics in the context of these socio-technical systems. Before joining Microsoft Research, he worked on search technologies at Bing for 15+ years. He is serving as the ACM SIGIR Community Relations Coordinator and on the NIST TREC program committee. He has co-organized several workshops (Neu-IR @ SIGIR 2016–2017 and HIPstIR 2019), shared evaluation tasks (TREC Deep Learning Track 2019–2023, TREC Tip-of-the-Tongue Track 2023, and MS MARCO ranking leader boards), and tutorials (WSDM 2017–2018, SIGIR 2017, and ECIR 2018).

Johanne Trippas is a Vice-Chancellor's Research Fellow at RMIT University, specializing in intelligent systems, focusing on digital assistants and conversational information seeking. Her research aims to enhance information accessibility through conversational systems, interactive information retrieval, and human-computer interaction. Additionally, Johanne is currently part of the NIST TREC program committee and is an ACM CHIIR steering committee member. She serves as vice-chair of the SIGIR Artifact Evaluation Committee, tutorial chair for ECIR'24, general chair of the ACM CUI'25, and ACM SIGIR-AP'23 proceedings chair. She has organized workshops (CHIIR'20–22), a TREC Track (CAsT'22), and tutorials (CHIIR'21, SIGIR'22, and WebConf'23).

Zhaochun Ren is an Associate Professor at Leiden University. His research interests focus on joint research problems in information retrieval and natural language processing, with an emphasis on conversational information seeking, question-answering, and recommender systems. He aims to develop intelligent systems that can address complex user requests and solve core challenges in both information retrieval and natural language processing towards that goal. In addition to his academic experience, he worked on e-commerce search and recommendation at JD.com for 2+ years. He has co-organized workshops at SIGIR (2020) and WSDM (2019, 2020).

The First International Workshop on Open Web Search (WOWS)

Sheikh Mastura Farzana[1], Maik Fröbe[2](✉), Michael Granitzer[3], Gijs Hendriksen[4], Djoerd Hiemstra[4], Martin Potthast[5], and Saber Zerhoudi[3]

[1] German Aerospace Center (DLR), Cologne, Germany
sheikh.farzana@dlr.de
[2] Friedrich-Schiller-Universität Jena, Jena, Germany
maik.froebe@uni-jena.de
[3] University of Passau, Passau, Germany
michael.granitzer@uni-passau.de
[4] Radboud University, Nijmegen, The Netherlands
djoerd.hiemstra@ru.nl
[5] Leipzig University and ScaDS.AI, Leipzig, Germany
martin.potthast@uni-leipzig.de

Abstract. We organize the first international Workshop on Open Web Search (WOWS) at ECIR 2024 with two calls for contributions. The first call targets scientific contributions on cooperative search engine development.This includes cooperative crawling of the web and cooperative deployment and evaluation of search engines. We specifically highlight the potential of enabling public and commercial organizations to use an indexed web crawl as a resource to create innovative search engines tailored to specific user groups, instead of relying on one search engine provider. The second call aims at gaining practical experience with joint, cooperative evaluation of search engine prototypes and their components using the Information Retrieval Experiment Platform (TIREx).

1 Introduction

Web search is a critical technology for the digital economy, dominated by a few gatekeepers [4]. These gatekeepers take care of and thus control every aspect of search: they crawl the web, index it, develop search engines, and provide search-based applications. Since these gatekeepers also control the market for search engine advertisements that are displayed alongside search results, they have incentives to sacrifice search result quality by promoting best-paying results over the most relevant results. The gatekeepers have every opportunity to dictate their will to their users: They even control the web browsers and operating systems that users need to access the web [15]. Additionally, these companies also have a profound influence on academic research, both indirectly and directly through funding of research facilities, researchers, and sponsorship of conferences [1].

https://opensearchfoundation.org/wows2024/.

N. Goharian et al. (Eds.): ECIR 2024, LNCS 14612, pp. 426–431, 2024.
https://doi.org/10.1007/978-3-031-56069-9_58

With these tech giants of web search controlling every aspect of search, web publishers need to optimize their content for the search engines instead of the search engines optimizing their results for the content. This has led to a closed ecosystem of search engines and risks compromising quality for publishers. Many search engine users are aware of this fact [11].

The first International Workshop on Open Web Search (WOWS) promotes and discusses ideas and approaches to opening up today's closed search ecosystem. We are particularly interested in approaches that allow organizations to provide search engines cooperatively, e.g., by specializing in one task and sharing their expertise, tools, and resources with others for mutual benefit. We are also interested in ideas and approaches that allow organizations to collaborate on these tasks and open standards for search and open source retrieval.

Beyond search, the web has demonstrated its critical role as a resource, e.g., for training large language models, for analyzing human behavioral data, or for studying the web as a structure itself. However, tapping into such a resource requires a large infrastructure, appropriate technical capabilities, and much data cleaning and pre-processing. While some large organizations can provide these resources, small research groups or young startups either lack the skills or would have to devote much time and resources outside their core area, which is usually beyond their budget. Since the open source landscape of web data processing is surprisingly small, we are interested in approaches where organizations share their data processing infrastructures and treat their data as open and accessible.

From a practical point of view, the workshop will also focus on the scientific evaluation of search engines and their components, using test collections from shared tasks. We will encourage participants to submit and evaluate new retrieval approaches or components of retrieval pipelines using TIRA/TIREx [7,8]. Modern web search engines use complex pipelines with many components (e.g., query/document understanding), so different organizations can bring different components to open search ecosystems. Therefore, we ask participants to implement retrieval pipelines and/or components. Using TIREx, these components can be systematically combined and run on many test collections, making their results publicly available, and comparing their effectiveness. If the outputs of standard components are available for multiple test collections, researchers can reuse them without having to rerun the software itself, promoting GreenIR [18] along the entire retrieval pipeline [9]. Therefore, the goals of the workshop are:

1. Sharing novel concepts, algorithms, and ideas for cooperatively building web search engines to open the web as a resource for researchers and innovators.
2. Gaining hands-on experience with joint, collaborative evaluation of search engines and/or their components with TIRA/TIREx.

2 Workshop Program

The workshop's intended audience includes researchers and practitioners in open web search and has a call for research contributions and a call for software. We see the Workshop on Open Web Search as a continuation of workshops on the topic of open source information retrieval that began with the first International Workshop on Open Source Web Information Retrieval in 2005, the second International Workshop on Open Source Information Retrieval in 2006 [22], the Workshop on Open Source Information Retrieval in 2012 [19], the Lucene for information access and retrieval research (LIARR) workshop in 2017 [3], and the workshops on replicability, including the Reproducibility, Inexplicability, and Generalizability of Results (RIGOR) workshop in 2016 [2] and the Open-Source Information Retrieval Replicability Challenge (OSIRRC) in 2019 [5].

2.1 Call for Research Contributions

We seek research contributions that address elements of a traditional web search pipeline and also incorporate recent developments such as (open source) large language models to interface with retrieval systems. Specifically, our focus is on contributions that address the importance of an open web search pipeline and the creation of an open web index as a basis for the development of search applications for specific purposes and communities. Interesting topics include, but are not limited to (1) crawling for an open web index (e.g., fast DNS resolution, distribution of crawl jobs, etc.). (2) pre-processing and enrichment (e.g., entity extraction/linking, geo-parsing, spam classification, etc.). (3) indexing and search architectures (e.g., with tools like Terrier [14], OldDog [17], GeeseDB [10], Spinque [6], CIFF [12], Anserini [21], PISA [16], JASSv2 [20]). (4) search interfaces and paradigms (e.g., novel search paradigms like conversational search, temporal argument search, or geospatial search with aspects like trust, privacy, bias, and ethics). Overall, we invite contributions in the form of research papers and resource papers along the following non-exclusive list of topics: standards for search and interoperability; deployment of search engines; collaborative crawling; open infrastructures for evaluation; open source search engines; open source replicability; ethical and legal aspects of web search; alternatives for query logs and click logs; vertical search engines; search engines for low-resource languages; web master control (robots.txt and more); energy efficiency of web search; and large scale web data pre-processing pipelines.

2.2 Call for Evaluation Components

Evaluation plays a critical role for every (web) search engine. Robust and insightful evaluation is essential to guide the development efforts of cooperative open web search engines, especially when multiple decoupled organizations with potentially incompatible goals contribute to the ecosystem. However, evaluations of general-purpose web search engines are difficult due to the size of the web and the immense technical effort required to implement a practical web search engine. Therefore, we request the submission of software to TIRA/TIREx [7,8] that implements search engine components that the community can declaratively assemble to evaluate retrieval pipelines composed of submitted components.

This call leverages the open infrastructure for evaluation provided by TIRA/TIREx that enables the construction of experimental retrieval pipelines that can be evaluated on many information retrieval test collections without having to own these collections or even invest in preprocessing or indexing them (e.g., by submitting re-rankers). Each step of a retrieval pipeline in TIREx is a Docker image, and submitted Docker images can be used in declarative PyTerrier [14] pipelines after the shared task has finished [7]. All software submissions to TIREx are immutable to allow caching of a component's output. Typical IR test collections are static, so that many steps in retrieval pipelines need to be executed only once in a lifetime. Thus, we will make all outputs of the submitted software publicly available where the dataset licenses allow this so that the community can build complex retrieval pipelines without re-executing many of its components. At the same time, TIREx provides a provenance "chain of trust" to users who reuse the output of a software component, as it documents which Docker image produced which output. Because TIRA runs each submitted Docker image in a sandbox not under the participant's control, reproducibility and replicability are improved while simplifying reusing components.

The outcomes of the shared task are the dockerized components submitted by the participants and the outputs produced by those docker images on standard IR benchmarks in TIRA/TIREx. The focus of our call for individual (potentially fine-grained) components of retrieval pipelines is in contrast to previous efforts that collected complete retrieval systems [2,5], as small components (or their outputs) can be more easily reused in a wide range of IR experiments than complete systems, especially in fast-changing research environments. All software submissions in TIREx are implemented against the interface provided by ir_datasets [13], which provides abstraction and standardization over typical data structures of IR experiments such as information needs and documents.

Our call for software asks for submissions that fall into one or more of the following classes of retrieval components: (1) query processing (e.g., query performance prediction, query reformulation, entity detection, query segmentation, query expansion, etc.), (2) document processing (e.g., document expansion, document reduction, spam classification, web genre classification, topic modeling, etc.), and (3) query–document processing (e.g., re-ranking, rank fusion, etc.).

3 Conclusion

In summary, the ECIR 2024 Workshop on Open Web Search (WOWS) addresses some of the most pressing concerns related to web search such as transparency, availability, and accessibility of web resources. We expect that WOWS will foster collaboration and innovation in building open search ecosystems. WOWS will explore new approaches that enable organizations to collaborate on various aspects of search engines, promote open standards, and challenge the current closed ecosystem of search. In addition, WOWS aims to facilitate data sharing and open access to web resources. The practical focus of the workshop is to promote scientific evaluation through shared tasks, collaboration in building search pipelines and retrieval test collections, with the vision that different organizations contribute to building a sustainable open search framework.

Acknowledgments. This work has received funding from the European Union's Horizon Europe research and innovation programme under grant agreement No. 101070014 (OpenWebSearch.EU, https://doi.org/10.3030/101070014).

References

1. Abdalla, M., Abdalla, M.: The grey hoodie project: big tobacco, big tech, and the threat on academic integrity. In: Proceedings of the 2021 AAAI/ACM Conference on AI, Ethics, and Society, pp. 287–297 (2021)
2. Arguello, J., Crane, M., Diaz, F., Lin, J., Trotman, A.: Report on the SIGIR 2015 workshop on reproducibility, inexplicability, and generalizability of results (RIGOR). SIGIR Forum **49**(2), 107–116 (2016)
3. Azzopardi, L., et al.: The Lucene for information access and retrieval research (LIARR) workshop at SIGIR 2017. In: Proceedings of the 40th International ACM SIGIR Conference on Research and Development in Information Retrieval, pp. 1429–1430 (2017)
4. Bahrke, J., Podesta, A., Regnier, T., Simonini, S.: Digital markets act: commission designates six gatekeepers. European Commission Press Release (2023). https://ec.europa.eu/commission/presscorner/detail/en/IP_23_4328
5. Clancy, R., Ferro, N., Hauff, C., Lin, J., Sakai, T., Wu, Z.Z.: The SIGIR 2019 open-source IR replicability challenge (OSIRRC 2019). In: Piwowarski, B., Chevalier, M., Gaussier, É., Maarek, Y., Nie, J., Scholer, F. (eds.) Proceedings of the 42nd International ACM SIGIR Conference on Research and Development in Information Retrieval, pp. 1432–1434 (2019)
6. Cornacchia, R.: Graph Ranking – part 1. Spinque (2021). https://spinque.com/blog/graph-ranking-part-1/
7. Fröbe, M., et al.: The information retrieval experiment platform. In: Proceedings of the 46th International ACM SIGIR Conference on Research and Development in Information Retrieval (2023)
8. Fröbe, M., et al.: Continuous integration for reproducible shared tasks with TIRA.io. In: Kamps, J., et al. (eds.) Advances in Information Retrieval, ECIR 2023. LNCS, vol. 13982, pp. 236–241. Springer, Cham (2023). https://doi.org/10.1007/978-3-031-28241-6_20
9. Granitzer, M., Voigt, S., et al.: Impact and development of an open web index for open web search. J. Assoc. info. Sci. Technol. **2023**, 1–9 (2023)
10. Kamphuis, C., de Vries, A.P.: GeeseDB: a Python graph engine for exploration and search. In: Proceedings of the 2nd International Conference on Design of Experimental Search and Information Retrieval Systems (DESIRES) (2021)
11. Lewandowski, D., Schultheiß, S.: Public awareness and attitudes towards search engine optimization. Behav. Inform. Technol. **42**, 1025–1044 (2022)
12. Lin, J., et al.: Supporting interoperability between open-source search engines with the common index file format. In: Proceedings of the 43rd International ACM SIGIR Conference on Research and Development in Information Retrieval, pp. 2149–2152 (2020)
13. MacAvaney, S., Yates, A., Feldman, S., Downey, D., Cohan, A., Goharian, N.: Simplified data wrangling with ir_datasets. In: SIGIR (2021)
14. Macdonald, C., Tonellotto, N., MacAvaney, S., Ounis, I.: PyTerrier: declarative experimentation in Python from BM25 to dense retrieval. In: Proceedings of the 30th ACM International Conference on Information & Knowledge Management (CIKM), pp. 4526–4533 (2021)

15. Mager, A., Norocel, C., Rogers, R. (eds.): The State of Google Critique and Intervention. Big Data Soc. **10** (2023)
16. Mallia, A., Siedlaczek, M., Mackenzie, J., Suel, T.: PISA: performant indexes and search for academia. In: Proceedings of the Open-Source IR Replicability Challenge co-located with 42nd International ACM SIGIR Conference on Research and Development in Information Retrieval, OSIRRC@SIGIR 2019, Paris, France, 25 July 2019, pp. 50–56 (2019). http://ceur-ws.org/Vol-2409/docker08.pdf
17. Mühleisen, H., Samar, T., Lin, J., De Vries, A.: Old dogs are great at new tricks: column stores for information retrieval prototyping. In: Proceedings of the 37th International ACM SIGIR Conference on Research & Development in Information Retrieval, pp. 863–866 (2014)
18. Scells, H., Zhuang, S., Zuccon, G.: Reduce, reuse, recycle: green information retrieval research. In: Amigó, E., Castells, P., Gonzalo, J., Carterette, B., Culpepper, J.S., Kazai, G. (eds.) The 45th International ACM SIGIR Conference on Research and Development in Information Retrieval, SIGIR 2022, pp. 2825–2837 (2022)
19. Trotman, A., Clarke, C.L., Ounis, I., Culpepper, S., Cartright, M.A., Geva, S.: Open source information retrieval: a report on the SIGIR 2012 workshop. SIGIR Forum **46**(2), 95–101 (2012)
20. Trotman, A., Lilly, K.: JASSjr: the minimalistic BM25 search engine for teaching and learning information retrieval. In: Huang, J.X., et al. (eds.) Proceedings of the 43rd International ACM SIGIR conference on Research and Development in Information Retrieval, SIGIR 2020, Virtual Event, China, 25–30 July 2020, pp. 2185–2188, ACM (2020). https://doi.org/10.1145/3397271.3401413
21. Yang, P., Fang, H., Lin, J.: Anserini: enabling the use of Lucene for information retrieval research. In: Kando, N., Sakai, T., Joho, H., Li, H., de Vries, A.P., White, R.W. (eds.) Proceedings of the 40th International ACM SIGIR Conference on Research and Development in Information Retrieval, Shinjuku, Tokyo, Japan, 7–11 August 2017, pp. 1253–1256. ACM (2017). https://doi.org/10.1145/3077136.3080721
22. Yee, W.G., Beigbeder, M., Buntine, W.: SIGIR06 workshop report: open source information retrieval systems (OSIR06). ACM SIGIR Forum **40**(2), 61–65 (2006)

Third Workshop on Augmented Intelligence in Technology-Assisted Review Systems (ALTARS)

Giorgio Maria Di Nunzio[1(✉)], Evangelos Kanoulas[2], and Prasenjit Majumder[3]

[1] Department of Information Engineering, University of Padova, Padua, Italy
giorgiomaria.dinunzio@unipd.it

[2] Faculty of Science, Informatics Institute, University of Amsterdam, Amsterdam, The Netherlands
E.Kanoulas@uva.nl

[3] DAIICT, Gandhinagar, India
prasenjit_t@isical.ac.in

Abstract. In this third edition of the workshop on Augmented Intelligence in Technology-Assisted Review Systems (ALTARS), we focus on the capacity of building test collections for the evaluation of High-recall Information Retrieval (IR) systems which tackle challenging tasks that require the finding of (nearly) all the relevant documents in a collection. During the workshop, the organizers as well as the participants will discuss the problems of how to build and evaluate these types of systems and prepare a set of guidelines for a correct evaluation of those systems according to the current and future available datasets.

Keywords: Technology-Assisted Review Systems · Augmented Intelligence · Evaluation · Systematic Reviews · eDiscovery

1 Introduction

Technology-assisted review (TAR) systems use a kind of human-in-the-loop approach where classification and/or ranking algorithms are continuously trained according to the relevance feedback from expert reviewers, until a substantial number of the relevant documents are identified. This approach has been shown to be more effective and more efficient than traditional e-discovery and systematic review practices, which typically consists of a mix of keyword searches and manual review of the search results.

The first edition of the ALTARS workshop was successfully held at ECIR 2022[1] [5]. During (and after) that workshop, there was a lively discussion about open questions that still need to be addressed and clarified in this research area. The second edition witnessed an even larger participation at ECIR 2023[2] [6]

[1] https://altars2022.dei.unipd.it/.

[2] https://altars2023.dei.unipd.it/.

N. Goharian et al. (Eds.): ECIR 2024, LNCS 14612, pp. 432–436, 2024.
https://doi.org/10.1007/978-3-031-56069-9_59

with around 20 participants on-site and 10 online. The aim of the second edition was to study innovative approaches to fathom the effectiveness of these TAR systems.

In this third edition of the workshop[3], we will invite researchers to discuss the issues related to the shortage of test collections for the evaluation of TAR systems and how to facilitate standardized evaluations and the creation of such datasets and to ensure that TAR systems are fine-tuned to meet the high-recall retrieval requirements of tasks like electronic discovery, systematic reviews, and investigations.

2 Workshop Goals and Objectives

The main goal of this edition of the workshop will be to focus on the evaluation of the different definitions of the effectiveness of TAR systems which is a research challenge itself. The idea is to go beyond a "traditional" retrieval approach and study the problem from different perspectives.

In the first edition of the workshop, we organized a special issue for the Intelligent Systems with Applications journal where extended versions of the papers presented at the workshop were published in open access [4].[4]

In the second edition of the workshop, we organized another special issue for the MDPI Information journal with the same aim (call for papers is still ongoing at present time).[5]

The desired outcome of this third workshop is having a collection of high-quality short papers to be presented at the workshop and a selected number of papers that will be invited in a new special issue and gather momentum in this interdisciplinary area together with researchers, stakeholders in the fields (such as the https://icasr.github.io/about.html and https://www.zylab.com/en/), and international projects (https://dossier-project.eu).

The goals of the workshop are threefold:

- To foster cross-discipline collaborations between researchers with different perspectives and research backgrounds in the TAR systems;
- To combine and analyze existing theoretical and empirical contributions in order to determine shared issues, and novel research questions;
- To create a set of shared datasets dedicated towards the evaluation of TAR systems, thus enabling a wider research community to benefit from the outcomes of the workshop.

2.1 Challenges and Issues

One last goal of the workshop is to discuss open issues and challenges and have as a final product a Horizon Europe/NSF proposal to strengthen the network of researchers in this topic.

[3] https://altars2024.dei.unipd.it/.

[4] https://www.sciencedirect.com/journal/intelligent-systems-with-applications/special-issue/10CLX27456Q.

[5] https://www.mdpi.com/journal/information/special_issues/74F9Z52LA4.

In particular, in the first two editions of the ALTARS workshop, the discussions have focused on the challenges that reflect the complexity and diverse applications of TAR systems in contexts where high recall is crucial. These issues show the multifaceted nature of research and development in this field:

- Scalability: Developing TAR systems that can efficiently handle large volumes of diverse data, especially in e-Discovery where various document types and formats need classification.
- Effectiveness: Achieving high recall without compromising precision remains a challenge, particularly in contexts like systematic reviews where the comprehensive gathering of relevant studies is crucial.
- Algorithmic Bias: Addressing biases in test collections due to missing relevant labels is essential. Developing methods to mitigate biases and ensure representative test collections for fair evaluation of retrieval models.
- Varied Data Types: Creating TAR systems that can effectively process and interpret different types of data, including text, images, and structured data, as seen in e-Discovery where correspondence, memos, and balance sheets need classification.
- Human-in-the-Loop: Designing systems that incorporate human expertise effectively, especially in contexts like systematic reviews where domain knowledge is crucial in identifying relevant studies.
- Generalizability: Ensuring the applicability and effectiveness of TAR systems across diverse domains, considering that systematic reviews, e-Discovery, and other contexts may have distinct requirements and nuances.
- Evaluation Metrics: Developing robust evaluation metrics and standards to assess the performance of TAR systems accurately, especially when dealing with complex tasks like systematically gathering evidence or classifying legal documents.
- Ethical and Legal Implications: Addressing ethical concerns and legal implications, particularly in e-Discovery, where the accuracy and fairness of document classification can impact legal proceedings.

3 Format and Structure

This workshop will be a full-day workshop and will be structured in four sessions: two sessions in the morning and two in the afternoon.

The call for papers of the workshop will include both full and short papers. All the authors of the accepted papers will have the possibility to give a talk; moreover, during an afternoon coffee break, we plan to organize a poster session to have some additional time for extra discussions that may arise during the day.

There will be an invited keynote speaker in the morning (to be confirmed) and a panel/general discussion in the afternoon after all the paper presentation.

4 Organizing Team

All the three organizing committee members have been active participants in the past editions of the TREC, CLEF and FIRE evaluation forum for the Total

Recall and Precision Medicine TREC Tasks, TAR in eHealth tasks, and AI for Legal Assistance. The committee members have strong research record with a total of more than 400 papers in international journals and conferences. They have been doing research in technology-assisted review systems and problems related to document distillation both in the eHealth and eDiscovery domain and made significant contributions in this specific research area.

Giorgio Maria Di Nunzio is Associate Professor at the Department of Information Engineering of the Universiryt of Padova. He has been the co-organizer of the ongoing Covid-19 Multilingual Information Access Evaluation forum,[6] in particular for the evaluation of high-recall systems and high-precision system tasks. He will bring to this workshop the perspective of alternative (to the standard) evaluation measures and multilingual terminological challenges [2,3,7,8,11].

Evangelos Kanoulas is Full Professor at the Faculty of Science of the Informatics Institute at the University of Amsterdam. He has been the co-organizer CLEF eHealth Lab and of the Technologically Assisted Reviews in Empirical Medicine task.[7] He has also been the co-organizer of a related workshop at ECIR 2023, the first international workshop on Legal Information Retrieval (LegalIR). He will bring to the workshop the perspective of the evaluation of the costs in eHealth TAR systems, in particular of the early stopping strategies as well as the conversational aspects of interactive systems [10,13,14].

Prasenjit Majumder is Associate Professor at the Dhirubhai Ambani Institute of Information and Communication Technology (DA-IICT), Gandhinagar and TCG CREST, Kolkata, India. He has been the co-organizer of the Forum for Information Retrieval Evaluation and, in particular, the Artificial Intelligence for Legal Assistance (AILA) task.[8] He will bring to the workshop the perspective of the evaluation of the costs of eDiscovery, in particular of the issues related to legal precedence findings and text representations in financial texts [1,9,12].

References

1. Bhattacharya, P., et al.: FIRE 2020 AILA track: artificial intelligence for legal assistance. In: Majumder, P., Mitra, M., Gangopadhyay, S., Mehta, P. (eds.) Forum for Information Retrieval Evaluation, FIRE 2020, Hyderabad, India, 16–20 December 2020, pp. 1–3. ACM (2020). https://doi.org/10.1145/3441501.3441510
2. Clipa, T., Di Nunzio, G.M.: A study on ranking fusion approaches for the retrieval of medical publications. Information **11**(2), 103 (2020). https://doi.org/10.3390/info11020103
3. Di Nunzio, G., Faggioli, G.: A study of a gain based approach for query aspects in recall oriented tasks. Appl. Sci. **11**(19), 9075 (2021). https://doi.org/10.3390/app11199075. https://www.mdpi.com/2076-3417/11/19/9075

[6] http://eval.covid19-mlia.eu.
[7] https://clefehealth.imag.fr.
[8] https://sites.google.com/view/aila-2021.

4. Di Nunzio, G.M., Kanoulas, E.: Special issue on technology assisted review systems. Intell. Syst. Appl. **20**, 200260 (2023). https://doi.org/10.1016/j.iswa.2023.200260. https://www.sciencedirect.com/science/article/pii/S2667305323000856
5. Di Nunzio, G.M., Kanoulas, E., Majumder, P.: Augmented intelligence in technology-assisted review systems (ALTARS 2022): evaluation metrics and protocols for ediscovery and systematic review systems. In: Hagen, M., et al. (eds.) ECIR 2022. LNCS, vol. 13186, pp. 557–560. Springer, Cham (2022). https://doi.org/10.1007/978-3-030-99739-7_69
6. Di Nunzio, G.M., Kanoulas, E., Majumder, P.: 2nd workshop on augmented intelligence in technology-assisted review systems (ALTARS). In: Kamps, J., et al. (eds.) Advances in Information Retrieval - 45th European Conference on Information Retrieval, ECIR 2023, Dublin, Ireland, 2–6 April 2023, Proceedings, Part III. LNCS, vol. 13982, pp. 384–387. Springer, Switzerland (2023). https://doi.org/10.1007/978-3-031-28241-6_41
7. Di Nunzio, G.M., Marchesin, S., Silvello, G.: A systematic review of automatic term extraction: what happened in 2022? Digit. Scholarsh. Humanit. **38**(Suppl._1), 41–47 (2023). https://doi.org/10.1093/llc/fqad030
8. Di Nunzio, G.M., Vezzani, F.: Did I miss anything? A study on ranking fusion and manual query rewriting in consumer health search. In: Barrón-Cedeño, A., et al. (eds.) Experimental IR Meets Multilinguality, Multimodality, and Interaction - 13th International Conference of the CLEF Association, CLEF 2022, Bologna, Italy, 5–8 September 2022, Proceedings. LNCS, vol. 13390, pp. 217–229. Springer, Cham (2022). https://doi.org/10.1007/978-3-031-13643-6_17
9. Gangopadhyay, S., Majumder, P.: Text representation for direction prediction of share market. Exp. Syst. Appl. **211**, 118472 (2023). https://doi.org/10.1016/j.eswa.2022.118472
10. Li, D., Kanoulas, E.: When to stop reviewing in technology-assisted reviews: sampling from an adaptive distribution to estimate residual relevant documents. ACM Trans. Inf. Syst. **38**(4), 41:1–41:36 (2020). https://doi.org/10.1145/3411755
11. Marchesin, S., Di Nunzio, G.M., Agosti, M.: Simple but effective knowledge-based query reformulations for precision medicine retrieval. Inf. **12**(10), 402 (2021). https://doi.org/10.3390/info12100402
12. Parikh, V., et al.: AILA 2021: shared task on artificial intelligence for legal assistance. In: Ganguly, D., Gangopadhyay, S., Mitra, M., Majumder, P. (eds.) Forum for Information Retrieval Evaluation, FIRE 2021, Virtual Event, India, 13–17 December 2021, pp. 12–15. ACM (2021). https://doi.org/10.1145/3503162.3506571
13. Zou, J., Aliannejadi, M., Kanoulas, E., Pera, M.S., Liu, Y.: Users meet clarifying questions: toward a better understanding of user interactions for search clarification. ACM Trans. Inf. Syst. **41**(1), 16:1–16:25 (2023). https://doi.org/10.1145/3524110
14. Zou, J., Kanoulas, E.: Towards question-based high-recall information retrieval: locating the last few relevant documents for technology-assisted reviews. ACM Trans. Inf. Syst. **38**(3), 27:1–27:35 (2020). https://doi.org/10.1145/3388640

2nd International Workshop on Geographic Information Extraction from Texts (GeoExT 2024)

Xuke Hu[1](✉), Ross Purves[2], Ludovic Moncla[3], Jens Kersten[1], and Kristin Stock[4]

[1] Institute of Data Science, German Aerospace Center (DLR), Jena, Germany
{xuke.hu,jens.kersten}@dlr.de

[2] Department of Geography, University of Zurich, Zurich, Switzerland
ross.purves@geo.uzh.ch

[3] University of Lyon, INSA Lyon, CNRS, UCBL, LIRIS, UMR 5205, Lyon, France
ludovic.moncla@insa-lyon.fr

[4] School of Mathematical and Computational Sciences, Massey University, Auckland, New Zealand
K.Stock@massey.ac.nz

Abstract. A wealth of unstructured textual content contains valuable geographic insights. This geographic information holds significance across diverse domains, including geographic information retrieval, disaster management, and spatial humanities. Despite significant progress in the extraction of geographic information from texts, numerous unresolved challenges persist, ranging from methodologies, systems, data, and applications to privacy concerns. This workshop will serve as a platform for the discourse of recent breakthroughs, novel ideas, and conceptual innovations in this field.

Keywords: Geographic information extraction · document geolocation · geoparsing · toponym recognition · toponym resolution

1 Motivation and Relevance to ECIR

Vast and growing volumes of semi- and unstructured text data, including news, scientific articles, historical records, travel blogs, and social media posts, are readily accessible. These texts often refer to geographic regions or specific locations on Earth and contain valuable geographic insights in the form of toponyms, location references, and complex location descriptions [7]. This information can contribute to both scientific studies, such as sociolinguistics and spatial humanities [10], and practical applications like geographic search [12], disaster management [13], urban planning [11], disease surveillance [1], tourism planning [4], and crime prevention [2]. Researchers from various fields, including information retrieval, natural language processing, and geographic information science, have

N. Goharian et al. (Eds.): ECIR 2024, LNCS 14612, pp. 437–441, 2024.
https://doi.org/10.1007/978-3-031-56069-9_60

strived to develop methods to deduce the geographical context of documents [9], extract geographic references from diverse texts [6], and unambiguously link them to places [5] or spaces like geographic grids [3,12] on the Earth's surface.

Despite significant progress in geographic information extraction, a myriad of challenges persist, including methods, systems, data, applications, and privacy concerns. For example, integrating Large Language Models (LLMs) such as GPT-4 offers promising advancements [8] but also faces significant hurdles, including lack of transparency, inherent bias in training data, high computational needs, and difficulties in maintaining consistency across diverse languages and cultures. Besides, there is a critical need for varied and sophisticated methodologies tailored to specific applications, which may include balancing accuracy with processing speed, offering multilingual support, or adapting to constraints in computational resources. The absence of comprehensive, high-quality datasets hampers the evaluation of existing methods. Gazetteers necessitate updates and enrichment to accommodate contemporary spatial changes, ancient toponyms, and their variations. Beyond, prioritizing the protection of location privacy for online users demands increased attention.

Geographic information extraction from texts serves as a crucial facet of information retrieval and has attracted considerable research attention. The relevance of this proposal to ECIR is evident, given its focus on this vital sub-task.

2 Workshop Goals

2.1 Objectives and Overall Vision

The second GeoExT Workshop is dedicated to fostering dynamic discussions and facilitating the exchange of knowledge surrounding cutting-edge advancements in the extraction of geographic information from unstructured texts. Central to our mission is the emphasis on methodological innovations. Our primary objective is to investigate these innovations and tackle the accompanying challenges. Nonetheless, we extend a warm invitation to submissions that explore various facets of geographic information extraction. Our scope of interests includes, but is not restricted to, the following areas:

- Document geocoding
- Toponym recognition and resolution
- Toponym matching
- Address matching
- Method generalizability (regions, languages, data sources)
- Multi-source data (e.g., text and image) fusion for method improvement
- Location description extraction and resolution
- Relation extraction and disambiguation
- Social media geolocating at hyper-local levels
- Fine-grained toponym resolving
- Historical toponym resolving
- Historical archive geovisualization

- Gazetteer enrichment
- Dataset annotation for method training and evaluation
- Location privacy protection for online users
- Novel applications of geographic information in texts

2.2 Desired Outcomes

We aim to support continued discussions and exchanges between researchers and practitioners, and to strengthen the existing network within the First GeoExT Workshop. In this regard, the following outcomes are desired:

- Collection of papers, data sets, and tools submitted by participants
- Compilation of overall findings contributed by participants in a summarizing paper
- Establish a common shared task in the future
- Enhance the established network/interest group of researchers and practitioners
- Foster new collaborations in national and international projects
- Paving the road towards follow-up ECIR workshops on this topic

3 Format and Structure

During ECIR 2023, we organized the first GeoExT workshop as a half-day event, which received 13 submissions. Regrettably, due to time constraints, we could only accommodate 7 of these submissions for presentation. With the upcoming 2nd GeoExT workshop, we anticipate an increase in submissions and participates and consequently, we plan to extend it to a full-day workshop format. This extension will allow us to accommodate a larger number of attendees and foster more in-depth discussions. The workshop will feature a blend of keynote talks, paper presentations, demonstrations, and open discussions which proved to enhance interactivity during GeoExT 2023.

To facilitate enriching dialogues, we strongly encourage submissions falling under the following categories:

- Full paper + oral presentation
- Short paper + oral presentation
- Demonstration paper + corresponding hands-on demo

The papers (original, unpublished contributions) will be included in an open-access proceedings volume of CEUR Workshop Proceedings (indexed by both Scopus and DBLP).

4 Potential Attendees

Building on the success of the previous workshop with around 20 attendees, we anticipate a larger participation of over 25 this year due to an expanding network from the first event. We will use our existing network to spread the call for workshop submissions. Our workshop extends a warm invitation to all researchers attending the conference, encouraging their active participation.

5 Organisers

Xuke Hu received his doctoral degree (Dr.-Ing.) in Geoinformatics from Universität Heidelberg, Germany. He is now a permanent senior researcher at the Data Science Institute of the German Aerospace Center (DLR). He organized/co-organized several workshops at ECIR, ACM SIGSPATIAL, and the ISPRS Geospatial Week. He has several years of working experience in geographic information extraction, especially toponym recognition and resolution in the context of disaster management.

Ross Purves is a professor of Geographic Information Science at the University of Zurich. He has a long history of research developing methods related to Geographic Information Retrieval, and with Chris Jones chaired a series of workshops on this theme. As well as research developing methods related to geographic extraction, he has in recent years applied such methods extensively to work exploring how the notion of place can be represented computationally, for example exploring how perception varies in landscapes.

Ludovic Moncla received his PhD in computer science from Université de Pau et des Pays de l'Adour (France) and Universidad de Zaragoza (Spain) in 2015. He is now an associate professor in computer science at National Institute of Applied Science of Lyon (INSA Lyon) and LIRIS laboratory (UMR 5205 CNRS). He co-organized the GeoHumanities workshop series at ACM SIGSPATIAL (from 2019 to 2023). His research interests are oriented towards pluridisciplinary aspects of Natural Language Processing (NLP), information retrieval, data mining, digital humanities and geographical information science (GIS).

Jens Kersten received his doctoral degree (Dr.-Ing.) in remote sensing and computer vision and his diploma in geodesy from the Technical University of Berlin, Germany. He is a senior researcher at the German Aerospace Center (DLR). His group at DLR's Institute of Data Science researches methods to acquire and analyze web- and social media-data for applications related to civil security, natural hazards and environmental impacts. Developing new methods for place name extraction and resolution are essential to address the groups overarching objective of supporting operational crisis response.

Kristin Stock is Director of the Massey Geoinformatics Collaboratory and Associate Professor in Computer Science and Information Technology. She has qualifications in surveying (and is a Registered Surveyor), urban and regional planning, computer science and geospatial science, and her PhD addressed the subject of semantic integration of geographic databases. She has worked as a mining, engineering and cadastral surveyor in private practice and in 1999 she designed a land information system for the Australian Capital Territory government that is still in use. Her current research focuses on extracting geospatial knowledge from text sources using natural language processing techniques in application areas including disaster management and biological collections.

References

1. Allen, T., et al.: Global hotspots and correlates of emerging zoonotic diseases. Nat. Commun. **8**(1), 1–10 (2017)
2. Arulanandam, R., Savarimuthu, B.T.R., Purvis, M.A.: Extracting crime information from online newspaper articles. In: Proceedings of the Second Australasian Web Conference, vol. 155, pp. 31–38 (2014)
3. Gritta, M., Pilehvar, M.T., Limsopatham, N., Collier, N.: What's missing in geographical parsing? Lang. Resour. Eval. **52**(2), 603–623 (2018)
4. Haris, E., Gan, K.H.: Mining graphs from travel blogs: a review in the context of tour planning. Inform. Technol. Tourism **17**(4), 429–453 (2017)
5. Hu, X., Sun, Y., Kersten, J., Zhou, Z., Klan, F., Fan, H.: How can voting mechanisms improve the robustness and generalizability of toponym disambiguation? Int. J. Appl. Earth Obs. Geoinf. **117**, 103191 (2023)
6. Hu, X., et al.: Location reference recognition from texts: A survey and comparison. arXiv preprint arXiv:2207.01683 (2022)
7. Hu, Y., Adams, B.: Harvesting big geospatial data from natural language texts. In: Werner, M., Chiang, Y.-Y. (eds.) Handbook of Big Geospatial Data, pp. 487–507. Springer, Cham (2021). https://doi.org/10.1007/978-3-030-55462-0_19
8. Hu, Y., et al.: Geo-knowledge-guided gpt models improve the extraction of location descriptions from disaster-related social media messages. Int. J. Geogr. Inf. Sci. **37**(11), 2289–2318 (2023)
9. Kinsella, S., Murdock, V., O'Hare, N.: " i'm eating a sandwich in glasgow" modeling locations with tweets. In: Proceedings of the 3rd International Workshop on Search and Mining User-generated Contents, pp. 61–68 (2011)
10. Melo, F., Martins, B.: Automated geocoding of textual documents: a survey of current approaches. Trans. GIS **21**(1), 3–38 (2017)
11. Milusheva, S., Marty, R., Bedoya, G., Williams, S., Resor, E., Legovini, A.: Applying machine learning and geolocation techniques to social media data (twitter) to develop a resource for urban planning. PLOS ONE **16**(2), 1–12 (02 2021). https://doi.org/10.1371/journal.pone.0244317
12. Purves, R.S., Clough, P., Jones, C.B., Hall, M.H., Murdock, V.: Geographic information retrieval: Progress and challenges in spatial search of text. Foundations and Trends® in Information Retrieval **12**(2-3), 164–318 (2018). https://doi.org/10.1561/1500000034
13. Scalia, G., Francalanci, C., Pernici, B.: Cime: context-aware geolocation of emergency-related posts. Geoinformatica **26**(1), 125-157 (2022). https://doi.org/10.1007/s10707-021-00446-x

Bibliometric-Enhanced Information Retrieval: 14th International BIR Workshop (BIR 2024)

Ingo Frommholz[1], Philipp Mayr[2](✉), Guillaume Cabanac[3,4], and Suzan Verberne[5]

[1] School of Engineering, Computing and Mathematical Sciences, University of Wolverhampton, Wolverhampton, UK
ifrommholz@acm.org
[2] GESIS – Leibniz-Institute for the Social Sciences, Cologne, Germany
philipp.mayr@gesis.org
[3] Computer Science Department, University of Toulouse, IRIT UMR 5505, Toulouse, France
guillaume.cabanac@univ-tlse3.fr
[4] Institut Universitaire de France (IUF), Paris, France
[5] Leiden University, LIACS, Leiden, The Netherlands
s.verberne@liacs.leidenuniv.nl

Abstract. The 14th iteration of the Bibliometric-enhanced Information Retrieval (BIR) workshop series takes place at ECIR 2024 as a full-day workshop. BIR addresses research topics related to academic search and recommendation, at the intersection of Information Retrieval, Natural Language Processing, and Bibliometrics. As an interdisciplinary scientific event, BIR brings together researchers and practitioners from the Scientometrics/Bibliometrics community on the one hand, and the Information Retrieval and NLP communities on the other hand. BIR is an ever-growing topic investigated by both academia and the industry.

Keywords: Academic Search · Information Retrieval · Digital Libraries · Bibliometrics · Scientometrics

1 Motivation and Relevance to ECIR

The aim of the BIR workshop series is to bring together researchers and practitioners from Scientometrics/Bibliometrics as well as Information Retrieval (IR). Scientometrics is a sub-field of Bibliometrics which, like IR, is in turn a sub-field of Information Science. Bibliometrics and Scientometrics are concerned with all quantitative aspects of information and academic literature [7], which naturally make them interesting for IR research, in particular when it comes to academic search, recommendation, and other domains in which citations play a central role, for example, legal and patent retrieval. In the early 1960s, Salton was already striving to enhance IR by including clues inferred from bibliographic

N. Goharian et al. (Eds.): ECIR 2024, LNCS 14612, pp. 442–446, 2024.
https://doi.org/10.1007/978-3-031-56069-9_61

Table 1. Overview of the BIR workshop series and CEUR proceedings

Year	Conference	Venue		Papers	Proceedings
2014	ECIR	Amsterdam,	NL	6	Vol-1143
2015	ECIR	Vienna,	AT	6	Vol-1344
2016	ECIR	Padua,	IT	8	Vol-1567
2016	JCDL	Newark,	US	10 + 10[a]	Vol-1610
2017	ECIR	Aberdeen,	UK	12	Vol-1823
2017	SIGIR	Tokyo,	JP	11	Vol-1888
2018	ECIR	Grenoble,	FR	9	Vol-2080
2019	ECIR	Cologne,	DE	14	Vol-2345
2019	SIGIR	Paris,	FR	16 + 10[b]	Vol-2414
2020	ECIR	Lisbon (Online),	PT	9	Vol-2591
2021	ECIR	Lucca (Online),	IT	9	Vol-2847
2022	ECIR	Stavanger,	NO	5	Vol-3230
2023	ECIR	Dublin,	IE	7	To appear

[a]With CL-SciSumm 2016 Shared Task; [b]with CL-SciSumm 2019 Shared Task

citations [8]. In the course of decades, both disciplines (Bibliometrics and IR) evolved apart from each other over time, leading to the two loosely connected fields we know today [10].

However, the exploding number of scholarly publications and the requirement to satisfy scholars' specific information needs led Bibliometric-enhanced IR to receive growing recognition in the IR as well as the Scientometrics communities. Challenges in academic search and recommendation became particularly apparent during the COVID-19 crisis as well as the ongoing trend of publishing on preprint servers first (and sometimes exclusively), for instance in the rapidly developing field of artificial intelligence and deep learning. This results, for instance, in the information overload researchers and practitioners are facing, as well as the need to ensure the timeliness and quality of published research. Tackling these challenges requires effective and efficient solutions for scholarly search, recommendation and discovery of high-quality publications and heterogeneous data. Bibliometric-enhanced IR tries to provide these solutions to the peculiar needs of scholars to keep on top of the research in their respective fields, utilising the wide range of suitable relevance signals that come with academic scientific publications, such as keywords provided by authors, topics extracted from the full-texts, co-authorship networks, citation networks, altmetrics, bibliometric figures, and various classification schemes of science. Bibliometric-enhanced IR systems must deal with the multifaceted nature of scientific information by searching for or recommending academic papers, patents, venues (i.e., conference proceedings, journals, books, manuals, grey literature), authors, experts (e.g., peer reviewers), references (to be cited to support an argument), and datasets.

A further discussion of the various research directions in bibliometric-enhanced IR can be found in [4].

To this end, the BIR workshop series was founded in 2014 [5] to tackle these challenges by tightening up the link between IR and Bibliometrics. We strive to bring the 'retrievalists' and 'citationists' [10] active in both academia and industry together. The success of past BIR events, as shown in Table 1, evidences that BIR@ECIR is a much-needed interdisciplinary scientific event that attracts researchers and practitioners from IR, Bibliometrics, and Natural Language Processing alike.

2 Workshop Goals and Objectives

Our vision is to bring together researchers and practitioners from Scientometrics/Bibliometrics on the one hand and IR + NLP, on the other hand, to create better methods and systems for instance for academic search and recommendation. Our view is to expose people from one community to the work of the respective other community and to foster fruitful interaction across communities. Submissions to the workshop should also address novel challenges and opportunities, for example, coming from Generative AI and Large Language Models [6,9] in a scholarly setting. Therefore, in the call for papers for the 2024 BIR workshop at ECIR, we address, but are not limited to, current research issues regarding 4 aspects of the academic search/recommendation process:

1. User needs and behaviour regarding scientific information, such as:
   - Finding relevant papers/authors for a literature review.
   - Identifying expert reviewers for a given submission.
   - Understanding information-seeking behaviour and HCI in academic search.
   - Filtering high-quality research papers, e.g., in preprint servers.
   - Measuring the degree of plagiarism in a paper.
   - Flagging predatory conferences and journals, or other forms of scientific misbehaviour [1].
2. Mining the scientific literature, such as:
   - Information extraction, text mining and parsing of scholarly literature.
   - Natural language processing (e.g., citation contexts).
   - Discourse modelling and argument mining.
3. Academic search/recommendation systems, such as:
   - Modelling the multifaceted nature of scientific information.
   - Building test collections for reproducible BIR.
   - System support for literature search and recommendation.
4. Generative AI and Large Language Models with bibliometric-enhanced IR
   - Retrieval-augmented LLMs [9] for academic search and recommendation.
   - LLM-enhanced retrieval and recommendation in scholarly settings [6].
   - Challenges with generative LLMs for scholarly texts and references.

3 Target Audience and Dissemination

The target audience of the BIR workshops is researchers and practitioners, junior and senior, from Scientometrics as well as IR and Natural Language Processing (NLP). These could be IR/NLP researchers interested in potential new application areas for their work as well as researchers and practitioners working with bibliometric data and interested in how IR/NLP methods can make use of such data. BIR 2024 will be open to anyone interested in the topic.

The BIR organisers (the authors of this paper) are well established in the IR and the Bibliometrics community, respectively and have a years-long experience as workshop organizers. We have sent the call for papers to major professional mailing lists in IR (ACM SIGIR, IR-List, JISC IR) and Bibliometrics (ASIS&T Sigmetrics and ISSI). We also sent it to the former BIR and BIRNDL participants (in the range of a few hundred people) and scientists who publish in both IR and Bibliometrics venues, based on the mining of the DBLP. We further advertised the call for papers through our social media channels.

The 10th-anniversary edition in 2020 ran online with an audience peaking at 97 online participants [2]. BIR 2022, the 12th edition, was run as a hybrid event with on-site and online participants. We were surprised by how successful this model went and how satisfied the speakers, audience, and organizers were with the hybrid workshop.

In December 2020, we published our third special issue emerging from the past BIR workshops [3].

4 Previous Workshops

The BIR workshop series has a long tradition of taking place along major IR conferences such as ECIR, SIGIR, and JCDL, as documented on the BIR overview page. Table 1 provides an overview of past events. BIR@ECIR2024 would be the continuation of a highly successful conference series.

5 Organisers

- Ingo Frommholz, Reader in Data Science at the University of Wolverhampton, UK.
- Philipp Mayr, team leader at the GESIS – Leibniz-Institute for the Social Sciences department Knowledge Technologies for the Social Sciences, Germany.
- Guillaume Cabanac, Professor of Computer Science at the University of Toulouse and research chair at the Institut Universitaire de France.
- Suzan Verberne, Professor of Natural Language Processing at the Leiden Institute of Advanced Computer Science (LIACS), Leiden University.

Acknowledgements. This work was funded by the European Union under the Horizon Europe grant OMINO - Overcoming Multilevel INformation Overload (grant number 101086321, http://ominoproject.eu). Ingo Frommholz is funded by UKRI (ref

EP/X040496/1). Views and opinions expressed are those of the authors alone and do not necessarily reflect those of the European Union or the European Research Executive Agency. Neither the European Union nor the European Research Executive Agency can be held responsible for them.

References

1. Cabanac, G.: Decontamination of the scientific literature. arXiv preprint arXiv:2210.15912 (2022)
2. Cabanac, G., Frommholz, I., Mayr, P.: Report on the 10th anniversary workshop on bibliometric-enhanced information retrieval (BIR 2020), SIGIR Forum **54**(1) (2020). https://doi.org/10.1145/3451964.3451974
3. Cabanac, G., Frommholz, I., Mayr, P.: Scholarly literature mining with information retrieval and natural language processing: preface. Scientometrics **125**(3), 2835–2840 (2020). https://doi.org/10.1007/s11192-020-03763-4
4. Frommholz, I., Cabanac, G., Mayr, P., Verberne, S.: Report on the 11th bibliometric-enhanced information retrieval workshop (BIR 2021). SIGIR Forum **55**(1) (2021). https://doi.org/10.1145/3476415.3476426
5. Mayr, P., Scharnhorst, A., Larsen, B., Schaer, P., Mutschke, P.: Bibliometric-enhanced information retrieval. In: de Rijke, M., et al. (eds.) Advances in Information Retrieval. ECIR 2014. LNCS, vol. 8416, pp. 798–801. Springer, Cham (2014). https://doi.org/10.1007/978-3-319-06028-6_99
6. Pride, D., Cancellieri, M., Knoth, P.: CORE-GPT: combining open access research and large language models for credible, trustworthy question answering. In: Alonso, O., Cousijn, H., Silvello, G., Marrero, M., Lopes, C.T., Marchesin, S. (eds.) Proceedings of the 27th International Conference on Theory and Practice of Digital Libraries (TPDL 2023). LNCS, vol. 14241, pp. 146–159. Springer, Cham (2023). https://doi.org/10.1007/978-3-031-43849-3_13
7. Pritchard, A.: Statistical bibliography or bibliometrics? [Documentation notes]. J. Document. **25**(4), 348–349 (1969). https://doi.org/10.1108/eb026482
8. Salton, G.: Associative document retrieval techniques using bibliographic information. J. ACM **10**(4), 440–457 (1963). https://doi.org/10.1145/321186.321188
9. Shuster, K., Poff, S., Chen, M., Kiela, D., Weston, J.: Retrieval augmentation reduces hallucination in conversation. In: Findings of the Association for Computational Linguistics: EMNLP 2021, pp. 3784–3803. Association for Computational Linguistics (2021). https://doi.org/10.18653/v1/2021.findings-emnlp.320
10. White, H.D., McCain, K.W.: Visualizing a discipline: an author co-citation analysis of information science, 1972–1995. J. Am. Soc. Inf. Sci. **49**(4), 327–355 (1998)

Conference and Labs of the Evaluation Forum (CLEF)

The CLEF-2024 CheckThat! Lab: Check-Worthiness, Subjectivity, Persuasion, Roles, Authorities, and Adversarial Robustness

Alberto Barrón-Cedeño[1](✉), Firoj Alam[2], Tanmoy Chakraborty[3], Tamer Elsayed[4], Preslav Nakov[5], Piotr Przybyła[6,7], Julia Maria Struß[8], Fatima Haouari[4], Maram Hasanain[2], Federico Ruggeri[9], Xingyi Song[10], and Reem Suwaileh[11]

[1] DIT, Università di Bologna, Forlì, Italy
a.barron@unibo.it
[2] Qatar Computing Research Institute, HBKU, Doha, Qatar
[3] Indian Institute of Technology Delhi, New Delhi, India
[4] Qatar University, Doha, Qatar
[5] Mohamed bin Zayed University of Artificial Intelligence, Abu Dhabi, UAE
[6] Univesitat Pompeu Fabra, Barcelona, Spain
[7] Institute of Computer Science, Polish Academy of Sciences, Warsaw, Poland
[8] University of Applied Sciences Potsdam, Potsdam, Germany
[9] DISI, Università di Bologna, Bologna, Italy
[10] University of Sheffield, Sheffield, UK
[11] HBKU, Doha, Qatar

Abstract. The first five editions of the `CheckThat!` lab focused on the main tasks of the information verification pipeline: check-worthiness, evidence retrieval and pairing, and verification. Since the 2023 edition, it has been focusing on new problems that can support the research and decision making during the verification process. In this new edition, we focus on new problems and —for the first time— we propose six tasks in fifteen languages (Arabic, Bulgarian, English, Dutch, French, Georgian, German, Greek, Italian, Polish, Portuguese, Russian, Slovene, Spanish, and code-mixed Hindi-English): Task 1 estimation of check-worthiness (the only task that has been present in all `CheckThat!` editions), Task 2 identification of subjectivity (a follow up of `CheckThat!` 2023 edition), Task 3 identification of persuasion (a follow up of SemEval 2023), Task 4 detection of hero, villain, and victim from memes (a follow up of CONSTRAINT 2022), Task 5 Rumor Verification using Evidence from Authorities (a first), and Task 6 robustness of credibility assessment with adversarial examples (a first). These tasks represent challenging classification and retrieval problems at the document and at the span level, including multilingual and multimodal settings.

General and task coordinators appear first, in alphabetical order.

N. Goharian et al. (Eds.): ECIR 2024, LNCS 14612, pp. 449–458, 2024.
https://doi.org/10.1007/978-3-031-56069-9_62

Keywords: disinformation · fact-checking · check-worthiness · subjectivity · political bias · factuality · authority finding · model robustness

1 Introduction

During its previous five editions, the CheckThat! lab has focused on developing technology to assist the *journalist fact-checker* during the main steps of the verification process [5–8, 11, 12, 30–34]. Given a document, or a claim, it first has to be assessed for check-worthiness, i.e. whether a journalist should check its veracity. If this is so, the system needs to retrieve claims verified in the past that could be useful to fact-check the current one. Further evidence to verify the claim could be retrieved from the Web, if necessary. Finally, with the evidence gathered from diverse sources, a decision can be made: whether the claim is factually true or not. This year, we propose six tasks:

Task 1 Check-Worthiness Estimation: to identify claims that could be important to verify on social- and mainstream media (the only task that has been organized during all editions of the lab; cf. Sect. 2).

Task 2 Subjectivity in News Articles: to spot text that should be processed with specific strategies [42]; benefiting the fact-checking pipeline [22, 24, 51] (cf. Sect. 3).

Task 3 Persuasion Techniques: to identify text spans in which a persuasion technique is being issued to influence the reader (cf. Sect. 4).

Task 4 Detecting Hero, Villain, and Victim from Memes: to predict the role of each entity: *hero*, *villain*, *victim*, or *other* in a given meme and a list of entities (cf. Sect. 5).

Task 5 Rumor Verification Using Evidence from Authorities: to retrieve evidence from trusted sources (authorities that have "real knowledge" on the matter) and determine if the rumor is supported, refuted, or unverifiable according to the evidence (cf. Sect. 6).

Task 6 Robustness of Credibility Assessment with Adversarial Examples: to discover examples indicating low robustness of misinformation detection models (cf. Sect. 7).

2 Task 1: Check-Worthiness Estimation

Motivation. Fact-checking is a complex process. Before assessing the truthfulness of a claim, determining if it can be fact-checked at all is essential. Given the time-consuming nature of this process, it is important to prioritize claims that are important to be fact-checked.

Task Definition. The aim of this task is to assess whether a statement, sourced from either a tweet or a political debate, requires fact-checking [1]. To make this decision, one must consider questions such as "Does it contain a verifiable factual claim?" and "Could it be harmful?" before assigning a final label for its check-worthiness.

Data. The dataset is comprised of multigenre content in Arabic, English, Dutch and Spanish. The Arabic and Dutch datasets consist of tweets that were collected using keywords related to COVID-19 and vaccines, following the annotation schema described in [2]. The dataset for English consists of transcribed sentences from candidates during the US presidential election debates and annotated by human annotators [4]. We use essentially the same dataset reported in [4], with some updates that reflect improved annotation accuracy. The Spanish dataset consists of tweets collected from Twitter accounts and transcriptions from Spanish politicians, which are manually annotated by professional journalists who are experts in fact-checking. These datasets include 8.9*k*, 1.9*k*, 23.9*k* and 30*k* instances in Arabic, Dutch, English, and Spanish, respectively [1,29]. We split them into training (∼74%), development (∼12%), and development-test (∼15%) sets (an average estimate from all languages) to facilitate training, parameter tuning, and to obtain initial results on the development-test set. For the evaluation of systems in this lab edition, new test sets containing ∼500 instances per language will be released.

Evaluation. This is a binary classification task and we evaluate it on the basis of the F_1-measure on the check-worthiness class.

3 Task 2: Subjectivity Detection

Motivation. Verifiable facts are not only communicated in objective and neutral statements, but can also be found in subjectively colored ones. While objective sentences can be directly considered for verification, subjective ones require additional processing steps, e.g., extracting an objective version of the contained claims.

Task Definition. Given a sentence from a news article, determine whether it is subjective or objective. This is a binary classification task and is offered in Arabic, English, German, Italian and in a cross-lingual setting.

Data. For training and validation we provide 1.9*k* sentences in Arabic, 1.3*k* in English, 1.3*k* in German, and 2.2*k* in Italian from last year's iteration [13]. About 300 new sentences are being collected and labelled for each language to be used as novel test sets. The dataset for the cross-lingual setting will be compiled from the individual datasets of the aforementioned languages.

Evaluation. We use macro-averaged F_1-measure as the official evaluation metric.

4 Task 3: Detection of Persuasion Techniques in News Articles

Motivation. A major characteristic of disinformation is that it is not just about lying, but also about convincing people to think or to act in a specific way.

Thus, it is conveyed using specific rhetorical devices: persuasion techniques (e.g., emotional appeals, logical fallacies, personal attacks). Here, we aim to detect the use of such techniques in news articles in various languages.

Task Definition. Given a set of news articles and a list of 23 persuasion techniques organized into a 2-tier taxonomy, including logical fallacies and emotional manipulation techniques that might be used to support flawed argumentation [36], the task consists of identifying the spans of texts in which each technique occurs. This is a multi-label multi-class sequence tagging task.

Data. We will use an existing corpus, consisting of $2k$ news articles in 9 languages annotated with $48K$ instances of persuasion techniques [37], as our training dataset. A new test dataset of $\sim$ 500 news articles in Arabic, Bulgarian, English, Portuguese, and Slovene will be provided.

Evaluation. The task is evaluated using an extension of the F_1-measure taking into account partial overlaps between predicted and golden spans [10], and an evaluation at both coarse- and fine-grained level with respect to the type of persuasion technique is envisaged.

5 Task 4: Detecting the Hero, the Villain, and the Victim from Memes

Motivation. Memes, characterized by their diverse multimodal nature, are frequently employed to communicate intricate concepts effortlessly on social media. However, this simplicity can sometimes oversimplify intricate concepts, leading to the potential delivery of harmful content, often wrapped in humor. While previous studies identified various types of harm caused by memes [25,39,44,48], they largely overlook nuanced analyses like "narrative framing", especially in resource-constrained settings. Current approaches have limitations in addressing multimodality and reasoning about visual and semantic elements in memes, as noted in prior findings [46]. Identifying narrative roles in memes is crucial for in-depth semantic analysis, especially when examining their potential connection to harmful content like hate speech, offensive material, and cyberbullying [45].

Task Definition. The task aims to determine the roles of entities within memes, categorizing them as "hero", "villain", "victim", or "other" through a multi-class classification setting that considers the systematic modeling of multimodal semiotics [46].

Data. We already have the `HVVMemes` dataset [47], including $6.9k$ labeled instances. Additionally, we will introduce a new test dataset of 500 instances for the following languages: Arabic, English, and code-mixing of Hindi and English.

Evaluation. The macro-averaged F_1-measure will primarily assess model performance. Two role-label experts will annotate each official test set, overseen by a consolidator following guidelines from previous work [47].

6 Task 5: Rumor Verification Using Evidence from Authorities

Motivation. Several existing studies addressed rumor verification in social media by exploiting evidence extracted from propagation networks or the Web [19,21,35]. Finding and incorporating authorities for rumor verification in Twitter was proposed recently [16–18]. In the previous edition of the lab, we offered the task of *Authority Finding in Twitter* [20]; this year, we offer a follow-up task with the objective of retrieving evidence from timelines of authorities, and, accordingly, deciding whether the rumors are supported, refuted, or unverifiable.

Task Definition. Given a rumor expressed in a tweet and a set of authorities (one or more authority Twitter accounts) for that rumor, represented by a list of tweets from their timelines during the period surrounding the rumor, the system should retrieve up to 5 evidence tweets from those timelines, and determine if the rumor is supported (true), refuted (false), or unverifiable (in case not enough evidence to verify it exists in the given tweets) according to the evidence. This task is offered in both Arabic and English.

Data. The dataset comprises 160 Arabic rumors expressed in tweets selected from the AuFIN [18,20] and AuSTR [16,17] datasets, and 693 timelines of authority Twitter accounts comprising about 34*k* annotated tweets in total. The same data will be automatically translated to English and validated manually. The data will be split into 60%, 20%, and 20% of the rumors for training, development, and testing respectively.

Evaluation. The official evaluation measure for evidence retrieval is Mean Average Precision (MAP). The systems get no credit if they retrieve any tweets for unverifiable rumors. Other evaluation measures to be considered are Recall@5 and Precision@5. For rumor classification, we use the F_1-measure. Additionally, we also consider a strict evaluation where the rumor label is considered correct **only if** at least one retrieved authority evidence is correct.

7 Task 6: Robustness of Credibility Assessment with Adversarial Examples

Motivation. The aim of the task is to assess the *robustness* of text classifiers in the misinformation detection domain, i.e. their resilience to input data that were purposefully prepared to elicit a misguided response, known as *adversarial examples* (AEs). The vulnerability of deep learning models to AEs has been initially shown for image classification [14,49], but such weaknesses exist for text as well, even though finding them is more challenging [52]. However, exploring this area is of paramount importance, especially in the case of misinformation detection challenges, where motivated adversaries are active [40].

Task Definition. The task is realized in five domains: style-based news bias assessment (HN), propaganda detection (PR), fact checking (FC), rumour detection (RD) and COVID-19 misinformation detection (C19). For each domain, the participants are provided with three victim models, trained for the corresponding binary classification task, as well as a collection of 400 text fragments. Their aim is to prepare adversarial examples, which preserve the meaning of the original examples, but are labelled differently by the classifiers.

Data. The task is based on the publicly available corpora with expert-annotated credibility used in the BODEGA framework [41]. HN uses news articles [38] gathered for SemEval-2019 Task 4 [26]; PR is based on the corpus accompanying SemEval-2020 Task 11 [9], with 14 propaganda techniques annotated in 371 newspapers articles by professional annotators; FC uses the claims-evidence pairs gathered for FEVER [50]; RD is based on the augmented dataset of rumors and non-rumors for rumor detection [15], created from Twitter threads. Additionally, C19 will use a previously unreleased dataset [23,28].

Evaluation. The quality of the adversarial examples will be assessed using the BODEGA score [41], which combines the change in the classifier's decision with the similarity between the original and modified example: character-level through Levenshtein distance [27] and semantic using *BLEURT* [43].

8 Conclusions

The seventh edition of the CheckThat! lab at CLEF provides a diverse collection of challenges to the research community interested in developing technology to support and understand the journalistic verification process. The tasks go from core verification tasks such as assessing the check-worthiness of a text to understanding the strategies used to influence the audience and identifying the stance of relevant characters on questionable affairs. For the first time, the lab looks at the impact of data purposefully shaped to disguise classifiers for different relevant tasks. As in every year, the evaluation framework for all tasks is freely released to the community in order to foster the development of technology against disinformation and misinformation.

Acknowledgments. The work of F. Haouari is supported by GSRA grant #GSRA6-1-0611-19074 from the Qatar National Research Fund. The work of T. Elsayed was made possible by NPRP grant #NPRP-11S-1204-170060 from the Qatar National Research Fund (a member of Qatar Foundation). The work of F. Alam, M. Hasanain and R. Suwaileh is partially supported by NPRP 14C-0916-210015 from the Qatar National Research Fund (a member of Qatar Foundation). The work of P. Przybyła is part of the ERINIA project, which has received funding from the European Union's Horizon Europe research and innovation programme under grant agreement No 101060930. Views and opinions expressed are however those of the author(s) only and do not necessarily reflect those of the funders. Neither the European Union nor the granting authority can be held responsible for them.

References

1. Alam, F., et al.: Overview of the CLEF-2023 CheckThat! lab task 1 on check-worthiness in multimodal and multigenre content. In: Aliannejadi et al. [3]
2. Alam, F., et al.: Fighting the COVID-19 infodemic: modeling the perspective of journalists, fact-checkers, social media platforms, policy makers, and the society. In: Findings of EMNLP 2021, pp. 611–649 (2021)
3. Aliannejadi, M., Faggioli, G., Ferro, N., Vlachos, M. (eds.): Working Notes of CLEF 2023 - Conference and Labs of the Evaluation Forum, CLEF 2023, Thessaloniki, Greece (2023)
4. Arslan, F., Hassan, N., Li, C., Tremayne, M.: A benchmark dataset of check-worthy factual claims. In: Proceedings of the International AAAI Conference on Web and Social Media, vol. 14, pp. 821–829 (2020)
5. Barrón-Cedeño, A., et al.: The CLEF-2023 CheckThat! lab: checkworthiness, subjectivity, political bias, factuality, and authority. In: Kamps, J., et al. (eds.) Advances in Information Retrieval. LNCS, vol. 13982, pp. 506–517. Springer, Cham (2023). https://doi.org/10.1007/978-3-031-28241-6_59
6. Barrón-Cedeño, A., et al.: Overview of the CLEF-2023 CheckThat! lab checkworthiness, subjectivity, political bias, factuality, and authority of news articles and their source. In: Arampatzis, A., et al. (eds.) Experimental IR Meets Multilinguality, Multimodality, and Interaction. Proceedings of the Fourteenth International Conference of the CLEF Association (CLEF 2023) (2023)
7. Barrón-Cedeño, A., et al.: CheckThat! at CLEF 2020: enabling the automatic identification and verification of claims in social media. In: Jose, J., et al. (eds.) ECIR 2020. LNCS, vol. 12036, pp. 499–507. Springer, Cham (2020). https://doi.org/10.1007/978-3-030-45442-5_65
8. Barrón-Cedeño, A., et al.: Overview of CheckThat! 2020: automatic identification and verification of claims in social media. In: Arampatzis, A., et al. (eds.) Experimental IR Meets Multilinguality, Multimodality, and Interaction. Proceedings of the Eleventh International Conference of the CLEF Association (CLEF 2020), pp. 215–236. LNCS, vol. 12260. Springer, Cham (2020). https://doi.org/10.1007/978-3-030-58219-7_17
9. Da San Martino, G., Barrón-Cedeño, A., Wachsmuth, H., Petrov, R., Nakov, P.: SemEval-2020 Task 11: detection of propaganda techniques in news articles. In: Proceedings of the Fourteenth Workshop on Semantic Evaluation (SemEval-2020), pp. 1377–1414 (2020)
10. Da San Martino, G., Yu, S., Barrón-Cedeño, A., Petrov, R., Nakov, P.: Fine-grained analysis of propaganda in news article. In: Proceedings of the 2019 Conference on Empirical Methods in Natural Language Processing and the 9th International Joint Conference on Natural Language Processing (EMNLP-IJCNLP), pp. 5636–5646. Association for Computational Linguistics, Hong Kong (2019)
11. Elsayed, T., et al.: CheckThat! at CLEF 2019: automatic identification and verification of claims. In: Advances in Information Retrieval - European Conference on IR Research (ECIR 2019), pp. 309–315 (2019)
12. Elsayed, T., et al.: Overview of the CLEF-2019 CheckThat!: automatic identification and verification of claims. In: Experimental IR Meets Multilinguality, Multimodality, and Interaction. LNCS, vol. 11696, pp. 301–321. Springer, Cham (2019). https://doi.org/10.1007/978-3-030-28577-7_25
13. Galassi, A., et al.: Overview of the CLEF-2023 CheckThat! lab task 2 on subjectivity in news articles. In: Aliannejadi et al. [3]

14. Goodfellow, I.J., Shlens, J., Szegedy, C.: Explaining and Harnessing Adversarial Examples. In: Bengio, Y., LeCun, Y. (eds.) 3rd International Conference on Learning Representations, ICLR 2015, San Diego, 7–9 May 2015, Conference Track Proceedings (2015). http://arxiv.org/abs/1412.6572
15. Han, S., Gao, J., Ciravegna, F.: Neural language model based training data augmentation for weakly supervised early rumor detection. In: Proceedings of the 2019 IEEE/ACM International Conference on Advances in Social Networks Analysis and Mining (ASONAM 2019), pp. 105–112. Association for Computing Machinery, Inc (2019)
16. Haouari, F., Elsayed, T.: Are Authorities Denying or Supporting? Detecting Stance of Authorities Towards Rumors in Twitter (2023). https://doi.org/10.21203/rs.3.rs-3383493/v1
17. Haouari, F., Elsayed, T.: Detecting stance of authorities towards rumors in Arabic tweets: a preliminary study. In: Kamps, J., et al. (eds.) ECIR 2023. LNCS, vol. 13981, pp. 430–438. Springer, Cham (2023). https://doi.org/10.1007/978-3-031-28238-6_33
18. Haouari, F., Elsayed, T., Mansour, W.: Who can verify this? finding authorities for rumor verification in Twitter. Inf. Process. Manag. **60**(4), 103366 (2023)
19. Haouari, F., Hasanain, M., Suwaileh, R., Elsayed, T.: ArCOV19-Rumors: Arabic COVID-19 Twitter dataset for misinformation detection. In: Proceedings of the Arabic Natural Language Processing Workshop (WANLP 2021), pp. 72–81 (2021)
20. Haouari, F., Sheikh Ali, Z., Elsayed, T.: Overview of the CLEF-2023 CheckThat! lab task 5 on authority finding in twitter. In: Aliannejadi et al. [3]
21. Hu, X., Guo, Z., Chen, J., Wen, L., Yu, P.S.: MR2: a benchmark for multimodal retrieval-augmented rumor detection in social media. In: Proceedings of the 46th International ACM SIGIR Conference on Research and Development in Information Retrieval (SIGIR 2023), pp. 2901–2912. Association for Computing Machinery, New York (2023)
22. Jerônimo, C.L.M., Marinho, L.B., Campelo, C.E.C., Veloso, A., da Costa Melo, A.S.: Fake news classification based on subjective language. In: Proceedings of the 21st International Conference on Information Integration and Web-Based Applications and Services, pp. 15–24 (2019)
23. Jiang, Y., Song, X., Scarton, C., Aker, A., Bontcheva, K.: Categorising fine-to-coarse grained misinformation: an empirical study of COVID-19 infodemic. arXiv preprint arXiv:2106.1 (2021)
24. Kasnesis, P., Toumanidis, L., Patrikakis, C.Z.: Combating fake news with transformers: a comparative analysis of stance detection and subjectivity analysis. Inf. **12**(10), 409 (2021)
25. Kiela, D., et al.: The hateful memes challenge: Detecting hate speech in multimodal memes. In: NeurIPS 2020 (2020)
26. Kiesel, J., et al.: SemEval-2019 Task 4: hyperpartisan news detection. In: Proceedings of the 13th International Workshop on Semantic Evaluation, pp. 829–839. Association for Computational Linguistics, Minneapolis (2019)
27. Levenshtein, V.I.: Binary codes capable of correcting deletions, insertions, and reversals. Sov. Phys. Doklady **10**, 707–710 (1966)
28. Mu, Y., et al.: A large-scale comparative study of accurate COVID-19 information versus misinformation. In: TrueHealth 2023: Workshop on Combating Health Misinformation for Social Wellbeing (2023)
29. Nakov, P., et al.: Overview of the CLEF-2022 CheckThat! lab task 1 on identifying relevant claims in tweets. In: Working Notes of CLEF 2022–Conference and Labs of the Evaluation Forum (CLEF 2022) (2022)

30. Nakov, P., et al.: Overview of the CLEF-2022 CheckThat! lab on fighting the COVID-19 infodemic and fake news detection. In: Proceedings of the 13th International Conference of the CLEF Association: Information Access Evaluation meets Multilinguality, Multimodality, and Visualization (CLEF 2022) (2022)
31. Nakov, P., et al.: The CLEF-2022 CheckThat! lab on fighting the COVID-19 infodemic and fake news detection. In: Advances in Information Retrieval - European Conference on IR Research (ECIR 2022), pp. 416–428 (2022)
32. Nakov, P., et al.: Overview of the CLEF-2018 lab on automatic identification and verification of claims in political debates. In: Working Notes of CLEF 2018 - Conference and Labs of the Evaluation Forum (CLEF 2018). (2018)
33. Nakov, P., et al.: Overview of the CLEF-2021 CheckThat! lab on detecting check-worthy claims, previously fact-checked claims, and fake news. In: Candan, K., et al. (eds.) Experimental IR Meets Multilinguality, Multimodality, and Interaction. Proceedings of the Twelfth International Conference of the CLEF Association. LNCS, vol. 12880, pp. 294–291. Springer, Cham (2021). https://doi.org/10.1007/978-3-030-85251-1_19
34. Nakov, P., et al.: The CLEF-2021 CheckThat! lab on detecting check-worthy claims, previously fact-checked claims, and fake news. In: Advances in Information Retrieval - 43rd European Conference on IR Research (ECIR 2021), vol. 12657, pp. 639–649 (2021)
35. Nielsen, D.S., McConville, R.: Mumin: a large-scale multilingual multimodal fact-checked misinformation social network dataset. In: Proceedings of the 45th International ACM SIGIR Conference on Research and Development in Information Retrieval (SIGIR 2022), pp. 3141–3153. Association for Computing Machinery, New York (2022)
36. Piskorski, J., et al.: News categorization, framing and persuasion techniques: annotation guidelines. Tech. Rep. JRC-132862, European Commission Joint Research Centre, Ispra (Italy) (2023)
37. Piskorski, J., Stefanovitch, N., Da San Martino, G., Nakov, P.: SemEval-2023 task 3: detecting the category, the framing, and the persuasion techniques in online news in a multi-lingual setup. In: Proceedings of the 17th International Workshop on Semantic Evaluation (SemEval-2023), pp. 2343–2361. Association for Computational Linguistics, Toronto (2023)
38. Potthast, M., Kiesel, J., Reinartz, K., Bevendorff, J., Stein, B.: A stylometric inquiry into hyperpartisan and fake news. In: Proceedings of the 56th Annual Meeting of the Association for Computational Linguistics (Volume 1: Long Papers), pp. 231–240. Association for Computational Linguistics (2018)
39. Pramanick, S., et al.: Detecting harmful memes and their targets. In: Findings of the Association for Computational Linguistics: ACL-IJCNLP 2021, pp. 2783–2796. Association for Computational Linguistics (2021)
40. Przybyła, P., Saggion, H.: ERINIA: evaluating the robustness of non-credible text identification by anticipating adversarial actions. In: NLP-MisInfo 2023: Workshop on NLP applied to Misinformation, held as part of SEPLN 2023: 39th International Conference of the Spanish Society for Natural Language Processing. CEUR-WS.org (2023)
41. Przybyła, P., Shvets, A., Saggion, H.: Verifying the Robustness of Automatic Credibility Assessment. arXiv preprint arXiv:2303.08032 (2023)
42. Riloff, E., Wiebe, J.: Learning extraction patterns for subjective expressions. In: Proceedings of the 2003 Conference on Empirical Methods in Natural Language Processing (EMNLP 2003), pp. 105–112 (2003)

43. Sellam, T., Das, D., Parikh, A.: BLEURT: learning robust metrics for text generation. In: Proceedings of the 58th Annual Meeting of the Association for Computational Linguistics, pp. 7881–7892. Association for Computational Linguistics (2020)
44. Shang, L., Zhang, Y., Zha, Y., Chen, Y., Youn, C., Wang, D.: Aomd: an analogy-aware approach to offensive meme detection on social media. Inf. Process. Manage. **58**(5) (2021)
45. Sharma, S., et al.: Detecting and understanding harmful memes: a survey. In: Proceedings of the Thirty-First International Joint Conference on Artificial Intelligence (IJCAI 2022), pp. 5597–5606. International Joint Conferences on Artificial Intelligence Organization (2022), survey Track
46. Sharma, S., et al.: Characterizing the entities in harmful memes: who is the hero, the villain, the victim? In: Proceedings of the 17th Conference of the European Chapter of the Association for Computational Linguistics, pp. 2149–2163. Association for Computational Linguistics, Dubrovnik (2023)
47. Sharma, S., et al.: Findings of the CONSTRAINT 2022 shared task on detecting the hero, the villain, and the victim in memes. In: Proceedings of the Workshop on Combating Online Hostile Posts in Regional Languages during Emergency Situations, pp. 1–11. Association for Computational Linguistics, Dublin (2022)
48. Suryawanshi, S., Chakravarthi, B.R.: Findings of the shared task on troll meme classification in Tamil. In: Proceedings of the First Workshop on Speech and Language Technologies for Dravidian Languages, pp. 126–132. Association for Computational Linguistics (2021)
49. Szegedy, C., et al.: Intriguing properties of neural networks. arXiv:1312.6199 (2013)
50. Thorne, J., Vlachos, A., Cocarascu, O., Christodoulopoulos, C., Mittal, A.: The fact extraction and VERification (FEVER) shared task. In: Proceedings of the First Workshop on Fact Extraction and VERification (FEVER) (2018)
51. Vieira, L.L., Jerônimo, C.L.M., Campelo, C.E.C., Marinho, L.B.: Analysis of the subjectivity level in fake news fragments. In: Proceedings of the Brazillian Symposium on Multimedia and the Web (WebMedia 2020), pp. 233–240. ACM (2020)
52. Zhang, W.E., Sheng, Q.Z., Alhazmi, A., Li, C.: Adversarial attacks on deep-learning models in natural language processing. ACM Trans. Intell. Syst. Technol. **11**(3) (2020)

ELOQUENT CLEF Shared Tasks for Evaluation of Generative Language Model Quality

Jussi Karlgren[1(✉)], Luise Dürlich[2], Evangelia Gogoulou[2], Liane Guillou[2], Joakim Nivre[2], Magnus Sahlgren[1,3], and Aarne Talman[1]

[1] Silo AI, Helsinki, Finland
jussi.karlgren@silo.ai
[2] RISE Research Institutes of Sweden, Stockholm, Sweden
[3] AI Sweden, Stockholm, Sweden

Abstract. ELOQUENT is a set of shared tasks for evaluating the quality and usefulness of generative language models. ELOQUENT aims to bring together some high-level quality criteria, grounded in experiences from deploying models in real-life tasks, and to formulate tests for those criteria, preferably implemented to require minimal human assessment effort and in a multilingual setting. The selected tasks for this first year of ELOQUENT are (1) probing a language model for topical competence; (2) assessing the ability of models to generate and detect hallucinations; (3) assessing the robustness of a model output given variation in the input prompts; and (4) establishing the possibility to distinguish human-generated text from machine-generated text.

Keywords: Generative language models · LLM · Multilinguality · Shared task · Quality benchmarks · CLEF

Assessing the Quality of Generative Language Models

Generative language models ("LLMs") have become a foundational technology for both research and information service development in the past year, and have prompted great public interest through some quite innovative application scenarios.

A generative language model as a component in an information service can handle a broad variety of input data robustly and elegantly, and can provide appropriately creative generated output to fit the varied application situations and the preferences of a diverse user population. An information service with a generative language model can be built to provide a flexible low threshold conversational interface for its users: there is considerable interest to put generative language models to use in productive practical applications, in industry, NGOs, and public sector, that cover one or multiple languages.

Deploying generative models, or selecting among the several recent models with various generative capacities, is challenged by run-time costs, training costs,

N. Goharian et al. (Eds.): ECIR 2024, LNCS 14612, pp. 459–465, 2024.
https://doi.org/10.1007/978-3-031-56069-9_63

and to a great extent by concerns about the quality, predictability, relevance, appropriateness, and reliability of the output from the generative component. A language model can provide a system with fluently generated output but without guarantees of relevance, veracity, or consistency. In research contexts, quality assurance, evaluation, and benchmarking of large language models is performed using tests that focus on fluency, and details about language understanding and some aspects of reasoning using language. There is a large number of tasks available for this purpose. However, many factors that are crucial for application development and deployment to third parties, such as veracity and robustness, are not addressed systematically through existing benchmarks. In addition, the current benchmarks are primarily Anglo-centric, which raises questions about the quality of output produced in other languages, or in multilingual settings.

The ELOQUENT evaluation lab will experiment with new evaluation methods for generative language models to meet some of the challenges in the path from laboratory to application. We will formulate a series of tasks to be developed iteratively over several years, each year drawing from experiences from the previous year. We recognise that organisations, whether commercial or not, will likely be unable and unwilling to expend great human effort on testing, benchmarking, sampling, and validation to perform quality assurance on a language module which oftentimes is a subsidiary component of a larger envisioned service [8]. We intend to develop tasks which do not rely on fixed gold standards and which involve minimal human effort expended on assessment, instead leveraging the intrinsic qualities of generative AI for test development and scoring.

The outcome of this lab is not only to score the performance of submitted models and to provide benchmarks for future reuse but to establish useful top-level quality criteria, which existing and new tests will contribute to assessing, and to more generally point out future directions for evaluation of generative and learning models [9]. Such top-level quality criteria include the directly linguistic notions of utterance-level correctness, clarity and fluency; of session-level discourse and conversational competence and awareness of social norms and conventions; of relevance, truthfulness, and topical quality, and of resilience and consistency in face of varying input. Such quality criteria are only partially covered by existing test suites. We intend to address some identifiable gaps.

The tasks are intended to involve comparatively little human effort, and instead leverage the capacity of the current generation of AI models to generate, process, and assess human language input. The organisers wish to encourage efforts in multilingual research and as hope to see participants work with models in many languages. Addressing the challenges of accommodating a variety of languages and cultural areas in a common evaluation framework is one of the intended focus topics of the workshop.

Task 1: Topical Competence

Is a generative model knowledgeable about some given topic?

A generative language model in practical application will in most envisioned use cases be expected to stay within topical boundaries, to generate material restricted to the domain it is employed to work within, and to have competence in the terminology and conventions of that domain. Examples of relevant topical domains could be business domains, such as finance [20] or healthcare [16], or even recreational activities such as sailing or basketball, ranging to differences in how a topic is treated differently across linguistic and cultural areas or in specific demographic groups. We do not intend to constrain the type of topic to be included and will in this task experiment with a wide range of topics and discourse types, welcoming suggestions and proposals from task participants.

The discourse control structures and domain knowledge for generative language models are in today's architectures implicit and learned via the acquisition of linguistic competence, but increasing interest has been shown towards a separate knowledge representation for domain information coupled to the input and output capacities of a language model.[1] The generality of language models, and their ability to work across domains with little or no fine-tuning is part of their appeal, and verification of their capacity for a specific domain under consideration is usually done as part of a fine-tuning process. The more general question, given a foundation model, whether it already is competent for some application domain of interest is not addressed in general terms.

This task will both answer to the need for verifying a model's understanding of an application domain of interest and will also in future iterations probe the capacity of a model to communicate with another model. This first year, the task will be for a model to generate and score topical tests, applied between participating models in a round-robin fashion. We expect this task to develop during the next years, with coming years including more challenges related to conversational competence with two or more models set up to engage in a conversation to solve some task which will involve making use of both topical and conversational competence.

Task 2: HalluciGen Detection

Can a model identify hallucinated content?

Detecting hallucinated or factually incorrect information may be difficult for humans. LLMs should therefore be rigorously tested for their ability to generate accurate output prior to deployment. One possible approach for detecting hallucinated content, that we will explore in this lab, is the use of LLMs to evaluate LLM output.

[1] https://wsl.iiitb.ac.in/cikm-2023-workshop-on-enterprise-knowledge-graphs-using-large-language-models/.

Previous work has framed the problem as one of cross-model evaluation where one model evaluates another's output [10,15], or self-evaluation where the model evaluates its own output [10,12]. Inspired by the *Build it, break it: Language Edition* task [6], we divide the task into a builders subtask which focuses on the development of model-evaluators, and a breakers subtask which focuses on the development of challenging evaluation datasets.

The focus of Year 1 will be on the *builder task*. Participants will develop multilingual and monolingual hallucination-aware models (*evaluators*) that are able to both detect and generate hallucinated content. Given a pair of submitted models, A and B, and within a conversational setting, model A will be prompted to generate two versions of the same input containing either the presence/absence of hallucinated content, and model B will be asked to detect whether the output contains hallucinated content. We will leverage pre-existing manually-compiled datasets, e.g. data from the SHROOM task on *Hallucinations and Related Observable Overgeneration Mistakes*[2] at SemEval 2024, to validate the ability of the models to generate and detect hallucinated content. For the generation step, we will measure the semantic similarity between model output and a human-authored gold standard, and for the detection step we will validate performance using a held-out test set, possibly machine-generated. This task will assess both model awareness of hallucination and the viability of cross-model evaluation.

The focus of Year 2 will be on the *breaker task*. Participants will develop novel evaluation datasets/benchmarks designed to *break* the hallucination detection capabilities of one or more baseline *evaluator* models. The evaluator model(s) could, for example, be the best-performing model(s) from Year 1. The aim of the task is to provide challenging examples, of human or near-human quality, that expose the strengths and weaknesses of the models in a monolingual or multilingual setting. This task is inspired by the WMT challenge set subtask [7] and the development of the Adversarial NLI dataset [13]. The focus will be on automated methods for dataset construction, for example leveraging pre-existing datasets and applying techniques such as data augmentation (e.g. using LLMs). This task will explore the use of automated methods for generating high-quality evaluation benchmarks.

Task 3: Robustness

Does a model generate equivalent content for varied but equivalent input?

Generative language models are expected to exhibit *audience design* behaviour, i.e. to fit their output to the preceding input [2]. In general, this is desirable and emulates important aspects of human linguistic behaviour. However, if this variation extends to content-related aspects of the output, this may have the unfortunate effect of systematically generating different material depending on

[2] https://helsinki-nlp.github.io/shroom/.

user group, if e.g. the system is sensitive to dialectal, sociolectal, cross-cultural, or otherwise significant variability in linguistic items in the input.

This task will assess the how elicited responses from generative language models vary by input prompt variation. The task will organise a set of input prompts that vary along some dimensions of style and form, and the resulting outputs form participating systems will be scored for some observable content-related variables of interest. Robustness has been identified as a quality criterion when models have positional biases in responses to multiple choice questions [21] and in the face of adversarial attacks [1, 11, 19]. We intend to explore and extend this notion further in this task.

This task will, besides assessing the quality of participating models, establish quantitative measures for the range of elicited responses from the generative language models and for the robustness vs. volatity of generative language models with respect to variation in input.

Task 4: Voight-Kampff

> Can human-authored text be reliably distinguished from machine-generated text?

Recent advances in generative language models have made it possible to automatically generate content for websites, news articles, social media, etc. Labelling AI-generated content, in order to tackle disinformation, is something that the EU has recently urged technology companies to do.[3]

Detecting automatically generated text, with the increased quality of generative AI, is becoming a task quite similar to human authorship verification, and recent experiments have shown detection of one type of model does not readily translate to cross-model detection, with generative models exhibiting characteristics peculiar to them. [14].

This task will be organised in collaboration with the PAN lab at CLEF, which allows us to build on years of experience from previous shared tasks organised by the PAN lab on authorship verification and closely related tasks. Recent editions of PAN have studied authorship verification from a *cross-domain* [4,5,18] and *cross-discourse type verification* perspective [17].

The Voight-Kampff task will be split into two subtasks, in the builder-breaker style, similar to the HalluciGen task described above. In the *builder* task PAN lab participants will build systems to discriminate between automatically-generated and human-authored texts [3]. The systems will be evaluated using the dataset(s) constructed by the *breakers*.

In the *breaker* task ELOQUENT lab participants will focus on constructing an evaluation dataset designed to challenge the discriminative capabilities of the *breaker* systems. A human-authored test set representing different styles and genres will be selected from existing texts, and the automated set will be

[3] https://www.theguardian.com/technology/2023/jun/05/google-and-facebook-urged-by-eu-to-label-ai-generated-content.

generated by first automatically summarising the human-authored texts, and then instructing an LLM to generate a text of equal length to the original based on the summary. Participants may employ different strategies to evade detection by the *breaker* systems, for example designing LLMs to output text in a similar in style to the human-authored texts, or to inject human-like errors in the texts.

This task is designed to explore whether automatically-generated text can be distinguished from human-authored text.

Invitation to Participate

Those interested in participating in this first edition of the ELOQUENT evaluation lab should register for the task through the CLEF registration site found through the conference pages[4] Defining and shaping new tasks for coming editions will be discussed during the course of experimentation with this year's tasks — anyone interested in suggesting new tasks, new target notions, or new methodologies is welcome to join the discussion!

References

1. Altinisik, E., Sajjad, H., Sencar, H.T., Messaoud, S., Chawla, S.: Impact of adversarial training on robustness and generalizability of language models. arXiv preprint arXiv:2211.05523 (2023)
2. Bell, A.: Language style as audience design. Lang. Soc. **13**(2) (1984)
3. Bevendorff, J., et al.: Overview of PAN 2024: multi-author writing style analysis, multilingual text detoxification, oppositional thinking analysis, and generative AI authorship verification. In: Advances in Information Retrieval: 46th European Conference on IR Research (ECIR) (2024)
4. Bevendorff, J., et al.: Overview of PAN 2021: authorship verification, profiling hate speech spreaders on twitter, and style change detection. In: Experimental IR Meets Multilinguality, Multimodality, and Interaction – 12th International Conference of the CLEF Association (2021)
5. Bevendorff, J., et al.: Overview of PAN 2020: authorship verification, celebrity profiling, profiling fake news spreaders on twitter, and style change detection. In: Experimental IR Meets Multilinguality, Multimodality, and Interaction – 11th International Conference of the CLEF Association (2020)
6. Ettinger, A., Rao, S., Daumé III, H., Bender, E.M.: Towards linguistically generalizable NLP systems: a workshop and shared task. In: Proceedings of the First Workshop on Building Linguistically Generalizable NLP Systems. Association for Computational Linguistics (2017)
7. Freitag, M., et al.: Results of WMT22 metrics shared task: stop using BLEU - neural metrics are better and more robust. In: Proceedings of the Seventh Conference on Machine Translation (WMT). Association for Computational Linguistics (2022)

[4] ELOQUENT lab page with more practical information can be found at https://eloquent-lab.github.io/.

8. Karlgren, J.: Adopting systematic evaluation benchmarks in operational settings. In: Information Retrieval Evaluation in a Changing World: Lessons Learned from 20 Years of CLEF (2019)
9. Karlgren, J., et al.: Evaluating learning language representations. In: Experimental IR Meets Multilinguality, Multimodality, and Interaction – 6th International Conference of the CLEF Association (2015)
10. Manakul, P., Liusie, A., Gales, M.J.F.: Selfcheckgpt: zero-resource black-box hallucination detection for generative large language models. arXiv preprint arXiv:2303.08896 (2023)
11. Moradi, M., Samwald, M.: Evaluating the robustness of neural language models to input perturbations. In: Proceedings of the 2021 Conference on Empirical Methods in Natural Language Processing (2021)
12. Mündler, N., He, J., Jenko, S., Vechev, M.: Self-contradictory hallucinations of large language models: evaluation, detection and mitigation. arXiv preprint arXiv:2305.15852 (2023)
13. Nie, Y., Williams, A., Dinan, E., Bansal, M., Weston, J., Kiela, D.: Adversarial NLI: a new benchmark for natural language understanding. In: Proceedings of the 58th Annual Meeting of the Association for Computational Linguistics. Association for Computational Linguistics (2020)
14. Sarvazyan, A.M., González, J.Á., Rosso, P., Franco-Salvador, M.: Supervised machine-generated text detectors: family and scale matters. In: Arampatzis, A., et al. (eds.) Experimental IR Meets Multilinguality, Multimodality, and Interaction. CLEF 2023. LNCS, vol. 14163, pp. 121–132. Springer, Cham (2023). https://doi.org/10.1007/978-3-031-42448-9_11
15. Saunders, W., et al..: Self-critiquing models for assisting human evaluators. arXiv preprint arXiv:2206.05802 (2022)
16. Singhal, K., et al.: Large language models encode clinical knowledge. Nature **620**(7972) (2023)
17. Stamatatos, E., et al.: Overview of the authorship verification task at PAN 2022. In: Faggioli, G., Ferro, N., Hanbury, A., Potthast, M. (eds.) CLEF 2022 Labs and Workshops, Notebook Papers. CEUR-WS.org (2022)
18. Stamatatos, E., Potthast, M., Pardo, F.M.R., Rosso, P., Stein, B.: Overview of the PAN/CLEF 2015 evaluation lab. In: Experimental IR Meets Multilinguality, Multimodality, and Interaction – 6th International Conference of the CLEF Association (2015)
19. Wang, B., et al.: InfoBERT: improving robustness of language models from an information theoretic perspective. In: International Conference on Learning Representations (2021)
20. Wu, S., et al.: Bloomberggpt: a large language model for finance. arXiv preprint arXiv:2303.17564 (2023)
21. Zheng, C., Zhou, H., Meng, F., Zhou, J., Huang, M.: Large language models are not robust multiple choice selectors. arXiv preprint arXiv:2309.03882 (2023)

Overview of Touché 2024: Argumentation Systems

Johannes Kiesel[1(✉)], Çağrı Çöltekin[2], Maximilian Heinrich[1], Maik Fröbe[3], Milad Alshomary[4], Bertrand De Longueville[5], Tomaž Erjavec[6], Nicolas Handke[7], Matyáš Kopp[8], Nikola Ljubešić[6], Katja Meden[6], Nailia Mirzhakhmedova[1], Vaidas Morkevičius[9], Theresa Reitis-Münstermann[12], Mario Scharfbillig[11], Nicolas Stefanovitch[5], Henning Wachsmuth[4], Martin Potthast[7,10], and Benno Stein[1]

[1] Bauhaus-Universität Weimar, Weimar, Germany
johannes.kiesel@uni-weimar.de, touche@webis.de
[2] University of Tübingen, Tübingen, Germany
[3] Friedrich-Schiller-Universität Jena, Jena, Germany
[4] Leibniz University Hannover, Hannover, Germany
[5] European Commission, Joint Research Centre (JRC), Ispra, Italy
[6] Jožef Stefan Institute, Ljubljana, Slovenia
[7] Leipzig University, Leipzig, Germany
[8] Charles University, Prague, Czechia
[9] Kaunas University of Technology, Kaunas, Lithuania
[10] ScaDS.AI, Leipzig, Germany
[11] European Commission, Joint Research Centre (JRC), Brussels, Belgium
[12] Arcadia Sistemi Informativi Territoriali, Milan, Italy
https://www.touche.webis.de

Abstract. Decision-making and opinion-forming are everyday tasks that involve weighing pro and con arguments. The goal of Touché is to foster the development of support-technologies for decision-making and opinion-forming and to improve our understanding of these processes. This fifth edition of the lab features three shared tasks: (1) Human value detection (ValueEval), where participants detect (implicit) references to human values and their attainment in text; (2) Multilingual Ideology and Power Identification in Parliamentary Debates, where participants identify from a speech the political leaning of the speaker's party and whether it was governing at the time of the speech (new task); and (3) Image retrieval or generation in order to convey the premise of an argument with visually. In this paper, we briefly describe the planned setup for the fifth lab edition at CLEF 2024 and summarize the results of the 2023 edition.

Keywords: Argumentation · Human values · Ideology · Image retrieval

N. Goharian et al. (Eds.): ECIR 2024, LNCS 14612, pp. 466–473, 2024.
https://doi.org/10.1007/978-3-031-56069-9_64

1 Introduction

Decision-making and opinion-forming are everyday tasks, for which everybody has the chance to acquire knowledge on the Web on almost every topic. However, conventional search engines are primarily optimized for returning *relevant* results, which is insufficient for collecting and weighing the pros and cons for a topic. To close this gap of technologies that support people in decision-making and opinion-forming, the Touché lab's shared tasks[1] (https://touche.webis.de) call for the research community to develop respective approaches. In 2024, we organize the three following shared tasks:

1. Human Value Detection (a continuation of ValueEval'23 @ SemEval [8][2]) features two subtasks in ethical argumentation of detecting human values in texts and their attainment, respectively.
2. Ideology and Power Identification in Parliamentary Debates features two subtasks in debate analysis of detecting the ideology and position of power of the speaker's party, respectively (new task).
3. Image Retrieval for Arguments (third edition, now joint task with ImageCLEF) is about the retrieval or generation of images to help convey an argument's premise.

After having organized four successful Touché labs on argument retrieval at CLEF 2020–2023 [1–4], we now organize a fifth lab edition to bring together researchers from the fields of information retrieval, natural language processing, and computational linguistics working on argumentation. During the previous Touché labs, we received 243 runs from 74 teams. We manually labeled the relevance and quality of more than 30,000 argumentative texts, web documents, and images for 200 search topics (topics and judgments are publicly available at the lab's web page, https://touche.webis.de).

This year's edition of Touché intends to widen its scope. After having explored causal questions in last year's edition, we now explore ethical argumentation in the task of human value detection. Compared to ValueEval'23 @ SemEval, this year's task features a larger dataset that also considers multiple languages and is created in a joint effort of over 70 value scholars. The second task targets deep linguistic analyses of debates, by analyzing the language of different ideologies and positions of power in parliamentary speeches. It is based on the multilingual ParlaMint corpus.[3] In addition, we have further developed the task of finding images for arguments, which this year focuses on finding images for specific arguments rather than topics. Moreover, for the first time we allows participants to use a text-to-image generative AI and submit generated images. As in the previous Touché editions, we will encourage participants to deploy their software in our cloud-based evaluation-as-a-service platform TIRA [11] for better reproducibility.

---

[1] 'touché' confirms "a hit in fencing or the success or appropriateness of an argument, an accusation, or a witty point." [https://merriam-webster.com/dictionary/touche].

[2] Demo of best-performing approach: https://values.args.me.

[3] https://www.clarin.eu/parlamint.

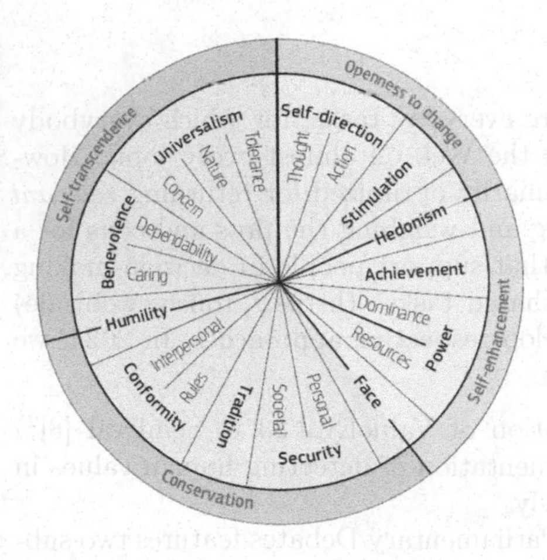

Inner circle: 19 human values
(see https://valueeval.webis.de)

Outer circle: four general directions
(not used in this task)

- **Openness to change**
  Being independent and exploring
- **Self-enhancement**
  Seeking pleasure, wealth, and esteem
- **Conservation**
  Preserving group cohesion, order, and security
- **Self-transcendence**
  Helping others, close ones, and nature

Fig. 1. The 19 values used in this task, shown in the Schwartz value taxonomy [16].

2 Task Definitions

Task 1: Human Value Detection (ValueEval). In argumentation, one has to consider that people have different beliefs and priorities of what is generally worth striving for (e.g., personal achievements vs. humility) and how to do so (e.g., being self-directed vs. respecting traditions), referred to as (human) values.

Overview. The task is to identify the values of the widely accepted value taxonomy of Schwartz [16] (cf. Fig. 1) and their attainment in long texts of eight languages (Bulgarian, Dutch, English, French, German, Hebrew, Italian, and Turkish). This taxonomy has been replicated in over 200 samples in 80 countries and is the backbone of value research [15]. A value can either be mentioned as something that is or should be attained (i.e., lead towards fulfilling the value) or something that is not attained or constrained. For example, for Security, (partial) attainment would mean that something is made safer or healthier. In contrast, an event can be stated in a way that thwarts or constrains safety or health. Participating teams can submit software in one or both of two sub-tasks: (1) Given a text, for each sentence, detect which human values the sentence refers to; and (2) Given a text, for each sentence and value this sentence refers to, detect whether this reference (partially) attains or constrains the value.

Data. The task employs a collection of 3000 human-annotated texts between 400 and 800 words (across eight languages) from news articles and political texts (excerpts of speeches, debates, and party manifestos). Texts are sampled to reflect diverse opinions (different parties; mainstream news and not; from 2019 to 2023). The data is annotated as part of the ValuesML project[4] by over 70 value

[4] https://knowledge4policy.ec.europa.eu/projects-activities/valuesml-unravelling-expressed-values-media-informed-policy-making_en.

scholars. Dedicated team leaders per language train the respective annotators, consolidate annotations, and discuss disagreement (measured continuously by the organizers) in their language teams. The team leaders discuss issues with the organizers in bi-weekly meetings. The test set covers 20% of this data.

Evaluation. Submissions are evaluated using macro F_1-score over all values. To facilitate quick develop-and-test cycles, the report facilities in TIRA provide participants with detailed feedback on the prediction errors in their submissions.

Task 2: Multilingual Ideology and Power Identification in Parliamentary Debates. Parliaments are one of the most important institutions in modern democratic states where issues with high societal impact are discussed. The impact of the decisions made in a parliament often goes beyond their borders, and may even have global effects. As a form of political debate, however, speeches in a parliament are often indirect and present challenges for automated systems for analyzing them.

Overview. This task is concerned with predicting *ideology* and *power* in (transcribed) parliamentary speeches from multiple national parliaments, recorded in multiple languages. Both subtasks are formulated as binary classification tasks. The first subtask is about predicting the political orientation (`left–right`) of speakers from their speeches. The second subtask is about predicting whether the speaker is a member of a governing party or the opposition.

Data. The data for both tasks is a sample of ParlaMint [6], a corpus of parliamentary speeches from 29 national or regional parliaments with varying amounts of instances. The time span of the data is from 2015 to 2022 across all parliaments. To ease participation and balance the dataset, this task uses a sample of ParlaMint (full data is up to 90 million words per parliament). The dataset for both tasks includes at least speeches from national parliaments of Belgium, Iceland, Italy, Poland, Slovenia, Spain, The Netherlands, Turkey and United Kingdom. ParlaMint contains machine translation of all data to English, which participants can use as supporting data.

Evaluation. Submissions are evaluated using macro F_1-score in both subtasks, for all languages. Even though the participants are encouraged to make use of multilingual data for improving results for individual languages, we do not evaluate zero- or few-shot settings separately.

Task 3: Image Retrieval/Generation for Arguments (Joint Task with ImageCLEF). Argumentation is a communicative activity of producing and exchanging reasons to support claims. Though mostly associated with the exchange of words, argumentation often involves also images, either for exemplification, illustration, or evoking emotions. This task investigates how images can be used to convey an argument. Whereas the first two editions of this task

Topic: Photo identification at polling stations

Claim: Legislation to impose restrictive photo ID requirements has the potential to block millions of American voters.

Premise: People will forget their ID cards and be denied their right to vote.

Submissions:

Images:			
Rationale:	**Woman who forgot her ID**	**Embarrassed man who forgot his ID**	**Retired nuns barred from voting**
Relevance:	**1**	**2**	**0**

Fig. 2. Three possible submissions for the specified argument. The first (retrieved) image could help to convey the "forget"-part of the premise but does not relate to voting, unlike the second image (which was generated) that is thus rated higher on relevance (1 vs. 2). The third image (which was generated) does not indicate that someone forgets their ID or is barred from voting, and is thus rated irrelevant (0).

followed the setup of Kiesel et al. [9] to retrieve images for a topic, this year's edition focuses on images for specific arguments.

Overview. Given a set of arguments, the task is to return for each argument several images that help convey the argument's premise. A suitable image could depict the argument or show a generalization or specialization. Participants can optionally add a short rationale that explains the meaning of the image.

Data. The task data consists of 50 arguments, each consisting of a claim and a premise (cf. Fig. 2). Premises are either facts or anecdotal. As document collection we provide a focused crawl of at least 1000 images per argument. Following the idea of the infinite index [5], we also provide an API for a Stable Diffusion image generator [14].

Evaluation. Images can be (1) retrieved from the focused crawl and (2) generated using the Stable Diffusion API. The task follows the classic TREC-style methodology of teams submitting ranked results to be judged by human assessors. For a metric, the task uses standard nDCG [7] to represent a user looking through a ranked list of images retrieved for the specific argument.

3 Touché at CLEF 2023: Brief Overview

In 2023, Touché at CLEF included the following four shared tasks [2]: (1) Retrieval of documents that contain arguments and opinions on some controversial topic. (2) Retrieval of documents that contain evidence on whether a causal relationship between two events exists. (3) Retrieval of images to visually corroborate textual arguments and to provide a quick overview of public opinions on controversial topics. (4) Stance classification of comments on proposals from the multilingual participatory democracy platform CoFE[5] to support opinion formation on socially important topics. Touché 2023 received 41 registrations, from which 7 teams actively participated in the tasks and submitted 30 results (runs; every team could submit up to 5 results). The three retrieval tasks followed the traditional TREC methodology: the participants received document collections and topics, and submitted their results (up to five runs) for each topic to be judged by human assessors. In the retrieval tasks, all teams used BM25 or BM25F [12,13] for first-stage retrieval. The final ranked lists (runs) were often created based on argument quality estimation and predicted stance (Task 1), based on the presence of causal relationships in documents (Task 2), and exploiting the contextual similarity between images and queries and using the predicted stance for images (Task 3). The participants trained feature-based and neural classifiers to predict argument quality or stance, and often used ChatGPT with various prompt-engineering methods. To predict the stance of multilingual texts in Task 4, the participants used transformer-based models exploiting a few-step fine-tuning, data augmentation, and label propagation techniques.

The corpora, topics, and judgments are available on the Touché website.[6] Parts of the data are also available in BEIR [17] and `ir_datasets` [10].

4 Conclusion

At Touché, we continue to foster research on argumentation systems, building respective test collections, and bringing the research community together. During the previous four years, the submitted approaches developed from sparse to dense retrieval and zero-shot models, combined with assessments of document "argumentativeness," argument quality, stance detection, and sentiment analysis.

Touché 2024 brings in new tasks and refines existing ones, targeting more subtle aspects of argumentation. With ethical argumentation (human value detection) and the identification of ideology and power in speeches we focus on deep linguistic analyses of argumentation, the former continuing a very successful task at SemEval'23. The third year of the image retrieval task explores a more specific task and the opportunity to submit generated images.

Acknowledgements. This work was partially supported by the European Commission under grant agreement GA 101070014 (https://openwebsearch.eu).

[5] https://futureu.europa.eu.

[6] https://touche.webis.de/.

References

1. Bondarenko, A., et al.: Overview of touché 2020: argument retrieval. In: Working Notes of CLEF 2020 - Conference and Labs of the Evaluation Forum (CLEF 2020), CEUR Workshop Proceedings, vol. 2696. CEUR-WS.org (2020). http://ceur-ws.org/Vol-2696/paper_261.pdf
2. Bondarenko, A., et al.: Overview of touché 2023: argument and causal retrieval. In: Arampatzis, A., et al. (eds.) 14th International Conference of the CLEF Association (CLEF 2023). LNCS, vol. 14163, pp. 507–530, Springer, Heidelberg (2023). https://doi.org/10.1007/978-3-031-42448-9_31
3. Bondarenko, A., et al.: Overview of Touché 2022: argument retrieval. In: Working Notes of CLEF 2022 - Conference and Labs of the Evaluation Forum (CLEF 2022), CEUR Workshop Proceedings, vol. 3180, pp. 2867–2903. CEUR-WS.org (2022). http://ceur-ws.org/Vol-3180/paper-247.pdf
4. Bondarenko, A., et al.: Overview of Touché 2021: argument retrieval. In: Working Notes of CLEF 2021 - Conference and Labs of the Evaluation Forum (CLEF 2021), CEUR Workshop Proceedings, vol. 2936, pp. 2258–2284. CEUR-WS.org (2021). http://ceur-ws.org/Vol-2936/paper-205.pdf
5. Deckers, N., et al.: The infinite index: information retrieval on generative text-to-image models. In: Gwizdka, J., Rieh, S.Y. (eds.) ACM SIGIR Conference on Human Information Interaction and Retrieval (CHIIR 2023), pp. 172–186. ACM (2023). https://doi.org/10.1145/3576840.3578327
6. Erjavec, T., et al.: The parlamint corpora of parliamentary proceedings. Lang. Resour. Eval. **57**, 415–448 (2022). https://doi.org/10.1007/s10579-021-09574-0
7. Järvelin, K., Kekäläinen, J.: Cumulated gain-based evaluation of IR techniques. ACM Trans. Inf. Syst. **20**(4), 422–446 (2002). https://doi.org/10.1145/582415.582418
8. Kiesel, J., et al.: SemEval-2023 Task 4: ValueEval: identification of human values behind arguments. In: Kumar, R., Ojha, A.K., Doğruöz, A.S., Martino, G.D.S., Madabushi, H.T. (eds.) 17th International Workshop on Semantic Evaluation (SemEval 2023), pp. 2287–2303. Association for Computational Linguistics, Toronto (2023). https://doi.org/10.18653/v1/2023.semeval-1.313
9. Kiesel, J., Reichenbach, N., Stein, B., Potthast, M.: Image retrieval for arguments using stance-aware query expansion. In: 8th Workshop on Argument Mining (ArgMining 2021). ACL (2021)
10. MacAvaney, S., Yates, A., Feldman, S., Downey, D., Cohan, A., Goharian, N.: Simplified data wrangling with `ir_datasets`. In: 44th International ACM SIGIR Conference on Research and Development in Information Retrieval (SIGIR 2022), pp. 2429–2436. ACM (2021). https://doi.org/10.1145/3404835.3463254
11. Potthast, M., Gollub, T., Wiegmann, M., Stein, B.: TIRA integrated research architecture. In: Information Retrieval Evaluation in a Changing World - Lessons Learned from 20 Years of CLEF, The Information Retrieval Series, vol. 41, pp. 123–160. Springer, Cham (2019). https://doi.org/10.1007/978-3-030-22948-1_5
12. Robertson, S.E., Walker, S., Jones, S., Hancock-Beaulieu, M., Gatford, M.: Okapi at TREC-3. In: 3rd Text REtrieval Conference (TREC 1994), NIST Special Publication, vol. 500–225, pp. 109–126. NIST (1994)
13. Robertson, S.E., Zaragoza, H., Taylor, M.J.: Simple BM25 extension to multiple weighted fields. In: 13th International Conference on Information and Knowledge Management (CIKM 2004), pp. 42–49. ACM (2004). https://doi.org/10.1145/1031171.1031181

14. Rombach, R., Blattmann, A., Lorenz, D., Esser, P., Ommer, B.: High-resolution image synthesis with latent diffusion models. In: IEEE/CVF Conference on Computer Vision and Pattern Recognition (CVPR 2022), pp. 10674–10685. IEEE (2022). https://doi.org/10.1109/CVPR52688.2022.01042
15. Scharfbillig, M., et al.: Values and Identities - a Policymaker's Guide. Tech. Rep. KJ-NA-30800-EN-N, European Commission's Joint Research Centre, Luxembourg (2021). https://doi.org/10.2760/349527
16. Schwartz, S.H., et al.: Refining the theory of basic individual values. J. Personal. Soc. Psychol. **103**(4), 663–688 (2012). https://doi.org/10.1037/a0029393
17. Thakur, N., Reimers, N., Rücklé, A., Srivastava, A., Gurevych, I.: BEIR: a heterogeneous benchmark for zero-shot evaluation of information retrieval models. In: Vanschoren, J., Yeung, S. (eds.) Neural Information Processing Systems (NeurIPS 2021). NeurIPS (2021)

eRisk 2024: Depression, Anorexia, and Eating Disorder Challenges

Javier Parapar[1(✉)], Patricia Martín-Rodilla[1], David E. Losada[2], and Fabio Crestani[3]

[1] Information Retrieval Lab, Centro de Investigación en Tecnoloxías da Información e as Comunicacións (CITIC), Universidade da Coruña, A Coruña, Spain
{javierparapar,patricia.martin.rodilla}@udc.es

[2] Centro Singular de Investigación en Tecnoloxías Intelixentes (CiTIUS), Universidade de Santiago de Compostela, Santiago, Spain
david.losada@usc.es

[3] Faculty of Informatics, Universitá della Svizzera italiana (USI), Lugano, Switzerland
fabio.crestani@usi.ch

Abstract. In 2017, we launched eRisk as a CLEF Lab to encourage research on early risk detection on the Internet. Since then, thanks to the participants' work, we have developed detection models and datasets for depression, anorexia, pathological gambling and self-harm. In 2024, it will be the eighth edition of the lab, where we will present a revision of the sentence ranking for depression symptoms, the third edition of tasks on early alert of anorexia and eating disorder severity estimation. This paper outlines the work that we have done to date, discusses key lessons learned in previous editions, and presents our plans for eRisk 2024.

1 Introduction

The eRisk Lab[1] is an ongoing project focused on evaluating early risk detection on the internet, with a particular emphasis on health and safety issues. Since its pilot edition in 2017 in Dublin [6], it has been a part of CLEF. Throughout its various editions [6–9,11–13], many collections and models have been presented under the eRisk umbrella, and the presented dataset construction approaches and evaluation strategies can be applied to different types of risks.

Our interdisciplinary Lab addresses tasks that require and combine from information retrieval, computational linguistics, machine learning, and psychology knowledge. Diverse experts have collaborated to design monitoring models for critical societal problems. These models could be used, for example, to alert when someone exhibits signs of suicidal thoughts on social media. Previous eRisk editions have addressed issues like depression, eating disorders, gambling, and self-harm detection.

[1] https://erisk.irlab.org.

N. Goharian et al. (Eds.): ECIR 2024, LNCS 14612, pp. 474–481, 2024.
https://doi.org/10.1007/978-3-031-56069-9_65

eRisk has introduced early alert, sentence ranking for risk symptoms, and severity estimation tasks. Early risk tasks (Sect. 2.1) involve predicting risks by analyzing a temporal text stream (e.g., social media posts) and accumulating evidence to make decisions about specific risks, such as the development of depression. In the severity estimation challenges (Sect. 2.2), participants use all user writings to compute a detailed estimate of symptoms of a specific risk, filling out a standard questionnaire as real users would do. Last year, we presented the sentence ranking for signs of depression (Sect. 2.3). This new type of task complements the other two by challenging participants to rank sentences from a collection of user writings according to their relevance to the symptoms of a specific risk.

2 A Brief History of eRisk

In the inaugural eRisk edition in 2017 [6], the sole pilot task focused on early depression risk detection. Data was released in weekly chunks, and participants had to submit predictions after each release. The demanding nature of this process led to only eight groups out of thirty completing the tasks by the deadline. Evaluation methods and metrics were based on those defined in [5].

In 2018 [7], the same setup was continued, featuring the early detection of depression and introducing a new task for early detection of anorexia. Task 1 on depression received 45 system submissions, while for anorexia we received 35.

In 2019 [8], a significant change occurred as the release of user posts became more fine-grained using a server. Two early risk detection tasks continued (anorexia and self-harm), and a new task on severity estimation for depression was introduced, involving clinically validated questionnaires. The number of submissions for these tasks was 54, 33, and 33 for tasks 1, 2, and 3, respectively. In 2020 [9], the early detection of self-harm and the estimation of depression symptom severity tasks persisted. There were 46 system submissions for the early risk task and 17 for the severity estimation task.

The year 2021 [11] saw the introduction of three tasks, with the third edition of the early self-harm detection and depression symptom severity tasks. Additionally, a new task focused on early detection in the domain of pathological gambling was introduced, receiving 115 runs from 18 teams out of 75 registered. In 2022 [12], the early risk detection of pathological gambling task continued, along with the first edition of early depression risk detection under the new fine-grained setup. Another severity estimation task was presented, focusing on eating disorders and using a standard questionnaire. In total, the proposed tasks received 117 runs from 18 teams.

In 2023 [13], a new task was introduced, focusing on locating markers in sentences for 21 depression symptoms as defined by the BDI-II questionnaire. This marked the first edition of the sentence ranking task. The three-year cycle for early alert of pathological gambling was closed and we also ran the second edition of the eating disorder severity estimation task under the EDE-Q questionnaire, with 105 runs submitted by 20 teams for the proposed tasks.

Over these seven years, eRisk has received a stable number of active participants, slowly placing the Lab as a reference forum for early risk research. We summarised the eRisk experience and the best models presented so far in our recent book [4].

2.1 Early Risk Prediction Tasks

The initial challenges revolved around early risk prediction in various domains (such as depression, anorexia, and self-harm). In each edition, the teams analyzed social media writings (posts or comments) sequentially in chronological order to detect signs of risk as early as possible. All shared tasks in different editions sourced their data from the social media platform Reddit.

Reddit users tend to write prolifically, often with posts spanning several years. Many communities (subreddits) on Reddit focus on mental health disorders, providing a valuable resource for obtaining the writing history of redditors to build eRisk datasets [5]. In these datasets, redditors are categorized into a positive class (e.g., individuals with depression) and a negative class (control group). To identify positive redditors, the methodology developed by Coppersmith and colleagues [3] was followed. For instance, in the case of identifying positive redditors for depression, the writings were searched for explicit strings (e.g., "Today, I was diagnosed with depression") indicating a diagnosis. Notably, phrases like "I am anorexic" or "I have anorexia" were not considered explicit affirmations of a diagnosis. This semi-supervised method has been used to extract information about patients diagnosed with different conditions since 2020, with the assistance of the Beaver tool [10] for labeling positive and negative instances.

Regarding the evaluation methodology, the first edition of eRisk introduced a new metric called ERDE (Early Risk Detection Error) to measure early detection [5]. ERDE differs from standard classification metrics as it takes into account prediction latency, considering both the correctness of the binary decision and the latency. The original ERDE metric quantified latency by counting the number of posts (k) processed before reaching a decision. In 2019, an alternative metric for early risk prediction, $F_{latency}$, was adopted, as proposed by Sadeque et al. [14]. Starting in 2019, with the release of user texts at the writing level, user rankings were generated based on participants' estimated degree of risk. These rankings have been evaluated using common information retrieval metrics, including P@10 and nDCG [8].

2.2 Severity Level Estimation Tasks

In 2019, we introduced a new task focused on estimating the severity of depression, a task that continued in 2020 and 2021. In 2022 we ran the severity estimation on the eating disorder domain, where participants were required to automatically complete the EDE-Q questionnaire. In these tasks, participants had access to the writing history of some redditors who volunteered to fill out the standard questionnaires. Their challenge was to develop models that could answer each of the questionnaire's questions based on the evidence found in the provided writings.

For depression assessment, we used Beck's Depression Inventory (BDI-II) [1], which consists of 21 questions related to the severity of depression signs and symptoms, each with four alternative responses corresponding to different severity levels (e.g., loss of energy, sadness, and sleeping problems). For eating disorders, we used questions 1–12 and 19–28 from the Eating Disorder Examination Questionnaire (EDE-Q) [2].

To create the ground truth, we compiled surveys completed by social media users, along with their writing history. Given the unique nature of the task, we introduced new evaluation measures to assess the participants' estimations. In the case of depression, four metrics were defined: Average Closeness Rate (ACR), Average Hit Rate (AHR), Average DODL (ADODL), and Depression Category Hit Rate (DCHR). Detailed descriptions of these metrics can be found in [8]. In the case of eating disorders, we adopted new metrics in the previous year, including Mean Zero-One Error (MZOE), Mean Absolute Error (MAE), Macroaveraged Mean Absolute Error (MAE_{macro}), Global ED (GED), and the corresponding Root Mean Square Error (RMSE) for four sub-scales: Restraint, Eating Concern, Shape Concern, and Weight Concern [12].

2.3 Sentence Ranking for Symptoms of Risk Tasks

In the 2023 edition, Task 1 presented a novel challenge, focusing on the creation of sentence rankings based on their relevance to specific symptoms of depression. Participants were instructed to rank sentences extracted from user writings based on their relevance to the 21 standardized symptoms as outlined in the BDI-II Questionnaire [1]. In this context, a sentence was considered relevant to a particular symptom if it provided information about the user's condition related to that symptom. It's crucial to highlight that a sentence could be considered relevant even if it conveyed positive information about the symptom. For example, a sentence like "I feel quite happy lately" should still be regarded as relevant for symptom 1, which is "Sadness" in the BDI-II. Using participants' results and *top-k* pooling as document adjudication model, we created the relevance judgements with expert assessors. The ranking-based evaluation was conducted using Mean Average Precision (MAP), mean R-Precision, mean Precision at 10, and mean nDCG at 1000.

2.4 Results

According to the CLEF tradition, Labs' Overview and Extended Overview papers compile the summaries and critical analysis of the participants' systems results [6–9,11–13].

Over the course of eleven editions of early detection tasks for four mental health disorders, we have seen a diverse array of models and methods, with most participants primarily focusing on improving classification accuracy on training data, rather than considering the accuracy-delay trade-off that's crucial for timely detection. Notably, we've observed varying system performance across different disorders over the years. For instance, anorexia and pathological gambling tasks appear to be more manageable than depression detection, possibly due to differences in available training data and the nature of the disorders.

Our observations suggest that the likelihood of patients leaving traces of their condition in their social media language may vary depending on the illness. Nonetheless, the results demonstrate a consistent pattern of participants improving detection accuracy from edition to edition, which encourages us to continue supporting research in text-based early risk detection on social media.

Furthermore, some participants have shown promising results in developing automatic or semi-automatic screening systems for predicting the severity of specific risks. The results also suggest that analyzing the entire user's writing history can be a complementary technique for extracting indicators or symptoms related to the disorder, particularly for depression, where some systems achieved a 40% hit rate in answering the BDI questions in the same way as real users. In the case of the eating disorder questionnaire, results in the second edition, with participants using training data, showed improvement over the previous edition.

In the inaugural sentence ranking task, performance varied among the 37 runs, but the best team achieved promising results (best nDCG: 0.596, best P@10: 0.861).

3 The Tasks of eRisk 2024

The outcomes of previous editions have encouraged us to continue the Lab in 2024 and examine the interaction between text-based screening from social media and risk prediction and estimation. The following is the task breakdown for our plans for this year:

3.1 Task 1: Search for Symptoms of Depression

As in the 2021 task, this challenge will involve ranking sentences from a collection of user writings based on their relevance to each of the 21 symptoms of depression outlined in the BDI-II questionnaire. A sentence will be considered relevant to a BDI symptom if it provides information about the user's condition related to that symptom. We will provide a dataset of tagged sentences along with the BDI-II questionnaire. Participants are free to choose their strategy for generating queries based on the BDI symptom descriptions in the questionnaire. Each system will submit 21 sentence rankings, one for each BDI item.

Once we receive submissions from the participating teams, we will create relevance judgments with the assistance of three human assessors using top-k pooling. These resulting *qrels* will be used to evaluate the systems using standard ranking metrics like MAP, nDCG, among others. This newly annotated corpus of sentences will be a valuable resource with numerous applications beyond eRisk.

In this second edition of the task, we will provide participants with the data from the previous year. Additionally, building on the experience gained in 2023, we plan to include some context to assess the relevance of the target sentences. Specifically, relevance will be determined based on the target sentence along with the preceding and following sentences from the original corpus.

3.2 Task 2: Early Detection of Anorexia

The challenge focuses on the sequential processing of evidence to detect early signs of anorexia as quickly as possible. Texts are processed in the order of their creation, allowing systems that excel at this task to be applied for sequential monitoring of user interactions in blogs, social networks, or other online media.

The task is divided into two stages. In the training stage, participating teams will have access to a training server, where we will release the complete history of writings for a set of training users. The training data will be derived from the 2018 and 2019 editions. The test stage involves a period during which participants must connect to our server[2] and iteratively retrieve user writings and submit their responses.

3.3 Task 3: Measuring the Severity of the Signs of Eating Disorders

The task involves assessing an individual's level of eating disorder based on their historical written submissions. Participants are required to use automated solutions to complete a standard eating disorder questionnaire based on the complete history of user's writings.

The Eating Disorder Examination Questionnaire (EDE-Q) is used to evaluate the range and severity of features associated with eating disorders. It consists of 28 items divided into four subscales: restraint, eating concern, shape concern, and weight concern, in addition to a global score [2]. Using the user's written history, algorithms must estimate the user's responses to each individual item.

We will gather questionnaires filled out by social media users, along with their written histories (collected immediately after the user completes the questionnaire). These user-filled questionnaires serve as the ground truth and will be used to evaluate the quality of responses provided by participating systems. Participants will have access to training data from 2022 and 2023.

4 Conclusions

The results achieved in eRisk and the engagement of the research community inspire us to keep introducing new challenges related to risk identification in Social Media. We extend our heartfelt gratitude to all participants for their contributions to the success of eRisk. We strongly encourage research teams to continue refining and developing new models for upcoming tasks and risks. Despite the time and effort required to create these resources, we firmly believe that the societal benefits far outweigh the associated costs.

[2] https://erisk.irlab.org/server.html.

Acknowledgements. The first and second authors thank the financial support supplied by the Consellería de Cultura, Educación, Formación Profesional e Universidades (accreditation 2019-2022 ED431G/01, ED431B 2022/33) and the European Regional Development Fund, which acknowledges the CITIC Research Center in ICT of the University of A Coruña as a Research Center of the Galician University System and the project PID2022-137061OB-C21 (Ministerio de Ciencia e Innovación, Agencia Estatal de Investigación, Proyectos de Generación de Conocimiento; suppported by the European Regional Development Fund). The third author thanks the financial support supplied by the Consellería de Cultura, Educación, Formación Profesional e Universidades (accreditation 2019-2022 ED431G-2019/04, ED431C 2022/19) and the European Regional Development Fund, which acknowledges the CiTIUS-Research Center in Intelligent Technologies of the University of Santiago de Compostela as a Research Center of the Galician University System. The third author thanks the financial support obtained from: i) project PID2022-137061OB-C22 (Ministerio de Ciencia e Innovación, Agencia Estatal de Investigación, Proyectos de Generación de Conocimiento; suppported by the European Regional Development Fund) and ii) project SUBV23/00002 (Ministerio de Consumo, Subdirección General de Regulación del Juego). The first, second, and third author also thank the funding of project PLEC2021-007662 (MCIN/AEI/10.13039/501100011033, Ministerio de Ciencia e Innovación, Agencia Estatal de Investigación, Plan de Recuperación, Transformación y Resiliencia, Unión Europea-Next Generation EU).

References

1. Beck, A.T., Ward, C.H., Mendelson, M., Mock, J., Erbaugh, J.: An inventory for measuring depression. JAMA Psychiatry **4**(6), 561–571 (1961)
2. Carey, M., Kupeli, N., Knight, R., Troop, N.A., Jenkinson, P.M., Preston, C.: Eating disorder examination questionnaire (EDE-Q): norms and psychometric properties in UK females and males. Psychol. Assess. **31**(7), 839 (2019)
3. Coppersmith, G., Dredze, M., Harman, C.: Quantifying mental health signals in Twitter. In: ACL Workshop on Computational Linguistics and Clinical Psychology (2014)
4. Crestani, F., Losada, D.E., Parapar, J. (eds.): Early Detection of Mental Health Disorders by Social Media Monitoring. Springer, Cham (2022). https://doi.org/10.1007/978-3-031-04431-1
5. Losada, D.E., Crestani, F.: A test collection for research on depression and language use. In: Proceedings Conference and Labs of the Evaluation Forum CLEF 2016, Evora, Portugal (2016)
6. Losada, D.E., Crestani, F., Parapar, J.: eRisk 2017: clef lab on early risk prediction on the internet: experimental foundations. In: Jones, G.J., et al. (eds.) CLEF 2017. LNCS, vol. 10456, pp. 346–360. Springer, Cham (2017). https://doi.org/10.1007/978-3-319-65813-1_30
7. Losada, D.E., Crestani, F., Parapar, J.: Overview of eRisk: early risk prediction on the internet. In: Bellot, P., et al. (eds.) CLEF 2018. LNCS, vol. 11018, pp. 343–361. Springer, Cham (2018). https://doi.org/10.1007/978-3-319-98932-7_30
8. Losada, D.E., Crestani, F., Parapar, J.: Overview of eRisk 2019 early risk prediction on the internet. In: Crestani, F., et al. (eds.) CLEF 2019. LNCS, vol. 11696, pp. 340–357. Springer, Cham (2019). https://doi.org/10.1007/978-3-030-28577-7_27

9. Losada, D.E., Crestani, F., Parapar, J.: Overview of of eRisk 2020: early risk prediction on the internet. In: Arampatzis, A., et al. (eds.) CLEF 2020. LNCS, vol. 12260, pp. 272–287. Springer, Cham (2020). https://doi.org/10.1007/978-3-030-58219-7_20
10. Otero, D., Parapar, J., Barreiro, Á.: Beaver: efficiently building test collections for novel tasks. In: Proceedings of the Joint Conference of the Information Retrieval Communities in Europe (CIRCLE 2020), Samatan, Gers, France, 6–9 July 2020 (2020). https://ceur-ws.org/Vol-2621/CIRCLE20_23.pdf
11. Parapar, J., Martín-Rodilla, P., Losada, D.E., Crestani, F.: Overview of eRisk 2021: early risk prediction on the internet. In: Candan, K.S., et al. (eds.) CLEF 2021. LNCS, vol. 12880, pp. 324–344. Springer, Cham (2021). https://doi.org/10.1007/978-3-030-85251-1_22
12. Parapar, J., Martín-Rodilla, P., Losada, D.E., Crestani, F.: Overview of eRisk 2022: early risk prediction on the internet. In: Barrón-Cedeño, A., et al. (eds.) CLEF 2022. LNCS, vol. 13390, pp. 233–256. Springer, Cham (2022). https://doi.org/10.1007/978-3-031-13643-6_18
13. Parapar, J., Martín-Rodilla, P., Losada, D.E., Crestani, F.: Overview of eRisk 2023: early risk prediction on the internet. In: Arampatzis, A., et al. (eds.) CLEF 2022. LNCS, vol. 14163, pp. 294–315. Springer, Cham (2023). https://doi.org/10.1007/978-3-031-42448-9_22
14. Sadeque, F., Xu, D., Bethard, S.: Measuring the latency of depression detection in social media. In: Proceedings of the Eleventh ACM International Conference on Web Search and Data Mining, WSDM 2018, pp. 495–503. ACM, New York (2018)

QuantumCLEF - Quantum Computing at CLEF

Andrea Pasin[1](✉), Maurizio Ferrari Dacrema[2], Paolo Cremonesi[2], and Nicola Ferro[1]

[1] University of Padua, Padua, Italy
andrea.pasin.1@phd.unipd.it, nicola.ferro@unipd.it
[2] Politecnico di Milano, Milan, Italy
{maurizio.ferrari,paolo.cremonesi}@polimi.it

Abstract. Over the last few years, *Quantum Computing (QC)* has captured the attention of numerous researchers pertaining to different fields since, due to technological advancements, QC resources have become more available and also applicable in solving practical problems. In the current landscape, *Information Retrieval (IR)* and *Recommender Systems (RS)* need to perform computationally intensive operations on massive and heterogeneous datasets. Therefore, it could be possible to use QC and especially *Quantum Annealing (QA)* technologies to boost systems' performance both in terms of efficiency and effectiveness. The objective of this work is to present the first edition of the QuantumCLEF lab, which is composed of two tasks that aim at:

- evaluating QA approaches compared to their traditional counterpart;
- identifying new problem formulations to discover novel methods that leverage the capabilities of QA for improved solutions;
- establishing collaborations among researchers from different fields to harness their knowledge and skills to solve the considered challenges and promote the usage of QA.

This lab will employ the QC resources provided by CINECA, one of the most important computing centers worldwide. We also describe the design of our infrastructure which uses Docker and Kubernetes to ensure scalability, fault tolerance and replicability.

1 Introduction

In the current challenging scenario where *Information Retrieval (IR)* and *Recommender Systems (RS)* systems face ever increasing amounts of data and rely on computational demanding approaches, *Quantum Computing (QC)* can be used to improve their performance. Although QC has already been applied in several domains, limited work has been done specifically for the IR and RS fields [6,10,13]. Indeed, despite there is an area of IR called Quantum IR [9,16,17], it consists of exploiting the concepts of quantum mechanics to formulate IR models and problems but it does not deal with implementing IR and RS models and algorithms via QC technologies.

N. Goharian et al. (Eds.): ECIR 2024, LNCS 14612, pp. 482–489, 2024.
https://doi.org/10.1007/978-3-031-56069-9_66

In this work we focus on *Quantum Annealing (QA)*, which exploits special-purpose devices able to rapidly find optimal solutions to optimization problems by leveraging quantum-mechanical effects. Our goal is to understand if QA can improve the efficiency and effectiveness of IR and RS systems. So, we present a new evaluation lab called *QuantumCLEF (qCLEF)*[1] [12], which aims at:

- evaluating the performance of QA with respect to traditional approaches;
- identifying new ways of formulating IR and RS algorithms and methods, so that they can be solved with QA;
- growing a research community around this new field in order to promote a wider adoption of QC technologies for IR and RS.

Working with QA does not require particular knowledge about how quantum physics works underneath it. There are in fact available tools and libraries that can be easily used to program and solve problems through this paradigm.

The paper is organized as follows: Sect. 2 introduces related works; Sect. 3 presents the tasks in the qCLEF lab; Sect. 4 considers some critical evaluation aspects; Sect. 5 shows the design of the infrastructure for the lab; finally, Sect. 6 draws some conclusions and outlooks some future work.

2 Related Works

What is Quantum Annealing. QA is a QC paradigm that is based on special-purpose devices (quantum annealers) able to tackle optimization problems. A quantum annealer represents a problem as the energy of a physical system and then leverages quantum-mechanical phenomena to let the system find a state of minimal energy, corresponding to the solution of the original problem.

These problems need to be formulated as minimization ones using the *Quadratic Unconstrained Binary Optimization (QUBO)* formulation, defined as follows:

$$\min \quad y = x^T Q x$$

where x is a vector of binary decision variables and Q is a matrix of constant values representing the problem to solve. The QUBO formulation is very general and can represent many problems [8]. Then, the *minor embedding* step maps the QUBO problem into the quantum annealer hardware, accounting for its topology. This can be done automatically, relying on some heuristics. A QUBO problem is usually solved by quantum annealers in few *milliseconds*.

Applications of Quantum Annealing. QA can have practical applications in several fields due to its ability to tackle *NP-Hard* integer optimization problems.

QA has been previously applied to tackle IR and RS tasks such as feature selection [10], showing feasibility and promising improvements in efficiency and effectiveness. QA has also been applied to *Machine Learning (ML)* tasks. For example, Willsch et al. [19] proposes a formulation of kernel-based *Support Vector Machine (SVM)* on a D-Wave 2000Q quantum annealer, while Delilbasic

[1] https://qclef.dei.unipd.it/.

et al. [4] proposes a quantum multiclass SVM formulation aiming to reduce the execution time for large training sets. Other works explore the application of QA to clustering; for example, Zaiou et al. [21] applies it to a balanced K-means method showing better performance according to the Davies-Bouldin Index.

3 Tasks

In the qCLEF lab there are two tasks, each with the following goals:

- find one or more possible QUBO formulations of the problem;
- evaluate the quantum annealer approach compared to a corresponding traditional approach to assess both its efficiency and its effectiveness.

In general, we expect QA to solve problems more quickly than traditional approaches, achieving results that are similar better in terms of effectiveness.

3.1 Task 1 - Quantum Feature Selection

This task focuses on formulating the well-known *NP-Hard* feature selection problem to solve it with a quantum annealer, similarly to other previous works [6,10].

Feature selection is a widespread problem for both IR and RS which requires to identify a subset of the available features (e.g., the most informative, less noisy etc.) to train a learning model. This problem is very impacting, since many IR and RS systems involve the optimization of learning models, and reducing the dimensionality and noise of the input data can improve their performance.

If the input data has n features, we can enumerate all the possible sets of input data having a fixed number k of features, thus obtaining $\binom{n}{k}$ possible subsets. Therefore, to find the best subset of k features the learning model should be trained on all the subsets of features, which is infeasible even for small datasets. So, in this task we want to understand if QA can be used to solve this problem more efficiently and effectively. Feature selection fits very well the QUBO formulation: there is one variable x per feature indicating whether it should be selected or not. The challenge lies in designing the objective function, i.e., matrix Q.

We have identified some possible datasets such as MQ2007 or MQ2008 [15] and The Movies Dataset[2] which have already been used in previous works [6, 10], LETOR4.0 and MSLR-WEB30K [15]. These datasets contain pre-computed features and the objective is to select a subset of these features to train a learning model, such as LambdaMART [3] or a content-based RS, and to achieve best performance according to metrics such as nDCG@10.

3.2 Task 2 - Quantum Clustering

This task focuses on the formulation of the clustering problem to solve it with a quantum annealer. Clustering is a relevant problem for IR and RS which involves grouping items together according to their characteristics.

[2] https://www.kaggle.com/datasets/rounakbanik/the-movies-dataset.

Clustering can be helpful for organizing large collections, helping users to explore a collection and providing similar results to a query. It can also be used to divide users according to their interests or build user models with the cluster centroids [20] boosting efficiency or effectiveness for users with limited data.

There are different clustering problem formulations, such as centroid-based Clustering or Hierarchical Clustering. In this task, each document or user can be represented as a vector in a similarity space and it is possible to cluster documents based on the similarity between each other.

Clustering fits very well with a QUBO formulation and various methods have already been proposed [1,2,18]. Most of them use variables x to represent the associated cluster to a datapoint, hence the number of points and clusters is the main limitation. There are ways to overcome this issue which can result in approximate solutions but allow the use of quantum annealers for large datasets.

For this task, we have identified MSMARCO [11] as a possible dataset, but due to the high number of documents in MSMARCO, we have identified an alternative smaller dataset such as 20 Newsgroups[3] or Wikipedia Movie Plots[4]. From the dataset we will produce embeddings using models such as BERT [5]. The cluster quality will be measured with user queries that undergo the same embedding process. These queries will match only the most representative embeddings of the clusters, avoiding computing similarities on the whole collection. For the recommendation task, we will generate user and item embeddings using state-of-the-art collaborative recommendation algorithms such as graph neural networks, on datasets Yelp and Amazon-Books. The cluster quality will be measured based on whether the centroids can be used to improve the efficiency and effectiveness of the user modeling similarly to what done in [20]. In this case the cluster quality will be measured according to the Silhouette coefficient and P@10.

3.3 Additional Challenges

Even though State-of-the-art quantum annealers nowadays have thousands of qubits (e.g., the D-Wave Advantage has $\sim$ 5000 qubits), one crucial limitation is that each qubit is physically connected only to a limited number of other qubits (15–20) in a graph of a certain topology. The process of *minor embedding* transforms the QUBO formulation in an equivalent one that fits in the particular hardware topology. This process may require to use multiple qubits to represent a single problem variable. Therefore, even if the quantum annealer has $\sim$ 5000 qubits, in practice it is possible to consider problems with at most hundreds of variables. If the problem does not fit, hybrid traditional-quantum methods exist to split the problem in smaller ones that can be solved on the quantum annealer and then combine the results. This is usually done in a general way, so a possible further challenge consists in finding better ways to split a problem in sub-problems exploiting its structure, as well as developing new problem formulations that account for the limited connectivity of quantum annealers.

[3] http://qwone.com/~jason/20Newsgroups/.

[4] https://www.kaggle.com/datasets/jrobischon/wikipedia-movie-plots.

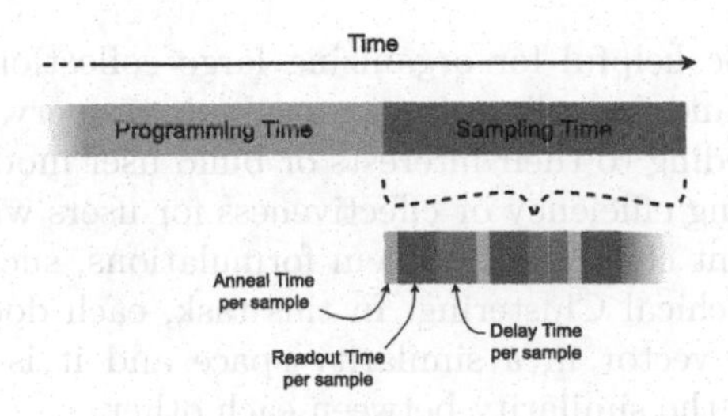

Fig. 1. The quantum annealer access time split in several steps.

4 Evaluation of Quantum Annealing

Using a quantum annealer requires several stages:

Formulation: compute the QUBO matrix Q;
Embedding: generate the *minor embedding* of the QUBO for the hardware;
Data Transfer: transfer the problem and the embedding to the datacenter that hosts the quantum annealer;
Annealing: run the quantum annealer itself.

Considering effectiveness, there are at least two layers of stochasticity. First, the embedding phase in which heuristic methods transform the QUBO formulation in an equivalent one that will fit in the hardware. This process is not deterministic: it could produce different embeddings for the same problem, that are in principle equivalent but in practice may affect the result. Second, the annealing phase that samples a low-energy solution. In some cases, many samples might be needed to get a reliable solution. Usually one selects the best solution found, but this may result in experiments with high variance. Therefore statistical evaluation measures are essential.

Considering efficiency, while the annealing phase in which the quantum annealer is actually used may last in the range of *milliseconds*, transferring the problem on the network introduces large delays and generating the minor embedding may require even minutes for particularly large problems. Furthermore, the runtime can be split in several phases, see Fig. 1: first the device needs to be programmed for the problem, then the quantum-mechanical annealing process is run and lastly the result is read. The annealing process is extremely fast, requiring *few microseconds*, but it is repeated multiple times due to the stochastic nature of the device. It is indeed necessary to consider the time requirements of all the steps involved to measure efficiency.

5 QuantumCLEF Infrastructure

We present our custom infrastructure that is required since participants cannot have direct access to quantum annealers and we want measurements to be as fair and reproducible as possible. As depicted in Fig. 2, it is composed of several components with specific purposes:

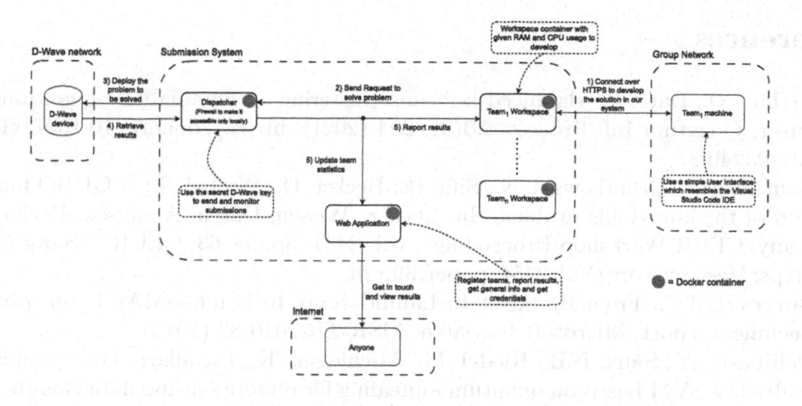

Fig. 2. High-level representation of the infrastructure.

- **Workspace**: each team has its own workspace which is accessible through the browser by providing the correct credentials. The workspace has a pre-configured git repository that is fundamental for reproducibility reasons.
- **Dispatcher**: it manages and keeps track of all the teams' submissions. It also holds the secret API Key that is used to submit problems to the quantum annealer. In this way, participants will never know the secret Key used.
- **Web Application**: it is the main source of information to the external users about the ongoing tasks. Moreover, it allows teams to view their quotas and some statistics through a dashboard. Also organizers have their own dashboard through which it is possible to manage teams and tasks.

The system can be deployed on cloud making use of different physical machines to handle several teams working together. Our infrastructure plays for QA a role similar to others, such as TIRA [14] or TIREx [7], for more general evaluation purposes. We will use the QC resources provided by CINECA that will make available D-Wave's cutting-edge quantum annealers thanks to an agreement already met.

6 Conclusions and Future Work

In this paper we have discussed the qCLEF lab, a new lab composed of two practical tasks aiming at evaluating the performance of QA applied to IR and RS. We have also discussed about the potential benefits that QA can bring to the IR and RS fields and we have highlighted how the evaluation of both efficiency and effectiveness should be performed. Finally, we have presented an infrastructure designed and implemented to satisfy both participants and organizers' needs.

qCLEF can represent a starting point for many researchers worldwide to know more about these new cutting-edge technologies that will likely have a big impact on the future of several research fields. Through this lab it will be also possible to assess whether QA can be employed to improve the current state-of-the-art approaches, hopefully delivering new performing solutions.

References

1. Arthur, D., Date, P.: Balanced k-means clustering on an adiabatic quantum computer. Quantum Inf. Process. **20**(9), 294 (2021). https://doi.org/10.1007/s11128-021-03240-8
2. Bauckhage, C., Piatkowski, N., Sifa, R., Hecker, D., Wrobel, S.: A QUBO formulation of the k-medoids problem. In: Lernen, Wissen, Daten, Analysen, Berlin, Germany, CEUR Workshop Proceedings, vol. 2454, pp. 54–63, CEUR-WS.org (2019). https://ceur-ws.org/Vol-2454/paper_39.pdf
3. Burges, C.J.C.: From RankNet to LambdaRank to LambdaMART: an overview. Technical report, Microsoft Research, MSR-TR-2010-82 (2010)
4. Delilbasic, A., Saux, B.L., Riedel, M., Michielsen, K., Cavallaro, G.: A single-step multiclass SVM based on quantum annealing for remote sensing data classification. arXiv preprint arXiv:2303.11705 (2023)
5. Devlin, J., Chang, M.W., Lee, K., Toutanova, K.: Bert: pre-training of deep bidirectional transformers for language understanding. arXiv preprint arXiv:1810.04805 (2018)
6. Ferrari Dacrema, M., Moroni, F., Nembrini, R., Ferro, N., Faggioli, G., Cremonesi, P.: Towards feature selection for ranking and classification exploiting quantum annealers. In: Proceedings of 45th Annual International ACM SIGIR Conference on Research and Development in Information Retrieval (SIGIR 2022), pp. 2814–2824. ACM Press, New York (2022)
7. Fröbe, M., et al.: The information retrieval experiment platform. In: Chen, H.H., Duh, W.J., Huang, H.H., Kato, M.P., Mothe, J., Poblete, B. (eds.) Proceedings of 46th Annual International ACM SIGIR Conference on Research and Development in Information Retrieval (SIGIR 2023), pp. 2826–2836. ACM Press, New York (2023)
8. Glover, F.W., Kochenberger, G.A., Du, Y.: Quantum bridge analytics I: a tutorial on formulating and using QUBO models. 4OR **17**(4), 335–371 (2019). https://doi.org/10.1007/s10288-019-00424-y
9. Melucci, M.: Introduction to Information Retrieval and Quantum Mechanics. The Information Retrieval Series, vol. 35. Springer, Heidelberg (2015). https://doi.org/10.1007/978-3-662-48313-8
10. Nembrini, R., Ferrari Dacrema, M., Cremonesi, P.: Feature selection for recommender systems with quantum computing. Entropy **23**(8), 970 (2021)
11. Nguyen, T., et al.: MS Marco: a human generated machine reading comprehension dataset. Choice **2640**, 660 (2016)
12. Pasin, A., Ferrari Dacrema, M., Cremonesi, P., Ferro, N.: qCLEF: a proposal to evaluate quantum annealing for information retrieval and recommender systems. In: Arampatzis, A., et al. (eds.) CLEF 2023. LNCS, vol. 14163, pp. 97–108. Springer, Cham (2023). https://doi.org/10.1007/978-3-031-42448-9_9
13. Pilato, G., Vella, F.: A survey on quantum computing for recommendation systems. Information **14**(1), 20 (2023). https://doi.org/10.3390/info14010020
14. Potthast, M., Gollub, T., Wiegmann, M., Stein, B.: TIRA integrated research architecture. In: Ferro, N., Peters, C. (eds.) Information Retrieval Evaluation in a Changing World. The Information Retrieval Series, vol. 41, pp. 123–160. Springer, Cham (2019). https://doi.org/10.1007/978-3-030-22948-1_5
15. Qin, T., Liu, T.Y.: Introducing LETOR 4.0 Datasets. arXiv org, Information Retrieval (cs.IR) arXiv:1306.2597 (2013)

16. van Rijsbergen, C.J.: The Geometry of Information Retrieval. Cambridge University Press, Cambridge (2004)
17. Uprety, S., Gkoumas, D., Song, D.: A survey of quantum theory inspired approaches to information retrieval. ACM Comput. Surv. (CSUR) **53**(5), 1–39 (2020)
18. Ushijima-Mwesigwa, H., Negre, C.F.A., Mniszewski, S.M.: Graph partitioning using quantum annealing on the d-wave system. CoRR abs/1705.03082 (2017). https://arxiv.org/abs/1705.03082
19. Willsch, D., Willsch, M., De Raedt, H., Michielsen, K.: Support vector machines on the d-wave quantum annealer. Comput. Phys. Commun. **248**, 107006 (2020)
20. Wu, Y., Cao, Q., Shen, H., Tao, S., Cheng, X.: INMO: a model-agnostic and scalable module for inductive collaborative filtering. In: SIGIR 2022: The 45th International ACM SIGIR Conference on Research and Development in Information Retrieval, Madrid, Spain, pp. 91–101. ACM (2022). https://doi.org/10.1145/3477495.3532000
21. Zaiou, A., Bennani, Y., Matei, B., Hibti, M.: Balanced k-means using quantum annealing. In: 2021 IEEE Symposium Series on Computational Intelligence (SSCI), pp. 1–7 (2021). https://doi.org/10.1109/SSCI50451.2021.9659997

BioASQ at CLEF2024: The Twelfth Edition of the Large-Scale Biomedical Semantic Indexing and Question Answering Challenge

Anastasios Nentidis[1,2](✉), Anastasia Krithara[1], Georgios Paliouras[1], Martin Krallinger[3], Luis Gasco Sanchez[3], Salvador Lima[3], Eulalia Farre[3], Natalia Loukachevitch[4], Vera Davydova[5], and Elena Tutubalina[5,6,7]

[1] National Center for Scientific Research "Demokritos", Athens, Greece
{tasosnent,akrithara,paliourg}@iit.demokritos.gr
[2] Aristotle University of Thessaloniki, Thessaloniki, Greece
[3] Barcelona Supercomputing Center, Barcelona, Spain
{martin.krallinger,lgasco,salvador.limalopez,eulalia.farre}@bsc.es
[4] Moscow State University, Moscow, Russia
[5] Sber AI, Moscow, Russia
[6] Artificial Intelligence Research Institute, Moscow, Russia
[7] Kazan Federal University, Kazan, Russia

Abstract. The large-scale biomedical semantic indexing and question-answering challenge (BioASQ) aims at the continuous advancement of methods and tools to meet the needs of biomedical researchers and practitioners for efficient and precise access to the ever-increasing resources of their domain. With this purpose, during the last eleven years, a series of annual challenges have been organized with specific shared tasks on large-scale biomedical semantic indexing and question answering. Benchmark datasets have been concomitantly provided in alignment with the real needs of biomedical experts, providing a unique common testbed where different teams around the world can investigate and compare new approaches for accessing biomedical knowledge.

The twelfth version of the BioASQ Challenge will be held as an evaluation Lab within CLEF2024 providing four shared tasks: (i) *Task b* on the information retrieval for biomedical questions, and the generation of comprehensible answers. (ii) *Task Synergy* the information retrieval and generation of answers for open biomedical questions on developing topics, in collaboration with the experts posing the questions. (iii) *Task MultiCardioNER* on the automated annotation of clinical entities in medical documents in the field of cardiology, primarily in Spanish, English, Italian and Dutch. (iv) *Task BioNNE* on the automated annotation of biomedical documents in Russian and English with nested named entity annotations. As BioASQ rewards the methods that outperform the state of the art in these shared tasks, it pushes the research frontier towards approaches that accelerate access to biomedical knowledge.

Keywords: Biomedical Information · Semantic Indexing · Question Answering

N. Goharian et al. (Eds.): ECIR 2024, LNCS 14612, pp. 490–497, 2024.
https://doi.org/10.1007/978-3-031-56069-9_67

1 Introduction

BioASQ[1] [27] is a series of international challenges and workshops on biomedical semantic indexing and question answering (QA). Each edition of BioASQ is structured into distinct but complementary tasks and sub-tasks relevant to biomedical information access. As a result, the participating teams can focus on particular tasks of interest to their specific area of expertise, including but not limited to machine learning, information retrieval, information extraction, and multi-document query-focused summarization. The BioASQ challenge has been running since 2012, with the participation of more than 100 teams from 30 countries, and a BioASQ workshop is organized annually [4,7,22,23]. This year, the BioASQ workshop is part of the fifteenth CLEF conference[2].

BioASQ allows multiple teams that work on biomedical information access systems around the world, to compete in the same realistic benchmark datasets and share, evaluate, and compare their ideas and approaches. Therefore, a key contribution of BioASQ is the benchmark datasets and the open-source infrastructure developed for running its tasks [6]. In particular, as BioASQ consistently rewards the most successful approaches in each task and sub-task, it eventually pushes toward systems that outperform previous approaches. Such successful approaches for semantic indexing and QA can eventually lead to the development of tools to support more precise access to biomedical knowledge and to further improve health services.

2 BioASQ Evaluation Lab 2024

The twelfth BioASQ challenge (BioASQ12) will consist of four tasks that are central to biomedical knowledge access and the question-answering process: (i) *Task b*[3] on the processing of biomedical questions, the generation of answers, and the retrieval of supporting material, (ii) *Task Synergy* on biomedical QA for developing problems under a scenario that promotes collaboration between biomedical experts and question-answering systems, (iii) *Task MultiCardioNER* on text mining and semantic indexing of diseases and medicantions in cardiology clinical case report documents in Spanish, including the annotation of concepts in unlabeled documents and the subsequent normalization of these concept annotations. *Task MultiCardioNER* can be considered as a follow-up task of the previous *MedcProcNER* [8] and *DisTEMIST* [13] tasks on mentions of diseases and medical procedures. (iv) *Task BioNNE* on the automated annotation of unlabelled biomedical documents, written in Russian and English, with nested named entities. As *Task b* and *Task Synergy* have also been organized in the context of previous editions of the BioASQ challenge [16,19], we refer to

[1] http://www.bioasq.org.
[2] https://clef2024.imag.fr/.
[3] Since the first BioASQ, the task on large-scale biomedical semantic indexing is called *Task a*, and the task on biomedical QA is called *Task b*, for brevity. After the completion of *Task a* [5], we keep this naming convention for *Task b* for consistency.

their current version, in the context of BioASQ12, as *task 12b*, *task Synergy 12* respectively.

2.1 Task 12b: Biomedical Question Answering

BioASQ *task 12b* takes place in two phases. In the first phase (Phase A), the participants are given questions in English formulated by biomedical experts and their systems have to retrieve relevant documents (from PubMed) and snippets (passages) of the documents. In the second phase (Phase B), some relevant documents and snippets identified by the experts (using the BioASQ tools [24]) are also provided, and 'exact' and 'ideal' answers are required by the systems. Depending on the type of question, the 'exact' answer can be a *yes* or *no* (yes/no), an entity name, such as a disease or gene (factoid), or a list of entity names (list). The 'ideal' answer is a paragraph-sized summary of the most important information for each question, regardless of its type.

About 500 new biomedical questions annotated with golden documents, snippets, and answers ('exact' and 'ideal'), will be developed by the BioASQ team of trained biomedical experts for testing and assessing the performance of the participating systems. In addition, a training set of about 5,050 biomedical questions, accompanied by answers, and supporting evidence (documents and snippets), will be available from previous versions of the tasks, as a unique resource for the development of question-answering systems [6]. The evaluation in *task 12b* is done manually by the experts and automatically by employing a variety of evaluation measures [12] as done in the previous version of the task [16]. The official measure for document retrieval is the Mean Average Precision and for snippet retrieval the F-measure. For the exact answers, the official measure depends on the type of question. For yes/no questions we use the macro-averaged F-Measure. For factoid questions, where up to five candidate answers can be submitted, we use the Mean Reciprocal Rank. For List questions, we use the mean F-Measure. Finally, for ideal answers, the official evaluation is based on manual scores assigned by experts assessing the readability, recall, precision, and repetition of each response.

2.2 Task Synergy 12: Question Answering for Developing Topics

In an effort to use the BioASQ question-answering ecosystem in order to promote research in developing biomedical topics, such as COVID-19, we introduced the BioASQ *task Synergy* in 2020 [7]. Contrary to the original *task b*, *task Synergy* is designed as a continuous dialog, that allows biomedical experts to pose unanswered questions for developing problems, for which they do not know beforehand whether a definite answer can be provided. The experts receive relevant material (documents and snippets) and potential answers to these questions and assess them, providing feedback to the systems, in order to improve their responses. This process repeats iteratively with new feedback and new system responses for the same questions, as well as with new questions that may have arisen. In each round of this task, new material is also considered based on the current version of the original document resource [18]. Since 2023, this evolving document resource is PubMed [16].

A training dataset of about 300 questions on COVID-19 and other developing topics is already available from previous versions of *task Synergy*, which took place in sixteen rounds over the last three years [17,20,21]. These questions are incrementally annotated with different versions of exact and ideal answers, as well as documents and snippets assessed by the experts as relevant or not. During the *task Synergy 12* this set will be extended with more than fifty new open questions on developing health topics. Meanwhile, any existing questions that remain relevant may be enriched with more updated answers and more recent evidence. Tn *task Synergy 12* we use the same evaluation measures with *task 12b*, considering only new material for the information retrieval part, an approach known as *residual collection evaluation* [26]. However, the focus of this task is to aid the experts in contributing to the incremental understanding of new developing health topics and the discovery of new solutions.

2.3 Task MultiCardioNER: Mutiple Clinical Entity Detection in Multilingual Medical Content

The extraction of clinical variables from unstructured content such as clinical case reports or EHRs is key to enable medical data analytics. Due to the highly specialized medical language, with considerable variation depending on the medical specialty or document type, more specialized automatic semantic annotation resources are needed, not only for English but also other languages. This is particularly true for clinical content related to the cardiovascular diseases (CVDs), which represent the leading cause of death globally, responsible for approximately 17.9 million death/year. Previous efforts to recognize clinical concepts, for instance in Spanish, have focused typically only on a single or limited number of entity types, using a general collection medical document, or focusing on clinical content written in a single language. This resulted in valuable datasets and resources, such as the DISTEMIST [13], SYMPTEMIST, PharmaCoNER [2], and Medprocner [9] corpora and systems, but the interplay and complementarity of multi-label entity extraction approaches were not targeted and evaluated. To address all these issues the novel *task MultiCardioNER* will focus on the automatic recognition, entity linking and indexing of key clinical variables or concept types, namely diseases and medications.

The *task MultiCardioNER* will focus on the recognition and indexing of these clinical entity types in cardiology clinical case documents in Spanish and other languages, by posing three subtasks on (1) automatic detection of mentions of diseases and medications in cardiology clinical case texts written in Spanish, (2) normalization of clinical entity mentions in Spanish to controlled vocabularies (clinical entity linking and concept normalization) and (3) automatic detection of mentions of diseases and medications in cardiology clinical case texts in English, Italian and Dutch.

The BioASQ *task MultiCardioNER* will rely on a collection of 1000 clinical case reports in Spanish annotated with disease and medication mentions, as well as a collection on 600 cardiology clinical case reports annotated also by clinical experts with diseases and medications. These mentions were manually mapped to their corresponding concept identifiers from SNOMED-CT. Moreover, a multilingual collection of 600 cardiology case reports translated into English, Italian and Dutch where concept mentions had been manually corrected by clinical experts will be used for the third sub-track. As a training set 1000 clinical case reports together with an additional collection of 400 cardiology case reports will be used, while the remaining 200 cardiology case reports will serve as a test set collection. The evaluation of systems for this task will use flat evaluation measures following the *task a* [3] track (mainly micro-averaged F-measure, MiF).

2.4 Task BioNNE: Nested NER in Russian and English

Most biomedical datasets and named entity recognition (NER) methods have been designed to capture flat (non-nesting) mention structures. To facilitate ongoing research on nested NER, we introduce a BioNNE shared task on nested NER in PubMed abstracts in Russian and English. The evaluation framework will be divided into three broad tracks (bilingual or language-oriented). The train/dev sets are based on the NEREL-BIO dataset [11] which extends an annotation scheme of the general-domain NEREL [10] and includes annotated mentions of disorders, anatomical structures, chemicals, diagnostic procedures, and biological functions. The NEREL-BIO dataset includes over 700 annotated PubMed abstracts in the Russian language and a small set of annotated English abstracts (app. 100). We strongly encourage participants to apply cross-language (Russian to English) and cross-domain (existing bio NER corpora to the BioNNE set) techniques in this task. Participants will be able to run their systems on novel test sets on the Codalab platform. The span-level macro-averaged F1 will be used as an evaluation metric.

In particular, the *Task BioNNE* will be structured into three tracks: **Track 1 - Bilingual:** participants in this track are required to train a single multi-lingual NER model using training data for both Russian and English languages. The model should be used to generate prediction files for each language's dataset. Please note that predictions from any mono-lingual model are not allowed in this track. **Tracks 2 & 3 - Language-oriented:** Participants in these tracks are required to train a model that works for one language. Participants are allowed to train any model architecture on any publicly available data in order to achieve best performance for the language of their choice.

2.5 BioASQ Datasets and Tools

During the eleven years of BioASQ, hundreds of systems from research teams around the world have been evaluated on the indexing, retrieval, and analysis of hundreds of thousands of biomedical publications and on answering thousands of biomedical questions. In this direction, BioASQ has developed a lively ecosystem of tools that facilitate research, such as the BioASQ Annotation Tool [24] for question-answering dataset development and a range of evaluation measures for automated assessment of system performance in all tasks. All BioASQ software[4] and datasets[5] are publicly available.

In particular, beyond the unique datasets described above for the four tasks running this year, BioASQ also provides: i) a training dataset of more than 16.2 million articles on large-scale biomedical semantic indexing with MeSH topics (task*task a*) [5], ii) the *task MESINEP* datasets [1,25] of more than 300K articles, on medical semantic indexing in Spanish, and iii) the *task DisTEMIST* [13] and *task MedProcNER* [8] datasets of 1,000 clinical cases in Spanish, on semantic indexing with SNOMED-CT entities for diseases and medical procedures.

3 Conclusions

BioASQ facilitates the exchange and fusion of ideas, providing unique realistic datasets and evaluation services for research teams that work on biomedical semantic indexing and question answering. Therefore, it eventually accelerates progress in the field, as indicated by the gradual improvement of the scores achieved by the participating systems [16,19]. An illustrative example is the Medical Text Indexer (MTI) [15], which achieved significant improvements [19] largely due to the adoption of ideas from the systems that compete in the BioASQ challenge [14], eventually reaching a performance level that allows the adoption of fully automated indexing in NLM [5].

Similarly, we expect that the new version of BioASQ will allow the participating teams to bring further improvement to the open tasks of biomedical question answering (*task 12b*), answering open questions for developing topics (*task Synergy 12*), multi-lingual named entity recognition in cardiology (*task MultiCardioNER*), and nested named entity recognition in Russian and English (*task BioNNE*). In conclusion, BioASQ aims to assist participating teams in their approach to the challenge's tasks, which represent key information needs in the biomedical domain.

Acknowledgments. Google was a proud sponsor of the BioASQ Challenge in 2023. The twelfth edition of BioASQ is also sponsored by Ovid Technologies, Inc., Elsevier, and Atypon Systems inc. The *task MultiCardioNER* is supported by the Spanish Plan for the Advancement of Language Technologies (Plan TL), the 2020 Proyectos de I+D+i-RTI Tipo A (Descifrando El Papel De Las Profesiones En La Salud De Los Pacientes A Traves De La Mineria De Textos, PID2020-119266RA-I00). This project has received funding from the European Union Horizon Europe Coordination and Support Action under Grant Agreement No 101058779 (BIOMATDB) and DataTools4Heart - DT4H, Grant agreement No 101057849. The work on the BioNNE task was supported by the Russian Science Foundation [grant number 23-11-00358].

[4] https://github.com/bioasq.
[5] http://participants-area.bioasq.org/datasets.

References

1. Gasco, L., et al.: Overview of BioASQ 2021-MESINESP track. Evaluation of advance hierarchical classification techniques for scientific literature, patents and clinical trials. CEUR Workshop Proceedings (2021)
2. Gonzalez-Agirre, A., Marimon, M., Intxaurrondo, A., Rabal, O., Villegas, M., Krallinger, M.: Pharmaconer: pharmacological substances, compounds and proteins named entity recognition track. In: Proceedings of The 5th Workshop on BioNLP Open Shared Tasks, pp. 1–10 (2019)
3. Kosmopoulos, A., Partalas, I., Gaussier, E., Paliouras, G., Androutsopoulos, I.: Evaluation measures for hierarchical classification: a unified view and novel approaches. Data Min. Knowl. Disc. **29**(3), 820–865 (2015)
4. Krallinger, M., Krithara, A., Nentidis, A., Paliouras, G., Villegas, M.: BioASQ at CLEF2020: large-scale biomedical semantic indexing and question answering. In: Jose, J., et al. (eds.) ECIR 2020. LNCS, vol. 12036, pp. 550–556. Springer, Cham (2020). https://doi.org/10.1007/978-3-030-45442-5_71
5. Krithara, A., Mork, J.G., Nentidis, A., Paliouras, G.: The road from manual to automatic semantic indexing of biomedical literature: a 10 years journey. Front. Res. Metrics Anal. **8** (2023). https://doi.org/10.3389/frma.2023.1250930
6. Krithara, A., Nentidis, A., Bougiatiotis, K., Paliouras, G.: BioASQ-QA: a manually curated corpus for Biomedical Question Answering. bioRxiv (2022)
7. Krithara, A., Nentidis, A., Paliouras, G., Krallinger, M., Miranda, A.: BioASQ at CLEF2021: large-scale biomedical semantic indexing and question answering. In: Hiemstra, D., Moens, M.F., Mothe, J., Perego, R., Potthast, M., Sebastiani, F. (eds.) ECIR 2021. LNCS, vol. 12657, pp. 624–630. Springer, Cham (2021). https://doi.org/10.1007/978-3-030-72240-1_73
8. Lima-López, S., et al.: Overview of MedProcNER task on medical procedure detection and entity linking at BioASQ 2023. In: CEUR Workshop Proceedings (2023)
9. Lima-López, S., et al.: Overview of medprocner task on medical procedure detection and entity linking at bioasq 2023. Working Notes of CLEF (2023)
10. Loukachevitch, N., et al.: NEREL: a Russian dataset with nested named entities, relations and events. In: Proceedings of the International Conference on Recent Advances in Natural Language Processing (RANLP 2021), pp. 876–885. INCOMA Ltd., Held Online (2021). https://aclanthology.org/2021.ranlp-1.100
11. Loukachevitch, N., et al.: NEREL-BIO: a dataset of biomedical abstracts annotated with nested named entities. Bioinformatics **39**(4), btad161 (2023)
12. Malakasiotis, P., Pavlopoulos, I., Androutsopoulos, I., Nentidis, A.: Evaluation measures for task b. Technical report, BioASQ (2018). https://participants-area.bioasq.org/Tasks/b/eval_meas_2018
13. Miranda-Escalada, A., et al.: Overview of DisTEMIST at BioASQ: automatic detection and normalization of diseases from clinical texts: results, methods, evaluation and multilingual resources. In: Working Notes of Conference and Labs of the Evaluation (CLEF) Forum. CEUR Workshop Proceedings (2022)
14. Mork, J., Aronson, A., Demner-Fushman, D.: 12 years on-Is the NLM medical text indexer still useful and relevant? J. Biomed. Semant. **8**(1), 8 (2017)
15. Mork, J., Jimeno-Yepes, A., Aronson, A.: The NLM Medical Text Indexer System for Indexing Biomedical Literature (2013)
16. Nentidis, A., et al.: Overview of BioASQ 2023: the eleventh BioASQ challenge on large-scale biomedical semantic indexing and question answering. In: Arampatzis, A., et al. (eds.) CLEF 2023. LNCS, vol. 14163, pp. 227–250. Springer, Cham (2023). https://doi.org/10.1007/978-3-031-42448-9_19

17. Nentidis, A., Katsimpras, G., Krithara, A., Paliouras, G.: Overview of BioASQ tasks 11b and Synergy11 in CLEF2023. In: CEUR Workshop Proceedings (2023)
18. Nentidis, A., et al.: Overview of BioASQ 2021: the ninth BioASQ challenge on large-scale biomedical semantic indexing and question answering. In: Candan, K.S., et al. (eds.) CLEF 2021. LNCS, vol. 12880, pp. 239–263. Springer, Cham (2021). https://doi.org/10.1007/978-3-030-85251-1_18
19. Nentidis, A., et al.: Overview of BioASQ 2022: the tenth BioASQ challenge on large-scale biomedical semantic indexing and question answering. In: Barrón-Cedeño, A., et al. (eds.) CLEF 2022. LNCS, vol. 13390, pp. 337–361. Springer, Cham (2022). https://doi.org/10.1007/978-3-031-13643-6_22
20. Nentidis, A., Katsimpras, G., Vandorou, E., Krithara, A., Paliouras, G.: Overview of BioASQ tasks 9a, 9b and synergy in CLEF2021. In: Proceedings of the 9th BioASQ Workshop A Challenge on Large-scale Biomedical Semantic Indexing and Question Answering. CEUR Workshop Proceedings (2021). https://ceur-ws.org/Vol-2936/paper-10.pdf
21. Nentidis, A., Katsimpras, G., Vandorou, E., Krithara, A., Paliouras, G.: Overview of BioASQ tasks 10a, 10b and Synergy10 in CLEF2022. In: CEUR Workshop Proceedings, vol. 3180, pp. 171–178 (2022)
22. Nentidis, A., Krithara, A., Paliouras, G., Farre-Maduell, E., Lima-Lopez, S., Krallinger, M.: BioASQ at CLEF2023: the eleventh edition of the large-scale biomedical semantic indexing and question answering challenge. In: Kamps, J., et al. (eds.) ECIR 2023. LNCS, vol. 13982, pp. 577–584. Springer, Cham (2023). https://doi.org/10.1007/978-3-031-28241-6_66
23. Nentidis, A., Krithara, A., Paliouras, G., Gasco, L., Krallinger, M.: BioASQ at CLEF2022: the tenth edition of the large-scale biomedical semantic indexing and question answering challenge. In: Hagen, M., et al. (eds.) ECIR 2022. LNCS, vol. 13186, pp. 429–435. Springer, Cham (2022). https://doi.org/10.1007/978-3-030-99739-7_53
24. Ngomo, A.C.N., Heino, N., Speck, R., Ermilov, T., Tsatsaronis, G.: Annotation tool. Project deliverable D3.3 (2013). https://www.bioasq.org/sites/default/files/PublicDocuments/2013-D3.3-AnnotationTool.pdf
25. Rodriguez-Penagos, C., et al.: Overview of MESINESP8, a Spanish Medical Semantic Indexing Task within BioASQ 2020 (2020)
26. Salton, G., Buckley, C.: Improving retrieval performance by relevance feedback. J. Am. Soc. Inf. Sci. **41**(4), 288–297 (1990). https://doi.org/10.1002/(SICI)1097-4571(199006)41:4⟨288::AID-ASI8⟩3.0.CO;2-H
27. Tsatsaronis, G., et al.: An overview of the BioASQ large-scale biomedical semantic indexing and question answering competition. BMC Bioinform. **16**, 138 (2015). https://doi.org/10.1186/s12859-015-0564-6

EXIST 2024: sEXism Identification in Social neTworks and Memes

Laura Plaza[1(✉)], Jorge Carrillo-de-Albornoz[1], Enrique Amigó[1], Julio Gonzalo[1], Roser Morante[1], Paolo Rosso[2,3], Damiano Spina[4], Berta Chulvi[2], Alba Maeso[2], and Víctor Ruiz[1]

[1] Universidad Nacional de Educación a Distancia (UNED), 28040 Madrid, Spain
{lplaza,jcalbornoz,enrique,julio,rmorant,victor.ruiz}@lsi.uned.es

[2] Universitat Politècnica de València (UPV), 46022 Valencia, Spain
prosso@dsic.upv.es, berta.chulvi@upv.es, amaeolm@inf.upv.es

[3] ValgrAI - Valencian Graduate School and Research Network of Artificial Intelligence, 46022 Valencia, Spain

[4] RMIT University, Melbourne 3000, Australia
damiano.spina@rmit.edu.au

Abstract. The paper describes the EXIST 2024 lab on Sexism identification in social networks, that is expected to take place at the CLEF 2024 conference and represents the fourth edition of the EXIST challenge. The lab comprises five tasks in two languages, English and Spanish, with the initial three tasks building upon those from EXIST 2023 (*sexism identification in tweets*, *source intention detection in tweets*, and *sexism categorization in tweets*). In this edition, two new tasks have been introduced: *sexism detection in memes* and *sexism categorization in memes*. Similar to the prior edition, this one will adopt the Learning With Disagreement paradigm. The dataset for the various tasks will provide all annotations from multiple annotators, enabling models to learn from a range of training data, which may sometimes present contradictory opinions or labels. This approach facilitates the model's ability to handle and navigate diverse perspectives. Data bias will be handled both in the sampling and in the labeling processes: seed, topic, temporal and user bias will be taken into account when gathering data; in the annotation process, bias will be reduced by involving annotators from different social and demographic backgrounds.

Keywords: sexism identification · sexism categorization · learning with disagreement · memes · data bias

1 Introduction

Sexism is rooted in beliefs concerning the inherent nature of women and men, dictating the roles they are expected to fulfill in society. These beliefs often materialize as gender stereotypes, potentially elevating one gender above the other in a hierarchical manner. This hierarchical mindset can manifest either

N. Goharian et al. (Eds.): ECIR 2024, LNCS 14612, pp. 498–504, 2024.
https://doi.org/10.1007/978-3-031-56069-9_68

consciously and overtly, or subtly and unconsciously, often appearing as unconscious bias. The impact of sexism is widespread, affecting individuals across the gender spectrum, with women bearing a significant brunt of its consequences.

Inequality and discrimination against women persist within society and are increasingly replicated online [1]. The internet has the power to perpetuate and normalize gender differences and sexist attitudes [2], especially among teenagers. Previous studies have shown how sexism towards women is perpetuated on social networks and their influence on the teenagers' health [3]. Although platforms like Twitter continually strive to identify and eradicate hateful content, they grapple with the overwhelming volume of user-generated data [4]. In this context, leveraging automatic tools becomes crucial not only for detecting and alerting against sexist behaviors and discourses, but also for estimating the prevalence of sexist and abusive situations on social media platforms, identifying common forms of sexism, and understanding how sexism is expressed through these channels.

EXIST 2024[1] will be the fourth edition of the sEXism Identification in Social neTworks challenge. EXIST is a series of scientific events and shared tasks that aim to capture sexism in a broad sense, from explicit misogyny to other subtle expressions that involve implicit sexist behaviors. The last edition of the EXIST shared task was held as a lab in CLEF 2023 [5,6], while the first two editions were held in the IberLEF Spanish evaluation forum [7,8]. The EXIST 2024 Lab will be one of the participating labs of the new MonsterCLEF initiative that focuses on the versatility of Large Language Models.[2]

In past editions, more than 75 teams participated from research institutions and companies from all around the world. While the three previous editions focused solely on detecting and classifying sexist textual messages, this new edition incorporates new tasks that center around images, particularly memes. Memes are images, typically humorous in nature, that are spread rapidly by social networks and Internet users. With this addition, we aim to encompass a broader spectrum of sexist manifestations in social networks, especially those disguised as humor. As indicated in [9], the consumption of memes and their increasing visual complexity is on the rise each day. Additionally, as indicated by [10], memes can propagate globally through repackaging in different languages and they can also be multilingual. Consequently, it becomes imperative to develop automated multimodal tools capable of detecting sexism in both text and images.

Similar to the approach in the 2023 edition, this edition will also embrace the Learning With Disagreement (LeWiDi) paradigm for both the development of the dataset and the evaluation of the systems. Our prior research demonstrated that perceiving sexism is significantly influenced by the demographic and cultural background of the individual. Consequently, when identifying sexist attitudes and expressions, and even when categorizing them into distinct sexist categories, disagreements among individuals are common. The LeWiDi paradigm doesn't rely on a single "correct" label for each example. Instead, the model is trained

[1] https://nlp.uned.es/exist2024.
[2] https://monsterclef.dei.unipd.it.

to handle and learn from conflicting or diverse annotations. This enables the system to consider various annotators' perspectives, biases, or interpretations, resulting in a fairer learning process.

In the following sections, we provide comprehensive information about the tasks, the dataset and the evaluation methodology that will be adopted in the EXIST 2024 challenge at CLEF.

2 EXIST 2024 Tasks

The last edition of EXIST focused on detecting sexist messages in Twitter as well as on categorizing these messages according to the type of sexist behavior they enclose and the intention of the source (Tasks 1 to 3). For the 2024 edition, we will address also multimedia content. In particular, we will aim to detect and categorize sexism in memes. Therefore, five tasks are proposed which are described below.

2.1 Sexism Identification

The first task is a binary classification task where systems must decide whether or not a given tweet is sexist. The following statements show examples of sexist and not sexist messages, respectively.

(1) **Sexist:** *Woman driving, be careful!.*
(2) **Non sexist:** *Just saw a woman wearing a mask outside spank her very tightly leashed dog and I gotta say I love learning absolutely everything about a stranger in a single instant.*

2.2 Source Intention Detection

This task aims to categorize the message according to the intention of the author. We propose a ternary classification task: (i) direct sexist message, (ii) reported sexist message and (iii) judgmental message. This distinction will allow us to differentiate sexism that is actually taking place in online platforms from sexism which is being suffered by women in other situations but that is being reported in social networks with the aim of complaining and fighting against sexism. The following categories are defined:

- **Direct** sexist message: the intention was to write a message that is sexist by itself or incites to be sexist, as in:
  (3) *A woman needs love, to fill the fridge, if a man can give this to her in return for her services (housework, cooking, etc.), I don't see what else she needs.*
- **Reported** sexist message: the intention is to report and share a sexist situation suffered by a woman or women in first or third person, as in:
  (4) *Today, one of my year 1 class pupils could not believe he'd lost a race against a girl.*
- **Judgemental** message: the intention is judgmental, since the tweet describes sexist situations or behaviors with the aim of condemning them.
  (5) *21st century and we are still earning 25% less than men #Idonotrenounce.*

2.3 Sexism Categorization

Many facets of a woman's life may be the focus of sexist attitudes including domestic and parenting roles, career opportunities, sexual image, and life expectations, to name a few. In this task, each sexist tweet must be categorized in one or more of the following categories:

- **Ideological and inequality**: this category includes all tweets that discredit the feminist movement in order to devalue, belittle and defame the struggle of women in any aspect of their lives. It also includes messages that reject inequality between men and women, or present men as victims of gender-based oppression. Some examples of this type of sexism discourse can be found in the following tweets:
  (6) *Think the whole equality thing is getting out of hand. We are different, that's how were made!*
- **Stereotyping and dominance**: this category includes messages that express false ideas about women that suggest they are more suitable or inappropriate for certain tasks. It also includes any claim that implies that men are somehow superior to women.
  (7) *Most women no longer have the desire or the knowledge to develop a high quality character, even if they wanted to.*
- **Objectification**: Objectification and physical stereotyping includes messages where women are presented as objects apart from their dignity and personal aspects. We also include messages that assume or describe certain physical qualities that women must have in order to fulfill traditional gender roles, for example, ideas that suggest that women should maintain a standard and ideal of beauty or attacks on a woman's body.
  (8) *No offense but I've never seen an attractive african american hooker. Not a single one.*
- **Sexual violence**: this category includes messages where sexual suggestions, requests or harassment of a sexual nature (rape or sexual assault) are made:
  (9) *I wanna touch your tits..you can't imagine what I can do on your body.*
- **Misogyny and non sexual violence**: this category includes expressions of hatred and violence towards women.
  (10) *Domestic abuse is never okay... Unless your wife is a bitch.*

2.4 Task 4: Sexism Identification in Memes

This is a binary classification task consisting on deciding whether or not a given meme is sexist. Figure 1 shows examples of sexist and non sexist memes, respectively.

2.5 Task 5: Sexism Categorization in Memes

This task aims to classify sexist memes according to the categorization provided for Task 3: (i) ideological and inequality, (ii) stereotyping and dominance, (iii) objectification, (iv) sexual violence and (v) misogyny and non-sexual violence.

(a) Sexist meme (b) Non sexist meme

Fig. 1. Examples of memes from the EXIST 2024 dataset

3 The EXIST 2024 Dataset

In recent years, a number of proposals have advocated for considering disagreement as informative content that can enhance task performance [11]. Evidence suggests that gold labels are an idealization and that irreconcilable disagreement is common, especially in tasks involving highly subjective judgments, such as sexism detection.

Following this trend, for the development of the dataset we will follow the learning with disagreement paradigm, so that rather than aggregating the (potentially) contradictory labels provided by the annotators, all of them will be kept and provided to the participating systems so that they may exploit disagreement in their systems. The LeWiDi paradigm may also help to mitigate bias and produce equitable NLP systems.

For the development of the EXIST 2024 dataset we will collect memes from different sources such as Google image search, Twitter, special forums such as Reddit or Forocoche. The downloaded images will be processed to remove duplicate items, as well as analyzed with an OCR software to extract the text following the methodology used in [12]. For the crawling and filtering, we will use the same methodology as in EXIST 2023.

The selection of annotators for the development of the EXIST 2024 dataset will take into account the heterogeneity necessary to avoid gender and age biases. Annotation will be performed by crowd-workers using for instance Prolific [13]. The Prolific crowdsourcing platform was specifically selected because of the features it provides to define participant criteria in the recruiting process - in our case, gender, age, and fluency in the different languages. Each tweet/meme will be annotated by six workers that will be selected according to their different demographic characteristics in order to minimize the label bias.

4 Evaluation Methodology and Metrics

As in EXIST 2023, we will carry out a **hard evaluation** and a **soft evaluation**. In the hard evaluation, both the system and the gold standard consist of one or more labels assigned to each instance in the dataset. In contrast, the soft evaluation is intended to measure the ability of the model to capture disagreement, by considering the probability distribution of labels in the output as a soft label and comparing it with the probability distribution of the annotations.

The LewiDi paradigm can be considered in both sides of the evaluation process:

- **The ground truth.** In a hard setting, the variability in the human annotations is reduced by selecting one and only one gold category per instance, the hard label. In a soft setting, the gold standard label for one instance is the set of all the human annotations existing for that instance. Therefore, the evaluation metric incorporates the proportion of human annotators that have selected each category (soft labels).
- **The system output.** In a hard traditional setting, the system predicts one or more categories for each instance. In a soft setting, the system predicts a probability for each category, for each instance. The evaluation score is maximized when the probabilities predicted match the actual probabilities in a soft ground truth.

Two different types of evaluation will be accomplished:

1. **Soft-soft evaluation**. This evaluation is intended for systems that provide probabilities for each category, rather than a single label. We will use a modification of the ICM metric (Information Contrast Measure [14]), ICM-Soft (see details in [5]), as the official evaluation metric in this variant and we will also provide results for the normalized version of ICM-Soft (ICM-Soft Norm). We will also provide results for Cross Entropy.
2. **Hard-hard evaluation**. This evaluation is intended for systems that provide a hard, conventional output. To derive the hard labels in the ground truth from the different annotators' labels, we will use a probabilistic threshold computed for each task. The official metric for this task will be the original ICM, as defined by Amigó and Delgado [14]. We will also report a normalized version of ICM (ICM Norm) and F1.

Acknowledgments. This work has been financed by the European Union (Next Generation EU funds) through the "Plan de Recuperación, Transformación y Resiliencia", by the Ministry of Economic Affairs and Digital Transformation and by the UNED University. It has also been financed by the Spanish Ministry of Science and Innovation (project FairTransNLP (PID2021-124361OB-C31 and PID2021-124361OB-C32)) funded by MCIN/AEI/10.13039/501100011033 and by ERDF, EU A way of making Europe, and by the Australian Research Council (DE200100064 and CE200100005).

References

1. Social Media and the Silencing Effect: Why Misogyny Online is a Human Rights Issue. NewStatesman. https://bit.ly/3n3ox68. Accessed 18 Oct 2023
2. Burgos, A., et al.: Violencias de Género 2.0, pp. 13–27 (2014)

3. Gil Bermejo, J.L., Martos, S.C., Vázquez, A.O., García-Navarro, E.B.: Adolescents, ambivalent sexism and social networks, a conditioning factor in the healthcare of women. Healthcare **9**(6), 721 (2021)
4. Twitter's Famous Racist Problem. The Atlantic. https://bit.ly/38EnFPw. Accessed 17 Oct 2023
5. Plaza, L., et al.: Overview of EXIST 2023 - learning with disagreement for sexism identification and characterization. Experimental IR meets multilinguality, multimodality, and interaction. In: Arampatzis, A., et al. (eds.) Proceedings of the Fourteenth International Conference of the CLEF Association (CLEF 2023), Thessaloniki, Greece (2023)
6. Plaza, L., et al.: Overview of EXIST 2023 - learning with disagreement for sexism identification and characterization (extended overview). In: Aliannejadi, M., Faggioli, G., Ferro, N., Vlachos, M. (eds.) Working Notes of CLEF 2023 - Conference and Labs of the Evaluation Forum (2023)
7. Rodríguez-Sánchez, F., et al.: Overview of EXIST 2021: sexism identification in social networks. Procesamiento del Lenguaje Natural **67**, 195–207 (2021)
8. Rodríguez-Sánchez, F., et al.: Overview of EXIST 2022: sexism identification in social networks. Procesamiento del Lenguaje Natural **69**, 229–240 (2022)
9. Valensise, C.M., Serra, A., Galeazzi, A., Etta, G., Cinelli, M., Quattrociocchi, W.: Entropy and complexity unveil the landscape of memes evolution. Sci. Rep. **11**(1), 1–9 (2021)
10. Sharma, S., et al.: Detecting and understanding harmful memes: a survey. In: Proceedings of the Thirty-First International Joint Conference on Artificial Intelligence, IJCAI 2022, pp. 5597–5606 (2022)
11. Basile, V., et al.: We need to consider disagreement in evaluation. In: Proceedings of the 1st Workshop on Benchmarking: Past, Present and Future, pp. 15–21, Online. Association for Computational Linguistics (2021)
12. Fersini, E., et al.: SemEval-2022 task 5: multimedia automatic misogyny identification. In: Proceedings of the 16th International Workshop on Semantic Evaluation (SemEval-2022), pp. 533–549. Association for Computational Linguistics (2022)
13. Prolific. https://www.prolific.com/. Accessed 18 Oct 2023
14. Amigó, E., Delgado, A.: Evaluating extreme hierarchical multi-label classification. In: Proceedings of the 60th Annual Meeting of the Association for Computational Linguistics, pp. 5809–5819 (2022)

Author Index

N. Goharian et al. (Eds.): ECIR 2024, LNCS 14612, pp. 505–508, 2024.
https://doi.org/10.1007/978-3-031-56069-9

Printed in the United States
by Baker & Taylor Publisher Services